D1087248

Proceedings
SPM 2006
ACM Symposium on Solid and Physical Modeling

Cardiff, Wales, United Kingdom
June 06 – 08, 2006

Honorary Chair

Jarek Rossignac (Georgia Tech)

Symposium Co-Chairs

Ralph Martin (Cardiff University)
Shi-Min Hu (Tsinghua University)

Program Co-Chairs

Leif Kobbelt (RWTH Aachen University)
Wenping Wang (Hong Kong University)

Proceedings Production Editor

Stephen N. Spencer, University of Washington

Sponsored by ACM SIGGRAPH

The Association for Computing Machinery, Inc.
1515 Broadway
New York, New York 10036

ACM ISBN:1-59593-358-1
Additional copies may be ordered prepaid from:

ACM Order Department
P.O. Box 11405
Church Street Station
New York, NY 10286-1405

Phone: 1-800-342-6626
(USA and Canada)
+1-212-626-0500
(All other countries)
Fax: +1-212-944-1318
E-mail: acmhelp@acm.org

ACM Order Number: 433069

Printed in the USA

Table of Contents

Table of Contents

Preface

The ACM Symposium on Solid and Physical Modeling is an annual international forum for the exchange of recent research results and applications of spatial modeling and computations in design, analysis and manufacturing, as well as in emerging biomedical, geophysical and other areas. Previous symposia in this series were held in Austin, Texas, 1991; Montreal, Canada, 1993; Salt Lake City, Utah, 1995; Atlanta, Georgia, 1997; Ann Arbor, Michigan in 1999 and 2001; Saarbrucken, Germany, 2002; Seattle, Washington, 2003; Genova, Italy, 2004; and Cambridge, Massachusetts, 2005. For additional information, please visit www.solidmodeling.org, the home page of The Solid Modeling Association that oversees this symposium series.

The SPM symposium series started initially with the name "ACM Symposium on Solid Modeling and Applications." To emphasize the fact that solid modeling entails not only handling their geometric shapes, but also their physical properties and behaviors, the name of the symposium was expanded to The ACM Symposium on Solid and Physical Modeling (abbreviated as SPM) in 2005.

SPM'06 was held in plenary sessions on the campus of Cardiff University, Wales, United Kingdom from Tuesday June 6 to Thursday June 8, 2006. Fifty six technical papers have been submitted and were reviewed by the international program committee and expert reviewers from around the world. At least three external reviewers and members of the program committee reviewed and discussed each submission. A total of 21 refereed papers have been selected for plenary presentation and publication in the proceedings. The symposium program also includes three invited presentations by Bruno Levy, Sara McMains and Jürgen Weese, all leading researchers in their fields. There were also two panels where recent trends and challenges were discussed. These panel sessions were organized by Kenji Shimada (Physical Modeling and Simulation) and Sara McMains (Design, Analysis, and Manufacturing).

We thank all the authors, volunteer reviewers and program committee members for their great efforts and contributions that make this symposium a success. Symposium chairs, Ralph Martin and Shimin Hu, deserve distinct recognition for their hard work in organizing the symposium. Stephan Bischoff ran the online review management system in an admirably efficient and professional manner. Last but not least, we are grateful to UGS Corporation for sponsoring the Best Paper Award, We also thank Cardiff University for hosting the symposium and ACM SIGGRAPH for sponsorship.

Leif Kobbelt and Wenping Wang
Program Co-Chairs, 2006 ACM Symposium on Solid and Physical Modeling

Segmenting Reliefs on Triangle Meshes

Shenglan Liu* Ralph R. Martin Frank C. Langbein Paul L. Rosin

School of Computer Science, Cardiff University
{Shenglan.Liu, ralph, F.C.Langbein, Paul.Rosin}@cs.cf.ac.uk

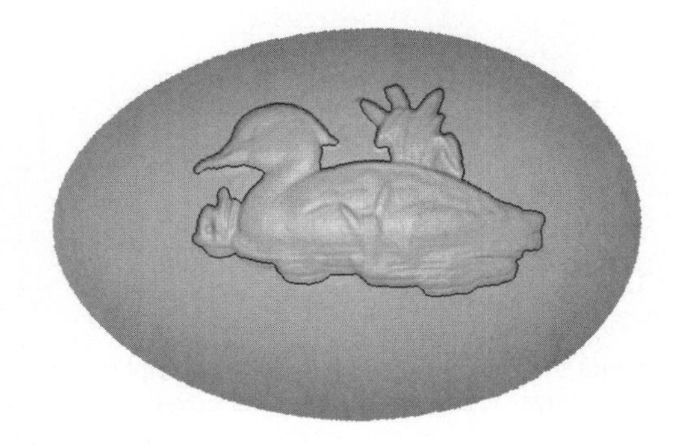

Figure 1: A porcelain relief and a corresponding segmented triangle mesh

Abstract

Sculptural *reliefs* are widely used in various industries for purposes such as applying brands to packaging and decorating porcelain. In order to easily apply reliefs to CAD models, it is often desirable to reverse-engineer previously designed and manufactured reliefs. 3D scanners can generate triangle meshes from objects with reliefs; however, previous mesh segmentation work has not considered the particular problem of separation of reliefs from background. We consider here the specific case of segmenting a simple relief delimited by a single outer contour, which lies on a smooth, slowly varying background. Generally, such reliefs meet the surrounding surface in a small step, enabling us to devise a specific method for such relief segmentation.

We find the boundary between the background and the relief using an adaptive *snake*. It starts at a simple user-drawn contour, and is driven inwards by a collapsing force until it matches the relief's boundary. Our method is insensitive to the choice of the initial contour. The snake's limiting position is controlled by a feature energy term designed to find a step. A refinement strategy is then used to drive the snake into concavities of the relief contour.

We demonstrate operation of our algorithm using real scanned models with different relief contour shapes and triangle meshes with different resolutions.

CR Categories: I.3.5 [Computer Graphics]: Computational Geometry and Object Modeling—Curve, surface, solid, and object representations

Keywords: Relief segmentation, snakes, mesh processing.

*Corresponding author

SPM 2006, Cardiff, Wales, United Kingdom, 06–08 June 2006.
© 2006 ACM 1-59593-358-1/06/0006 $5.00

1 Introduction

Complex decorative *reliefs* are often added to CAD models in such application areas as sign-making, packaging, and ceramics to make product designs more interesting, more characteristic of the company, or of higher intrinsic value. A typical example is the Wedgwood item shown on the left in Figure 1. New reliefs can be created from 2D artwork using software such as Delcam's ArtCAM, or can be hand-crafted, typically on a planar surface, and reverse-engineered using a 3D scanner. Extracting a relief from a planar background surface is a trivial task. However, in other cases it is desired to scan an existing relief lying on a curved object, rather than generating a new design. As in general such reliefs lie on freeform surfaces, reverse engineering methods are needed to separate the relief from the non-planar background, and to assist in the application of the relief to a different base surface.

A real world example of the need for reverse engineering reliefs occurs in the porcelain industry. When extending an existing range of porcelain, it is essential for any new item to exactly match existing designs on old items. At present it is necessary for a sculptor to hand-copy existing relief designs, a process that is time consuming and tedious for the artist, and expensive for the manufacturer—if indeed he can find suitably skilled workers. Clearly, the ability to automatically extract the relief, flatten it and then apply it to a new object would greatly reduce production time, and has the potential to produce better results.

To be more precise, in this paper, we define a *relief* to be extra material added locally to some underlying surface. This added material forms a surface with sculpted features clearly different from the underlying surface. In particular, we assume that the relief has a small height relative to the characteristic size of features on the underlying surface, which allows us to distinguish between the two surface regions. (There is nothing in principle, however, to prevent our methods being applied to 'negative reliefs', i.e., *embossing*, too, after making suitable minor changes).

There are various kinds of reliefs, and they can be imposed on diverse backgrounds. The simplest kind of relief is an *isolated* relief, which is bounded by a single outer contour (and which may also contain inner contours). See, for example, the duck in Figure 1. Other reliefs may be *cyclic* and wrap right around the object, such

Figure 2: Relief on a textured background

as the frieze around the rim in Figure 1. A more complex kind of relief is a *backgroundless* relief where the relief covers all or most of the surface of the model—we may envisage the model as having some smooth underlying surface over which a relief is applied everywhere. Relief processing also needs to take into account the nature of the *background* surface: the underlying surface maybe relatively simple and smooth, as in Figure 1, or more intricate—in more complex cases, the background can itself be textured, as in Figure 2. We could consider this object to be covered with a backgroundless relief, but alternatively, we might be interested in segmenting the bird and branch from the underlying pattern, which would then be considered to be a textured background.

In this paper we address a particular problem in reverse engineering of reliefs; other cases are to be considered in future work. We assume we are given as input a triangular mesh created from data points captured by a 3D scanner. Here we *only* consider the simple problem of segmenting the part of the mesh delimited by the outer contour of an *isolated* relief which lies on a *smooth* and slowly varying background. There may be multiple isolated reliefs present on the original object; we assume that the user draws a rough *closed contour* on the mesh around one relief to indicate which relief is to be segmented. This is done by specifying a few points on the mesh which are then automatically joined across the mesh; these may be relatively far from the relief. We can also handle the case where the relief is bounded by an *open contour*, as seen for example in Figure 7(c); in this case the user must indicate on which side of the initial contour the relief lies. We hope to extend our results to more complex reliefs and backgrounds in future, and so the approach taken has future generalisability in mind.

Our method uses a *snake* which starts from the user-drawn contour, and evolves until it matches the boundary of the relief. We adapt the snake to suit the mesh resolution, and allow most of the parameters to be determined automatically, minimising the need for non-technical users to have to understand and specify control parameters, whilst also providing high performance.

In Sections 2 and 3 we review closely related work: we briefly discuss the original idea of snakes as used for segmenting images, and their generalisation to 3D surfaces. In Section 4 we explain the novel contributions of this paper. The following Sections give details of our method: Section 5 explains the energy terms we use to control the snake and Section 6 describes the evolution process for the snake, Section 7 gives further implementation details, especially of the parameters used to control the method. Experiments using such snakes show that they can quickly and effectively determine isolated reliefs, as discussed in Section 8. Section 9 concludes the paper and gives various ideas for future extensions of our approach to other relief types.

2 Snakes on 2D Images

Snakes (or, more formally, *active contour models*) were originally proposed by Kass et al. [1988], and have been widely used for image segmentation and motion tracking in computer vision and image analysis. The basic idea is to deform a starting curve to an energy-minimising position under the influence of internal and external forces, in order to detect or follow features in an image or image sequence. Internal forces coming from the curve itself shrink and smooth the curve. External forces derived from the image help to drive the curve toward the desired features of interest using an iterative process. The snake evolves until an equilibrium of all forces is reached, which is equivalent to a minimum of the energy function.

There are two main limitations of the original snake approach. Firstly, the initial contour needs to be quite close to the object, otherwise the snake may converge to the wrong result, getting stuck in a local minimum. Secondly, when the object being sought is nonconvex, snakes have difficulties in following the object's boundary in concavities. Both of these difficulties can also be encountered when using snakes for relief contour extraction.

If the snake is too sensitive to choice of initial contour, this forces the user to trace a starting contour which reasonably follows the complicated concavities and convexities of the relief's boundary, which is tedious and error-prone. In practice, concavities are very common on relief contours, and often they can be very deep and narrow. Various methods have been proposed to address the problem of initial contour sensitivity by modifying the external forces. The 'balloon' model [Cohen 1991] adds an additional pressure force (which can be inflationary or deflationary) to make the contour behave like a balloon. This helps avoid the contour getting stuck at weak features caused mainly by noise, and aiding convergence onto a more distant strong boundary. However, this introduces the new problem of choosing an appropriate pressure force.

Cohen and Cohen [1993] suggests employing an attractive force derived from the edge map of the image to provide a larger capture range for the snake in order to address the limitation on placement of the initial contour. Xu and Prince [1998] employs a gradient vector flow, an external force computed as a diffusion of the gradient vectors of a gray-level or binary edge map derived from the image, to address both limitations.

3 Snakes on 3D Surfaces

When extending the ideas of snakes from 2D images to 3D mesh surfaces, various problems arise. The first is the greater complexity caused by the connectivity of the mesh not being regular, unlike a 2D grid of pixels. Secondly, the internal energy terms caused by the internal forces must be defined in local surface tangent planes, rather than a global plane, and care must be taken when forces are added. Thirdly, the extra forces need to be calculated based on mesh features, instead of intensity image features. Generally, we require more complex energy terms to describe mesh features, and certain extended extra forces proposed for image snakes are difficult to directly adapt to meshes. Finally, the snake must be restricted to lie on the mesh when the energy function is minimised, thus leading to a constrained rather than unconstrained minimisation problem.

Milroy et al. [1997] extended snakes to 3D surfaces to perform surface segmentation. Their approach links curvature extrema points on an *orthogonal cross-section model*, a surface mesh that traces an object's contours with closed curves running in the x, y, and z directions. They extended the definition of the internal energy to support a discretised snake with irregular segment lengths in 3D. Two extra forces, an inflation force and an attraction force, were introduced to inflate contours, and attract small, closed, user-defined contours to the curvature extrema points. A greedy algorithm was used to minimise the energy.

Snakes on triangular meshes have been used to detect features where a certain property such as curvature changes drastically, allowing the detection of sharp edges and peak vertices [Lee and Lee 2002], for example. The authors define the external feature energy on the mesh in terms of the variation of normal directions between neighbouring faces. They parameterise the surface region surrounding the snake, then minimise the energy function and compute the motion of the snake in 2D, and finally remap the snake back onto the 3D mesh.

Other more recent research has also considered snakes for triangular meshes. Jung and Kim [2004] use snakes to find features related to Gaussian curvature. They move snakes step-by-step, from one mesh vertex to another, and give them the ability to change topology, allowing a snake to split into multiple separate snakes as appropriate. Based on their framework of parameterisation-free active contour models [2004], Bischoff et al. [2005] present a new representation and method for evolving snakes on triangular meshes. Their method enables collision detection and supports topological controls such as snake merging and splitting, in this case constraining the vertices of the snakes to move on mesh edges.

4 Novelty

Many methods have been proposed for segmentation of meshes; some of these try to cut a mesh into natural pieces (such as the fingers on a hand), whereas others try to find surface features such as sharp edges. As just one example, we cite the work by Funkhouser et al. [2004], which uses a least-cost path for mesh segmentation to find the natural seams of the mesh. Although snakes have been previously used for problems such as feature detection and segmentation of meshes, previously reported work is not particularly tailored to detecting reliefs on an underlying surface. Careful choice of appropriate forces for this specific problem will lead to better results than applying a general method, or a method which looks for, e.g., sharp edges. Note that a relief may meet the underlying surface in a sharp step, or the relief may blend with the underlying surface over a short distance. This is a different kind of feature from a sharp edge and warrants a specific type of feature energy term particular to relief segmentation. We describe our specific energy term for detecting a relief boundary in Section 5.3.

As mentioned above, other general issues are that snakes can be sensitive to the choice of initial contour, and can have particular difficulties with finding concavities in contours such as those found near the duck's neck in Figure 1. Although sensitivity to initial contours has been tackled in previous work through the use of an external pressure force [Cohen 1991; Milroy et al. 1997], the force strength has to be set carefully if the equilibrium contour is to end up in the desired place, rather than overrunning, or ceasing to move inward too soon. The method proposed in Xu and Prince [1998] is able to move snakes into concavities in images, but is not directly applicable to 3D surface meshes. Our approach, on the other hand, works with a crudely specified initial contour which does not have to be close to the relief contour—it simply has to separate the desired relief from other possible reliefs.

Furthermore, previous methods based on snakes have required several parameters to be carefully controlled so that the snake's movement leads to the desired results. In many cases appropriate parameter choices are data-dependent and can require considerable tuning by users to get good results. However, we expect many users of relief segmentation will not have a technical background, making it hard for them to understand how to tune the method for best results. Much time can be wasted in making a series of trials to find the best parameter settings. We thus have devised an approach which keeps such tuning to a minimum, and, where possible, we derive parameter settings from the model directly, without user interaction. Other remaining parameters have intuitively obvious meanings and effects.

Taking these points into account, we introduce a novel adaptive snake-based approach suitable for relief contour detection on a triangular mesh. Our main contributions are:

1. A feature energy term specifically designed for relief contour segmentation.
2. A deflation force with strength determined in such a way as to make the snake insensitive to choice of initial contour.
3. A snake refinement phase designed to make the snake explore relief contour concavities.
4. A denoising term in the deflation energy and use of bilateral filtering of feature properties to enable the snake to robustly traverse a noisy background.

The next several Sections present the details of our approach to using snakes for relief contour detection on triangular meshes.

5 Energy Terms

We now introduce the snake used; in this Section we define all the energy terms used and explain the resulting energy minimisation problem. Section 5.1 describes the energy functional used and approach to solving the energy minimisation problem. Details of the internal energy terms and external energy terms are explained separately in Sections 5.2 and 5.3.

5.1 Energy Functional

A parametric snake in 3D may be represented by a curve $v(s) = (x(s), y(s), z(s))$, where s represents arc-length. Following previous researchers' ideas [Williams and Shah 1992; Milroy et al. 1997], the snake on the triangular mesh proposed in this paper is approximated using connected straight line segments, joined by an ordered list of discrete vertices (*snaxels*):

$$v(s) = [v_i], \qquad i = 0, \ldots, n-1 \tag{1}$$

where v_i is a point on the mesh lying either on a mesh vertex, an edge or a face. The list is a circular list for a closed relief contour: v_0 and v_{n-1} denote the same snaxel, or it may be an ordinary list for an open relief contour. The distances between snaxels need not be equal, even approximately. The energy functional representing the energy of the snake can be written as

$$E = \int E_{int}(v(s)) + E_{ext}(v(s))\, ds, \tag{2}$$

and depends on an *internal energy* term and an *external energy* term. The *internal energy* E_{int} depends on the snake itself. It is made up of two energy terms, representing *tension energy* and *bending energy*; minimising E_{int} makes the snake shrink and straighten (which are to some extent conflicting requirements).

These drive the snake inwards, and also prevent it from locally bending in and out too much. The *external energy* E_{ext} depends on the mesh. It creates extra forces which move and deform the snake to make it best fit the features sought; these forces are described in detail later. The forces arising from these various energy terms drive the snake to a position of lower overall energy. Ultimately, this should make the snake stabilise at the desired contour, located at a local minimum of energy.

In image processing, *variational calculus* [Kass et al. 1988] is widely used to determine the minimum energy of Eqn. 2. However, previous formulations cannot immediately be applied to the mesh case as this formalism does not permit the inclusion of the necessary *hard constraint* that the snake must lie on the mesh and not move off it. Lee and Lee [2002] handles this problem by parameterising the mesh surrounding the snake and then minimising the energy in 2D. However, parametrisation introduces an additional complexity, and may lead to unstable or inaccurate results in highly curved regions. Note that such regions at the boundary of the relief are precisely those we wish to find in our application!

A greedy algorithm for image snakes proposed by Williams and Shah [1992] provides a formulation for incorporating hard constraints quickly and easily. In this approach, the energy function is computed for every snaxel and each of its neighbouring pixels in the image, and the pixel (neighbour or current snaxel) which has the lowest energy is chosen as the new position of the snaxel. The energy of the snake is minimised one snaxel at a time and the whole snake stabilises when every snaxel is located at a local minimum of energy. Such a greedy method has also been used by Milroy et al. [1997] on a 3D mesh. The energy is evaluated at two adjacent locations on the surface perpendicular to the current snaxel boundary direction: this is justified in that any force parallel to the tangent direction of the snake affects the distribution of snaxels along the snake but does not change the snake's shape. The new positions of the snaxels are computed on a quadratic surface which locally approximates the orthogonal cross-section surface at every vertex. A greedy algorithm was also used by Jung and Kim [2004] on triangle meshes; they consider as candidate new snaxel positions the 1-ring neighbourhood vertices for each snaxel (which in their algorithm lie at mesh vertices).

We adopt Milroy's method, and use a greedy algorithm. For each snaxel, we consider two new candidate snaxel positions, at fixed distances from the snake and approximately perpendicular to it. In practice these candidate locations are chosen on the mesh along the directions which bisect the two edges linking the current snaxel, and the previous and next ones along the snake. We allow snaxels to be anywhere on the mesh, not at mesh vertices or on mesh edges. The snaxel v_i and the two candidate points, v_i^* and v_i^{**}, are shown in Figure 3. Snaxel v_i either moves to one of these two points or remains still depending on which of these three points has the lowest energy.

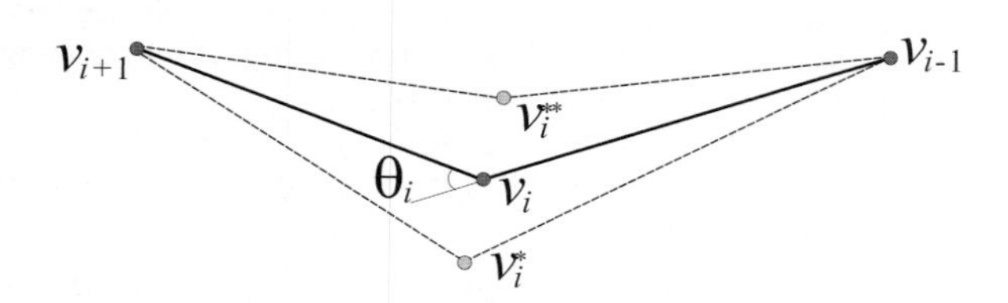

Figure 3: Candidate new snaxel positions

5.2 Internal Energy

The internal energy imposes internal shrinking and straightening forces on the snake which make the snake act both as a spring and as a flexible rod. It includes a *tension energy* term, E_t, and a *bending energy* term, E_b: $E_{int} = E_t + E_b$. We follow Milroy et al. [1997] in using the following energy terms associated with v_i: a first-order term which encourages the snake to be short,

$$E_t(i) = \frac{1}{4}\alpha\left(|v_{i-1}v_i| + |v_iv_{i+1}|\right), \tag{3}$$

and a second-order term representing the curvature of the snake, which discourages a large angle between snaxel segments, keeping it locally smooth,

$$E_b(i) = \beta\frac{1-\cos\theta_i}{\left(|v_{i-1}v_i| + |v_iv_{i+1}|\right)^2}. \tag{4}$$

The constants α and β are used to normalise the energy terms, giving each energy a value between 0 and 1. θ_i is the angle between $v_{i-1}v_i$ and v_iv_{i+1} as shown in Figure 3.

5.3 External Energy

The external energy's purpose is to drive the snake towards the relief boundary and localise the snake at it. Hence, the external energy should have a minimum at the boundary such that the snake converges towards it. For this we utilise a *feature energy* term E_f and a *deflation energy* term E_d. The external energy is: $E_{ext} = E_f + E_d$. The deflation energy helps the snake to contract from the user-drawn initial contour which may be at a large distance from the relief boundary and also to elongate the snake into concavities. The feature energy helps to creates a counter-force to the deflation force to stop the snake at the relief boundary.

We first consider the feature energy E_f designed to locate the relief contour. We do so by presenting three versions of the feature energy. A general representation is given first, and then we discuss how a revised one based on *bilateral filtering* can overcome local noise. Finally we present the version using in our program which is specifically tailored to finding the particular feature at the edge of a relief, which is at a higher level the underlying surface. We finish this section by considering the deflation energy term.

5.3.1 General Definition of Feature Energy

In order to detect general geometric features in a mesh, energies can be defined in terms of properties such as planarity, variation in normal direction, or curvature, measured at vertices of the mesh. The planarity $P(v)$ of a vertex v is defined as the signed distance between v and some plane fitted to its neighbours, where a positive value represents a vertex in a concave region and a negative value represents a vertex in a convex region. Given the planarity at mesh vertices, the planarity of points inside mesh faces can be determined by linear interpolation. Planarity is a simple measure to compute and provides an easy way to find such features as sharp edges.

For such general features, based on the idea of image feature energy, we may define E_f at a snaxel v_i as

$$E_f^1(i) = -\gamma\left(P(v_i)\right)^2, \tag{5}$$

where γ is a normalisation parameter. This gives a large negative energy where the surface is locally non-planar (as might occur near the edge of a relief).

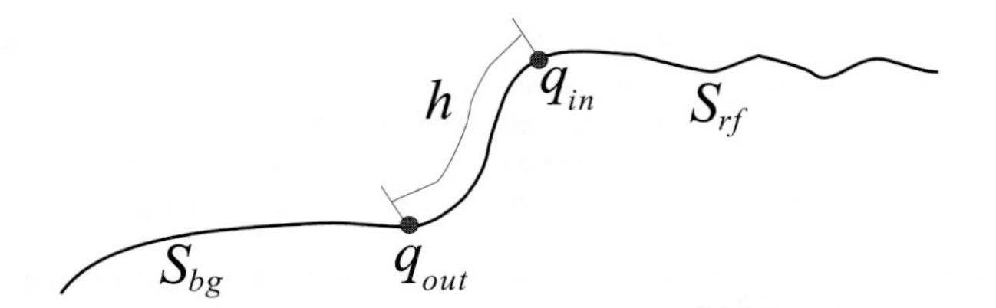

Figure 4: Cross-section of mesh near relief contour

5.3.2 Feature Energy for Noisy Surfaces

To reduce the effects of noise, mesh smoothing generally needs to be done before the properties like planarity are calculated. *Bilateral filtering*, proposed for images, is a one-sided filter derived from Gaussian blurring, which on the one hand smooths similar data while avoiding averaging dissimilar values, preserving steps in values [Tomasi and Manduchi 1998]. Extending this idea to meshes, Fleishman et al. [2003] yields a successful mesh-smoothing method which removes noise while preserving features. In the particular case of planarity, the bilaterally filtered value of planarity, $\hat{P}(v)$, at point v is defined by

$$\hat{P}(v) = \frac{\sum_{v' \in N(v)} W_c(|v - v'|) W_s(P(v) - P(v')) P(v)}{\sum_{v' \in N(v)} W_c(|v - v'|) W_s(P(v) - P(v'))}, \tag{6}$$

where $N(v)$ is a neighbourhood of v, W_c is a Gaussian filter with standard deviation σ_c, given by $W_c(x) = \exp\left(-x^2/(2\sigma_c^2)\right)$, and W_s is a similarity weight function, with standard deviation σ_s that penalises large variation in planarity, given by $W_s(x) = \exp\left(-x^2/(2\sigma_s^2)\right)$. Settings for these parameters are discussed in Section 7.2.1.

We may now revise the feature energy and replace Eqn. 5 by

$$E_f^2(i) = -\gamma\left(\hat{P}(v_i)\right)^2, \tag{7}$$

which will reduce the tendency of the snake to get stuck at local minima, caused by noise in the background region due to measurement errors.

5.3.3 Relief Step Feature Energy

When applied to a relief contour, Eqn. 7 can readily detect a sharp step boundary between a relief and the background, but does not work so well if the relief meets the background in a more gentle step. Consider how a cross section of the mesh might appear near the boundary of a relief; an example is shown in Figure 4. S_{bg} represents the background surface and S_{rf} is the relief surface. A small step joins the two surfaces and causes two local feature energy extrema at the boundary of the relief: the outer edge of the step located at q_{out} is represented by a local *maximum* of positive planarity in the sectional view, while the inner edge of the step located at q_{in} has a local *minimum* of negative planarity. This leads to E_f^1 or E_f^2 having two minima in the vicinity of the edge of the relief. Thus, we make a further modification to Eqn. 7 to give the final version of feature energy used in our algorithm. The key idea is to take advantage of the planarity signature expected near the edge of a relief. Although the relief itself may have quite a variable height, usually it will have a characteristic *step height* above the background surface *where it meets the background.* Suppose this height is h—see Figure 4. We assume that it can be measured, or estimated by the user. We use this as a typical distance in the further modified definition of the feature energy function:

$$E_f^3(i) = \begin{cases} -\gamma(\hat{P}(v_i) - \hat{P}(q_i))^2 & \text{if } \hat{P}(v_i) > \hat{P}(q_i); \\ 0 & \text{otherwise.} \end{cases} \tag{8}$$

Suppose the point q_i is a point on the mesh which is located inside the snake at geodesic distance h from snaxel v_i, measured perpendicular to the snake at v_i. When the snake is at some point of the background surface, v_i and q_i are expected to have nearly the same planarity, leading to a feature energy close to zero. When the snake is at the outer contour of the relief boundary, v_i will be the point having the local maximum planarity and q_i will have the local minimum planarity, leading to a very large negative feature energy.

However, in practice the relief may have a complicated cross-section—for example, it may rapidly go down again just inside the contour. Thus, just choosing one particular position for q_i may not lead to stable results. Instead, we use a small set of candidate positions for q_i at distances $(1-\varepsilon)h$ to $(1+\varepsilon)h$ from v_i, and select the location of q_i as the point with minimum planarity value for use in determining the energy. Generally, ε is given a fixed value 0.2 in our algorithm. Note that even if a model has large variations of relief step height, satisfactory results can be obtained by setting h as the minimum relief step height: note that point q_{out} in Figure 4 will still be the first location of local minimum energy encountered by the snake.

In practice we observe that using the function E_f^3 as the feature energy locates the relief contour very well when either the bottom or the top of the step is a sharp edge, and still produces good results when the relief blends into the background. It works well in practice as shown by the examples in Figures 6–8.

5.3.4 Deflation Energy

For the *external energy*, in addition to the feature energy, we also use a *deflation energy* E_d producing an extra force which tries to force the snake inwards. If global optimisation were used, the tension energy term in Eqn. 3 would make the snake shorter, but unfortunately this does not occur in a greedy algorithm which moves one snaxel at a time and only considers its immediate neighbours. The tension energy in this case simply tends to make the snake flat rather than shrink.

Our deflationary force is similar to that used in the *balloon model* [Cohen 1991], but it dynamically balances the internal energy taking into account the length of each snake segment and the speed of movement of the snake. Generally, the contour of a relief, when considered as a curve on an underlying surface, is composed of convex regions (e.g. around the beak of the duck in Figure 1), concave regions (e.g. around the neck of the duck in Figure 1) and flat regions (neither especially convex nor concave). Let us reconsider E_{int}. Snaxel v_i gets a smaller internal energy if it is moved towards the line between its neighbours. The result is that the snake does not move further inwards in concave regions, or even in flat regions, if we only use the internal energy alone: it would be moving to a position of higher energy. We must add a deflationary energy term to overcome this problem.

The deflationary energy E_d is defined to produce a corresponding force which acts inwards. The minimum deflationary force at a given point v_i is required when v_{i-1}, v_i and v_{i+1} lie in a straight line, and in such cases the force should be just sufficient to disrupt the flat status, while not being so large as to cause the snake to overrun the relief boundary. If correctly balanced, the snake will continue to move until it meets the relief boundary, but will not enter concavities. (We discus later how the snake is forced into concavities). For conciseness, let l_i be the length $0.5(|v_{i-1}v_i| + |v_iv_{i+1}|)$; we start evolution with all snaxels having values $l_i = l$, some desired initial value, and add or remove extra snaxels as evolution proceeds to keep snaxels about l apart, as explained in detail later. Suppose the movement step size for the snaxel v_i is τl_i. The just sufficient deflationary energy required can be derived from Eqns. 3 and 4. Taking into account that the background may be somewhat noisy, we define E_d at the inside candidate point for snaxel v_i to be:

$$E_d(i) = -\frac{1}{2}\alpha(\sqrt{1+\tau^2}-1)l_i - 2\beta\frac{\tau^2}{(1+\tau^2)^2 l_i^2} - \gamma\sigma^2, \quad (9)$$

where σ is the standard deviation of the planarity of the background surface, which can be estimated using points on the initial contour on the background as indicated by the user, and γ is the same parameter as in Eqn. 7. This equation comprises three terms. The first two resist the internal energy and the term $-\gamma\sigma^2$ eliminates the influence of any noise on the background surface.

6 Evolution

We now discuss how the snake moves from its initial contour to the final relief contour, which we call its *evolution*. The evolution of the snake occurs in three phases: firstly, *coarse evolution*, secondly, *contour refinement*, and thirdly, *final stabilisation*. Coarse evolution is used to approach the relief contour quickly. The snake stabilises close to most of the relief boundary except near concavities. Coarse evolution is carried out using large snake segments for efficiency. Contour refinement is then used both to explore the concavities, and to better capture fine detail of the contour. Smaller snake segments are used, and the internal energy is decreased. Final stabilisation is then used to accurately locate the snake at the relief contour and produce a smooth result.

In one evolutionary step, every snaxel is moved inwards or outwards a step, or kept still, depending on which of these three positions has minimum energy. Then a postprocessing process is applied before the next step to keep the snake regular, as done by Milroy et al. [1997]. First, any sharp notches which are smaller than a given sharpest angle are removed. Snaxels can be randomly distributed so without this it would be possible for the snake to fold over itself: there is no explicit control of the topology of the snake in our algorithm. Secondly, the length of each snake segment is kept within a desired range by merging any snaxels which are too short, and subdividing ones which are too long. Evolution stops after the snake has converged to a position when no snaxels move (or may be aborted if the iteration count exceeds some large number).

6.1 Coarse Evolution

Because the initial contour is far from the relief contour, we initially aim to move quickly across the background triangles on the mesh, while avoiding becoming stuck on local noise. During this phase, we define energy to be the internal energy, plus the feature energy and deflation energy. After the snake has stabilised, it will have reached a position which matches the relief boundary expect near some of the concavities, as shown for example in Figure 6(b).

Long snake segments are used at this stage, causing the snake to move quickly and capture the relief contour coarsely. Given a relief with step height h, we use a discrete curve with segment lengths about h to coarsely represent the relief boundary.

For efficiency, the status of whether each snaxel has moved or not during the current iteration is stored. If a given snaxel and its two nearest neighbours did not move on the previous iteration, we know that the centre snaxel must remain still during the current iteration, so we can save time by not bothering to compute its energy.

It is possible for small oscillations to occur in the positions of snaxels. We check after every 10 iterations whether the distance moved by each snaxel is less than a small amount (we use $2\tau l$, twice the movement step size), and if so, the snaxel is locked in position, to avoid it oscillating forwards and backwards.

6.2 Contour Refinement

After the snake has stabilised at the coarse contour, we now refine it, both to more accurately capture the details of the model, and to drive it into any concavities. We change the energy terms used, and the snake segment length, in this step.

Using the energy terms used for the coarse contour does not permit the snake to explore concavities, as its internal energy term prevents it from locally bending or expanding enough. Thus, we now modify the energy used to drive evolution.

Firstly, we adjust the deflation energy given in Eqn. 9, replacing the term $-\gamma\sigma^2$ by 0. This term is useful when coarsely finding the contour, as it allows us to step over noise in the background. However, when we are near the relief contour, we deactivate this term to prevent the snake from crossing the relief boundary. This is important to prevent snake entering the relief where its boundary is weakly defined.

We also add weights to the energy terms. Initially, during the coarse phase, each energy term had unit weight. Now the energy is redefined as

$$E = W_{int}(E_t + E_b) + W_f E_f + W_d E_d. \quad (10)$$

Decreasing W_{int} reduces the internal energy, allowing the snake to bend more, while having less tendency to shrink. Increasing W_f helps to force the snake to stop at the relief boundary. Increasing W_d helps to drive the snake into concavities. Clearly, it is the relative sizes of these weights which are important. In practice we fix W_f at 1. W_{int} is chosen by the user, as described later. W_d is then set to be larger than one, keeping the sum of the weights $2W_{int} + W_f + W_d = 4$.

During the refinement step, we set the initial snake segment length appropriate to the resolution of the mesh: half the average triangle edge length. The inserted snaxels are computed by resampling the snake on the mesh between the coarse snaxels.

6.3 Final Stabilisation

After the the refined snake has converged near the relief boundary, using the approach described above, we perform a final adjustment step to restabilise the snake, to remove unnecessary noise and yield a smoother boundary. We turn off the deflationary force by setting W_d to 0, to stop pushing the snake inwards—we have now driven the snake into the concavities, and now it is important not to overrun the boundary. W_f is again set to 1, to localise the snake at the relief boundary. We set $W_{int} = 1$ to help make the final snake smooth. The snake is re-adjusted until its position converges.

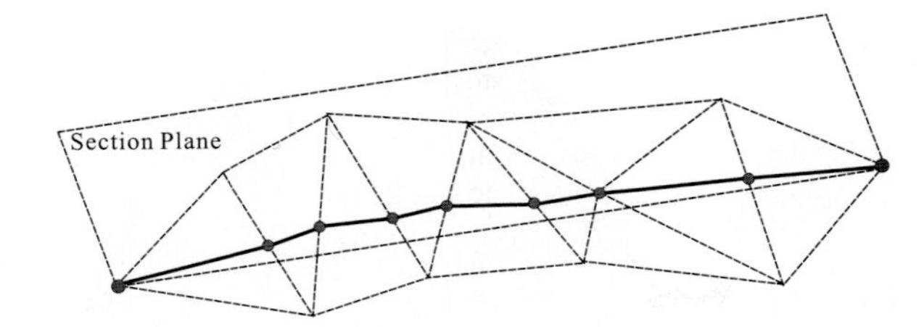

Figure 5: Intersecting the mesh with a section plane

7 Implementation Details

We now give various further implementation details of our algorithm.

7.1 Initial Contour

The user selects a few (say 4, or 6) mesh points around the outside of the relief, which are then joined across the mesh to give the initial snake. Normally this is a closed curve, but an open curve can also be used, for example if the relief runs to the edge of the mesh. We could use a shortest path algorithm to construct the initial contour [Kanai and Suzuki 2001; Surazhsky et al. 2005]. However, our snake is insensitive to the initial contour, so any cheap and simple method for constructing a reasonable connecting path between each pair of vertices can be used. We use the simple method (illustrated in Figure 5) of creating a section plane through the two vertices and the average of their normals, and intersecting the plane with the mesh. This is simple, fast and effective.

These intersection curves are sampled as described earlier to get the initial snake segments. Though the snake segments initially have equal lengths, there is no need to keep this restriction during evolution as our energy expressions properly take into account the lengths of individual snake segments. However, we do merge or divide any that become too long or too short.

7.2 Parameters

Snake-based methods are usually sensitive to choice of parameters. We have tried to make our method depend on few parameters, or at least to operate in a way which is insensitive to precise choice. The main parameter required is a user estimate for the relief step height h at the boundary of the relief; if this is variable, an estimate for its minimum value should be given.

Other factors affecting the discretised snake are adjusted according to the mesh model, as described next.

7.2.1 Parameters for Planarity Calculation

The number of neighbours of a given point to use when calculating planarity at a point depends on the resolution of the input mesh. We use all neighbours within a geodesic distance equal to the relief step height for this planarity calculation, as this is a length scale appropriate to the problem. The fast-marching method [Kimmel and Sethian 1998] is used to compute the geodesic distance to the neighbours of every vertex. For planarity bilateral filtering, the Gaussian filter W_c in Eqn. 6 decreases quickly as the distance increases. As further neighbours do not affect the filtering result much, just the 2-ring neighbours are used. Following Fleishman et al. [2003], σ_c is set to half of the maximum distance to any of the two-ring neighbours, and σ_s is set to the variance of the planarity.

7.2.2 Parameters for Snaxel Control

The relevant parameters here are, l, the initial length of each snake segment, the range of permitted snake segment lengths, and the sharpest permissible angle between segments (used to avoid self-intersections). The length of the snake segments affects speed of movement towards the contour, as well as its ability to precisely locate details of the contour and to explore concavities. As noted earlier, at the start of the coarse evolution phase, l is set to h, while at the start of the refinement phase, l is set to half of the average triangle edge length.

Segment lengths can vary in length as evolution proceeds, and although small changes are unimportant, they may change greatly wherever the snake shrinks or lengthens significantly. Redundant snaxels slow the algorithm down, while snaxels which are too sparse will not capture enough details of the contour. Thus, during both phases, extra snaxels are added, or snaxels are removed, if the snaxel spacing varies outside the range $[0.5l, 1.5l]$.

If two neighbouring segments form a sharp angle, self-intersection of the snake may occur after a few more steps. Thus, snaxels are removed if θ_i (see Figure 3) is larger than some maximum angle, set to 160° in our program.

7.2.3 Parameters for the Energy Terms

Various parameters are used to control the relative importance of the different energy terms. These are three normalisation parameters α, β, γ (see Eqns. 3, 4 and 7) and three weights W_{int}, W_f and W_d (see Eqn. 10).

The parameters α, β and γ are used to normalise the different energy terms. The snake segment length is initialised to l and it ranges between $l_{min} = 0.5l$ and $l_{max} = 1.5l$. Thus E_t lies in the range $0.5\alpha l_{min} \le E_t \le 0.5\alpha l_{max}$. From this, it is easy to choose α to normalise E_t to lie in the range $[0,1]$. Note, however, that different values are used in the coarse evolution and refinement phases. β and γ can be determined in a similar way.

The deflation energy is dynamically defined by snaxel length and speed of movement according to the internal energy at concavities, and is not normalised.

The weights W_{int}, W_f and W_d are used during refinement and stabilisation to adjust the relative importance of the energy terms as explained earlier. As noted, the only free parameter is W_{int} which is chosen by the user. Typically, the user only has to try a *very* small number of values, each of which should be half the previous value, before satisfactory results are obtained.

7.2.4 Other Parameters

Our method is relatively insensitive to choice of other parameters, and these are fixed in the algorithm. These include:

τ: the movement step size: 0.2;

ε: for locating candidate points for feature energy: 0.2;

length tolerance used for oscillation detection: $2\tau l$.

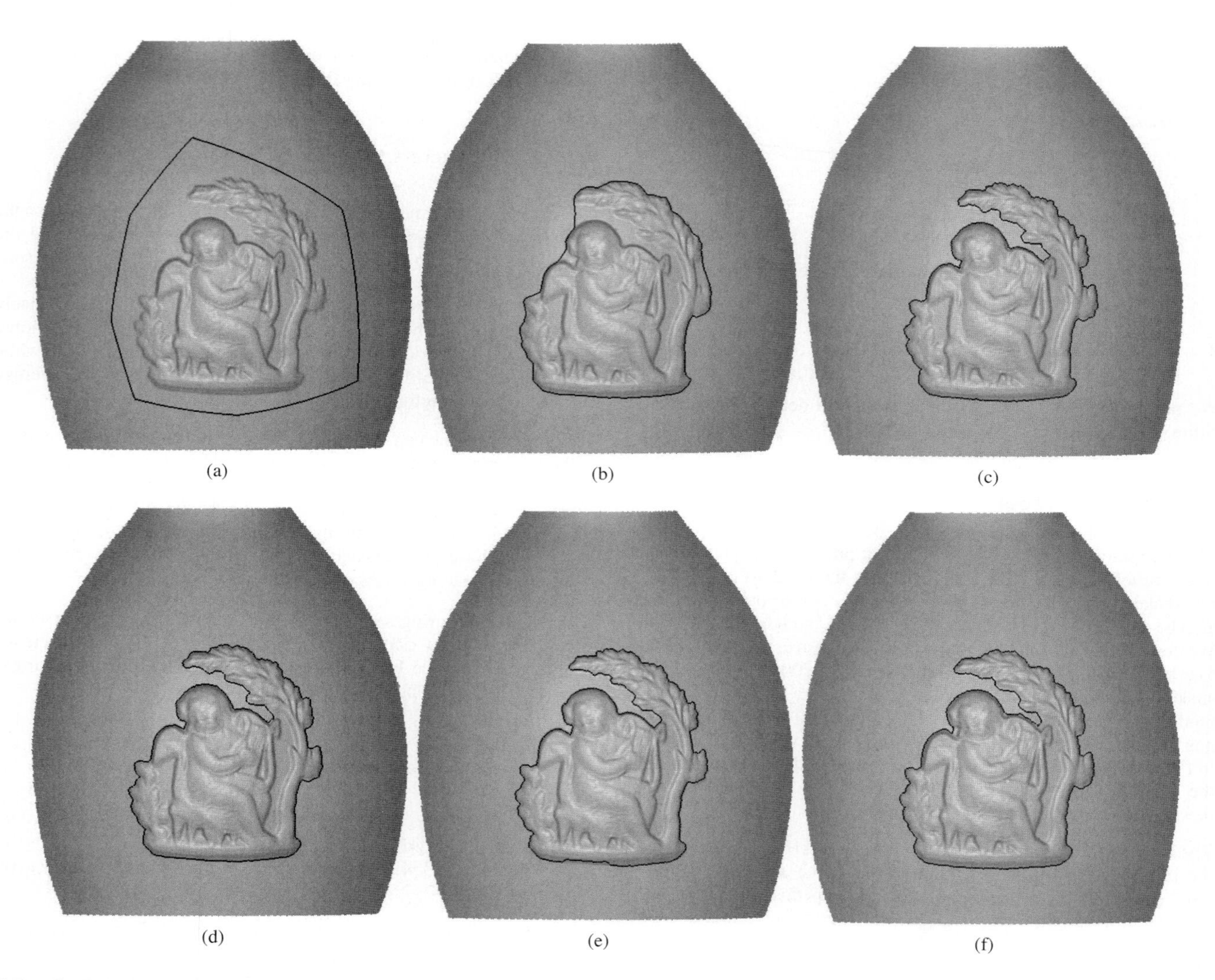

Figure 6: Relief segmentation of a model with a deep concavity: (a) initial contour, (b) coarse result, (c) refined result, (d) final contour ($h = 1.0$), (e) final contour with $h = 0.6$, (f) final contour with $h = 1.4$

The step height of the relief h need not be estimated very accurately by the user. In our experiments with the duck model (see Figure 8), good results were obtained using estimates in the range from 0.4mm–1.0mm. We provide the user with an interactive tool to calculate the geometric distance between two user-defined vertices on the mesh, allowing the user to easily estimate h.

Overall, our algorithm simply requires the user to estimate the parameter h and to tune the parameter W_{int}. All other parameters are fixed, automatically calculated or adaptively change according to the model and the snake evolution process. It is not hard for a non-technical user to produce good results.

8 Experimental Results

A variety of real scanned geometric models have been tested, having different relief contour shapes and with different resolutions and characteristics. All examples were tested on a computer with a 2.4GHz CPU and 1GB RAM.

8.1 Examples

Figure 6 shows the process of relief segmentation for a relief whose contour has a deep concavity. The model has 287446 triangles, with average edge length 0.21mm. Figure 6(a) shows the initial contour as created from a typical set of 7 user-specified points. Figures 6(b–d) illustrate the snake evolution process. A coarse relief contour with 175 snaxels is obtained after 131 iterations of energy minimisation during the coarse step, taking 21 seconds, as shown in Figure 6(b). Figure 6(c) shows the result after the refinement step; in this case the weight W_{int} was set to 0.25. It took 60 seconds using 459 iterations to produce the contour which comprises 1136 snaxels. Figure 6(d) is the result after final stabilisation. This process took 10 seconds and 94 iterations. Overall, the total process took 91 seconds. Throughout Figures 6(a–d), the parameter h was set to 1.0mm. Figures 6(e) and 6(f) show other final results when h was set to 0.6mm and 1.4mm respectively. Note that the final result does not change greatly.

Relief segmentation results for a variety of reliefs on a variety of backgrounds are shown in Figure 7. Figure 7(a) shows a model where the initial contour has deliberately been drawn as a zigzag

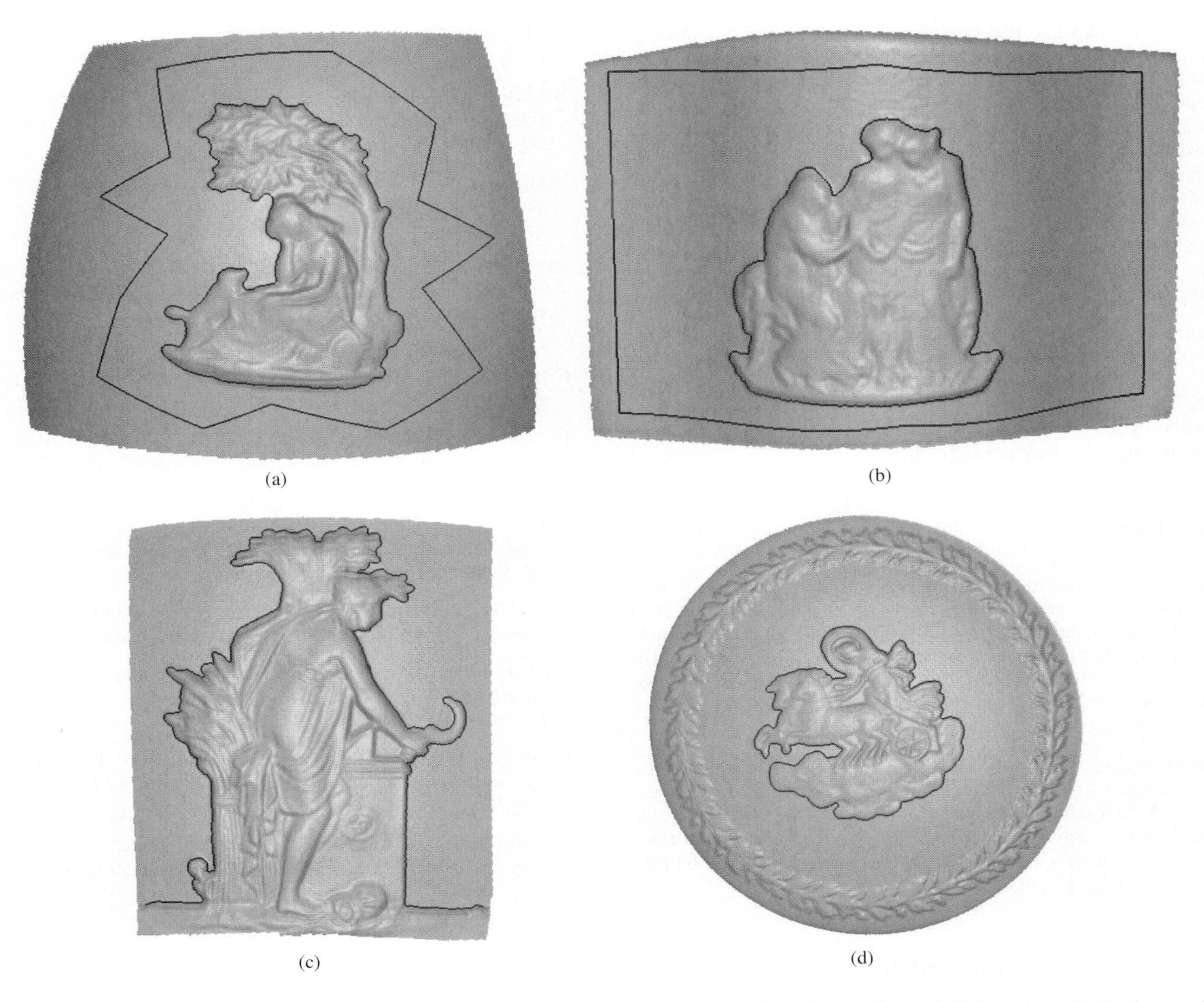

(a) (b)

(c) (d)

Figure 7: Relief segmentation for various relief and background shapes: (a) snake evolved from a zigzag initial contour, (b) background of varying curvature, (c) an open snake, (d) horse relief

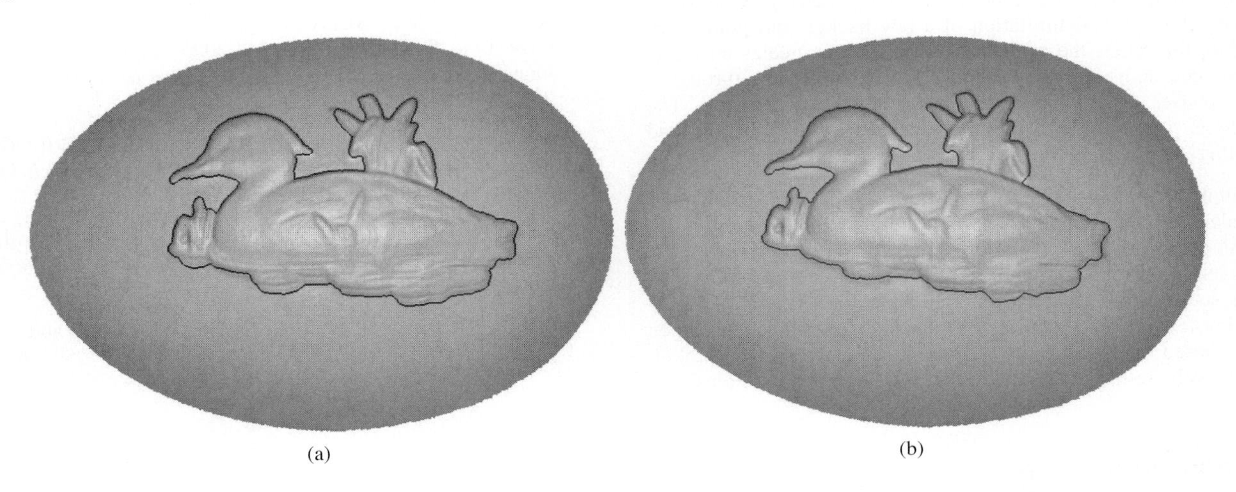

(a) (b)

Figure 8: Segmentation results for duck relief with different mesh resolutions: (a) scanned by Minolta VI-910, (b) using decimated mesh

shape to demonstrate that our method is insensitive to choice of initial contour. The snake both explores the concavities very well, and also represents the shape of the convex part in detail. For this model, it took 20 seconds to get the coarse contour using 194 snaxels and another 66 seconds to obtain the final results using 1525 snaxels. Figure 7(b) shows an example where the initial contour is far from the relief, and where the background surface has both positive and negative Gaussian curvature. In Figure 7(c), the relief boundary is an open contour, ending at the mesh boundary; for this example the background is also very noisy. Again we get a good result. Figure 7(d) is another example which shows that the snake goes into deep concavities while not overrunning the convex parts.

Figure 8 shows segmentation results for a duck relief using 2 different mesh resolutions. The mesh used in Figure 8(a) was scanned using a Minolta VI-910, and has 168700 triangles with average length 0.227mm. The initial model for Figure 8(b) was produced by decimation of the mesh used in Figure 8(a), and has 47327 triangles with average edge length 0.555mm.

8.2 Discussion

Our testing has shown that the coarse evolution phase readily produces reasonable and stable results however the initial contour is chosen. During the coarse phase, sensitivity to h and other parameters is low, but it is higher in the later phases. For example, for Figure 6, h can be successfully set anywhere in the range $[0.3, 1.6]$ during the coarse phase, but must be in the range $[0.6, 1.4]$ during the refinement phase.

Our method can deal with most concavities except those having very narrow entrances. Adjusting the method to force the snake into such concavities also generally results in the snake overrunning the relief contour elsewhere. Generally, the snake cannot enter concavities whose widths at their mouths are less than 4 to 6 times the average edge length of the mesh.

9 Conclusions and Future Work

In this paper, we have shown how to uses snakes for segmenting isolated reliefs lying on a smooth and slowly varying background surface. After user delimitation of a few background points surrounding the relief, the algorithm automatically creates an initial contour and actively evolves it to the relief boundary through a coarse phase, a refinement phase and a stabilisation phase. The process is controlled by a user estimate of relief step height, and one other parameter.

For future work, we intend to deal with reliefs with small internal holes, as well as coping with concavities with narrow mouths, by first estimating a continuation of the background surface. We also intend to extend our algorithm to cope with more complicated reliefs such as those having textured backgrounds, and cyclic reliefs. For textured background, we hope to drive the snake using an energy based on classification of the mesh into texture and relief. Closed cyclic reliefs, necessitating analysis of repeating items.

Acknowledgements

The authors wish to acknowledge the support of Delcam plc, including many helpful discussions with Richard Barratt and Steve Hobbs, and the support of EPSRC grant GR/T24425, for this work.

References

BISCHOFF, S., AND KOBBELT, L. 2004. Parameterization-free active contour models with topology control. *The Visual Computer 20*, 4, 217–228.

BISCHOFF, S., WEYAND, T., AND KOBBELT, L. 2005. Snakes on triangle meshes. http://www-i8.informatik.rwth-aachen.de/publications/publications.html.

COHEN, L. D., AND COHEN, I. 1993. Finite element methods for active contour models and balloons for 2d and 3d images. *IEEE Trans. Pattern Analysis and Machine Intelligence 15*, 11, 1131–1147.

COHEN, L. D. 1991. On active contour models and balloons. *CVGIP: Image Understanding 53*, 2, 211–218.

FLEISHMAN, S., DRORI, I., AND COHEN-OR, D. 2003. Bilateral mesh denoising. *ACM Transactions on Graphics 22*, 3, 950–953.

FUNKHOUSER, T., KAZHDAN, M., SHILANE, P., MIN, P., KIEFER, W., TAL, A., RUSINKIEWICZ, S., AND DOBKIN, D. 2004. Modeling by example. *ACM Transactions on Graphics 23*, 3, 652–663.

JUNG, M., AND KIM, H. 2004. Snaking across 3d meshes. *Proceedings of the Computer Graphics and Applications 12*, 87–93.

KANAI, T., AND SUZUKI, H. 2001. Approximate shortest path on a polyhedral surface and its applications. *Computer-Aided Design 33*, 11, 801–811.

KASS, M., WITKIN, A., AND TERZOPOULOS, D. 1988. Snakes: Active contour models. *International Journal of Computer Vision 1*, 4, 321–331.

KIMMEL, R., AND SETHIAN, J. A. 1998. Computing geodesic paths on manifolds. *Proceedings of National Academy of Sciences 1995 15*, 8431–8435.

LEE, Y., AND LEE, S. 2002. Geometric snakes for triangular meshes. *Computer Graphics Forum 21*, 3, 229–238.

MILROY, M. J., BRADLEY, C., AND VICKERS, G. W. 1997. Segmentation of a wrap-around model using an active contour. *Computer-Aided Design 29*, 4, 299–320.

SURAZHSKY, V., SURAZHSKY, T., KIRSANOVAND, D., GORTLER, S., AND HOPPE, H. 2005. Fast exact and approximate geodesics on meshes. *Proceedings of ACM SIGGRAPH 2005 24*, 553–560.

TOMASI, C., AND MANDUCHI, R. 1998. Bilateral filtering for gray and color images. *Proceedings of the 1998 IEEE International Conference on Computer Vision*, 839–846.

WILLIAMS, D. J., AND SHAH, M. 1992. A fast algorithm for active contours and curvature estimation. *CVGIP: Image Understanding 55*, 1, 14–26.

XU, C., AND PRINCE, J. L. 1998. Snakes, shapes, and gradient vector flow. *IEEE Trans. Image Processing 7*, 3, 359–369.

Feature Sensitive Mesh Segmentation

Yu-Kun Lai
Tsinghua University
Beijing, China *

Qian-Yi Zhou
Tsinghua University
Beijing, China †

Shi-Min Hu
Tsinghua University
Beijing, China‡

Ralph R. Martin
Cardiff University
Cardiff, UK§

Abstract

Segmenting meshes into natural regions is useful for model understanding and many practical applications. In this paper, we present a novel, automatic algorithm for segmenting meshes into meaningful pieces. Our approach is a clustering-based top-down hierarchical segmentation algorithm. We extend recent work on feature sensitive isotropic remeshing to generate a mesh hierarchy especially suitable for segmentation of large models with regions at multiple scales. Using integral invariants for estimation of local characteristics, our method is robust and efficient. Moreover, statistical quantities can be incorporated, allowing our approach to segment regions with different geometric characteristics or textures.

1 Introduction

Triangular meshes are now widely used in computer graphics. They are easy to acquire and widely available. The demands for techniques of analysis, processing, storage, transmission and rendering of triangular meshes are ever increasing. However, due to their irregular connectivity and lack of high-level semantic structures, the automatic analysis of meshes is challenging. Cutting mesh models into meaningful pieces is one important step towards surface understanding, and has a wide range of applications.

Segmentation is a crucial step in reverse engineering of CAD models: it divides the mesh into regions, each of which is fitted using a single analytical surface [Várady et al. 1997]. In computer graphics, various applications use segmentation as a preprocessing step. Mesh simplification can be improved by constraining contraction to take place within segmented regions, leading to improved quality of simplified models [Zuckerberger et al. 2002]. Li et al. [Li et al. 2001] demonstrated the use of well chosen segmentation in improving the performance of collision detection. Segmentation is also useful in morphing [Shlafman et al. 2002; Zuckerberger et al. 2002] and skeleton-driven animation [Katz and Tal 2003].

Ideas from cognitive science give a useful basis for model segmentation. Hoffmann and Richards [Hoffmann and Richards 1984] proposed the so-called *minimal rule*: the human visual system perceives region boundaries along negative minima of principal curvature, or concave creases. Later, Hoffmann and Singh [Hoffmann and Singh 1997] pointed out that the depth of the concavity directly affects the salience of region boundaries. Thus, concave feature regions, as well as other features, are crucial for segmentation [Katz and Tal 2003; Liu and Zhang 2004]. In addition, we wish to take geometric texture information into account—texture segmentation is widely used in image processing and should also be useful here.

Recently, the *regularized isophotic metric* was proposed in [Pottmann et al. 2004]; this distance function depends both on position and normal information. Going further, using the idea of an image manifold from image processing [Kimmel et al. 2000], each point $\mathbf{x}$ on the surface can be mapped to a corresponding point $\mathbf{x}_f = (\mathbf{x}, w\mathbf{n})$ in $\mathbb{R}^6$, where $\mathbf{n}(\mathbf{x})$ is the unit normal vector corresponding to $\mathbf{x}$ and w is a user specified constant controlling feature sensitivity [Lai et al. 2006]. This maps the surface $\Phi \subset \mathbb{R}^3$ to the corresponding 2-manifold $\Phi_f \subset \mathbb{R}^6$. Then, the *feature sensitive distance* for a curve c may be defined to be the Euclidean length of the image curve c_f. A method for feature sensitive remeshing was proposed in [Lai et al. 2006] using isotropic remeshing of Φ_f. The characteristics of the meshes produced were studied in that paper, leading to the suggestion that they could be a useful tool for robust feature extraction.

Our segmentation approach here is a clustering-based segmentation algorithm, like other state-of-the-art mesh segmentation methods such as [Katz and Tal 2003]. It uses locally defined integral invariants [Manay et al. 2004] to estimate local properties of the surface, which is much more robust than simply computing dihedral angles or estimating discrete curvatures. Feature sensitive remeshing [Lai et al. 2006] is a useful tool for efficiently computing integral invariants for segmentation. Only normal information is used, avoiding direct estimation of higher-order differential quantities.

We use feature sensitive remeshing to produce a *hierarchy* of meshes, allowing us to efficiently construct a *hierarchical segmentation*. In this way, our method performs segmentation of models using the most appropriate level in the mesh hierarchy. It can thus can handle larger models than previous k-means clustering based methods, and it can also segment coarse and fine details of complicated model using the same hierarchy. Moreover, by using statistical measures of integral invariants, we can achieve segmentation based on large-scale variation of local surface characteristics or variation in geometric texture. In this way, our method can separate regions without clear boundaries.

In Section 2, recent work on surface segmentation is discussed. We then outline our segmentation algorithm in Section 3. The two key steps, hierarchical feature sensitive remeshing, and hierarchical k-means clustering based segmentation, are detailed in Sections 4 and 5 respectively. Experimental results are presented in Section 6, while conclusions and discussions for future work are given in Section 7.

2 Related Work

Image segmentation is a key step in image analysis and understanding, and has received much attention. Its counterpart for 3D surfaces has been studied only much more recently.

Based on different aims, segmentation methods can generally be classified into two types: *patch-type* segmentation and *part-type* segmentation [Shamir 2004]. The former methods mostly aim to producing patches that satisfy certain geometric properties, often being similarity of geometric properties. For example, Sander et al. [Sander et al. 2003] segment a mesh model into a set of charts,

*laiyk@cg.cs.tsinghua.edu.cn
†zhouqy@cg.cs.tsinghua.edu.cn
‡shimin@tsinghua.edu.cn
§ralph@cs.cf.ac.uk

SPM 2006, Cardiff, Wales, United Kingdom, 06–08 June 2006.
© 2006 ACM 1-59593-358-1/06/0006 $5.00

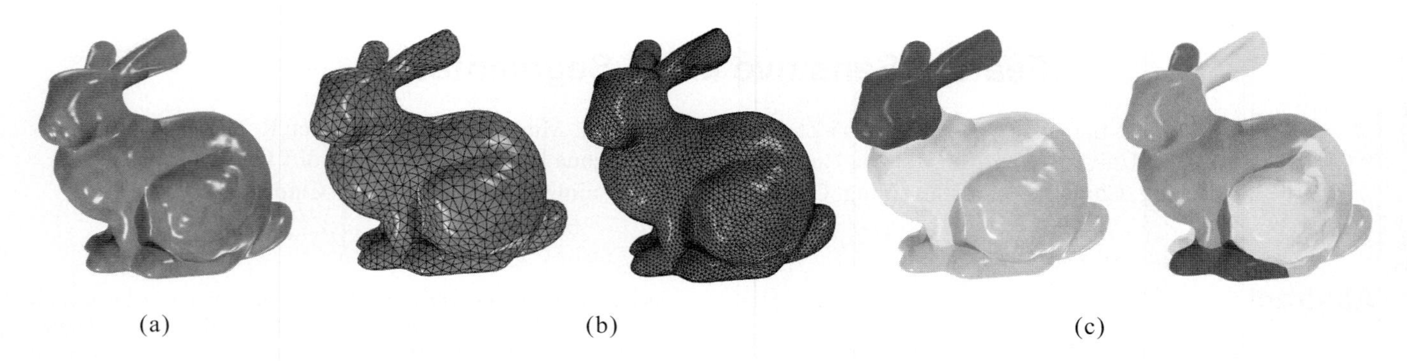

Figure 1: Steps of our segmentation algorithm

each of which is almost planar; such segmentation is suitable for parameterization as done in multi-chart geometry images. Trying to produce planar patches is too restrictive, and later improvements by Juliu et al. [Julius et al. 2005] try to segment models into quasi-developable patches. Part-type segmentation on the other hand tries to segment models into meaningful pieces, usually based on significant features. Our method is of the latter type, and we will mainly consider this type of segmentation.

Segmentation methods generally fall into two classes: *region based*, and *boundary based*. Segmentation methods rely on estimating local properties. Boundary based approaches use special values of these local properties as candidate locations for boundaries, and regions are deduced from the located boundaries. Region based methods, however, look for areas having similar properties, which define the regions, and the boundaries are deduced from them.

Various operators have been used for estimating properties. Some methods use discrete curvature estimators (e.g. [Mangan and Whitaker 1999; Page et al. 2003; Srinark and Kambhamettu 2003; Zhang et al. 2002]). Other work [Katz and Tal 2003; Liu and Zhang 2004; Shlafman et al. 2002] uses a combination of geodesic and angular distances for similarity measurement, angular distances being a function of dihedral angle between adjacent triangles. Gelfand and Guibas [Gelfand and Guibas 2004] proposed a rather different shape descriptor, using slippage analysis, which is more suitable for segmenting mechanical components than computer graphics models.

Some segmentation methods are boundary based. For example, Zhang et al. [Zhang et al. 2002] give an algorithm which explicitly locates boundaries using discrete curvatures. Generally, such approaches depend on accurate discovery of boundary loops.

However, many segmentation methods are region based. Mangan and Whitaker [Mangan and Whitaker 1999] extend the bobsledding watershed algorithm to triangular meshes. Page et al. [Page et al. 2003] use an alternative hill climbing algorithm for watershed segmentation. They compute a directional height map and use impeded climbing up negative principal curvature hills. Srinark et al. [Srinark and Kambhamettu 2003] classify local surface regions using curvature estimates, and then use region growing to segment the model, starting from certain seeds. Though fast, this approach does not seem to handle noise well, and can also result in over-segmentation. Gelfand and Guibas [Gelfand and Guibas 2004] use local slippage analysis and a multi-pass region growing approach based on slippage signatures to separate different regions. This approach also requires good quality models.

Other region-based approaches use iterative *clustering* as a tool (as we also do here). Use of global optimization allows such approaches to be more robust to variations in local properties caused by noise, for example. Shlafman et al. [Shlafman et al. 2002] use k-means clustering to provide a meaningful segmentation. However, the regions produced have jagged boundaries. This work was later improved in [Katz and Tal 2003], using hierarchical segmentation, fuzzy clustering and minimal boundary cuts to produce smoother boundaries. (In contrast, we use feature sensitive smoothing which, like the use of geometric snakes, tends to produce smoothed boundaries snapped to features). Spectral clustering was suggested by Liu et al. [Liu and Zhang 2004]; the authors claim it gives superior results for clean mesh models. However, these methods depend on dihedral angles whose computation is sensitive to noise. Geodesic and angular distances between all pairs of triangles must be precomputed and stored for efficiency; this limits the size of input mesh for which this is practical, even if done in a hierarchical manner. Another drawback is that the results depend on mesh connectivity. Cohen-Steiner et al. [Cohen-Steiner et al. 2004] use a similar Lloyd's-type clustering algorithm to that used in this work; however, their goal is surface approximation.

Unsupervised clustering techniques like the mean shift method can also be applied to mesh segmentation. Shamir et al. [Shamir et al. 2004] extend mean shift analysis to mesh models using local parameterization. Yamauchi et al. [Yamauchi et al. 2005] apply mean shift clustering to surface normals, and then use a method similar to that in [Sander et al. 2003] to compute the segmentation, based on the clustered normals. The number of clusters is computed during the segmentation process; however, the method presented in [Yamauchi et al. 2005] is likely to segment a model into more pieces than the desired number of meaningful parts.

Other approaches to segmentation also exist. Li et al. [Li et al. 2001] use edge contraction based skeletonization and space sweeping for mesh decomposition. Their method provides visually appealing results; however, it tends to capture large-scale shapes rather than features, making some decompositions impossible. Mitani and Suzuki [Mitani and Suzuki 2004] proposed a technique for making paper models from meshes. This can be considered as involving a special segmentation scheme that guarantees each region is developable. Wu and Levine [Wu and Levin 1997] proposed a method that simulates the distribution of electrical charges on the surface. Boundaries of regions are locations with minimal charge. The aim is to capture sharp concave features which are usually perceived as natural boundaries. Unlike most other methods, this approach does not depend on differential quantity estimation and is thus more stable. However, assumptions are needed about the nature of the input object, which should generally have an even distribution of concave feature regions for boundaries, and the mesh should also be closed. Katz et al. [2005] propose a segmentation algorithm based on feature points and core extraction. Pose-invariant results are reported in this paper. However, an expensive optimization method is used to find feature points, which limits the complexity of models that can be efficiently handled after simplification.

Other work has considered interactive mesh segmentation. As well as extracting features automatically, the approach proposed in [Lee et al. 2004] also gives the user tools to to close and optimize boundary loops separating different parts of the object. Funkhouser et al. [2004] present a modeling system based on searching for and stitching parts from a database. An intuitive interactive segmentation tool is given to find optimal cuts guided by user-drawn strokes, formulated as a constrained least-cost path problem.

Our approach is automatic and region-based. As in the method in [Katz and Tal 2003], we can produce a hierarchy of segmentation, of particular use in certain applications. We use hierarchical feature sensitive isotropic remeshing to efficiently and robustly segment large models at several levels. Integral invariants and statistical quantities are used as local properties in a k-means clustering approach. These provide more robustness than dihedral angles, are insensitive to small fluctuations in surfaces, and are also capable of segmentation using certain types of geometric textures.

3 Algorithm Overview

Given an input model represented as a triangular mesh, and a few user specified parameters, our method produces a set of disjoint, constituent regions, whose union is identical to the input mesh.

For some input models, preprocessing may be desirable. Though our algorithm can be applied to manifold models with or without holes, we do assume that any necessary topological correction has been carried out to remove unwanted gaps, tiny handles or other deficiencies (*topological noise*). For very large models (more than 10^6 triangles, using a current desktop PC), it may be desirable to apply mesh simplification [Garland and Heckbert 1997; Hoppe 1996] in order to obtain results in a reasonable time (a few minutes).

We now map the mesh from $\mathbb{R}^3$ to its counterpart in $\mathbb{R}^6$, as described earlier. This is done vertex by vertex while keeping the connectivity unaltered. As part of this process, the normal at each vertex needs to be estimated. For models that are not too noisy, normals can be reliably estimated using 1-ring neighbors. For noisier models, however, improved results can be achieved by estimating normals using local planar or quadratic surface fitting to a neighborhood—see [Lai et al. 2006] for further details.

Next, the new mesh is subjected to hierarchical feature sensitive remeshing, as explained in detail in Section 4. This process generates a hierarchy of feature sensitive (FS) meshes, of increasing resolution with successive levels of detail, and with clear correspondences between adjacent levels. This is accomplished by first constructing the coarsest level of remeshing; for each finer level, every triangle in the coarser level is subdivided into 4 triangles and newly inserted vertices are repositioned in nearby locations to optimize isotropic sampling in $\mathbb{R}^6$. The hierarchical output is conceptually similar to an Gaussian image pyramid. In the first step of segmentation, the coarsest level of remeshing is used, reducing the computational complexity and ensuring stability with respect to small scale fluctuations. As hierarchical segmentation proceeds, each area to be segmented at lower levels contains fewer triangles. When an area contains too few triangles on the current mesh so that the corresponding finer mesh contains fewer than a certain prespecified number of triangles, we move to a finer mesh.

A k-means clustering algorithm is used to segment a given area (at the top level, the remeshed surface, or part of it at lower levels). This clusters triangles to form regions as detailed in Section 5. The number of clusters, k, can be specified by the user, or can derived by optimization as in [Katz and Tal 2003].

A *metric* measuring distances between triangles is needed to perform clustering. We use a definition incorporating geodesic distance, integral invariants related to averaged normal curvature, and statistical measures of these invariants characterizing local properties such as geometric texture. Given a user specified number of regions to be generated at the current level, several initial regions are selected, which are then improved iteratively. This segmentation process is performed at several levels (if desired), giving a hierarchical segmentation.

Input: *a mesh model; certain user-specified parameters.*
Output: *a set of disjoint, contiguous regions representing meaningful parts of the input model.*

1. Preprocessing (optional);
2. Compute hierarchical feature sensitive remeshing.
3. Perform hierarchical segmentation with k-means clustering and a new distance metric.
4. Map the result back to the original model (optional).
5. Perform feature sensitive boundary smoothing.

Figure 2: Algorithm Overview

The segmentation results on the FS mesh can be mapped to the original input mesh by projection, in a similar way to the approach used in [Katz and Tal 2003]. Alternatively, the segmented FS mesh itself, which has better properties, may be used in downstream applications.

If the initial region boundaries are too jagged, they can be improved by *feature sensitive smoothing* as in [Lai et al. 2006] or by use of *geometric snakes* [Lee and Lee 2002].

Our algorithm is summarised in Figure 2.

4 Hierarchical Feature Sensitive Remeshing

Clustering-based segmentation algorithms need to compute distances between pairs of triangles on the mesh. In practice, distances between *most* pairs of triangles will be required. These pairwise distances need to be randomly accessed during k-means clustering and should be kept in main memory. They may be found using Dijkstra's shortest-path algorithm in $O(N^2 \log N)$ time, where N is the number of triangles; $O(N^2)$ storage is required. This is expensive for large N, and previous methods (e.g. [Katz and Tal 2003]) have used a simplified mesh as a means to provide a segmentation for large models. Simplification ideally would done in such a way as to ensure consistency of the segmentation with the original model.

Hierarchical segmentation was first introduced by Katz et al. [Katz and Tal 2003]. It is able to represent a decomposition of a model at different levels, mimicking the way people think. To achieve hierarchical segmentation, they simply segment each region at each level of detail into further regions, recursively. However, at the higher levels, the simplified triangulation count is low for each region, and inadequate detail may be present due to mesh simplification. This has an impact on the correctness and accuracy of the segmentation of the original mesh.

Our alternative approach is suited to large meshes: we use multiresolution, hierarchical, feature sensitive remeshing to reduce the size of the computational problem, while avoiding the loss of significant detail in the reduced size meshes. In particular, the input model is remeshed into a hierarchy of models with different resolutions with clear correspondences between adjacent levels in the hierarchy. The coarsest remeshing is used for the initial segmentation. In earlier stages of hierarchical segmentation, *details* of models are usually of little use, and the global shape at a coarse resolution is important. Later, areas of the model are segmented using more detailed meshes. At finer levels of detail, only single already-segmented areas of the mesh need by processed at a given time, not the whole mesh. By using hierarchical remeshing in conjunction with hierarchical segmentation, our method is capable of handling larger models, and segmenting them into more levels.

Thus, instead of using mesh simplification, we use *feature sensitive, isotropic, remeshing* [Lai et al. 2006]. Typically, we might make from one to three levels of such FS meshes, each with 1/4 the number of triangles of the previous level. An FS mesh in general has almost equilateral, equally sized triangles in $\mathbb{R}^6$. However, it also has the desirable property that triangles are elongated along sharp features. This makes it possible to efficiently represent models. Moreover, a topological disk on an FS mesh is a good approximation to a geodesic disk in $\mathbb{R}^6$. Let α represent the mapping between Φ and Φ_f. It has been shown in [Lai et al. 2006] that the affine first derivative mapping $D\alpha^{-1}$ maps the unit circle $\mathbf{k}_f$ in $\mathbb{R}^6$, centered at some point $\mathbf{x}_f$ on Φ_f, to an ellipse $\mathbf{k}$ in $\mathbb{R}^3$, in the corresponding tangent plane of Φ at the point $\mathbf{x}$. This mapping distorts the local shape. The *principal distortions*, corresponding to principal curvatures, are the extremal distortions. Thus, the distances of the vertices of the ellipse to its center are the corresponding principal distortions $1/\lambda_i$, $i = 1,2$, which satisfy

$$\lambda_i^2 = 1 - w^2 K + 2w^2 H \kappa_i = 1 + w^2 \kappa_i^2, \quad (1)$$

where κ_i are the two principal curvatures, respectively [Lai et al. 2006]. Thus the principal distortions, which can be estimated as an integral quantity, are closely related to the local surface curvatures. We explain how we use them for segmentation in Section 5.

FS remeshing can be computed by extending an isotropic remeshing algorithm (e.g. [Alliez et al. 2003; Surazhsky et al. 2003; Witkin and Heckbert 1994]) which works in $\mathbb{R}^3$ to the feature sensitive metric in $\mathbb{R}^6$. We use the method in [Lai et al. 2006]. It uses the iterative method in [Witkin and Heckbert 1994] to optimize the sampling, and to further improve the results, geodesic distances computed by the method in [Surazhsky et al. 2005], rather than Euclidean distances, are used in an energy function which is a sum of spring energies designed to cause vertices to repel one another.

As geodesic distances are used, an FS mesh contains almost equilateral triangles with almost identical size in terms of the feature sensitive metric on the input model. Given such a mesh, if each triangle is split into four smaller ones, by inserting a vertex at some appropriate point near the mid-point of each edge, the resulting refined model is still nearly isotropic in this sense. The refined model can be computed as follows. Insert a vertex at the mid-point of each edge of the coarse level mesh, and project these newly inserted vertices onto the input model in $\mathbb{R}^3$. (The projection can be done in $\mathbb{R}^3$ or $\mathbb{R}^6$. However, it appears to be more robust to do the projection in $\mathbb{R}^3$). The connectivity and positions of the vertices from the coarser level mesh are kept unaltered. This ensures that vertices do not move globally, and the correspondence between the coarse and finer levels is simply given by the above one-to-four mapping. The new vertices are repositioned using spring energy optimization. The neighborhood used for computing spring energy functions can be easily found from the topological neighbors based on subdivision. The optimization carried out is similar to the one used for remeshing at the coarsest level. Armijo rule [Kelley 1999] step-size control can be incorporated to make the result more stable. The initial positions are usually quite close to the required solution, and it takes just a few iterations to achieve acceptable remeshing results.

As noted earlier, it is usually sufficient to remesh models at from one to three levels, depending on detail in the input model and degree of segmentation required. Fig. 3 shows the Armadillo model (originally with 345,944 triangles) remeshed with 11,756 (left) and 47,024 triangles (right).

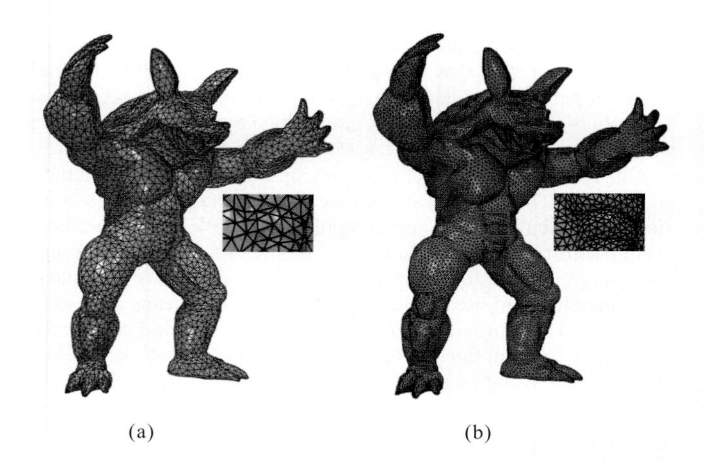

(a) (b)

Figure 3: FS meshes of the Armadillo model at coarser and finer resolutions

5 Hierarchical Segmentation

Use of remeshed models makes segmentation tractable, and estimation of geometric properties efficient. The triangles in the model as clustered into k meaningful regions using k-means clustering, which assigns triangles to clusters according to distances from an iteratively updated *representative* triangle for each cluster. Section 5.1 gives our basic approach to clustering-based hierarchical segmentation, Section 5.2 discusses the key issue of distance computation, and Section 5.3 briefly considers how to ensure that each region has a smooth boundary.

5.1 Hierarchical Segmentation Approach

Our hierarchical segmentation approach is similar to those in [Katz and Tal 2003] and [Shlafman et al. 2002]. The main differences lie in the distance computation (see Section 5.2) and the use of multiresolution remeshing. The number of clusters, k, can be specified by the user, or can derived automatically by optimization as in [Katz and Tal 2003].

The algorithm proceeds from coarse to fine segmentation of the input mesh. Segmentation is performed on the appropriate FS mesh and mapped back to the original mesh if desired. Initially, the entire lowest resolution FS level mesh is segmented into regions. Subsequently, if further segmentation is required, the segmentation process is applied to the individual regions identified at the previous level of segmentation. A finer FS mesh is used if the region has a manageable size in this mesh, i.e. contains fewer than a certain prespecified maximum number (say 10,000) of triangles Otherwise, the coarser FS mesh is still used.

In order to segment a target (the whole object or a region) into smaller regions, k-means clustering is used as follows:

1. **Precompute distances between triangles.** The distance between each pair of triangles is computed using a metric which combines geodesic, curvature-related and geometric texture-based information.

2. **Pick seeds.** Seed triangles may be chosen by the user. Otherwise a seed triangle is randomly selected for the first cluster, and seeds are found for the other clusters one by one, by choosing the triangle that has the largest average distance to all other seeds found so far.

3. **Assign triangles to the nearest cluster.** Each triangle is assigned to the cluster to whose representative is closest.

4. **Update the representative of each cluster.** The representative r is updated to be the one minimizing

$$\sum_{f \in \text{Region}(r)} D(f,r);$$

here $D(f,r)$ is the distance between triangles f and r.

Steps 3 to 4 are iterated; in practice, just a few iterations suffice to converge to adequate results.

Note that the precomputed pairwise distances can be used again for further levels of segmentation if the same resolution of FS mesh is used, leading to efficiency.

5.2 Distance Computation

Distance computation is the key step in any k-means clustering algorithm. It affects the clustering outcome and a suitable metric must be carefully chosen dependent on the problem. For clustering-based segmentation, geodesic distance and angular distance have been used [Katz and Tal 2003; Shlafman et al. 2002; Liu and Zhang 2004].

Geodesic distance favors segmentation of equal sized regions, whereas angular distance tries to force region boundaries to lie where surface direction changes quickly, e.g. at sharp edges. This works well for clean models, producing regions typically separated by (ideally deep) valleys. However, for real, noisy, scanned models, angular distances are less useful. Noisy meshes can contain triangles with relatively large dihedral angles between them. In the extreme, a small spike on a relatively smooth surface may be segmented as a region if its triangles are far from all its neighbors in angular distance. Moreover, the angular distance approach usually applies a nonlinear mapping (e.g. the cosine function) to the dihedral angles, in order to reduce this distance in flatter regions. The overall effect for a specific portion of surface depends on whether it is represented by a few large triangles, or a larger number of smaller triangles.

Like previous authors, we first define the distance between an *adjacent* pair of triangles. The distance between *any* pair of triangles on the mesh can then be computed by following the shortest path on the dual graph of the mesh, using Dijkstra's algorithm.

We define the distance function differently to previous authors, in terms of integral and statistical information about local features. In addition to using geodesic distance D_{geod} like previous work, we add two further terms, namely the *curvature distance* D_{curv}, derived from the mapping distortion of $\mathbb{R}^6$ geodesic disks, and the *texture distance* D_{texture} that measures changes in geometric texture or other statistical surface properties. The distance between an adjacent pair of triangles f_i and f_j is overall defined as

$$\begin{aligned} D(f_i,f_j) &= c_1 \cdot D_{\text{geod}}(f_i,f_j)/\bar{D}_{\text{geod}} \\ &+ c_2 \cdot D_{\text{curv}}(f_i,f_j)/\bar{D}_{\text{curv}} \\ &+ (1-c_1-c_2) \cdot D_{\text{texture}}(f_i,f_j)/\bar{D}_{\text{texture}} \end{aligned} \quad (2)$$

where the averages $\bar{D}_*$ are over all pairs of adjacent triangles. We typically use $0.1 \le c_1 \le 0.2$ and $0.7 \le c_2 \le 0.9$.

The geodesic distance $D_{\text{geod}}(f_i,f_j)$ between two adjacent triangles is defined as the sum of the distances from the barycenters of two triangles to the center of the edge that is shared by the two triangles.

To efficiently compute the other two distance terms, we compute the mapping distortion at each vertex of the FS mesh. For a vertex v, we use an r-ring of neighbours (typically $r = 1$ to 3) as an approximation to a $\mathbb{R}^6$ geodesic disk. The freedom to choose r differently provides flexibility to selectively ignore small scale features, allowing a tradeoff between accuracy and robustness. The principal distortions λ_{min} and λ_{max}, and corresponding directions $\mathbf{d}_{\text{min}}$ and $\mathbf{d}_{\text{max}}$, can be approximated using the geodesic disk. We find the shortest and longest distance from v to any point on the boundary of this neighborhood using a fast geodesic distance computation [Surazhsky et al. 2005]. Let these distances be Δ_{min} and Δ_{max} respectively, with corresponding vectors (projected on the tangent plane of v) $\mathbf{d}_{\text{min}}$ and $\mathbf{d}_{\text{max}}$. We can estimate the mapping distortion as $\lambda_{\text{min}} = \Delta_{\text{min}}/l$ and $\lambda_{\text{max}} = \Delta_{\text{max}}/l$, where l is the radius of the geodesic disk in the feature sensitive metric. We use $\mathbf{d}_{\text{min}}$ and its orthogonal direction in the tangent plane of v as approximate principal directions

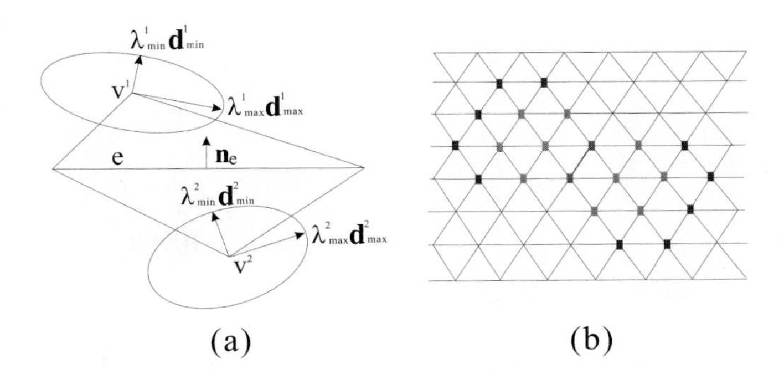

Figure 4: Distortion estimation and sampling pattern for geometric statistics

We now compute D_{curv} in Equation 2. For an adjacent triangle pair f_i and f_j, we consider the mapping distortion λ_e in the direction orthogonal to their common edge e, as follows. Taking the two vertices v^1, v^2 opposite to e in these triangles, we use their principal distortions to estimate the mapping distortion λ_e in the direction $\mathbf{n}_e$ orthogonal to e:

$$\lambda_e = \frac{1}{2} \sum_{i=1,2} (\lambda^i_{\text{min}} \mathbf{d}^i_{\text{min}} + \lambda^i_{\text{max}} \mathbf{d}^i_{\text{max}}) \cdot \mathbf{n}_e. \quad (3)$$

(see Fig. 4(a)). D_{curv} can then be defined as

$$D_{curv}(f_i,f_j) = \eta G(\frac{1}{\lambda_e} - 1). \quad (4)$$

where η is a coefficient controlling the relative importance of convex and concave regions. Cognitive theory emphasises the importance of segmentation at concave regions, so η should be a small number (e.g. $0.1 \le \eta \le 0.2$) for convex regions and 1.0 for concave ones. G is a sigmoidal nonlinear function used to reduce the effect of low responses. Its use is very similar to application of the cosine function to dihedral angles when computing angular distances. We use

$$G(x) = 1 - \cos(\frac{\pi}{c} \min(x,c)) \quad (5)$$

where c is a threshold; $c = 0.5$ works well in practice. All the examples presented in this paper use this setting for c.

Most previous methods for mesh segmentation do not take into account differences in geometric textures when performing segmentation. While computation of local similarity is possible, it tends to be expensive, and sensitive to noise. We therefore compute statistical properties of integral invariants (e.g. λ_{min}) for use as descriptors of local surface properties. Another suitable integral quantity, the radius ratio ρ, is given by

$$\rho = \sqrt{\text{Area in } \mathbb{R}^3/\text{Area in } \mathbb{R}^6}. \quad (6)$$

Small ρ corresponds to at least one of the principal curvatures being large. It can also be seen from Equation 1 that λ_{min} has a close

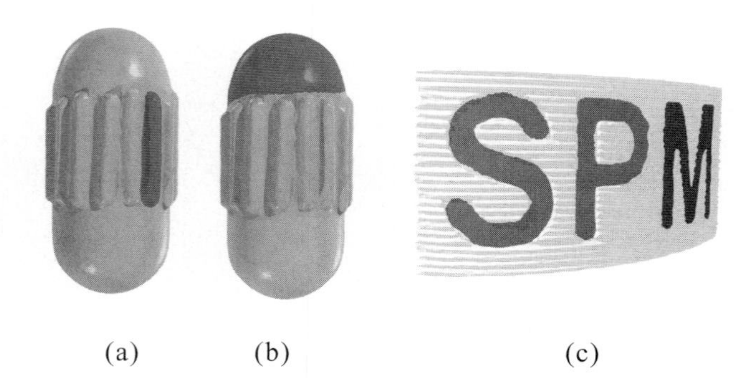

Figure 5: Segmentation by texture

relationship to the local average curvature. To use these descriptors for local shapes, given an edge e between two adjacent triangles f_i and f_j, we sample a specific neighborhood on either side of the edge (see Fig. 4(b)). The average and standard deviation of $\lambda_{\min}$ and ρ are computed and placed in a vector V (the standard deviation is multiplied by a weight to control relative importance of averages and standard deviations), and $D_{texture}$ is thus defined as:

$$D_{texture} = \|V_i - V_j\|^2. \quad (7)$$

By balancing the weights of constituent parts of the distance function, our method can produce varying results, as desired by the user. Fig. 5(b) shows an example where D_{texture} has been emphasised to segment a textured object into three regions; in this case we used $c_1 = c_2 = 0.1$. The result in Fig. 5(a) was produced with a weight for D_{texture} of 0.

Fig. 5(c) shows another example that segments the flat letters *SPM* from a surface covered with a grooved geometric texture. Because a large neighborhood has been used for geometric texture statistics, the locations of the boundaries are not very accurate and there also exist some rounding effects at corners due to the use of smoothing. A separate boundary optimization process is likely to be necessary in practice to further improve the results.

5.3 Patch Boundary Smoothing

Though our results tend to produce more robust boundaries near sharp features than previous methods, by utilising integral quantities rather than dihedral angles or discrete differential quantities, in some cases, the segmentation results are still jagged. We use *feature sensitive smoothing* [Lai et al. 2006] to improve this. The basic idea is to optimize discretized spline-in-tension energy in the feature sensitive metric. Segmentation boundaries tend to pass through feature regions, and such smoothing has the ability to snap the boundary to the features. It bears some similarity to geometric snakes [Lee and Lee 2002], but is much simpler to implement and avoids local parameterization which includes unavoidable mapping distortions.

Region boundary smoothing needs to be done carefully. As there are *branching* vertices on boundaries where three or more regions meet, we cannot move each boundary loop independently. Such vertices are detected, and the boundary is split into segments where any such vertices exist. Each segment is smoothed separately while keeping the locations of the branching vertices fixed (the positions of the latter could also be optimized too for further improvement).

The remeshing results can be mapped back to the original input model if desired using a projection method similar to the one in [Katz and Tal 2003].

Boundary smoothing should be done after projection.

6 Experimental Results

The weight w used for mapping into $\mathbb{R}^6$ should be chosen according to the scale of the input mesh [Lai et al. 2006] In the experiments reported later, we scaled the models to fit into a bounding box of size 1. The choice of w is not critical, and we typically used $0.05 \le w \le 0.1$.

Fig. 1 illustrates the main steps of our algorithm using the 'bunny' model with 25,000 triangles, remeshed using two levels of FS mesh with 17,336 and 4,334 triangles, respectively. Two levels of segmentation are shown, using the coarser and finer FS mesh respectively.

Fig. 6 shows segmentation of various objects with our method. The results are better for models with sharp features, especially concave features separating different parts (e.g. Figs. 6(a-d)). The FS meshes contain elongated triangles which tend to follow features, so that the computed region boundaries do not need smoothing. For models with fewer features, or whose features do not form closed loops, slightly more jagged boundaries result, and the segmentation is not as robust. For example, in Fig. 6(f), the front leg is cut higher than others, which in some sense is also reasonable, because the cut boundary takes into account certain creases on the model. Our method is robust in the presence of small fluctuations, as illustrated by Fig. 6(j). By varying the neighborhood size when computing integral invariants, the scale of features being considered can be altered. This is a useful property when small scale, sharper features coincide with large scale, smoother features. The various scanned models segmented in this Figure show the insensitivity of our method to noise.

Fig. 5 shows an example of segmentation by texture. The size of the statistical neighborhood used affects the texture captured. Our method can efficiently separate certain kinds of geometric textures, but due to the use of *simple* statistical measures, limitations exist; we wish to extend it to handle more complex kinds of geometric texture.

Fig. 7 shows two examples of segmenting models into a larger number of levels. The original 'eagle' model contains 33,072 triangles, and we used a simplified version of the 'Lucy' model containing 237,278 triangles. Segmentation was done to three levels. Using hierarchical FS meshes, detail can be retained while keeping the computation time and main memory manageable. Using the finest FS mesh, even the hand of 'Lucy' and the foot of the 'eagle' can be reliably segmented.

We tested our method on a Pentium IV 2.4GHz computer with 1GB RAM. Remeshing a model with 200K triangles to about 10K triangles takes under a minute. Segmentation time is directly related to the size of the input FS mesh. For models with 4K triangles, local property estimation takes 0.1s, pairwise distance computation 15.4s, and clustering 1.3s. For models with $11K$ triangles, the times are 0.2s, 134.8s, and 7.9s respectively. The time is dominated by computing pairwise distances. Note that smaller models require significantly less time to compute. By using *hierarchical* FS meshes, the time for pairwise distance computation is greatly reduced.

Compared to previous work, our approach is capable of handling complicated models with high efficiency, due to the use of feature sensitive hierarchical remeshing. Generally, our method is relatively insensitive to choices of the parameters in the algorithm. However, by using an appropriate neighborhood size in the computation of integral invariants, we believe more robust results might be achieved. Note that the ability to separate certain kinds of geometric *texture* means our method is particularly useful for certain types of application.

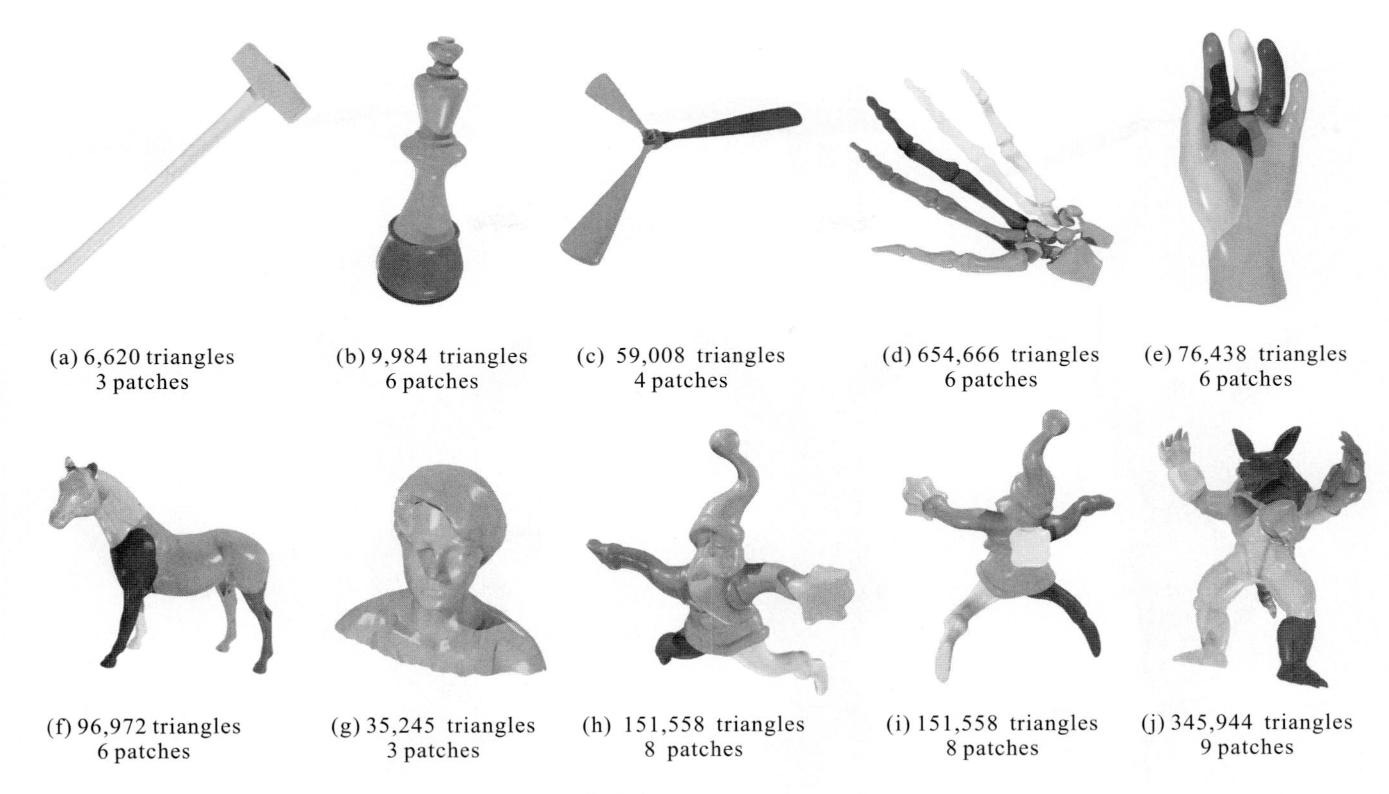

(a) 6,620 triangles 3 patches | (b) 9,984 triangles 6 patches | (c) 59,008 triangles 4 patches | (d) 654,666 triangles 6 patches | (e) 76,438 triangles 6 patches

(f) 96,972 triangles 6 patches | (g) 35,245 triangles 3 patches | (h) 151,558 triangles 8 patches | (i) 151,558 triangles 8 patches | (j) 345,944 triangles 9 patches

Figure 6: Various segmentation results.

7 Conclusions

In this paper, a top-down hierarchical mesh surface segmentation algorithm was presented. As it is based on isotropic remeshing, it is insensitive to the input triangulation of the mesh. By using integral and statistical properties, problems due to noise are avoided, and textures can also be captured and used to drive the segmentation. Hierarchical FS remeshing not only provides an efficient tool for computation, but can handle larger models with more accuracy than earlier methods. Use of finer models leads to better following of features, and better region boundaries. Improved results are obtained when more levels of segmentation are used.

We hope to address remaining limitations in the future. The algorithm is dependent on the initial seed points used, and better approaches for their placement needs to be explored. Choosing the number of regions automatically and reliably also needs further work. Our statistical approach needs extension to handle more complicated textures.

Acknowledgements

Models in this paper are courtesy of Cyberware, Stanford University, Georgia Institute of Technology and the Polhemus Corporation. This work was partially supported by the Natural Science Foundation of China (Projects 60225016, 60333010, 60321002) and the National Basic Research Project of China (Project 2002CB312101).

References

ALLIEZ, P., DE VERDIÈRE, É. C., DEVILLERS, O., AND ISENBURG, M. 2003. Isotropic surface remeshing. In *Proceedings of Shape Modeling International*, 49–58.

COHEN-STEINER, D., ALLIEZ, P., AND DESBRUN, M. 2004. Variational shape approximation. In *Proceedings of SIGGRAPH*, 905–914.

FUNKHOUSER, T., KAZHDAN, M., SHILANE, P., MIN, P., KIEFER, W., TAL, A., RUSINKIEWICZ, S., AND DOBKIN, D. 2004. Modeling by example. In *Proceedings of SIGGRAPH*, 652–663.

GARLAND, M., AND HECKBERT, P. 1997. Surface simplification using quadric error metrics. In *Proceedings of SIGGRAPH*, 209–216.

GELFAND, N., AND GUIBAS, L. J. 2004. Shape segmentation using local slippage analysis. In *Proceedings of Eurographics Symposium on Geometry Processing*, 219–228.

HOFFMANN, D. D., AND RICHARDS, W. A. 1984. Parts of recognition. *Cognition 18*.

HOFFMANN, D. D., AND SINGH, M. 1997. Salience of visual parts. *Cognition 63*, 29–78.

HOPPE, H. 1996. Progressive meshes. In *Proceedings of SIGGRAPH*, 27–36.

JULIUS, D., KRAEVOY, V., AND SHEFFER, A. 2005. D-charts: Quasi-developable mesh segmentation. *Computer Graphics Forum 24*, 3, 581–590.

KATZ, S., AND TAL, A. 2003. Hierarchical mesh decomposition using fuzzy clustering and cuts. In *Proceedings of SIGGRAPH*, ACM Press, vol. 22(3), 954–961.

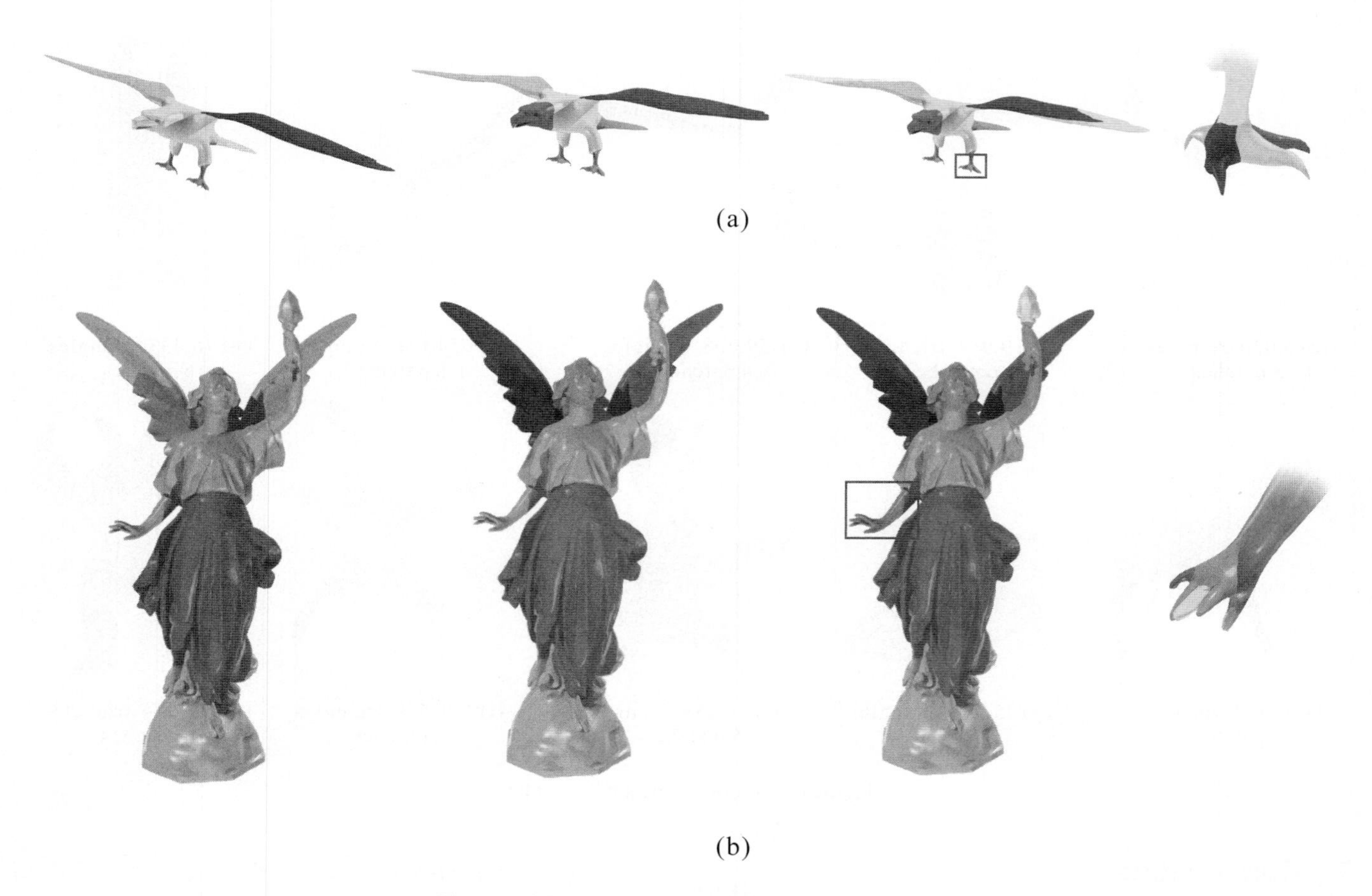

Figure 7: Hierarchical segmentation of larger models.

KATZ, S., LEIFMAN, G., AND TAL, A. 2005. Mesh segmentation using feature point and core extraction. *The Visual Computer 21*, 8–10, 865–875.

KELLEY, C. T. 1999. *Iterative Methods for Optimization*. SIAM.

KIMMEL, R., MALLADI, R., AND SOCHEN, N. 2000. Images as embedded maps and minimal surfaces: movies, color, texture and volumetric images. *Intl. J. Computer Vision 39*, 111–129.

LAI, Y.-K., ZHOU, Q.-Y., HU, S.-M., WALLNER, J., AND POTTMANN, H. 2006. Robust feature classification and editing. *IEEE Transactions on Visualization and Computer Graphics*. to appear.

LEE, Y., AND LEE, S. 2002. Geometric snakes for triangular meshes. *Computer Graphics Forum 21*, 3, 229–238.

LEE, Y., LEE, S., SHAMIR, A., COHEN-OR, D., AND SEIDEL, H.-P. 2004. Intelligent mesh scissoring using 3d snakes. In *Proceedings of Pacific Graphics*, 279–287.

LI, X., WOON, T. W., TAN, T. S., AND HUANG, Z. 2001. Decomposing polygon meshes for interactive applications. In *Proceedings of ACM Symp. on Interative 3D Graphics*, 35–42.

LIU, R., AND ZHANG, H. 2004. Segmentation of 3D meshes through spectral clustering. In *Proceedings of 12th Pacific Graphics*, 298–305.

MANAY, S., HONG, B.-W., YEZZI, A. J., AND SOATTO, S. 2004. Integral invariant signatures. In *Proceedings of ECCV*, Springer, 87–99.

MANGAN, A. P., AND WHITAKER, R. T. 1999. Partitioning 3D surface meshes using watershed segmentation. *IEEE Trans. on Visualization and Computer Graphics 5*, 4, 308–321.

MITANI, J., AND SUZUKI, H. 2004. Making papercraft toys from meshes using strip-based approximate unfolding. In *Proceedings of SIGGRAPH*, 259–263.

PAGE, D. L., KOSCHAN, A. F., AND ABIDI, M. A. 2003. Perception-based 3D triangle mesh segmentation using fast marching watershed. In *Proceedings of the IEEE Conf. on Computer Vision and Pattern Recognition*, 27–32.

POTTMANN, H., STEINER, T., HOFER, M., HAIDER, C., AND HANBURY, A. 2004. The isophotic metric and its application to feature sensitive morphology on surfaces. In *Proceedings of ECCV 2004, Part IV*, Springer, 560–572.

SANDER, P. V., WOOD, Z. J., GORTLER, S. J., SNYDER, J., AND HOPPE, H. 2003. Multi-chart geometry images. In *Proceedings of Eurographics Symposium on Geometry Processing*, 146–155.

SHAMIR, A., SHAPIRA, L., COHEN-OR, D., AND GOLDENTHAL, R. 2004. Geodesic mean shift. In *Proceedings of the 5th Korea Israel conference on Geometric Modeling and Computer Graphics*, 51–56.

SHAMIR, A. 2004. A formulation of boundary mesh segmentation. In *Proceedings of Second International Symposium on 3D Data Processing, Visualization and Transmission*, 82–89.

SHLAFMAN, S., TAL, A., AND KATZ, S. 2002. Metamorphosis of polyhedral surfaces using decomposition. *Computer Graphics Forum 21*, 3, 219–228.

SRINARK, T., AND KAMBHAMETTU, C. 2003. A novel method for 3D surface mesh segmentation. In *Proceedings of the 6th Intl. Conf. on Computers, Graphics and Imaging*, 212–217.

SURAZHSKY, V., ALLIEZ, P., AND GOTSMAN, C. 2003. Isotropic remeshing of surfaces: a local parameterization approach. In *Proceedings of 12th Intl. Meshing Roundtable*, 215–224.

SURAZHSKY, V., SURAZHSKY, T., KIRSANOV, D., GORTLER, S., AND HOPPE, H. 2005. Fast exact and approximate geodesics on meshes. In *Proceedings of SIGGRAPH*, 553–560.

VÁRADY, T., MARTIN, R. R., AND COX, J. 1997. Reverse engineering of geometric models - an introduction. *Computer Aided Design 29*, 4, 255–268.

WITKIN, A., AND HECKBERT, P. 1994. Using particles to sample and control implicit surfaces. In *Proceedings of SIGGRAPH*, 269–277.

WU, K., AND LEVIN, M. D. 1997. 3D part segmentation using simulated electrical charge distributions. *IEEE Transaction on Pattern Analysis and Machine Intelligence 19*, 11, 1223–1235.

YAMAUCHI, H., LEE, S., LEE, Y., AND OHTAKE, Y. 2005. Feature sensitive mesh segmentation with mean shift. In *Proceedings of Shape Modeling International*, 236–243.

ZHANG, Y., PAIK, J., KOSCHAN, A., ABIDI, M. A., AND GORSICH, D. 2002. A simple and efficient algorithm for part decomposition of 3-D triangulated models based on curvature analysis. In *Proceedings of Intl. Conf. on Image Processing*, III, 273–276.

ZUCKERBERGER, E., TAL, A., AND SHLAFMAN, S. 2002. Polyhedral surface decomposition with applications. *Computers & Graphics 26*, 5, 733–743.

Identifying Flat and Tubular Regions of a Shape by Unstable Manifolds

Samrat Goswami *
CS and ICES, U. Texas at Austin,
Austin, TX 78712

Tamal K. Dey †
CSE, Ohio State U.,
Columbus, OH 43210

Chandrajit L. Bajaj ‡
CS and ICES, U. Texas at Austin,
Austin, TX 78712

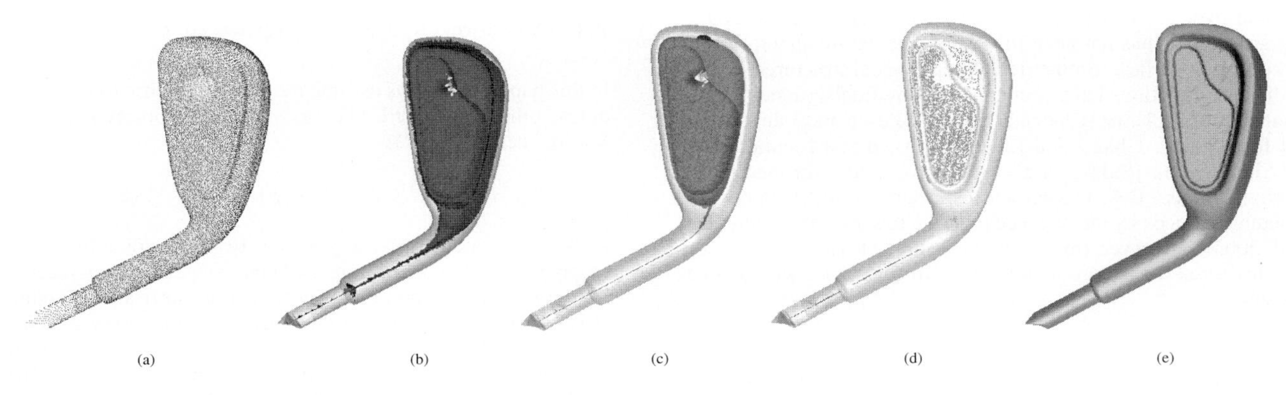

Figure 1: The steps of the algorithm are shown on an example dataset CLUB. Starting with an input set of points sampled from the surface (a), the medial axis in the interior of the shape is computed (b). The algorithm then detects the set of index 1 and index 2 saddle points lying on the interior medial axis and computes the unstable manifold of these saddle points (c). The unstable manifold of an index 1 saddle point is two dimensional (green) and the unstable manifold of an index 2 saddle point is one dimensional (red). The algorithm then collects the local maxima lying on the boundaries of these two types of unstable manifolds and tag them as falling into two different categories. The stable manifolds of these maxima are then used to map the 2-dimensional and 1-dimensional part of the medial axis back to the surface. The flat portion on the surface is colored cyan and the tubular region is colored golden (e).

Abstract

We present an algorithm to identify the flat and tubular regions of a three dimensional shape from its point sample. We consider the distance function to the input point cloud and the Morse structure induced by it on $\mathbb{R}^3$. Specifically we focus on the index 1 and index 2 saddle points and their unstable manifolds. The unstable manifolds of index 2 saddles are one dimensional whereas those of index 1 saddles are two dimensional. Mapping these unstable manifolds back onto the surface, we get the tubular and flat regions. The computations are carried out on the Voronoi diagram of the input points by approximating the unstable manifolds with Voronoi faces. We demonstrate the performance of our algorithm on several point sampled objects.

CR Categories: F.2.2 [Nonnumerical Algorithms and Problems]: Geometrical problems and computations; I.3.5 [Computational Geometry and Object Modeling]: Curve, surface, solid and object representations

Keywords: Point cloud, Voronoi diagram, Delaunay triangulation, Unstable Manifold.

*e-mail:samrat@ices.utexas.edu
†e-mail: tamaldey@cse.ohio-state.edu
‡e-mail:bajaj@cs.utexas.edu

1 Introduction

Problem and motivation. Many applications in shape modeling require to identify the salient features of a given shape. Some of them such as assembly planning, feature tracking, animations, structure elucidation of bio-molecules, human-body modeling benefit from a semantic annotation of the features. One such natural annotation is achieved by classifying the features as 'tubular' and 'flat'. Obviously, this annotation is ambiguous since the feature-space is a continuum resulting into features that cannot be simply classified as tubular or flat. Nevertheless, many designed and organic shapes have pronounced features that are perceived to be tubular and flat. We seek to identify these features using a topological method. The unstable manifolds induced by a shape distance function identify some one- and two-dimensional subsets of the medial axis. The preimage of a function that maps the points on the surface to the medial axis provides an association of the shape to these one- and two-dimensional subsets. The preimage of the one-dimensional subset is called tubular whereas that of the two-dimensional subset is called flat. Our experimental result shows that this classification can be effectively approximated for many datasets in practice.

Previous results. Because of the significance of the problem, quite a few work spanning various approaches have been reported

SPM 2006, Cardiff, Wales, United Kingdom, 06–08 June 2006.
© 2006 ACM 1-59593-358-1/06/0006 $5.00

in the literature. To mention a few, we refer to the curvature based methods of [Várady et al. 1997] and [Mortara et al. 2004a; Mortara et al. 2004b], the fuzzy clustering method of [Katz and Tal 2003], the method based on PCA of surface normals by [Pottmann et al. 2004], the hybrid variational surface approximation by [Wu and Kobbelt 2005] and the Reeb graph approach of [Shinagawa et al. 1996] and [Verroust and Lazarus 2000]. Remarkably the distance function over $\mathbb{R}^3$ which is defined by the distance to the boundary of the shape has not been fully used for feature annotation. In the context of surface reconstruction, topological structures induced by distance functions have been analyzed by Edelsbrunner [Edelsbrunner 2002], Chaine [Chaine 2003] and Giesen and John [Giesen and John 2003]. Chazal and Lieutier [Chazal and Lieutier 2004] and Siddiqi et al. [Siddiqi et al. 1998] have used it for medial axis approximations. Dey, Giesen and Goswami used the topological structures induced by the distance function to segment a shape [Dey et al. 2003]. However, this work stops short of using the topological structures for feature annotations. In this paper we complete this step.

Results. Given a compact surface Σ smoothly embedded in $\mathbb{R}^3$, a distance function h_Σ can be assigned over $\mathbb{R}^3$ that assigns to each point its distance to Σ.

$$h_\Sigma : \mathbb{R}^3 \to \mathbb{R}, \; x \mapsto \inf_{p \in \Sigma} \|x - p\|$$

In applications, Σ is often known via a finite set of sample points P of Σ. Therefore it is quite natural to approximate the function h_Σ by the function

$$h_P : \mathbb{R}^3 \to \mathbb{R}, \; x \mapsto \min_{p \in P} \|x - p\|$$

which assigns to each point in $\mathbb{R}^3$ the distance to the nearest sample point in P.

In this paper, we start with a finite sample P of Σ and identify the index 1 and index 2 saddle points of h_P from the Voronoi diagram $\mathrm{Vor}\,P$ and its dual Delaunay triangulation $\mathrm{Del}\,P$ of P. We then select only the saddle points of both indices which lie on the interior medial axis of Σ and compute their unstable manifolds. The unstable manifold of index 1 saddle points (U_1) are two dimensional whereas those of index 2 (U_2) are one dimensional. Exact computations of U_1 is prone to numerical error. So, we present an algorithm to compute them approximately. We then map the points belonging to U_1 and U_2 back to Σ. The image of U_1 under the mapping gives the flat regions of Σ and that of U_2 gives its tubular regions.

Thus, the main contributions of this paper are:

- Algorithms to compute the unstable manifolds of the index 2 saddles points of h_P exactly and those of the index 1 saddle points approximately,
- Identification of the tubular and flat features of Σ from its point sample P via the unstable manifolds of the saddle points,
- Experimental results exhibiting the performance of our algorithm on several point sampled objects.

The paper is organized as follows. In Section 2 we state some definitions and explain the terms such as Voronoi-Delaunay diagram, induced flow, stable/unstable manifolds etc. In Section 3 we describe the relation between the Voronoi-Delaunay diagram of the point set P and the induced flow. In Section 4 we describe the structure of the unstable manifolds of index 1 and index 2 saddle points and present an algorithm to compute them. In Section 5 we give an algorithm to map the unstable manifolds back to the surface to identify its flat and tubular features. In Section 6 we demonstrate the results of our algorithm on several models ranging from CAD objects to protein molecules. We conclude in Section 7.

2 Preliminaries

2.1 Voronoi-Delaunay Diagram of P

In this paper we always assume the distance metric to be Euclidean unless otherwise stated. For a finite set of points P in $\mathbb{R}^3$, the Voronoi cell of $p \in P$ is

$$V_p = \{x \in \mathbb{R}^3 : \forall q \in P - \{p\}, \|x - p\| \leq \|x - q\|)\}.$$

If the points are in general position, two Voronoi cells with non-empty intersection meet along a planar, convex Voronoi facet, three Voronoi cells with non-empty intersection meet along a common Voronoi edge and four Voronoi cells with non-empty intersection meet at a Voronoi vertex. A cell decomposition consisting of the *Voronoi objects*, that is, Voronoi cells, facets, edges and vertices is the Voronoi diagram $\mathrm{Vor}\,P$ of the point set P.

The dual of $\mathrm{Vor}\,P$ is the Delaunay diagram $\mathrm{Del}\,P$ of P which is a simplicial complex when the points are in general position. The tetrahedra are dual to the Voronoi vertices, the triangles are dual to the Voronoi edges, the edges are dual to the Voronoi facets and the vertices (sample points from P) are dual to the Voronoi cells. We also refer to the Delaunay simplices as *Delaunay objects*.

2.2 Induced Flow

The distance function h_P induces a flow at every point $x \in \mathbb{R}^3$. This flow has been characterized earlier [Giesen and John 2003]. See also [Edelsbrunner 2002]. For completeness we briefly mention it here.

Critical Points. The critical points of h_P are those points where h_P has no non-zero gradient along any direction. These are the points in $\mathbb{R}^3$ which lie within the convex hull of its closest points from P. It turns out that the critical points of h_P are the intersection points of the Voronoi objects with their dual Delaunay objects.

- *Maxima* are the Voronoi vertices contained in their dual tetrahedra,
- *Index 2 saddles* lie at the intersection of Voronoi edges with their dual Delaunay triangles,
- *Index 1 saddles* lie at the intersection of Voronoi facets with their dual Delaunay edges, and
- *Minima* are the sample points themselves as they are always contained in their Voronoi cells.

In this discrete setting, the index of a critical point is the dimension of the lowest dimensional Delaunay simplex that contains the critical point.

Flow. For every point $x \in \mathbb{R}^3$, let $V(x)$ be the lowest dimensional Voronoi object that contains x and $D(x)$ be its dual. Now *driver* of x, denoted as $d(x)$, is defined as

$$d(x) = \mathrm{argmin}_{y \in D(x)} \|x - y\|$$

The direction of steepest ascent can be uniquely determined by a unit vector in the direction of $x - d(x)$. The critical points coincide with their drivers. Now one can assign a vector v at every x with a zero vector assigned at the critical points. The resulting vector field is not necessarily continuous. Nevertheless, it induces a *flow* in $\mathbb{R}^3$. This flow tells how a point x moves in $\mathbb{R}^3$ along the steepest ascent of h_P and the corresponding path is known as the *orbit* of x. We can also define an *inverted orbit* of x where x moves in the direction of steepest descent.

Stable and Unstable Manifolds. For a critical point c its stable manifold is the set of points whose orbits end at c. The unstable manifold of a critical point c is the set of points whose inverted orbits end at c. The structure and computation of stable manifolds of the critical points of h_P were described in [Giesen and John 2003]. They can be computed from the Delaunay triangulations of the given point sets though they may not be subcomplexes of the Delaunay triangulations. For computational advantages they are also approximated by Delaunay subcomplexes as in [Dey et al. 2003].

We are interested in computing unstable manifolds and their approximations. As the Delaunay and Voronoi diagrams, the structures of stable and unstable manifolds have a duality. Interestingly, one can compute the unstable manifolds and their approximations from the Voronoi diagrams. Here we state some of the facts about the unstable manifolds of the critical points.

1. MAXIMA. The unstable manifold is the local maximum itself.
2. INDEX 2 SADDLES. The unstable manifold of an index 2 saddle point is a polyline starting at the saddle point and ending at a maximum.
3. INDEX 1 SADDLES. The unstable manifold of an index 1 saddle point is a two dimensional surface patch which is bounded by the unstable manifold of index 2 saddle points.
4. MINIMA. The unstable manifold of a local minimum is a three dimensional polytope bounded by the unstable manifold of critical points with higher indices.

In Section 4 the computation of the unstable manifold of index 1 and index 2 saddle points is described.

3 Flow on Voronoi Objects

Before we state the connection between the flow induced by h_P and the Vor-Del diagram of P, we would like to state some facts about the relative position of Voronoi and Delaunay objects. These relative positions can describe the nature of flows in the Voronoi objects. These facts were clearly explained in [Edelsbrunner 2002] for a more general setting of power distance.

Fact 1 *The unoriented normal to the supporting plane of a Voronoi facet is along its dual Delaunay edge and the plane passes through the midpoint of the edge. The Delaunay edge, though, may or may not intersect the dual Voronoi face.*

Figure 2 illustrates the two possibilities that may arise. The left figure corresponds to the situation that results in an index 1 saddle point.

Fact 2 *The supporting line of a Voronoi edge always intersects the plane of the dual Delaunay triangle at its circumcenter and is along its unoriented normal. The Voronoi edge may or may not intersect the interior of the Delaunay triangle.*

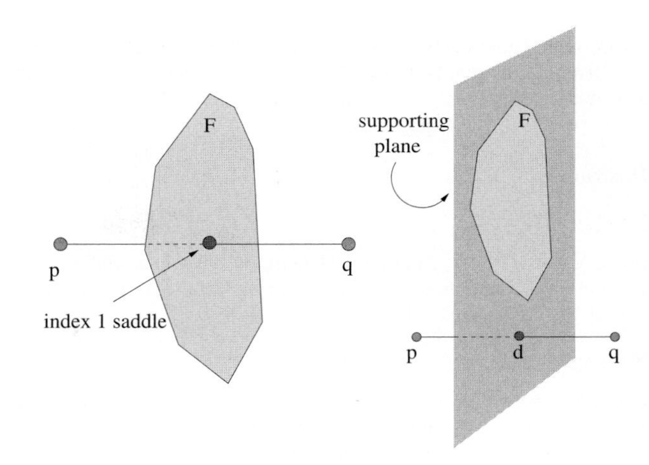

Figure 2: Relative position of a Voronoi facet F with respect to its dual Delaunay edge pq. The left picture shows the creation of an index 1 saddle point. The right picture shows the position of the driver d of F.

Figure 3 lists the four possible scenarios. The bottom right corresponds to the generation of an index 2 saddle point.

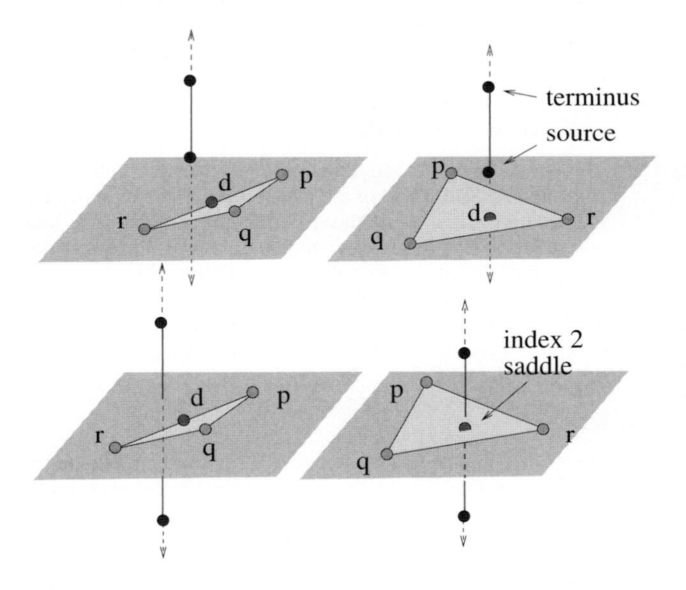

Figure 3: Relative position of a Voronoi edge e with respect to its dual Delaunay triangle pqr. The blue circles denote the two Voronoi vertices defining e. The driver of e is marked d and the supporting plane of triangle pqr is drawn in cyan.

We have already seen that the critical points of h_P can be computed from $\mathrm{Vor}\,P$ and $\mathrm{Del}\,P$. Also, the driver of a point x comes from the Delaunay object dual to the Voronoi object x lies in. In this context we would like to state the following lemma which is key to the further computations.

Lemma 1 *All interior points of a Voronoi object have the same driver.*

This result can be easily proved by considering all the different cases regarding the dimension of the Voronoi object and its position with respect to its dual Delaunay object.

By Lemma 1 and Facts 1 and 2 we can list the possible position of the drivers of the points lying in the interior of a certain dimensional Voronoi object.

Position of Drivers

Voronoi Cell

For a Voronoi cell V_p, the dual Delaunay object is a singleton set containing the sample point p and therefore all points x in the interior of V_p has p as their driver.

Voronoi Facet

Consider a Voronoi facet in the intersection of V_p and V_q. The dual Delaunay edge is pq and the midpoint of pq is the driver of all x lying in the interior of the Voronoi facet (Figure 2(right)).

Voronoi Edge

Next, consider a Voronoi edge in the intersection of V_p, V_q, V_r. As Fact 2 and Figure 3 indicate, the infinite line segment containing the Voronoi edge may or may not intersect the convex hull of p, q, r leading to two different possibilities

Case 1.1 In case of intersection, the circumcenter of pqr is the driver. Such Voronoi edges will be termed *non-transversal* edges as the flow is along the edge itself. The Voronoi edge has two Voronoi vertices as its endpoints. If both of them are in the same half-space defined by pqr, the closer Voronoi vertex is called *source* and the further one is called *terminus* of the Voronoi edge because the flow is directed from the closer to the further vertex. Figure 3 (top right) illustrates this case.

Case 1.2 If the Voronoi edge does not intersect the affine hull of $p, q,$ and r, the midpoint of the edge opposite to the largest angle of pqr is the driver. These Voronoi edges will be termed as *transversal*. If any point x moving along its orbit hits one such edge, the position of the driver implies that it will enter the Voronoi facet dual to the Delaunay edge opposite to the largest angle in pqr. Such Voronoi facet will be termed *acceptor* facets of that *transversal* Voronoi edge. Figure 4 illustrates the situation.

Voronoi Vertex

The case of Voronoi vertex again requires the analysis of two different cases. We assume, that it is outside its dual tetrahedron because otherwise it is a local maximum and hence is its own driver. Let v be a Voronoi vertex with the dual tetrahedron σ whose four neighbors are $\sigma_i, i = 1 \ldots 4$. Further, let the corresponding shared triangles between σ and σ_i be $t_i, i = 1 \ldots 4$ where $w_i, i = 1, \ldots 4$ is its opposite vertex in σ.

Case 2.1 There is only one triangle t_i of σ for which the Voronoi vertex v and the opposite vertex w_i lie in two different half-spaces defined by t_i. Let e_i be the Voronoi edge between the duals of σ and σ_i. Then, the driver for v (dual to σ) is same as the driver of e_i. In such cases, e_i is termed as the *outgoing* Voronoi edge of v. See top row of Figure 5 for an illustration.

Case 2.2 There are two triangles t_i, t_j of σ for which the Voronoi vertex v and the opposite vertex (w_i and w_j) lie in two different half-spaces defined by the corresponding triangles. Let e_i, e_j be the Voronoi edges defined as in Case 2.1. Note, in this case, both e_i, e_j are the *outgoing* Voronoi edges of v. There are two possibilities that we need to consider further.

Case 2.2.1 Both e_i, e_j are *transversal*: In this case the *acceptors* of both of them is dual to the Delaunay edge $t_i \cap t_j$ and the corresponding driver is the midpoint of $t_i \cap t_j$. See bottom-left subfigure of Figure 5.

Case 2.2.2 One of e_i, e_j is *transversal*: The driver is same as that of the non-transversal Voronoi edge. See bottom-right subfigure of Figure 5.

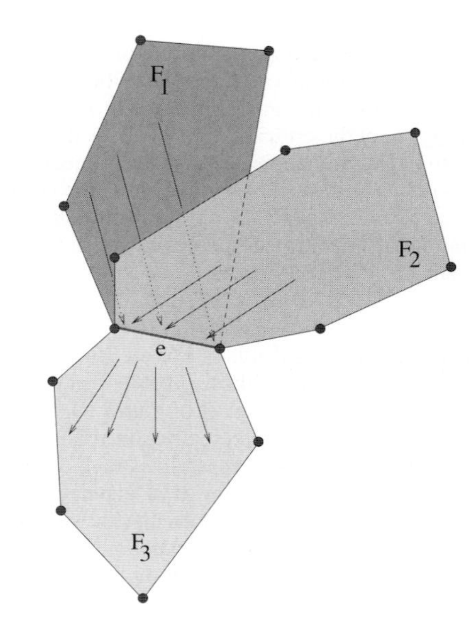

Figure 4: Transversal Voronoi edge e is shown in red with three incident Voronoi facets. Flow direction is shown with arrows. Flow from either of F_1 or F_2 hits e and enters F_3, the acceptor of e.

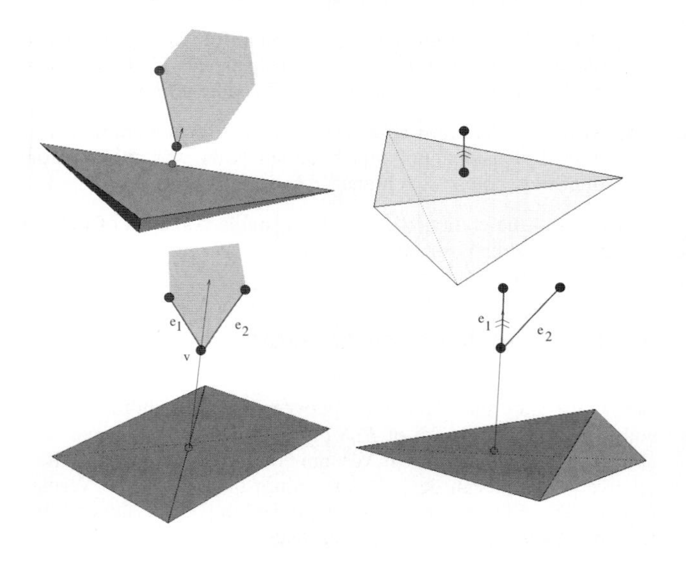

Figure 5: Possible driver positions of a Voronoi vertex v according to the cases 2.1 and 2.2.$(1-2)$. The acceptor Voronoi facet is shown in pink. The flow along a non-transversal Voronoi edge is shown with a double arrow. The driver is shown in red circle.

In this context we state another lemma that is important for subsequent developments.

Lemma 2 *Let F be an acceptor Voronoi Facet for the transversal Voronoi edges $e_1 = (v_1, v_2) \ldots e_k = (v_k, v_{k+1})$ around it.*

1. *The Voronoi edges $e_1 \ldots e_k$ form a continuous chain around F.*

2. *The Voronoi vertices $v_2 \ldots v_k$ fall in the category 2.2.1. The Voronoi vertices v_1 and v_{k+1} fall in the category 2.2.2.*

3. *F, $e_1 \ldots e_k$, $v_2 \ldots v_k$ have same driver which is the midpoint of the Delaunay edge dual to F.*

We omit the proofs of all of the above claims.

4 Computing Unstable Manifolds

4.1 Unstable Manifold of Index-2 Saddle Points

In this section we describe the structure and computation of the unstable manifolds of index 2 saddle points.

The unstable manifold of an index 2 saddle point is one dimensional. In our discrete setting it is a polyline with one endpoint at the saddle point and the other endpoint at a local maximum. The polyline consists of segments that are either subsets of non-transversal Voronoi edges or lie in the Voronoi facets. Due to the later case, the polyline may not be a subcomplex of Vor P.

Let us consider an index 2 saddle point, c, at the intersection of a Delaunay triangle t with a Voronoi edge e. Let the two tetrahedra sharing f be σ_1, σ_2. The edge e has the endpoints at the dual Voronoi vertices of σ_1 and σ_2, denoted as v_1, v_2 respectively. The unstable manifold $U(c)$ of c, has two intervals - one from c to v_1 and the other from c to v_2. We look at the structure of one of them, say the one from c to v_1, and the other one is similar.

At any point on the subsegment cv_1, the flow is toward v_1 from c. Once the flow reaches v_1, the subsequent flow depends on the driver of v_1. Instead of just looking at v_1, we consider a generic step, where the flow reaches at a Voronoi vertex v and we enumerate the possible situations that might occur depending on the position of driver of v. If v is a local maximum, the flow stops there, as the driver of v is v itself. Otherwise there are two cases to consider.

- v **falls into Case 2.1:** Let the dual tetrahedron be σ and the driver of v is same as that of the Voronoi edge e which is between the dual of σ and one of its neighbors, say σ'. If e is non-transversal, the flow will be along the Voronoi edge e till it hits the Voronoi vertex at the other endpoint (dual to σ'). Otherwise, the flow enters the acceptor Voronoi facet F of e. Due to Lemma 2, the driver of F is same as the driver of e. Therefore the next piece of the unstable manifold can be uniquely determined by the driver of e, say d and the Voronoi vertex v. It is the segment between v and the point where the ray $\overrightarrow{dv}$ intersects a Voronoi edge of F.

- v **falls under Case 2.2.x:** This situation is similar to the one described above. In case of both of the Voronoi edges being *transversal* (Case 2.2.1), the flow enters the acceptor Voronoi facet. In the other case (Case 2.2.2), the flow follows the non-transversal Voronoi edge.

Some segments of $U(c)$ are not along the Voronoi edges. Wherever the flow encounters a transversal Voronoi edge, it seizes to follow the Voronoi edge and enters a Voronoi facet which is acceptor for that Voronoi edge. This calls for the analysis of the flow when it crosses an acceptor Voronoi facet and hits a Voronoi edge. We have already characterized the position of the driver for a Voronoi edge and thereby classified those edges as either transversal or non-transversal. If the current edge intersected by the ray from the driver to v is a non-transversal edge, the flow will follow that Voronoi edge and hit a Voronoi vertex. Otherwise, it will enter the acceptor Voronoi facet of the Voronoi edge again. There is a technical difficulty we need to point out. Unless the acceptor for this Voronoi edge is different from the Voronoi facet the flow came from, we may encounter a cycle. The following lemma saves us from this awkward situation.

Lemma 3 *Let F be a Voronoi facet and let d be its driver. Let e be a Voronoi edge for which F is acceptor and x be any point on e. Also assume the ray from d to x intersects a Voronoi edge e'. If e' is transversal, the acceptor of e' is different from F.*

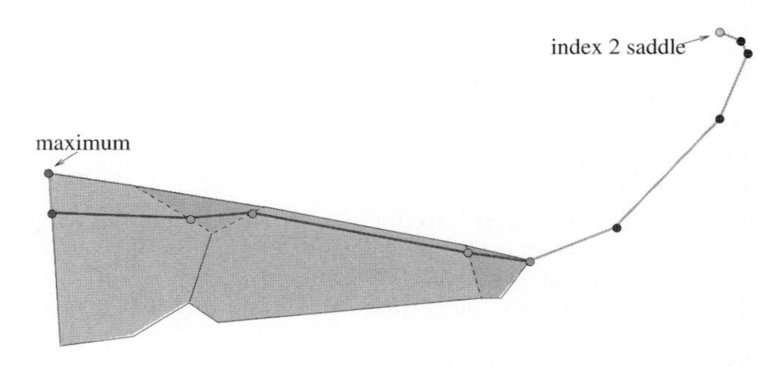

Figure 6: Unstable manifold $U(c)$ of an index 2 saddle point c. c is drawn with a cyan circle. The portion of $U(c)$ which is a collection of Voronoi edges is drawn in green with intermediate Voronoi vertices drawn in blue. The pink circle is a Voronoi vertex on $U(c)$ where the flow enters a Voronoi facet. The portion of $U(c)$ which lies inside the Voronoi facets is drawn in magenta. The transversal Voronoi edges intersected by this portion of $U(c)$ are dashed. $U(c)$ ends at a local maximum which is drawn in red.

Figure 6 shows an example of the unstable manifold of an index 2 saddle point.

Following the above discussion on the structure of $U(c)$ we devise the algorithm to compute the unstable manifold of an index 2 saddle point c. We assume, the saddle point c carries the information about the two neighboring tetrahedra σ_1, σ_2 and additionally we have access to Del P which is used to evaluate the utility routines like acceptor() , terminus() etc. The pseudo-code of the algorithm is given in Figure 7.

4.2 Unstable Manifold of Index-1 Saddle Points

Unstable Manifold of index 1 saddle points are two dimensional. Due to hierarchical structure, they are bounded by the unstable manifold of index 2 saddle points. In this section we first describe the structure of the unstable manifolds and then describe an algorithm that computes an approximation of the unstable manifold of an index 1 saddle point.

Let us consider an index 1 saddle point, c. This point lies at the intersection of a Voronoi facet F and a Delaunay edge. For any point $x \in F \setminus c$, the driver is c. For all such x, if they are allowed to move in the direction of flow, they will move radially outward and hit the Voronoi edges bounding F. Thus F is in $U(c)$. Now we analyze the flow when a point hits a Voronoi edge.

We have characterized the position of the drivers for a Voronoi edge and we have also seen that depending on the driver, one can classify the Voronoi edges into two categories - transversal and non-

UM_INDEX_2(c)

1 $U_1 = cv_1$ and $U_2 = cv_2$
2 $v = v_1$
3 end$(U_1) = v_1$
4 while (v is not a maximum) do
5 if(v is not a Voronoi vertex)
6 $e =$ Voronoi edge containing v
7 if(e is non-transversal)
8 end$(U_1) =$ terminus(e)
9 $U_1 = U_1 \cup$ segment$(v, \text{end}(U_1))$
10 $v =$ terminus(e)
11 else
12 $F =$ acceptor(e)
13 $d =$ driver$(F) =$ driver(e)
14 $x = \overrightarrow{dv} \cap e' \neq \emptyset$, e' is a Voronoi edge of F
15 end$(U_1) = x$
16 $U_1 = U_1 \cup$ segment$(v, \text{end}(U_1))$
17 $v = x$
18 else
19 if(v falls under Case 2.1)
20 $e =$ *outgoing* Voredge (v)
21 repeat steps 7-17.
22 else if(v falls under Case 2.2)
23 $F =$ acceptor(v)
24 repeat steps 13-17.
25 endwhile
26 Similarly compute U_2.
27 return $U_1 \cup U_2$.

Figure 7: Pseudo-code for computation of unstable manifold of an index 2 saddle point.

transversal. For a non-transversal Voronoi edge, the flow is along the Voronoi edge. Such Voronoi edges lie on the boundary of $U(c)$. On the other hand, $U(c)$ grows via the acceptor facets of transversal Voronoi edges. Depending on the position of the driver, which by Lemma 2 is same for both the edge and the acceptor facet, a *truncated cone* defines the extension of $U(c)$ into the acceptor Voronoi facet. Consider the cone defined by the two rays emanating from the driver and passing through the endpoints of the transversal Voronoi edge. The intersection of the acceptor facet with the cone defines the truncated cone. The truncated cone hits a continuous chain of Voronoi edges in the acceptor facet. Some of them are completely contained in the truncated cone and some of them are intersected by the two rays and hence are partially contained in it. This chain of edges defines the new boundary of $U(c)$ through some of which $U(c)$ can be extended further recursively. Figure 8 shows an example truncated cone in a Voronoi facet F by the driver d and the end Voronoi vertices of the transversal Voronoi edge (green).

To compute $U(c)$ accurately, one therefore needs to compute the intersection of a ray and a line segment in three dimension. Such computations are prone to numerical errors. Therefore, we rely on an approximation algorithm that computes a superset of $U(c)$. The algorithm works as follows.

Starting from the Voronoi facet F containing c, we maintain a list of Voronoi facets which are already in $U(c)$ and a list of active Voronoi edges which are transversal edges and lie on the boundary of the current approximation of $U(c)$. Through these transversal edges we collect their acceptor facets and grow $U(c)$. Instead of computing the new set of active edges by an expensive numerical calculation of ray-segment intersection, we collect all the transversal edges of this new acceptor Voronoi facets. This way we grow $U(c)$ recursively

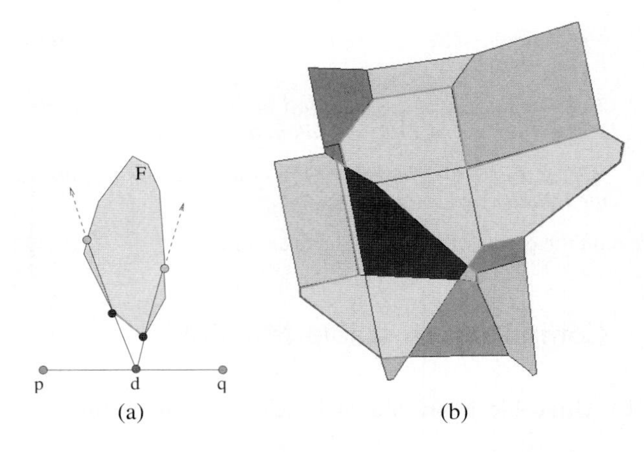

Figure 8: (a) Truncated Cone. Accurate computation selects only the pink region from the yellow Voronoi facet as part of unstable manifold of an index 1 saddle point c (not shown). (b) Snapshot of approximate computation of $U(c)$ at a generic stage.

till we have a set of Voronoi facets which are bounded by only a set of transversal Voronoi edges.

Figure 8(b) illustrates an intermediate stage of this computation. The index 1 saddle point c is contained in the blue Voronoi facet. The yellow Voronoi facets are already in $U(c)$. The red edges designate the static boundary as they are non-transversal and the green edges designate the active boundary through which the pink facets are included in $U(c)$ in the later stage of the algorithm. Following is the pseudo-code for this algorithm. Given an index 1 saddle point c it computes an approximation of $U(c)$. We assume c also has information about the Voronoi facet F it is contained in.

APPROX_UM_INDEX_1(c)

1 $U = F$
2 $B =$ Voronoi edges of F
3 while ($B \neq \emptyset$) do
4 $e =$ pop(B)
5 if (e is transversal)
6 $U = U \cup$ acceptor(e)
7 $B = B \cup$ unvisited edges of acceptor(e)
8 endwhile
9 return U.

Figure 9: Pseudo-code for approximate computation of unstable manifold of an index 1 saddle point.

4.3 Classification of Medial Axis

In the previous two subsections we have described the structures of the unstable manifolds of an index 1 and index 2 saddle points. We have also given an accurate and an approximate algorithm to compute them. Our goal is to identify the unstable manifolds near the medial axis of Σ. Ultimately these manifolds are mapped back to Σ for the feature annotation. For this we first compute a Voronoi subcomplex that approximates the medial axis M_Σ and then identify different regions of this approximate medial axis as the unstable manifolds computed by the two subroutines UM_INDEX_2 and APPROX_UM_INDEX_1.

Before we describe our approach, we briefly mention a recent result by Dey, Giesen, Ramos and Sadri [Dey et al. 2005] where they proved that under sufficient sampling of Σ by P, the critical points of h_P lie either close to Σ or close to M_Σ. This motivates our approach. Applying the same result, we filter out only the index 1 and index 2 saddle points near M_Σ instead of Σ. Further, we consider only the components of M_Σ which lie in the interior of the solid bounded by Σ. For this purpose we use the TIGHTCOCONE algorithm by Dey and Goswami [Dey and Goswami 2003]. The implementation of this algorithm is freely available in the public domain [Cocone] along with the software for medial axis approximations which is computed as a Voronoi subcomplex according to the algorithm by Dey and Zhao [Dey and Zhao 2004]. For the purpose of reconstruction, any other reconstruction algorithm also could be used [Bernardini et al. 1999; Bajaj et al. 1995]. Applying TIGHTCOCONE followed by medial axis approximation we get the approximate interior medial axis of Σ. We perform the critical point detection only within the Voronoi subcomplex that approximates this medial axis. Let us call this set of index 1 saddle points C_1 and that of index 2 saddle points C_2. We then apply UM_INDEX_2(c) for all $c \in C_2$ and APPROX_UM_INDEX_1(c) for all $c \in C_1$. $U(c \in C_1)$ is two dimensional and $U(c \in C_2)$ is one dimensional. Therefore, by restricting the unstable manifold computation only within M_Σ we obtain two subsets of M_Σ. In the next section, we describe how this classification can be mapped back to Σ for automatic identification of its flat and tubular regions.

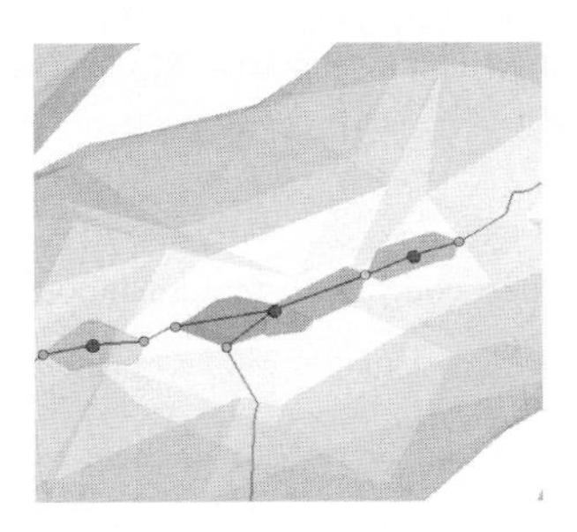

Figure 10: Removal of small patches in the tubular region via starring. Magenta circles indicate the centroids of these patches, green circles are the boundary vertices which connect a patch with a linear portion (red line) and cyan circle indicates where two different patches join at a common vertex. Blue lines are the replacements of these small patches obtained by the starring process.

Because of sampling artifacts, sometimes the interior medial axis in the tubular regions have a few index 1 saddle points. The unstable manifold of these saddle points need to be detected and approximated by lines. We partition the set C_1 based on the connectivity of their unstable manifolds via a common edge and every partition creates a patch which is the union of the unstable manifolds of all the index 1 saddle points falling into that partition. We further assign an *importance* value based on the area of the patch and sort the patches according to their *importance*. One could also employ other attributes like diameter, width etc. to evaluate the importance. The small clusters are then detected either by a user-specified threshold value or by simply selecting the k-smallest clusters where k is also a user-supplied parameter. These insignificant planar regions are then approximated by a set of straight lines emanating from the centroid of the patch to the boundary points which are connected to either a polyline (green circles in Figure 10) or another patch (cyan circle in Figure 10). We call this process *starring*.

The resulting one dimensional and two dimensional subsets of the interior medial axis is shown in Figure 11. Left column shows the approximate medial axis computed by [Dey and Zhao 2004]. The right column shows the subset of medial axis captured by $U(C_2)$ and $U(C_1)$.

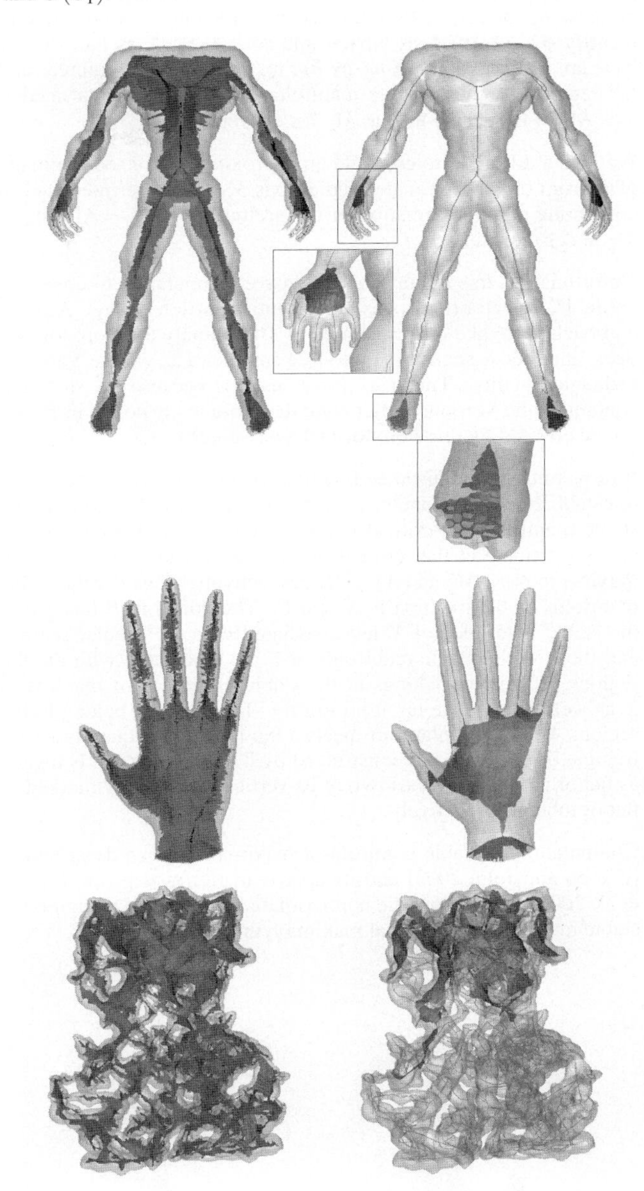

Figure 11: Results of Medial Axis classification. Top row shows the result for HEADLESS MAN. Two closeups have been shown to highlight the planar clusters in the palm of the hand and the feet. The closeup of hand has been rotated for visual clarity. The middle row shows the result on HAND dataset and the bottom row shows the result on a molecule data 1BVP.

5 Feature Annotation Algorithm

5.1 Mapping of Unstable Manifolds to Σ

There is a natural association between the medial axis M_Σ and Σ via the map $\phi : \Sigma \rightarrow M_\Sigma$ where $\phi(x)$ is the center of the medial ball touching Σ at x. Following this map, any subset $A \subseteq M_\Sigma$ can be associated with $\phi^{-1}(A) \subseteq \Sigma$. Let A_1 and A_2 be the closure of the unstable manifolds of index 2 and index 1 saddles in M_Σ defined by the distance function h_Σ. Recall that, generically, A_1 is one-dimensional and A_2 is two-dimensional. Ideally, we would like to identify $\phi^{-1}(A_1) \subseteq \Sigma$ as tubular and $\phi^{-1}(A_2) \subseteq \Sigma$ as flat. As we have an approximation of h_Σ by h_P, we compute these tubular and flat regions for the unstable manifolds in the approximate medial axis which we denote also as M_Σ for convenience.

We face a difficulty to compute an approximation of the preimage of ϕ from the approximate medial axis M_Σ. We are interested in computing an approximation of the preimage of $M'_\Sigma = A_1 \cup A_2 \subseteq M_\Sigma$ under the map ϕ.

Unfortunately, this requires an expensive computation to cover the entire M'_Σ which often spans a substantial portion of M_Σ. A naive approach is to take only a sample of M'_Σ, namely the Voronoi vertices, and then associate them to P, a sample of Σ, via the Voronoi-Delaunay duality. This also proves useless because M'_Σ does not contain all the Voronoi vertices and therefore many points in P cannot be covered by this Voronoi-Delaunay duality.

It turns out that the distance function h_P again proves to be useful to establish a correspondence between Σ and M'_Σ. Recall that, the stable manifold of a critical point is a collection of points whose orbits terminate at that critical point. Let X and Y be the set of maxima in $A_1 \subseteq M'_\Sigma$ and $A_2 \subseteq M'_\Sigma$ respectively. Consider the stable manifolds of the maxima in X and Y. The points in P that are in the stable manifolds of X are associated with the tubular regions and those in the stable manifolds of Y are associated with the flat regions. If a point belongs to the stable manifolds of maxima in X as well as in Y, we tag it arbitrarily. These points belong to the regions where a tubular part meets a flat part. Subsequently, every triangle of the surface reconstructed by TIGHT COCONE is tagged as flat or tubular if at least two of its vertices are already marked as flat or tubular respectively.

Computation of stable manifold of maxima has been described in [Giesen and John 2003] and its approximation was given in [Dey et al. 2003]. We follow the approximate algorithm to compute the stable manifolds of the local maxima lying on M'_Σ.

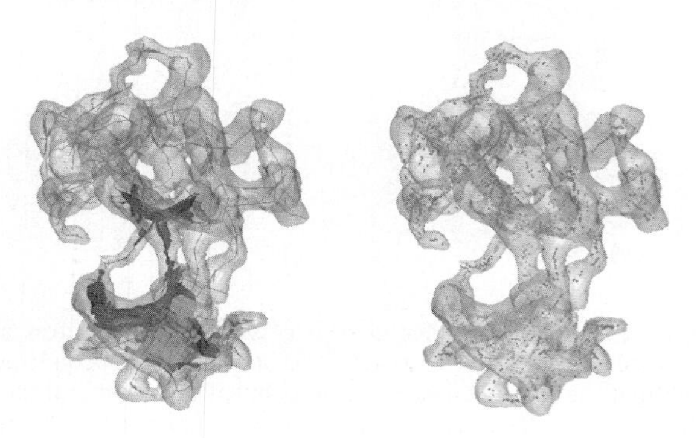

Figure 12: One dimensional subset of the interior medial axis is drawn in red and the two dimensional subset of the medial axis is drawn in green for the molecule data 1IRK. The right subfigure shows the selection of local maxima of the distance function in those two parts, colored accordingly.

Figure 12 shows the set M'_Σ of the molecule data 1IRK, and the set of maxima belonging to that set and identified as linear or planar. The corresponding flat and tubular portions of the surface captured by the mapping via stable manifold of these maxima - colored golden and cyan respectively - are shown in Figure 14. We collected the protein from Protein Data Bank [Berman et al. 2000] and blurred the molecule at a resolution 8 angstrom. Further we took the vertex set of a suitable level set as the input to our program. We verified the result with the existing literature in structural biology and we have seen that the flat regions identified by our algorithm correspond to the β-sheets of the protein molecule.

5.2 Annotation Algorithm

The modules described in the previous sections and subsections can thus be combined to devise an algorithm for automatic feature annotation of Σ. We give the pseudo-code of this annotation algorithm here.

```
IDENTIFY_FLAT_AND_TUBULAR_REGIONS(P)
 1   Compute Vor P and Del P.
 2   Compute the interior Medial Axis M_Σ
       by TIGHTCOCONE_AND_MA (P)
 3   C_1 = set of index 1 saddle points lying on M_Σ and
     C_2 = set of index 2 saddle points lying on M_Σ,
 4   A_1 = A_2 = ∅
 5   for all c ∈ C_2
 6      A_1 = A_1 ∪ UM_INDEX_2(c)
 7   for all c ∈ C_1
 8      A_2 = A_2 ∪ APPROX_UM_INDEX_1(c)
 9   X = maxima in A_1
10   Y = maxima in A_2
11   Σ_Tubular = MAPPING_VIA_STABLE_MANIFOLD(A_1)
12   Σ_Flat = MAPPING_VIA_STABLE_MANIFOLD(A_2)
13   return Σ_Tubular and Σ_Flat.
```

Figure 13: Pseudo-code of the feature annotation algorithm.

6 Results

6.1 Implementation Issues

The algorithm works on the Voronoi-Delaunay diagram of the set of sample points lying on the surface. To robustly compute the Delaunay triangulation and its dual Voronoi diagram for the input set of points we use the library CGAL [CGAL Consortium] which is freely available.

Even in CGAL-framework, we sometimes face the degenerate case of five or more points being cospherical. This case has to be handled with special care because only one Voronoi vertex is repeated and therefore the flow along the Voronoi edges is not well-defined anymore. To deal with such situations, we modify the algorithm slightly. At the start of the algorithm we collect the sets of tetrahedra which are cospherical. While computing the unstable manifold of index 2 saddle points, if the polyline hits a Voronoi vertex whose dual is a member of one such cospherical cluster, the algorithm automatically advances through the non-degenerate Voronoi edges which are dual to the triangles bounding the cospherical lump. This degeneracy poses a more serious threat to the computation of unstable manifold of index 1 saddle points and at this stage, we do not extend the manifold through any Voronoi edge whose dual Delaunay triangle is shared by two cospherical tetrahedra.

There are some parameters involved in the full feature annotation process. For surface reconstruction and medial axis approximation we used the software [Cocone]. The parameters for these routines are described in [Dey and Goswami 2003], [Dey and Zhao

2004]. For noisy inputs we replace TIGHT COCONE by ROBUST COCOCNE and the parameters for that step are again described in [Dey and Goswami 2004]. The rest of the algorithm requires only one parameter k which is the number of flat regions to be output.

6.2 Performance

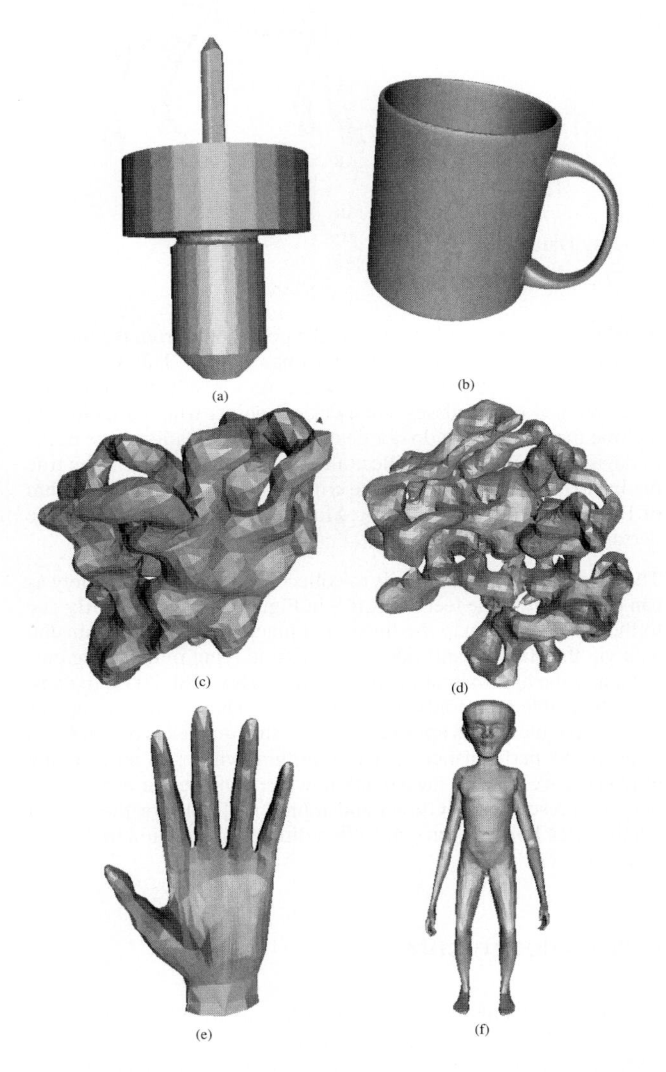

Figure 14: Performance of the feature annotation algorithm. The models are (a) PIN, (b) MUG, (c) molecule 1CID (d) molecule 1IRK, (e) HAND, (f) ALIEN

Figure 14 shows the performance of the annotation algorithm on six datasets. The datasets have been chosen to represent different domains this algorithm can possibly be applied in. PIN is a CAD dataset which has two tubular parts joined in the middle through a flat portion. The algorithm can identify them correctly. Similarly the method can correctly identify the handle as the tubular and the body as the flat region for the MUG dataset. In the second row we show the performance of our method on two protein molecules obtained from Protein Data Bank [Berman et al. 2000]. We took the crystal structure of these two molecules (PDB ID 1CID and 1IRK) and blurred them with Gaussian kernel. We further took a level set which represents a molecular surface and used the vertex set of that isosurface as the input to our algorithm. The flat features identified by our method correspond to the β-sheets of the secondary structure of those two proteins. In the last row we show the result on two free form objects containing both flat and tubular features. As we can see, the palm of the HAND has been detected as flat whereas the fingers have been detected as tubular. Our method can also capture the major flat and tubular features of ALIEN.

We purposefully show the performance of the algorithm on ALIEN as it brings forth the limitations of our algorithm. We see that a portion of the arm has been identified as flat. This is because the initial reconstruction phase could not separate the beginning of the arm from the torso due to lack of sampling. Secondly, one of the feet could not be fully identified as flat by our algorithm. This is because the approximate medial axis, that we started with, is not a close approximation of the true medial axis in that region, again due to lack of sampling. Because of that, our method fails to collect sufficiently many index 1 saddle points leading to incomplete identification of flat features in that region.

Figure 15 shows the performance of our method on noisy dataset HORSE. Instead of applying TIGHT COCONE, we first mark the interior and exterior of the closed surface from its noisy point sample by ROBUST COCONE ([Dey and Goswami 2004]) and then obtain the interior medial axis and proceed further with the unstable manifold computation and feature identification. Originally there were some thin flat regions due to the unstable manifold of some index 1 saddle points near the hind legs which we filtered out by thresholding in order to get a clean skeleton of the HORSE. In the rightmost picture we see some white triangles near the ears. These triangles appear as the mapping via stable manifold misses some points on the surface in that portion.

6.3 Timings

The time and space complexity of the algorithm is dominated by the complexity of Delaunay triangulation. We report the timings of the entire execution into four major steps

1. Step 1: Building the Voronoi-Delaunay diagram of the point set (Line 1 of Figure 13).
2. Step 2: Computation of interior medial axis. (Line 2 of Figure 13).
3. Step 3: Computation of unstable manifold of index 1 and index 2 saddle points lying on the interior medial axis. (Line 3-8 of Figure 13).
4. Step 4: Mapping the maxima in the planar and linear portion of the medial axis back to the surface. (Line 9-13 of Figure 13).

We built the code using CGAL [CGAL Consortium] and gnu C++ libraries. The code is compiled at an optimization level $-O2$. We run the experiments in a machine with INTEL XEON processor with 1GB RAM running at 1GHz cpuspeed. Table 1 reports the time taken in the four steps of the algorithm for a number of datasets. It is clear from the breakup of timing that the first two steps of building the Delaunay triangulation and then computing the interior medial axis are the two most expensive steps. For noisy datasets, additionally ROBUST COCONE is used to obtain an initial in-out marking. This step is comparatively inexpensive. For example, for NOISY HORSE (48,000 points) this step only adds 10 sec to the whole computation time which is approximately 100 sec.

Figure 15: Results on Noisy Data.

object	# points	Step 1 (sec.)	Step 2 (sec.)	Step 3 (sec.)	Step 4 (sec.)
1CID	5170	7.59	15.63	6.69	0.39
1IRK	13940	29.88	43.93	15.63	1
HEADLESS MAN	16287	18.63	51.30	16.01	1.26
PIN	15530	15.73	41.4	21.53	0.92
CLUB	16864	20.54	47.3	19.83	1.24
MUG	27109	37.68	83.28	47.14	2.19
HAND	40573	53.48	120.16	40.67	2.69
P8	48046	33.46	136.59	39.97	3.22
1BVP	53392	148.18	159.52	62.19	3.53
ALIEN	78053	102.62	242.33	64.11	5.4

Table 1: Timings

7 Conclusions

In this work, we first described the structure of the unstable manifold of index 1 and index 2 saddle points of the distance function induced by a set of points sampled from a surface. We further used this analysis to compute the unstable manifold of an index 2 saddle point exactly and the unstable manifold of an index 1 saddle point approximately. We then used the unstable manifold of index 1 and index 2 saddle points near the medial axis of the surface to automatically detect the flat and tubular features of the shape.

We believe that this work will be useful in many areas of science and engineering. One natural connection to structural biology is the elucidation of secondary structural properties of protein molecules. Secondary structure of a protein molecule is made up of a collection of α-helices and β-sheets. α-helices are tubular and β-sheets are flat. The results of our algorithm on protein molecules have been verified against the true structural information obtained from Protein Data Bank [Berman et al. 2000] and the existing literatures in structural biology. We have seen that our method can identify the secondary structural motifs correctly. Often the applications in structural biology require to elucidate the secondary structural information in the absence of atomic level representation of a protein molecule. This is particularly the case when a protein molecule is present in a larger assembly such as in virus capsid and the input is obtained only at a resolution coarser than 4 angstrom via electron microscopy. This method will prove fruitful in analyzing the secondary structural properties in such situations.

This work has triggered several questions. We have collected the initial set of index 1 and index 2 saddle points only from the interior medial axis computed by [Dey and Zhao 2004]. This is only an approximation of the true medial axis. As a result, the collection thus obtained often misses some critical points which are close to the true medial axis but do not lie on the approximation. One needs to devise an algorithm to collect all the critical points near the true medial axis efficiently, say by the critical point separation algorithm of Dey et al. [Dey et al. 2005]. Most likely this will improve the performance of our algorithm.

The algorithm sometimes fails to collect a flat region completely as can be seen from the feet of ALIEN in Figure 14. This is partly due to the fact that we map the linear and planar regions of the medial axis via the stable manifold of the maxima lying in those regions. We apply the approximation algorithm of [Dey et al. 2003] to compute the stable manifolds of the maxima. This method computes these stable manifolds approximately as subcomplexes of Del P. To improve the performance of our algorithm, while still maintaining efficiency, we plan to investigate how we can use the exact computation presented in [Giesen and John 2003] only for those local maxima that border a one- and a two-dimensional region in the medial axis.

Acknowledgments

First and third authors are supported in part by NSF grants ITR-EIA-0325550, CNS-0540033 and NIH grants P20 RR020647, R01 GM074258-021 and R01-GM073087. The second author is supported in part by NSF CARGO grant DMS-0310642 and ARO grant DAAD19-02-1-0347. We thank the Jyamiti group at The Ohio State University for providing the Tight Cocone and Medial software. We also thank Zeyun Yu for providing the molecular datasets.

References

BAJAJ, C., BERNARDINI, F., AND XU, G. 1995. Automatic reconstruction of surfaces and scalar fields from 3D scans. In *ACM SIGGRAPH*, 109–118.

BERMAN, H. M., WESTBROOK, J., FENG, Z., GILLILAND, G., BHAT, T., WEISSIG, H., SHINDYALOV, I., AND BOURNE, P. 2000. The protein data bank. *Nucleic Acids Research*, 235–242.

BERNARDINI, F., BAJAJ, C., CHEN, J., AND SCHIKORE, D. 1999. Automatic reconstruction of 3d cad models from digital scans. *Int. J. on Comp. Geom. and Appl. 9*, 4-5, 327–369.

CGAL CONSORTIUM. CGAL: Computational Geometry Algorithms Library. *http://www.cgal.org*.

CHAINE, R. 2003. A geometric convection approach of 3-d reconstruction. In *Proc. Eurographics Sympos. on Geometry Processing*, 218–229.

CHAZAL, F., AND LIEUTIER, A. 2004. Stability and homotopy of a subset of the medial axis. In *Proc. 9th ACM Sympos. Solid Modeling and Applications*, 243–248.

COCONE. Tight Cocone Software for surface reconstruction and medial axis approximation. *http://www.cse.ohio-state.edu/~tamaldey/cocone.html*.

DEY, T. K., AND GOSWAMI, S. 2003. Tight cocone: A water-tight surface reconstructor. In *Proc. 8th ACM Sympos. Solid Modeling and Applications*, 127–134.

DEY, T. K., AND GOSWAMI, S. 2004. Provable surface reconstruction from noisy samples. In *Proc. 20th ACM-SIAM Sympos. Comput. Geom.*, 330–339.

DEY, T. K., AND ZHAO, W. 2004. Approximating the Medial axis from the Voronoi diagram with convergence guarantee. *Algorithmica 38*, 179–200.

DEY, T. K., GIESEN, J., AND GOSWAMI, S. 2003. Shape segmentation and matching with flow discretization. In *Proc. Workshop Algorithms Data Strucutres (WADS 03)*, F. Dehne, J.-R. Sack, and M. Smid, Eds., LNCS 2748, 25–36.

DEY, T. K., GIESEN, J., RAMOS, E., AND SADRI, B. 2005. Critical points of the distance to an epsilon-sampling of a surface and flow-complex-based surface reconstruction. In *Proc. 21st ACM-SIAM Sympos. Comput. Geom.*, 218–227.

EDELSBRUNNER, H. 2002. Surface reconstruction by wrapping finite point sets in space. In *Ricky Pollack and Eli Goodman Festschrift*, B. Aronov, S. Basu, J. Pach, and M. Sharir, Eds. Springer-Verlag, 379–404.

GIESEN, J., AND JOHN, M. 2003. The flow complex: a data structure for geometric modeling. In *Proc. 14th ACM-SIAM Sympos. Discrete Algorithms*, 285–294.

KATZ, S., AND TAL, A. 2003. Hierarchical mesh decomposition using fuzzy clustering and cuts. In *Trans. on Graphics*, vol. 3, 954–961.

MORTARA, M., PATANÈ, G., SPAGNUOLO, M., FALCIDIENO, B., AND ROSSIGNAC, J. 2004. Blowing bubbles for the multi-scale analysis and decomposition of triangle meshes. *Algorithmica 38*, 227–248.

MORTARA, M., PATANÈ, G., SPAGNUOLO, M., FALCIDIENO, B., AND ROSSIGNAC, J. 2004. Plumber: a multi-scale decomposition of 3d shapes into tubular primitives and bodies. In *Proc. 9th ACM Sympos. Solid Modeling and Applications*, 139–158.

POTTMANN, H., HOFER, M., ODEHNAL, B., AND WALLNER, J. 2004. Line geometry for 3d shape understanding and reconstruction. *Computer Vision - ECCV 3021*, 1, 297–309.

SHINAGAWA, Y., KUNNI, T., BELAYEV, A., AND TSUKIOKA, T. 1996. Shape modeling and shape analysis based on singularities. *Internat. J. Shape Modeling 2*, 85–102.

SIDDIQI, K., SHOKOUFANDEH, A., DICKINSON, J., AND ZUCKER, S. 1998. Shock graphs and shape matching. *Computer Vision*, 222–229.

VÁRADY, T., MARTIN, R., AND COX, J. 1997. Reverse engineering of geometric models - an introduction. *Computer Aided Design 29*, 255–268.

VERROUST, A., AND LAZARUS, F. 2000. Extracting skeletal curves from 3d scattered data. *The Visual Computer 16*, 15–25.

WU, J., AND KOBBELT, L. 2005. Structure recovery via hybrid variational surface approximation. *Computer Graphics Forum 24*, 3, 277–284.

Constructive topological representations

Srinivas Raghothama
UGS Corp
10824 Hope St, Cypress, CA 90630
E-mail: raghotha@ugs.com

Abstract

Constructive representations, such as Constructive Solid Geometry (CSG) and its various feature-based extensions are inherently parametric in nature and are well suited for defining parametric family of solids. On the other hand, cell complex representations contain explicit shape elements (cells) and also their topology. However they are non-constructive, difficult to parameterize, and it is extremely difficult to enforce continuity in the usual cell complex topology (considered as a sub-space of Euclidean space). When cell complexes are used in conjunction with constructive representations, even if we can enforce continuity in limited cases, it is impossible to relate cellular operations with global semantics of constructive operations (such as Boolean and feature operations).

Using the framework developed in our earlier work for defining part families on any solid representation, we propose constructive topological representations by identifying every constructive representation with its corresponding unique spatial decomposition. By applying the proposed definitions to spatial CSG representations we will systematically develop algorithms for topologizing a CSG and also enforcing continuity between two given topologized CSG representations. These algorithms have been implemented in a prototype system about which we will briefly discuss. Finally, we will illustrate some interesting applications of the proposed constructive topological representations: specification of constructive families and the enforcement of global semantics of feature operations.

1 Introduction

1.1 Part families

Part families naturally arise in many engineering design and manufacturing applications. The notion of part family is defined in [S.Raghothama and V.Shapiro 2003] as objects and operations on objects, where the operations have to obey certain mathematical rules such as existence of identity, associativity and ability to compose operations. It is proposed that engineering part families need to be defined using topological spaces as objects and continuous maps as the operations. Under this proposal the mathematical model for engineering part families are the solid families in the usual Euclidean topology, since they capture the intuitive properties we desire in any family. As solids are computed using their representations such as CSG, cell complex and feature based representations, we can define families using particular representations defined in their respective topologies. But we need to be able to compare the properties of the representation families with the corresponding solid family that is unique (similar in spirit to the relationship between solids and their representations proposed by Requicha [Requicha 1980]).

Depending on the two predominant ways the solid representations are defined, we can broadly classify families as constructive and non-constructive. In what follows, we will illustrate the key properties and the fundamental differences between these two types of families.

1.2 Constructive representations and families

A constructive family is defined as the function of constructions globally parameterized by a solid representation. All objects in the family are defined by the same construction function and the operations of the family are continuous maps in the parametric space induced by a constructive solid representation. Such families are also known as parameter space families [S.Raghothama and V.Shapiro 2003]. Constructions are essentially global operations on a given set of primitives or features in the solid representations. Constructive operations are transformations and global set operations such as Booleans as well as Minkowski operations (these are used to define the blending and offsetting/shelling operations [Rossignac 1985]). Some well known examples of constructive solid representations are the CSG and feature representations, which induce constructive CSG and feature representation families respectively.

Figure 1 shows a constructive parametric family defined by a CSG representation that is a union of three cylinders A, B and C, each parameterized by its natural size (diameter and height) and also by their relative positions. Generating the members (or instances) of this CSG family is just a matter of changing parameters $P_i \in \mathbf{P}$ constrained by the Boolean expression of the CSG $F(\mathbf{P}) = (A \cup B \cup C)$.

Constructive parametric families defined in the parameter space have some nice properties such as global parametric control, correspond to (global) shape constructions, and have persistent references to "good" constructions as long as the references are unique. Typically the operations in a representation are defined on *instances* and in order to define a family, the operations should be definable on *all* the instances in the family. Without loss of generality, we assume that we can establish *correspondence* between the primitives in a constructive representation. The Booleans and other global set operations in constructive representations by the virtue of their algebraic nature (i.e, a set theoretic expression exists even when instances correspond to empty sets) possess this ability to act on all the instances of solids and hence easier to establish correspondence between the instances. On the other hand, constructive parametric representations are non-unique in general, harder to compute as well as represent any of the intuitive properties (for example, connectedness, number of holes and voids) of the solids, making them harder to relate to the Euclidean space. As can be seen from the solids in figure 1, the constructive family contains solids that are disconnected (figure 1(c)) and also have different connectivity (figure 1(b)) from the other solids.

Figure 2 shows another example of a constructive parametric family in which the connectivity of the solid changes for a parametric edit. Initially the solid has a "blind" hole at the top of the block

SPM 2006, Cardiff, Wales, United Kingdom, 06–08 June 2006.
© 2006 ACM 1-59593-358-1/06/0006 $5.00

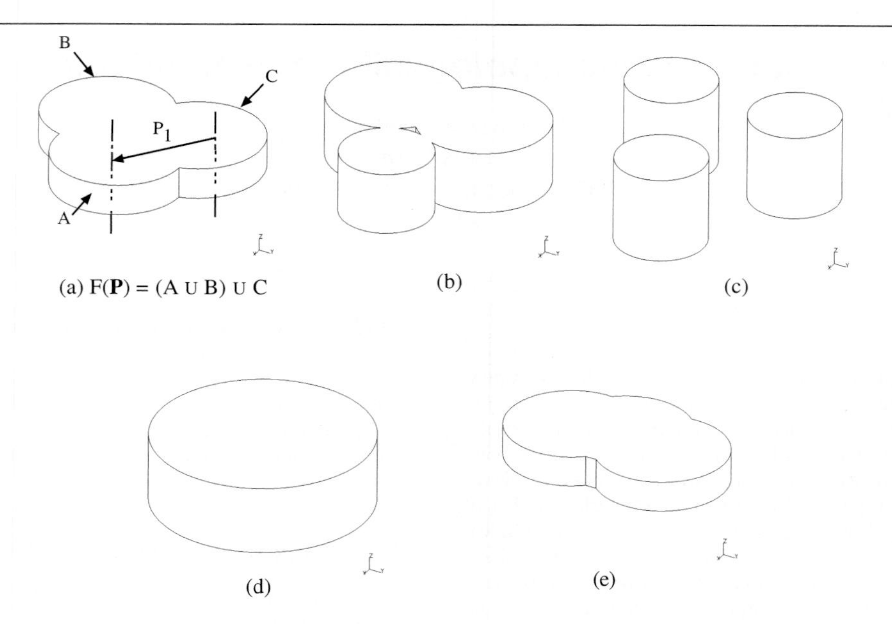

Figure 1: A constructive family defined by a CSG representation

and it becomes a "through" hole when the location of the cylinder is moved over the rectangular cutout at the bottom of the block. In feature based representations the cutout could be represented as a slot or a pocket feature.

Finally, figure 3 shows another constructive parametric family with the initial solid being one connected component, but the updated one (corresponding to increase in the depth of the slot feature) has two components. In fact, some CAD systems will not allow such edits and will flag them as errors due to the representational limitation of allowing only one component per solid. However note that such parametric families do have applications (such as shape optimization) where such global topological changes are desired.

1.3 Non-constructive representations and families

Unlike constructive solid representations, cellular representations such as b-reps possess more direct control of the shape and carry explicit geometric as well as topological information with them. A cellular family is defined using an *instance* of a cell complex along with its topology defined by adjacency structure of the cells and continuous transformations[1] on the instances in the cellular topology. Such families are also known as representation space families [S.Raghothama and V.Shapiro 2003]. Cell complex families are intuitive, possess local control ('tweaks' and other local shape operations such as blends, offsets, etc can be specified), and provide persistent spatial address. However cell complex families are not globally parametric, we have to deal with the harder problem of persistent naming [Kripac 1997; Capoyleas et al. 1996; Shapiro and Vossler 1995; S.Raghothama and V.Shapiro 1998] and most importantly are non-constructive i.e, do not naturally support history or construction procedure for replaying or editing the representation[2].

While cells provide a persistent spatial address within an instance (via explicit geometric information), they are not persistent in a family. The reason is due to the way the topology is defined in cell complexes in terms of adjacent cells of a cell. As a consequence, it is very difficult to establish correspondence between cells of various instances in a family and we typically require more information than given by the geometry of the carriers in a cell complex (for example, relative orientation of the lower dimensional cells with respect to higher dimension cells [S.Raghothama and V.Shapiro 1998]). Even in cases where we can establish the correspondence and enforce continuity using adjacency and relative orientation of the cells [Kripac 1997; Capoyleas et al. 1996; S.Raghothama and V.Shapiro 1998], it is difficult to enforce global semantics (say for Boolean operations) through cellular continuity. For example, the b-rep in figure 4(a) is derived from the union of 3 cylinders. If we locally transform (using the commonly available 'tweak' functionality of a commercial solid modeler) one of the front cylindrical faces of the cell complex, we get the result shown in figure 4(b). This result is semantically incorrect because it is not a union of 3 cylinders anymore, though the transformation is continuous in cell complex topology[Raghothama 2000]. Unfortunately, this semantic problem cannot be verified using a b-rep structure alone and in an earlier paper [S.Raghothama and V.Shapiro 2000] we showed how such inconsistencies as a result of boundary based tweaks can be verified by performing a consistent update on a corresponding CSG representation and verifying if the CSG is consistent with its corresponding b-rep.

A procedural or hybrid representation also contains a history of operations that define the resulting object in a procedural way (such representations are also called as generative representations [Rossignac et al. 1988; Chen and Hoffmann 1995; Rappoport 1995]). Unlike constructive representations, procedural representations contain fully evaluated b-reps at every node or step and permit direct shape modifications locally. In fact, most modern commercial CAD systems use a procedural representation that attempts to capture the history of operations along with partial or whole b-reps at each node. A classical example of a procedural representation

[1] Different types of cell complex families can be defined based on how we consider them and what type of continuity is desired on them, refer to [S.Raghothama and V.Shapiro 1998] for more details on such families.

[2] Though Euler operators [Mantyla 1988] provide a way to construct b-reps using local operators that preserve Euler characteristic, they are only useful for simple constructions and topologies.

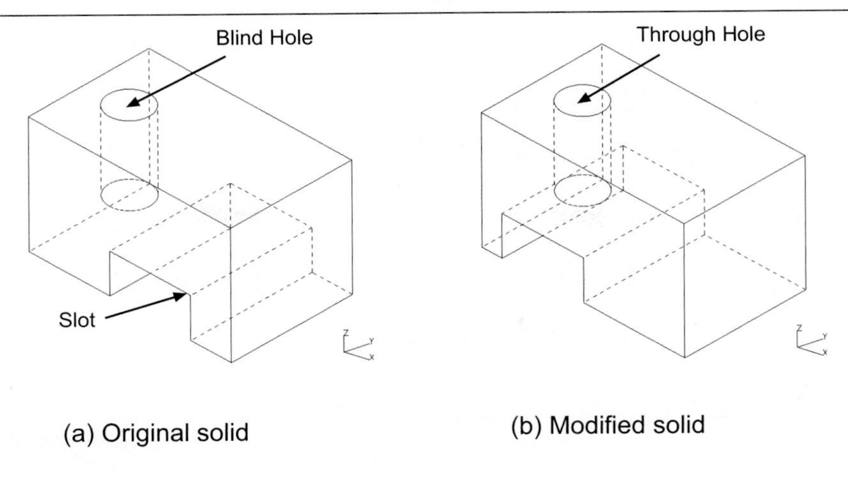

Figure 2: Blind hole changes to a through hole

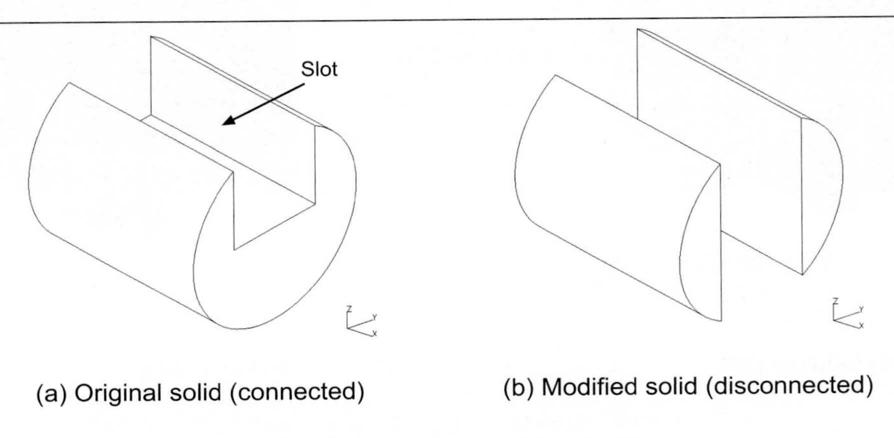

Figure 3: Update results in multiple solids

is the sweep representation [Requicha 1980; Chen and Hoffmann 1995], in which a given profile consisting of an open or closed set of 2D (curves/edges) or 3D entities (faces/surfaces) is transformed to create a surface or solid in 3D. The linearly transformed sweeps (extrusions) are currently a popular technique for creating complex primitives used in parametric feature-based systems [Pro-Engineer 1998; SolidWorks 2000; Unigraphics 2004]. In addition to global operations such as Booleans, several localized feature operations such as blend, offset, draft, etc, are provided in such procedural representations for adding intricate details to the shape to be designed.

Figure 5(a) shows a complex parametric feature-based solid model built in a contemporary CAD system. When Boss27's diameter is increased, Extrude72 which is nowhere near Boss27 jumps close to the Boss27 as shown in figure 5(b). The extrude feature was created using an edge of the b-rep of the solid prior to its creation and it has not be able to match the correct edge after editing the Boss27's diameter, leading to the wrong placement of the extrude feature. Unfortunately, the CAD system's procedural representation has not been able to detect the wrong location of Extrude72 and, due to symmetry even simple checks such as volume/mass properties of the solids (before and after an edit) will not be able to detect such semantic problems.

Though procedural families attempt to combine the advantages of the constructive and cell complex families, as the localized operations always refer to the b-rep entities (cells or their collections), they share the same problems inherent in both types of families. As we can see from the above examples, the input entities (not explicitly defined) used in the feature operations in a procedural representation are not always persistent, as they correspond to point-set instances of a b-rep and hence it is much harder to establish correspondence between them when the procedural representations are re-evaluated (due to parametric edit). Further, it is also harder to establish correspondence between parameters and the cells except in simpler domains (for example, in 2D). Besides, procedural representations have to deal with ill-defined feature constructions – especially the localized operations as illustrated earlier. In summary, procedural representations are non-unique, in general do not reliably support the basic solid modeling algorithms such as point membership classification (PMC), lead to non-constructive families and above all, are harder to topologize them for computing with families.

1.4 Outline

A topological framework for part families was proposed by the author in an earlier work [S.Raghothama and V.Shapiro 2003] based on category theory [Raghothama 2000]. According to the frame-

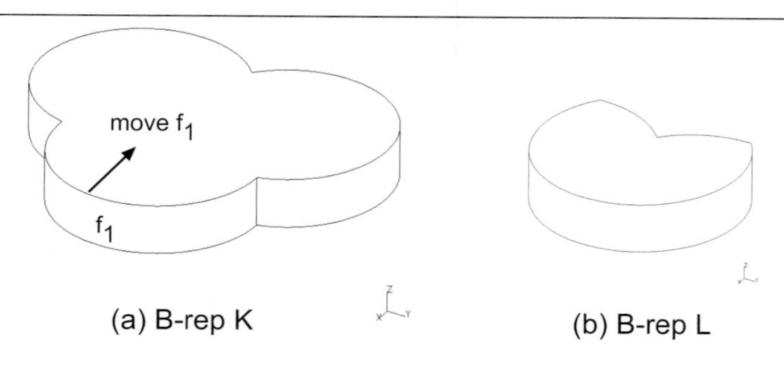

Figure 4: Incorrect result from a b-rep tweak

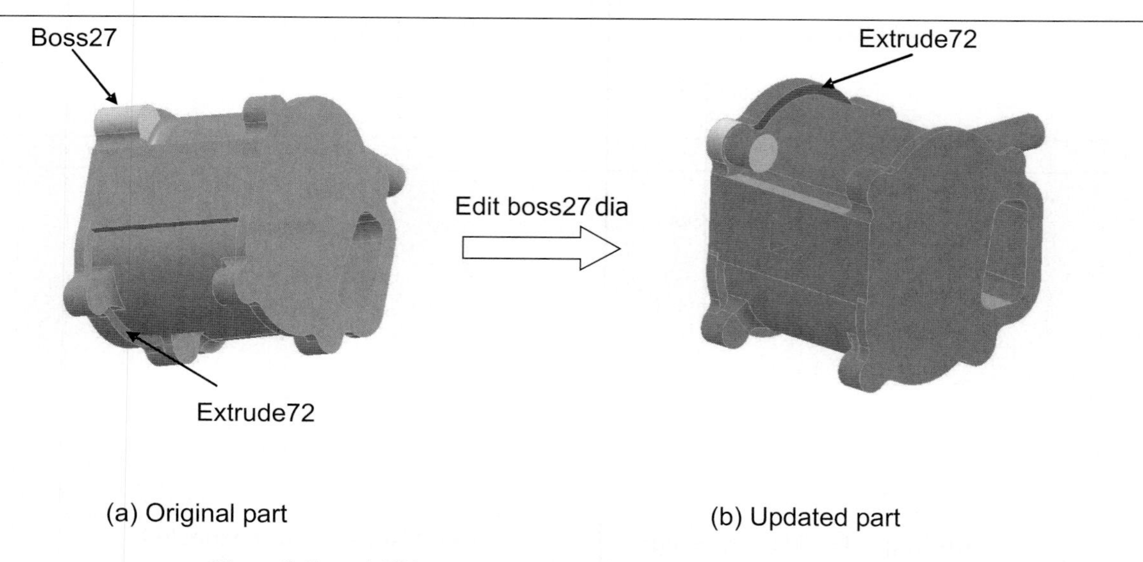

Figure 5: Extrude72 jumps to a different location when boss27 is edited

work, any part family can be defined by either using the parametric and/or representation space of the part representation. The objects in a family are topologized in parameter and/or representation space and the operations are continuous maps in the appropriate space. It was also shown that the two approaches can be combined using constructive part representations that are topologized in parameter space as well as in representation space using d-dimensional decompositions in E^d. This d-dimensional decomposition will be represented using a unique *constructive topological representation* $\mathcal{D}$, that is induced by a given *constructive parametric representation* $F(\mathbf{P})$. Every constructive topological representation is a collection of d-dimensional atoms (we will simply refer them as d-atoms) with a topology specified. Each d-atom is a closed but not necessarily connected pointset and is uniquely represented by a disjunctive canonical representation [Shapiro 1997]. This characterization of an atom in a constructive topological representation has the following advantages over the cells in a cell complex:

- we can establish a unique correspondence between d-atoms in not just instances, but across families;
- easier as well as reliable to compute, topologize and maintain the d-atoms as we don't have to deal with lower dimensional cells.

In this paper, we will formalize the general concept of constructive topological representations and illustrate how to define a topology on any constructive representation. We will focus on the CSG representation (which is a special case of constructive representation and considered as a decomposition of E^d) that has been topologized in both parameter as well as representation spaces. In keeping with the framework and terminology proposed in [S.Raghothama and V.Shapiro 2003], we will call the topologized representation of a CSG as a CSG complex. Further we will develop the algorithms for inducing a topology on a CSG representation and also enforcing continuity between two given CSG complexes computed from the same parametric CSG. Constructive topological representations are not only useful in defining constructive families, but they are also useful in specifying the semantics of features and to a larger extent (compared to parameter space continuity) preserve the intuitive properties of the solids in Euclidean space.

The rest of the paper is organized as follows. In section 2 we will define constructive topological representations and in section 3 we will develop an algorithm for computing the adjacency of atoms and induce a topology on a CSG complex. We will also develop two algorithms: brute-force and incremental, for enforcing continuity between two given CSG complexes. Finally, in section 4 we will conclude with a summary, discuss some applications and extensions of this work.

2 Formulation

2.1 Constructive representation

As a Boolean function with finite number of variables (carriers), every constructive representation of a solid can be written in a unique disjunctive canonical form as a union of intersection terms [Shapiro and Vossler 1991a]. The regularized intersection terms are homogeneously d-dimensional subsets of Euclidean space E^d that we will call *atoms*. Each *atom* is a closed, regular and (possibly) disconnected set that could contain holes or voids. For a fixed set of carriers, the *atoms* form a decomposition of the Euclidean space [Shapiro 1997], where each *atom* classifies as either IN or OUT with respect to the solid. From [Shapiro 1997] we know that the atoms in such a decomposition are finite in number. A carrier h represents a set of points in E^d such that: $h \equiv \{p \in E^d | f(p) * 0\}$, where $* \in \{\geq, \leq, =\}$ and f denotes any point classification function. In general this function is a ternary PMC (point membership classification) function with respect to the point (p) returning its classification of IN, ON or OUT with respect to the carrier h. However in case of constructive representations (CR) dealing with only d-atoms in E^d, we need only a binary classification function returning IN or OUT. When a given set of carriers $\mathbf{H} = \{h_1, h_2, \ldots, h_n\}$ is sufficient to represent a solid using a constructive representation, the solid is said to be describable by $\mathbf{H}$ and the regularized Boolean operations [Shapiro and Vossler 1991a]. If a solid is not describable with the given set of natural carriers, then we need additional separating carriers [Shapiro and Vossler 1991b]. As we assumed that we have a parametric representation $F(\mathbf{P})$ of a solid S, it is always describable with the carriers $\mathbf{H}$ present in F.

We define a *constructive representation* to be any collection of d-atoms such that the union of atoms that classify IN corresponds to the solid and the union of atoms that classify OUT corresponds to the complement of the d-dimensional solid in E^d. We will denote a constructive representation as D and the solid corresponding to the IN atoms as S. Note that a constructive representation does not satisfy the usual axioms of cell complexes, because the d-atoms are quasi-disjoint (they either do not intersect or intersect along their boundaries) sets of the same dimension d. The resulting *atomized* or stratified structure on the solids implies a 'natural' topology for constructive representations that is quite different from the usual Euclidean sub-space topology when the same solids are considered as cell complexes, and in the next subsection we will see how this 'natural' topology is defined for spatial constructive representations.

Figure 6 shows an example of simple 2-dimensional constructive representation D defined using the primitive set $\mathbf{H} = \{h_1, h_2, h_3\}$. The solid S is represented by the parametric CSG representation: $F(\mathbf{P}) = \{h_1 \cup h_2 \cap h_3\}$ and the constructive representation D is given by the union of the six atoms $\{C_1, C_2, \ldots, C_6\}$, where $\{C_1, C_2, C_3\}$ classify as IN and $\{C_4, C_5, C_6\}$ classify as OUT with respect to S.

Note that the parametric representation $F(\mathbf{P})$ of a solid is not unique, but a constructive representation D is unique since every atom in D is represented by a *unique* disjunctive canonical form Boolean expression. We will call this unique Boolean expression as a **characteristic map** of an atom. The characteristic map is key to the definition as well as computation of a constructive family. As the characteristic map of an atom is composed of all the carriers in the constructive representation, if we know the correspondence between the carriers we can use that to establish a correspondence between the atoms. We assumed that the carriers of two constructive representations are in general position, and we can always establish a correspondence between them (this is usually the case when they are obtained from the same parametric family given by a parametric representation $F(\mathbf{P})$).

Given a constructive representation D with its corresponding solid S and another constructive representation D' with solid S' (presumably obtained by a parametric update of the parametric representation $F(\mathbf{P})$), in order to establish a correspondence between two atoms in D and D', we require the following two conditions to be satisfied between the corresponding atoms:

1. characteristic map (C_i) = characteristic map ($C_j{}'$);
2. classify(C_i, S) = classify($C_j{}'$, S').

The latter requirement would not be necessary if the constructive representation is a union of the atoms in its corresponding solid (i.e, the atoms that classify as IN). But as the constructive representation contains atoms that are in the complement of a solid, we require this additional *partition-preserving* condition.

2.2 Constructive topological representation

A topological space $\mathcal{T}$ of a set X is usually defined as a collection of subsets of X called as open sets satisfying three conditions [Munkres 1975]:

1. the sets X and $\emptyset$ are always present in a topological space,
2. arbitrary union of the open sets is an open set,
3. finite intersection of the sets is an open set.

A topology $\mathcal{T}_1$ is said to be *coarser* than another topology $\mathcal{T}_2$, if $\mathcal{T}_1 \subset \mathcal{T}_2$. The inverse relationship between $\mathcal{T}_1$ and $\mathcal{T}_2$ implies a *finer* topology.

Several equivalent definitions of topology on a set exist [Munkres 1975; Jänich 1983], and one such definition that is of relevance to solid modeling is based on the notion of nearness or neighborhood [Requicha 1980; Kinsey 1993]. In this definition, a topological space is a set X with a collection $\mathcal{B}$ of subsets $N \subseteq X$, called neighborhoods, such that:

1. every point is in some neighborhood,
2. the intersection of any two neighborhoods of a point contains a neighborhood of the point.

The set $\mathcal{B}$ of all neighborhoods is called a basis for the topology on X. A subset $O \subseteq X$ is an open set if for each $x \in O$, there is a neighborhood $N \in \mathcal{B}$ such that $x \in N$ and $N \subseteq O$. The set $\mathcal{T}$ of all open sets is a topology on the set X. Note that the second definition of a topology is usually easier to work with, due to a smaller collection of neighborhoods than open sets. Also it should be clear that any neighborhood is itself an open set based on these definitions.

A topology can be defined for every constructive representation by defining neighborhood to each d-dimensional atom of the representation and thus define a constructive topological representation (CTR). As the constructive representation D also contains atoms in the complement of a solid S, we will be defining a topology for the whole of E^n and not just the solid S. Recall that constructive representations deal with regular sets in E^d and accordingly in the CR topology we need to use regularized intersection in order to satisfy the intersection condition of a neighborhood. This assumption in CR topology is consistent with the topology of closed regular sets [Requicha and Tilove 1978]. A notion of neighborhood can be defined using the star (St) of an atom in set-theoretic terms, where star (St) denotes adjacency of an atom.

Definition 1 (Star of an atom) *Given a constructive representation D, the star neighborhood of an atom C_i is defined as:*

$$St(C_i) = \{C_j | C_i \cap C_j \neq \emptyset\}. \tag{1}$$

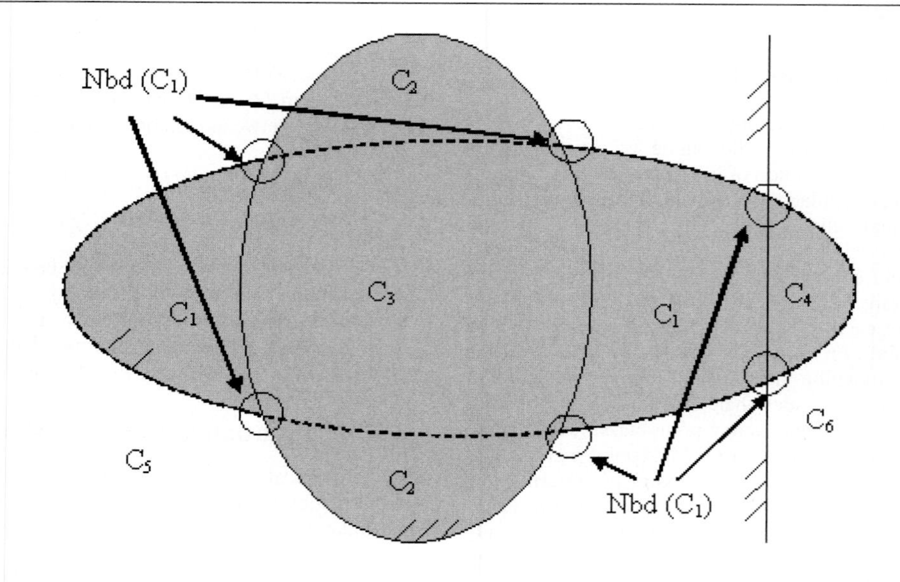

Figure 6: A 2-dimensional constructive representation corresponding to the solid $h_1 \cup h_2 \cap h_3$ is a union of 2-atoms

By definition the neighborhood of every atom can also be represented using a constructive representation, but is not necessarily canonical. As is usually common in constructing new spaces, we can use this finer topology to define a subspace topology on S. We will denote a given constructive representation D as the *nominal* representation and formally define the topology on a constructive representation.

Definition 2 (CR topology) *Given a constructive representation D and a notion of star (St), a CR topology is a collection of sets defined below:*

$$\mathcal{D} = \{|D|, \emptyset, C_1, C_2, \ldots, C_n, St(C_1), St(C_2), \ldots, St(C_n), \ldots\} \tag{2}$$

The trailing dots in the above definition denote arbitrary unions and intersection of the atoms. By definition, the closed sets in a constructive representation are also open and hence are also present in the topology. The distinction between D and $\mathcal{D}$ should be clear now: while D is a representation of an *instance* of a set, $\mathcal{D}$ is a representation of a *collection* of sets. Note that it should be clear that CR topology is very coarse compared to the usual Euclidean topology, for at least two reasons:

- we consider only the d-dimensional atoms and ignore the lower dimensional ones;
- ignored the connectivity (connectedness) characteristics of the atoms and also of the corresponding solid.

However due to the presence of atoms in the complement of the solid S, the above definition of a CR topology is a finer topology than the topology we can define using only the atoms in S.

We will illustrate how to define a CR topology and enumerate its topology through the constructive representation in Figure 6. We can define a topology for this constructive representation D by defining a neighborhood of each 2-atom in D. Even though this constructive representation does not have vertices or edges, as we will show later on that it is possible to define the vertex as well as edge neighborhoods for each atom and define a topology using these neighborhoods. For example, figure 6 shows all the vertex neighborhoods of the atom C_1. Thus all the atoms in one that intersect this neighborhood are in $St(C_1)$. The topology for D is given by:

$$\mathcal{D} = \{|D|, \emptyset, C_1, C_2, C_3, \ldots\ St(C_1), St(C_2), St(C_3), \ldots\}.$$

Once we have defined the notion of neighborhood on a constructive representation D, we have a constructive topological representation $\mathcal{D}$. Now we need to define precisely what we mean by neighborhood-preserving or continuous map between two constructive topological representations. As every constructive representation is a collection of atoms, given two constructive topologized representations $\mathcal{D}$ and $\mathcal{D}'$ we can define continuity from $\mathcal{D}$ to $\mathcal{D}'$ atom-by-atom.

Definition 3 (Continuous CTR map) *Given two constructive topologized representations $\mathcal{D}$ and $\mathcal{D}'$, a (possibly) many-to-one map from $\mathcal{D}$ to $\mathcal{D}'$ is a continuous CTR map if it takes every element of $\mathcal{D}$ into its image in $\mathcal{D}'$ such that neighborhood of every atom is mapped into its corresponding image atom's neighborhood.*

By definition, every continuous CTR map is a neighborhood-preserving map. Further, it can be shown that the following star condition holds for every continuous CTR map g:

$$g(St(C_i)) \subseteq St(g(C_i)). \tag{3}$$

Note that the continuity of a CTR map should be applicable to the whole of $\mathcal{D}$. One way to extend the map to all the elements of $\mathcal{D}$ is by checking the star condition on the atoms recursively i.e, $St(St(...(C_i))^3$.

In the following discussions we will refer to a continuous CTR map as just CTR map and implicitly assume the requirement of continuity of the map. An important insight behind a CTR map is:

[3]This notion was referred as the extended adjacency of cell in a b-rep [Chen and Hoffmann 1995].

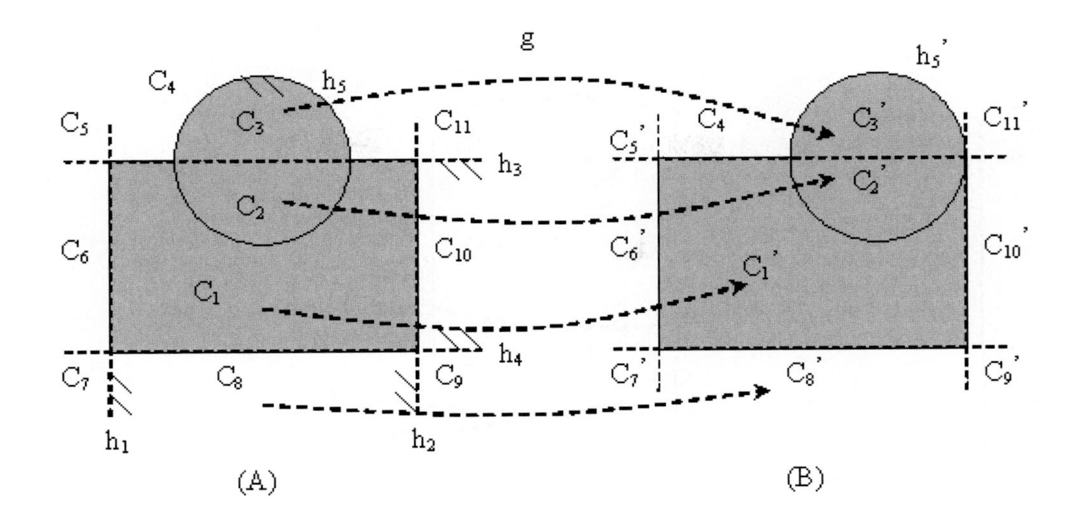

Figure 7: Continuous CTR map from the CTR in (a) to the CTR in (b)

new atoms may exist in constructive topological representation $\mathcal{D}'$ as long as all neighborhoods in $\mathcal{D}$ are mapped into their corresponding neighborhoods in $\mathcal{D}'$.

Let us illustrate the properties of CTR maps through the constructive representations in Figure 7. The one-to-one map g takes the 2-atom C_1 to C_1', C_2 to C_2' and C_n to C_n' (for all other atoms C_n, where $n = 3, \ldots 11$). Further, the star of atoms C_1, C_2 and their image atoms are given by:

$$
\begin{aligned}
St(C_1) &= \{C_1, C_2, \ldots C_{11}\},\\
St(C_2) = St(C_3) &= \{C_1, C_2, C_3, C_4\},\\
St(C_1') &= \{C_1', C_2', \ldots C_{11}'\},\\
St(C_2') = St(C_3') &= \{C_1', C_2', C_3', C_4', C_{10}', C_{11}'\},\\
g(St(C_1)) = g(C_1, C_2, \ldots C_{11}) &= \{C_1', C_2', \ldots C_{11}'\},\\
St(g(C_1)) = St(C_1') &= \{C_1', C_2', \ldots C_{11}'\}\\
&\Rightarrow g(St(C_1) = St(g(C_1)),\\
g(St(C_2)) = g(C_1, C_2, C_3, C_4) &= \{C_1', C_2', C_3', C_4'\},\\
St(g(C_2)) = St(C_2') &= \{C_1', C_2', C_3', C_4', C_{10}', C_{11}'\}\\
&\Rightarrow g(St(C_2)) \subset St(g(C_2)).
\end{aligned}
$$

In a similar fashion we can show that the star or neighborhood-preserving condition holds for all other 2-atoms in the constructive representations. In other words, g is a CTR map.

On the other hand, consider the mapping g between $\mathcal{D}$ and $\mathcal{D}'$ in Figure 8. The complete mapping g is given by:

$$
\begin{aligned}
g(C_1) = C_1',\ g(C_2) = \emptyset,\ g(C_3) = C_2',\\
g(C_4) = C_3',\ g(C_5) = C_4',\ g(C_6) = C_5',\\
g(C_7) = C_6',\ g(C_8) = C_7',\ g(C_9) = C_8',\\
g(C_{10}) = C_9',\ g(C_{11}) = C_{10}'.
\end{aligned}
$$

Now we need to check whether g also takes the star of each atom into its image atom's star. For instance:

$$
\begin{aligned}
St(C_2) = St(C_3) &= \{C_1, C_2, C_3, C_4\},\\
g(St(C_2)) = g(C_1, C_2, C_3, C_4) &= \{C_1', C_2', C_3'\}
\end{aligned}
$$

and $St(g(C_2)) = St(\emptyset) = \emptyset$. Thus $g(St(C_2)) \not\subseteq St(g(C_2))$. It can also be shown that $g(St(C_3)) \not\subseteq St(g(C_3))$ and hence g is not a CTR map.

2.3 Defining constructive families

Our goal is to define a family using both the parametric representation $F(\mathbf{P})$ and its corresponding nominal CTR $\mathcal{D}$, the latter is obtained by evaluating $F(\mathbf{P})$ and topologizing it using the star topology defined above. As illustrated in examples of section 1, we know that defining a parametric family using $F(\mathbf{P})$ in the parametric topology is quite straight-forward, and the members in this family can be easily generated by varying the parameter values in the set $\mathbf{P}$. We will denote the updated parametric representation as $F(\mathbf{P}')$. Now the constructive family (with CTR maps as the morphisms) can be defined using one of two paradigms:

- Classification: for every parametric edit of $F(\mathbf{P})$ and its corresponding CTR $\mathcal{D}$, re-evaluate another CR D' from $F(\mathbf{P}')$, topologize it and then classify $\mathcal{D}'$ against $\mathcal{D}$.
- Generation: for every parametric edit of $F(\mathbf{P})$, given the corresponding CTR $\mathcal{D}$ and a CTR map g, generate another CTR $\mathcal{D}'$ such that it is consistent with the parametric edit.

Clearly the generation paradigm is the ideal approach for defining a family, as is the case with the parametric family induced by $F(\mathbf{P})$. However in case of constructive families, generation appears to be harder than classification, since it implicitly assumes that we can parameterize the CTR maps and compute $\mathcal{D}'$ independently of $F(\mathbf{P})$. On the other hand, we can easily establish a correspondence between two given CTR (re-evaluated from a parametric edit of $F(\mathbf{P})$) and verify that it is a CTR map or not. Note that in either case, the parametric representation $F(\mathbf{P})$ is necessary to induce a CR. So far we haven't explained the algorithmic details of computing a CTR and enforcing the CTR map using the classification paradigm. In section 3, we will describe a brute force and also an incremental algorithm for accomplishing this task using CSG

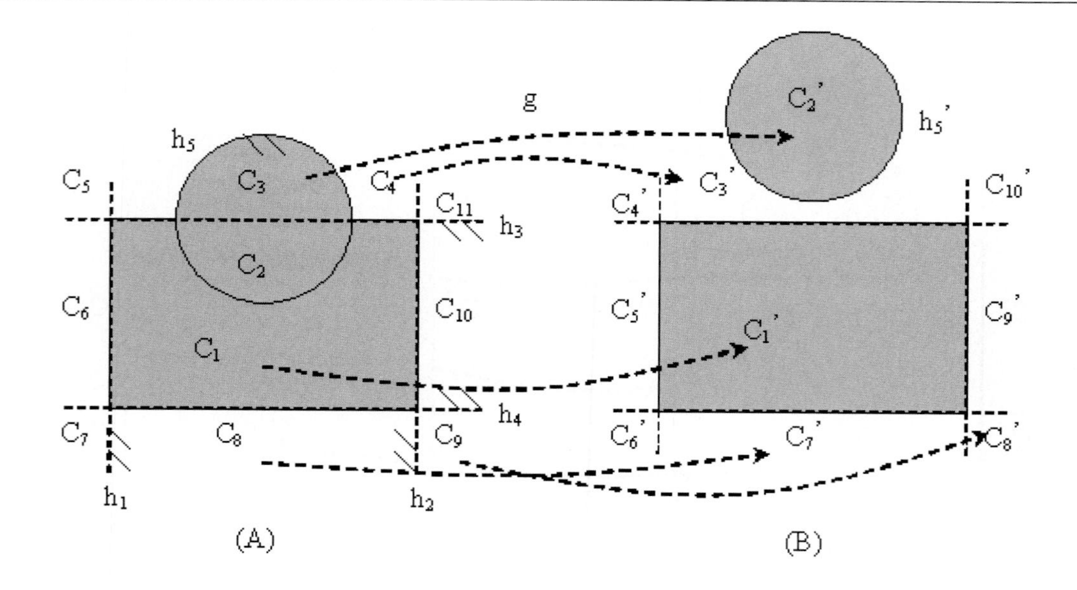

Figure 8: Correspondence between CTR (a)$\mathcal{D}$ and (b) $\mathcal{D}'$ is not a CTR map

representations (considered as a spatial decomposition), which is a special case of constructive representation and a CTR using CSG representation will be called as CSG complex in keeping with the terminology proposed in [S.Raghothama and V.Shapiro 2003].

3 Algorithms

3.1 Computing the CSG representation

By definition, every atom in a CSG representation D is sign-invariant with respect to the set of carriers $\mathbf{H}$ and is adequately represented by a single point from that atom [Shapiro and Vossler 1991a; Shapiro and Vossler 1993]. The set of all such points is called a characteristic point set (CPS) of a CSG representation [S.Raghothama and V.Shapiro 2000]. We generate points in the sign-invariant d-atoms using the same method that was used in brep-to-CSG conversion [Shapiro and Vossler 1991a; Shapiro and Vossler 1991b] and was also used for consistency verification between a CSG and b-rep under dual updates [S.Raghothama and V.Shapiro 2000]. Essentially this method consists of locally offsetting (by a small value ϵ) one or more carriers in $\mathbf{H}$, testing for intersections and generating the characteristic set of points for $\mathbf{H}$ in the sufficiently small vertex/edge neighborhoods. Note that the intersection of carriers will lead to more than one point per atom. But this redundancy of points for every atom is useful in computing the adjacency of a d-atom in E^d and hence induce a topology on a CSG representation directly without computing or representing the lower dimensional atoms.

The set of carriers that generate a point $s_i \in$ CPS will be called as *contributing* carriers of s_i. The contributing carriers are the minimum set of carriers necessary to represent the underlying pointset corresponding to an atom, but not necessarily describable by them (separators could be required [Shapiro 1997]). As we shall see soon, the contributing carriers are the key in computing the adjacency of an atom. If the boundary of an atom does not contain any vertex or edge (for instance the volume obtained by subtracting two concentric spheres), then we can use any point on the offset carrier. This implies that every point in the CPS has *at least* one contributing carrier.

The CSG D and its corresponding solid S (represented by the IN atoms in D) are represented by the CPS, where each point $s_i \in$ CPS has the following information:

1. Euclidean coordinates of the point $s_i = (x, y, z)$.
2. Source from where s_i came from (*contributing carriers*).
3. Classification with respect to the carriers $\mathbf{H}$ (*characteristic map*).
4. Classification with respect to the solid S (IN or OUT).

Once we compute the coordinates of s_i, we can easily compute the classification of s_i against the solid as well as the carriers using the usual PMC procedure. The contributing carriers can be deduced from the offsetting procedure: the original carriers which were offset to generate a given point $s_i \in$ CPS, are the contributing carriers. Thus the CPS is an adequate representation for the d-dimensional atoms of a CSG representation D, where all points contained in an atom have the same characteristic map and same classification with respect to the solid. But they need not be having the same set of contributing carriers.

Next we will see how to topologize a CSG representation D thus defining a CSG complex $\mathcal{D}$, and then see how to enforce the continuous CSG map between two given CSG complexes. This in turn requires establishing a (possibly) one-to-one map between the atoms of two CSG complexes and verify if the given map satisfies the star condition 3.

3.2 Inducing a CSG topology

Given a CSG representation D, we need to convert it into a CSG complex $\mathcal{D}$, by specifying the neighborhood of each atom in D. The key idea behind this neighborhood computation is directly based on the definition of star of an atom defined in section 2, i.e, the collection of all the atoms that intersect an atom. As each atom is defined using the carriers, the star of an atom can be computed using the fact that two adjacent atoms share one or more carriers which are common between the atoms.

From the CPS of a CSG representation D computed above, we could retain just one point per atom and discard the rest of the points in the atom for computing the adjacency. Then by removing one or more carriers of an atom we can determine its adjacent atoms. This is similar to certain steps in the Boolean minimization procedure of brep-to-CSG conversion [Shapiro and Vossler 1991a]. While this may be an efficient approach in terms of space requirements (due to the fact that there are lesser atoms than points), it may not be algorithmically simple, since it is not clear how many carriers need to be removed to determine the adjacency of an atom in general and we may need to do a multi-level minimization.

As we compute more than one point per atom anyway to start with, we can take advantage of the additional information available in the CPS to devise a simpler algorithm. Refer to the 2D CSG representation in figure 6. We find that in each of the neighborhoods shown, the four atoms are adjacent to each other and are separated by by the two elliptical carriers h_1 and h_2. These two carriers are essentially the contributing carriers for the four points in the CPS in this neighborhood. We see that if we remove the contributing carriers from the characteristic maps of two adjacent atoms, then they are identical (i.e, have the same canonical Boolean expression). This observation generalizes and holds true regardless of whether the carriers in a CSG intersect or not. For instance, consider the atom bounded by two parallel planes, two concentric spheres or two co-axial cylindrical surfaces. The inside atom bounded by the two carriers is contributed by both the carriers, but the innermost (for spheres and cylinders) and outer atoms are contributed by only one carrier. In all cases, the non-intersecting carriers are considered as contributing to an atom in the CSG representation as long as they bound the atom. Thus the algorithm for computing star of an atom in a CSG representation D follows from these observations. Atom C_j is in the star of atom C_i if they have:

1. at least one point each with the same contributing carriers; and,
2. same characteristic map with respect to all carriers except the contributing carriers.

The star of an atom C_i is the collection of all such atoms obtained by using the same two steps for all the characteristic points within C_i. In general, points in CPS that belong to adjacent atoms are separated [Shapiro and Vossler 1991a] by the contributing carriers and on the other hand, points belonging to non-adjacent atoms are separated by non-contributing carriers.

Using the CPS of a CSG representation D, we can build the adjacency structure for computing the star of an atom and thus impose a topology on D. Assuming the adjacency list representation of a graph for representing the star of each atom, we are lead to the following algorithm for computing the star of an atom.

When we are done with all the points of CPS, algorithm 1 would have obtained the adjacency information necessary for computing the star of each atom.

To better understand the adjacency computation, let us consider the 2-dimensional constructive representation in figure 9 (the constructive representation shown in figure 6 along with its CPS). The points in the CPS is shown labeled from s_1 through s_{24}, where each point s_i is generated by offsetting pairs of the carriers from the carrier set $\mathbf{H} = \{h_1, h_2, h_3\}$ and intersecting them. The adjacency of the 2-atoms C_i can be deduced by marching through the points in CPS and applying the algorithm 1 given above.

For instance, in order to compute the star of the 2-atom C_1 in figure 9, we start with the point $s_2 \in C_1$ and find that each of the three points (s_1, s_3, s_4) in CPS in its neighborhood is contributed

procedure ComputeAdjacency(D);
  //CHMAP denotes characteristic map
  for each point s_i in CPS of an atom C_j,
    get its contributing carriers $H(s_i)$,
    for each point s_j in CPS not in C_j,
      if $(H(s_i) = H(s_j))$ and

      CHMAP(C_j)-$H(s_i)$ = CHMAP(C_k)-$H(s_i)$,
      then
        $St(C_j) = AddToAdjacencyList(C_k)$
        $St(C_k) = AddToAdjacencyList(C_j)$
    end-for;
  end-for;
end-procedure;

Algorithm 1: Adjacency algorithm

by the two ellipses. In addition, their characteristic maps are equivalent if we exclude the two ellipses, and thus the atoms C_3, C_5 and C_2 corresponding to the three points s_1, s_3 and s_4 are adjacent to C_1. Next we consider the point $s_6 \in C_1$ and find that the set of points (s_5, s_7, s_8) in its neighborhood also satisfy the adjacency requirement. But we find their corresponding atoms already in the adjacency list and so we proceed to the next point in the CPS. Similarly for the points $s_{10} \in C_1$ and also $s_{16} \in C_1$, the atoms corresponding to the neighboring points s_{10} and s_{16} are already in C_6's adjacency list. We proceed to $s_{17} \in C_1$, and consider its neighboring set (s_{18}, s_{19}, s_{20}). From this, we set the atom C_4 and C_6 into the adjacency list of C_1. Finally we proceed to the last point $s_{21} \in C_1$ and it has three other points (s_{22}, s_{23}, s_{24}) in the neighborhood of s_{21}. But we find that their corresponding atoms C_4, C_5 and C_6 are already present in the adjacency list of C_1. Thus we obtain the star of C_1 as: $St(C_1) = \{C_1, C_2, C_3, C_4, C_5, C_6\}$. We repeat this procedure for each of the remaining 2-atoms to obtain their corresponding star as enumerated in section 2.

In general, the running time of the adjacency algorithm can be prohibitive for arbitrary spaces in d dimensions. But in practice for low degree (≤ 3) polynomial carriers in E^3 it is reasonable. We know that there can be $O(n^3)$ atoms in a CSG decomposition induced by n carriers [Shapiro 1997]. Though we use more than one point per atom, we can still consider it be $O(n^3)$ since the number of points per atom will be far less than n. Hence the adjacency algorithm described above can be estimated to be of $O(n^6)$. However note that this adjacency computation can be done incrementally when computing the decomposition carrier-by-carrier. Also we can use more efficient data structures (such as hash tables) and additional speedups for such graph-based computations. Once we know how to compute the star of every atom in a CSG representation D, we have the CSG complex $\mathcal{D}$ that represents the topology of D.

3.3 Enforcing CSG maps

3.3.1 Brute force approach

In the brute force approach to enforcing CSG maps, for any parametric edit on $F(\mathbf{P})$, we need to re-evaluate a new CSG representation D' from $F(\mathbf{P}')$, topologize it $\mathcal{D}'$, establish a correspondence between atoms of D and D', and finally verify if the implied correspondence is a CSG map from $\mathcal{D}$ to $\mathcal{D}'$ (the star condition 3).

The key computation lies in establishing a correspondence between the atoms of two CSG complexes. We know that every atom carries the characteristic map represented by a unique disjunctive canonical Boolean expression and is composed of all the carriers in the CSG complex. This implies that if we know the correspondence between the carriers, which is true always in our case, we can

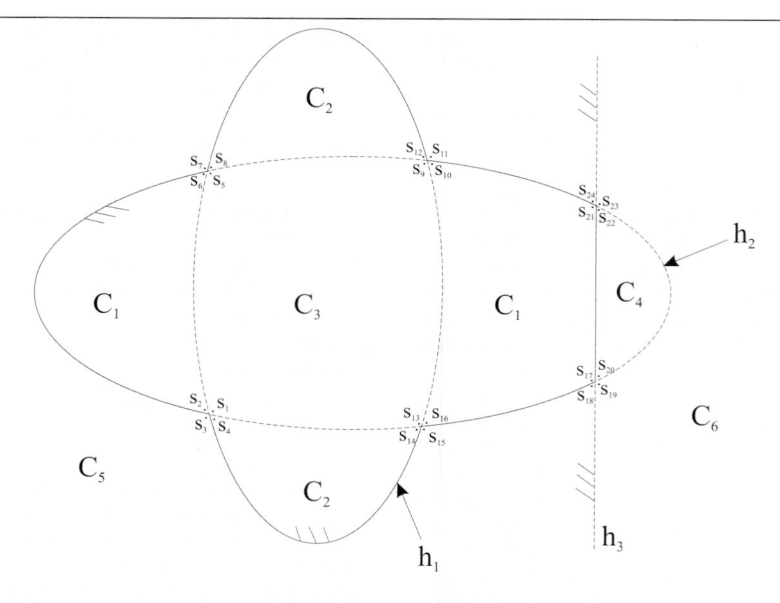

Figure 9: Computing the star of a 2-atom in a CSG complex by marching through its CPS

use that to establish a correspondence between the atoms. The two correspondence conditions in section 2 intuitively imply that the corresponding atoms need to have the same characteristic map and also same classification with respect to the corresponding solids.

These observations lead to the algorithm for verifying the star condition 3, given a pre-edit CSG complex $\mathcal{D}$, post-edit CSG complex $\mathcal{D}'$ and a correspondence g between the atoms. We will assume that a correspondence g is a function which takes an atom $C_i \in D$ and return its corresponding atom $C_j{}' \in D'$ (using the two conditions given above).

procedure VerifyCSGMAP($\mathcal{D}$, $\mathcal{D}'$, g);
  for each atom C_i in $\mathcal{D}$,
    $C_j' = g(C_i)$,
    if $g(St(C_i) \not\subseteq St(C_j')$, **then**
      error 'update resulted in discontinuity';
      return (false);
  end-for;
  return (true);
end-procedure;

Algorithm 2: Algorithm for verifying a CSG map

Using the same examples from section 2 we will see how to establish the correspondence implied there and enforce a CSG map. Refer to the 2-dimensional CSG complex shown in figure 7(a). The characteristic map of the atom C_1 is given by $(h_1 \cap h_2 \cap h_3 \cap h_4 \cap \bar{h}_5)$, and the characteristic map of the atom C_1' is given by $(h_1 \cap h_2 \cap h_3 \cap h_4 \cap \bar{h}_5')$. Given the fact that h_5 corresponds h_5' and the rest of the carriers are identical in the two CSG complexes in figure 7(a) and 7(b), and also the atoms have the same classification (IN) with their respective solids, it is easy to establish that $C_1 \in \mathcal{D}$ corresponds to $C_1' \in \mathcal{D}'$. Similarly we can show that C_2 corresponds to C_2' and so on. In this way we obtain the correspondence $g(C_i) = C_i'$, which can be further verified to obey the star condition 3 and hence g is a CSG map.

In case of the example in figure 8, the atom $C_2 \in \mathcal{D}$ has a characteristic map given by $(h_1 \cap h_2 \cap h_3 \cap h_4 \cap h_5)$. It is easy to see that C_2 does not have a corresponding atom in $\mathcal{D}'$ and thus it corresponds to $\emptyset$. The characteristic map of atoms C_1 and C_3 is given by $(h_1 \cap h_2 \cap h_3 \cap h_4 \cap \bar{h}_5)$ and $(h_1 \cap h_2 \cap \bar{h}_3 \cap h_4 \cap h_5)$ respectively. On the other hand, the characteristic map of atoms C_1' and C_2' is given by $(h_1 \cap h_2 \cap h_3 \cap h_4 \cap \bar{h}_5')$ and $(h_1 \cap h_2 \cap \bar{h}_3 \cap h_4 \cap h_5')$ respectively. The fact that $h_5 \in D$ corresponds to $h_5' \in D'$, and also the two atoms classify as IN with their respective solids implies that C_1 corresponds to C_1', and C_3 corresponds to C_2'. It is easy to find the corresponding atoms to the rest of the atoms in D. As the atoms C_1 and C_3 in C_2's neighborhood, map to the their corresponding atoms C_1' and C_2', but C_2 maps to $\emptyset$, the implied map in this case does not satisfy the star condition and consequently it cannot be a CSG map.

So far we haven't explained about the updates shown in figures 2, 3 and 5. The CSG representation in each of these examples were defined using standard primitives and features (blocks, cylinders, slot and hole). Both the updates shown in figures 2(b) and 3(b) can be shown to be continuous in CSG topology. There is a one-to-one mapping between their carriers and the atoms in both cases, though in one case the connectivity (figure 2) and in another case the number of connected components of the corresponding atom (figure 3) does not match. However if the CSG representation was defined using halfspaces, the update can be shown (though non-trivial to illustrate in a 2D figure) as a discontinuous map. Also it is easy to see that the update in figure 5(b) is not continuous, since Extrude72 has a completely different neighborhood (not labelled) in figure 5(a) (including Boss27) while its original neighborhood features are still in the same location.

These examples show that we can detect certain class of ambiguities in existing parametric CAD systems via CSG complexes and CSG continuity using some form of topology (either induced by halfspaces or primitives/features in the representation). In other words, we can augment non-constructive representations with a CSG complex as long as they use Boolean operations to define global feature operations, and use CSG continuity as a semantic checker for global validity that is dependent on the existence/non-existence as well as adjacency relationship between d-dimensional volumes.

3.3.2 Incremental approach

In the incremental approach to enforcing CSG maps, given a parametric update on $F(\mathbf{P})$, and a nominal CSG complex $\mathcal{D}$, we need to incrementally update $\mathcal{D}$ and enforce the CSG map locally (by keeping the parametric and CSG complex consistent). The key computational task in this paradigm involves constructing a new CSG representation D' by inducing a consistent classification to any new atoms or components of existing atoms in D' and by inducing a topology from $\mathcal{D}$ to D', and verifying if the star condition holds between $\mathcal{D}$ and $\mathcal{D}'$. If there are new atoms in D', we need to induce a consistent classification, based on the parametric representation $F(\mathbf{P})$ and its updated version $F(\mathbf{P}')$ respectively. This keeps $F(\mathbf{P})$ and D consistent. It is important to note that if an atom disappears after an update, it does not automatically lead to discontinuity. But if an atom disappears, then all of its neighborhood atoms should disappear in its corresponding atom's neighborhood for continuity. In other words, *the star condition of continuity permits neighborhoods to grow, but not shrink.*

As we have the task of computing a new CSG complex incrementally, we can rely on localized updates to determine and enforce the continuity between the CSG complexes, similar in principle to the incremental approach described in verifying consistency between a CSG and b-rep [S.Raghothama and V.Shapiro 2000].

Figure 10 shows a example of incrementally updating a CSG complex (shown in figure 10(a)) of a CSG complex family. The original CSG complex has three atoms and the updated CSG complex in figure 10(b) also has three atoms, each of which map one-to-one with the CSG complex in figure 10(a). However note that the new CSG complex has two new connected components corresponding to C_2' and C_3' that will be assigned the same classification as their corresponding atoms C_2 and C_3 in the original CSG complex.

In the above example there were no new atoms in the updated CSG complex. Figure 11 shows an example in which there is a new atom C_{10} created in the updated CSG complex shown in figure 11(b). We need to induce a consistent solid classification based on the parametric representation used to represent the original solid in figure 11(a). If the parametric representation is given by $F(\mathbf{P}) = ((((h_1 \cup h_2) \cup h_3) - h_4) - h_5)$, then the C_{10} will be classified as OUT; on the other hand if the parametric CSG representation is given by $F(\mathbf{P}) = ((((h_1 \cup h_2) - h_5) - h_4) \cup h_3)$, then the C_{10} will be classified as IN, for ensuring consistency between the parametric representation and its corresponding CSG representation as well as ensuring continuity between the CSG complexes.

3.4 Implementation

The brute force algorithm for enforcing a CSG map including the algorithm for computing the adjacency or star of an atom, has been implemented using Parasolid with HOOPS3D providing the GUI on the Windows platform. Parametric representation $F(\mathbf{P})$ is provided as input through a file and the system evaluates its b-rep for display purposes. Users can interactively edit (through a dialog box) one or more parameters of $F(\mathbf{P})$ and the system enforces continuity on the CSG complex computed internally. It throws an error message (in a dialog box) if the parametric edit does not lead to a continuous map in CSG topology and thus maintaining a parametric and CSG complex family.

4 Conclusions

4.1 Summary

In this paper we introduced constructive topological representations as an alternative to cell complexes. We further developed algorithms for computing the CSG topology and enforcing continuity in a CSG complex, which is a special case of CTR. Each atom in a CSG complex carries a unique characteristic map that is represented by a disjunctive canonical Boolean representation (with respect to all the carriers and operations in it). Also every atom contains a subset of the carriers which are called contributing for that atom and these contributing carriers along with the characteristic map is critical to the computation of adjacency or star of an atom. The specification of a star for every atom enables us to topologize a CSG representation and thus define a CSG complex. The characteristic map is also key in establishing a correspondence between atoms of any two CSG complexes, which is in turn necessary for verifying continuity between two given CSG complexes induced by a given parametric representation.

Using these algorithms we can define CSG complex family, which is a special and well-known constructive family. Constructive families defined this way have both global parametric as well as shape control. This control comes naturally to constructive families because they are defined using the CTR as the objects of these families. In addition, constructive families have two important advantages over non-constructive families in the following ways:

- ability to enforce continuity in both parametric **and** representation (shape) space;
- ability to support divide-and-conquer as well as incremental algorithms.

As demonstrated in this paper, while parametric continuity is quite easy to enforce, the representation space continuity in CSG is non-trivial. But it is the unique relationship between decompositions and representations that helped us develop systematic algorithms for defining a CSG topology as well as enforcing the CSG continuity.

4.2 Extensions

Though the algorithms in section 3 were developed for CSG representations, the definitions in section 2 are applicable to more general constructive representations including feature representations, assuming that these feature representations also rely on regularized Boolean and global operations (which permit canonical representations of their corresponding atoms). We can define constructive feature families using any feature based representation that relies on well defined constructive operations and represents the atoms uniquely. However as the decomposition and hence the topology is coarser for the primitives and feature-based constructive representations compared to the halfspace-based representations, the notion of continuity in their respective topologies loses the intuitiveness of Euclidean topology as illustrated in the examples of section 1.

Other extensions to CSG complex family are possible, for instance by modifying the decomposition by considering lower dimensional atoms and also the CSG topology. Details on this particular extension can be found in the first author's thesis[Raghothama 2000], where the *generic cell complex* is proposed as one generalization to CSG complex. This leads to more guarantees in the CSG continuity, but at the cost of restricting the range of family (i.e, the number of objects in the family) and also introducing the robustness problems that exist in cell complexes due to the requirement of lower dimensional cells.

Finally, it appears that CSG complexes with connected atoms can be used to compute the Betti numbers and also the Euler characteristic of a solid without using b-reps or cell complexes. The topology in such complexes coincides with the Euclidean sub-space topology and thus the properties of the solid are the same. While we know that the atoms are not describable or representable by a unique Boolean expression, in this case it does not matter as we do

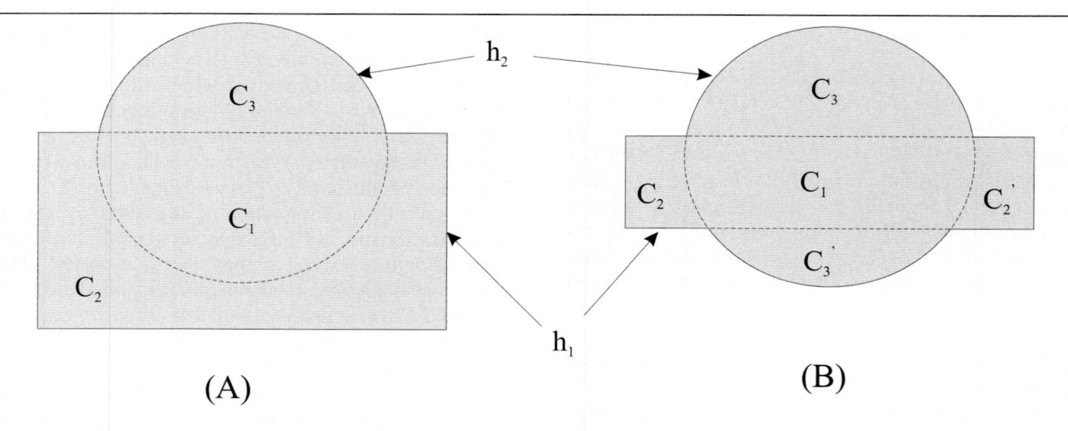

Figure 10: Incremental approach of computing a CSG complex

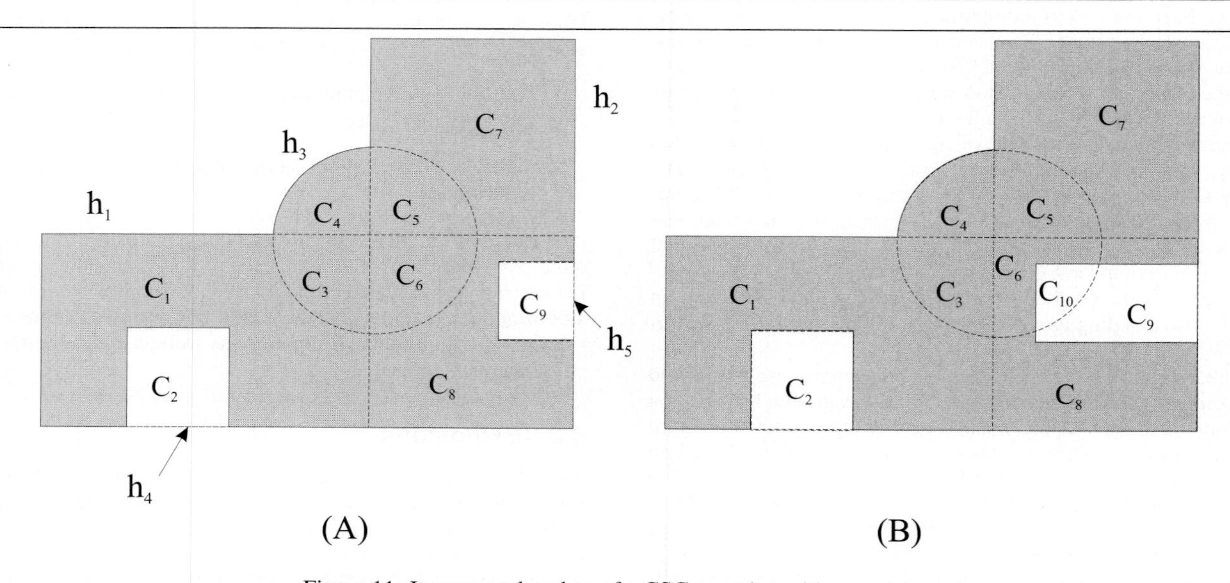

Figure 11: Incremental update of a CSG complex with new atoms

not have to deal with families and only with instances. One interesting application of a CSG complex with connected atoms is in improving the robustness of Boolean operations (of any two solids represented either using a CSG or a b-rep), for instance as a preprocessing step to Booleans for determining the validity and type of intersection (manifoldness, disconnected, etc) between two bodies. This is useful in particular to b-rep based Booleans involving freeform surfaces bounding the solids, where the numerical intersection algorithms can be unstable for degenerate cases and fail with imprecise and vague error messages.

ACKNOWLEDGMENTS

I thank Vadim Shapiro for his encouragement and constructive comments on several aspects of this paper's early versions.

References

CAPOYLEAS, V., CHEN, X., AND HOFFMANN, C. M. 1996. Generic naming in generative, constraint-based design. *Computer-Aided Design 28*, 1, 17–26.

CHEN, X., AND HOFFMANN, C. M. 1995. Towards feature attachment. *Computer-Aided Design 27*, 675–702.

JäNICH, K. 1983. *Topology*. Springer-Verlag, New York.

KINSEY, C. 1993. *Topology of Surfaces*. Springer-Verlag, New York.

KRIPAC, J. 1997. A mechanism for persistently naming topological entities in history-based parametric solid models. *Computer-Aided Design 29*, 2, 113–122.

MANTYLA, M. 1988. *An Introduction to Solid Modeling*. Computer Science Press, Maryland, USA.

MUNKRES, J. R. 1975. *Topology A First Course*. Prentice Hall, Englewood Cliffs, New Jersey.

PRO-ENGINEER. 1998. *Pro-Engineer Documentation, Version 20*. PTC, Waltham , MA.

RAGHOTHAMA, S. 2000. *Models and Representations for Parametric Family of Parts*. PhD thesis, Spatial Automation Laboratory, University of Wisconsin-Madison.

RAPPOPORT, A. 1995. Geometric modeling: A new fundamental framework and its practical implications. In *3rd ACM Symposium on Solid Modeling and Applications, Salt Lake City, Utah*.

REQUICHA, A. A. G., AND TILOVE, R. B. 1978. Mathematical foundations of constructive solid geometry: General topology of closed regular sets. Tech. rep., TM-27a, PAP, University of Rochester, Rochester, NY, June.

REQUICHA, A. A. G. 1980. Representations for rigid solids: Theory, methods and systems. *ACM Computing Surveys 12* (Dec.), 437–464.

ROSSIGNAC, J. R., BORREL, P., AND NACKMAN, L. R. 1988. Interactive design with sequences of parameterized transformations. In *Proc. of 2nd Eurographics Workshop in Intelligent CAD Systems: Implementation Issues*.

ROSSIGNAC, J. R. 1985. *Blending and offseting solid models*. PhD thesis, Rochester Institute of Technology.

SHAPIRO, V., AND VOSSLER, D. L. 1991. Construction and optimization of CSG representations. *Computer-Aided Design 23*, 1 (Jan.), 4–20.

SHAPIRO, V., AND VOSSLER, D. L. 1991. Efficient CSG representations of two dimensional solids. *Transactions of the ASME, Journal of mechanical design 113* (Sept.), 292–305.

SHAPIRO, V., AND VOSSLER, D. L. 1993. Separation for boundary to CSG conversion. *ACM Transactions on Graphics 12*, 1 (Jan.), 35–55.

SHAPIRO, V., AND VOSSLER, D. L. 1995. What is a parametric family of solids? In *3rd ACM Symposium on Solid Modeling and Applications, Salt Lake City, Utah, May*.

SHAPIRO, V. 1997. Maintenance of geometric representations through space decompositions. *International Journal of Computational Geometry and Applications 7*, 4, 383–418.

SOLIDWORKS. 2000. *SolidWorks 2000 Documentation*. SolidWorks Corp, Concord, MA.

S.RAGHOTHAMA, AND V.SHAPIRO. 1998. Boundary representation deformation in parametric solid modeling. *ACM Transactions on Graphics 17*, 4 (Oct.), 259–286.

S.RAGHOTHAMA, AND V.SHAPIRO. 2000. Consistent updates in dual representation systems. *Computer-Aided Design 32*, 8-9 (Aug.), 463–477.

S.RAGHOTHAMA, AND V.SHAPIRO. 2003. Topological framework for part families. *ASME Journal of Computing and Information Sciences 32*, 8-9 (Mar.).

UNIGRAPHICS. 2004. *NX4.0 Documentation*. UGS Corp, Cypress, CA.

Controlled-Topology Filtering

Yotam I. Gingold[*]
New York University

Denis Zorin[†]
New York University

Abstract

Many applications require the extraction of isolines and isosurfaces from scalar functions defined on regular grids. These scalar functions may have many different origins: from MRI and CT scan data to terrain data or results of a simulation. As a result of noise and other artifacts, curves and surfaces obtained by standard extraction algorithms often suffer from topological irregularities and geometric noise.

While it is possible to remove topological and geometric noise as a post-processing step, in the case when a large number of isolines are of interest there is a considerable advantage in filtering the scalar function directly. While most smoothing filters result in gradual simplification of the topological structure of contours, new topological features typically emerge and disappear during the smoothing process.

In this paper, we describe an algorithm for filtering functions defined on regular 2D grids with controlled topology changes, which ensures that the topological structure of the set of contour lines of the function is progressively simplified.

CR Categories: I.3.5 [Computing Methodologies]: Computer Graphics—Computational Geometry and Object Modeling

Keywords: computational topology, critical points, filtering, isosurfaces

1 Introduction

Many types of data are defined as scalar functions on unstructured or structured meshes. Such scalar fields are produced by MRI and CT scanners, scientific computing simulations, extracted from databases, or obtained by sampling distance functions to pointsets or surfaces. Quite often it is necessary to extract geometric information from such scalar fields, most commonly contour lines and isosurfaces, to which we will refer as *contours*. Contours often have to be extracted for multiple scalar function values, which motivates considering the topology of the complete set of contours, rather than that of an individual contour.

A variety of applications perform various types of processing either on the original scalar function data or on individual extracted isosurfaces. For example, the scalar field of an extracted contour can be smoothed to eliminate noise or to obtain a simplified representation of the object of interest, or enhanced to emphasize features of interest. The advantage of applying such processing operations to the scalar function, rather than to an extracted contour represented by a mesh, is that all contours are processed simultaneously and topology modification is possible. For example, spurious small-scale blobs due to noise in the scalar data can be eliminated.

While certain types of topology changes are desirable, other changes may have to be avoided. For example, if a blood vessel network is extracted from an image, breaking connected components of the isosurface is highly undesirable. Unfortunately, topological changes resulting from the application of a filter are difficult to control: even a simple Laplacian smoothing filter can result in undesirable disconnected components emerging (Figure 3).

It is desirable to be able to control the topology changes occurring during the filtering process. In the extreme case, all changes can be disallowed, resulting in topology-preserving filtering; in other cases, certain types of topology changes are allowed while other changes are not. For example, if topological simplification is desired, merging components is acceptable while creating components is not.

One possible approach to the problem is to perform a complete topology analysis using contour trees or discrete Morse-Smale complexes, construct a topology hierarchy when topology simplification is desired, and design filters respecting the constraints (for example, maximal descent paths in the Morse-Smale complexes). The advantage of this approach is complete and entirely predictable control over topology. At the same time, the filter construction is far more complicated, as relatively complex constraints need to be imposed (cf. [Bremer et al. 2004]). The other approach is to augment a filtering technique with topology control by detecting and preventing topological changes by local modifications to the filter. With the latter approach, one hopes that differences with the uncontrolled process can be minimized.

In this paper we describe an algorithm of the second type. Our algorithm adds topology control to flow-type filters which define a parametric family of results $p(t), t = 0 \dots t_1$ where $p(0)$ is the vector of initial values of a scalar function defined on a two-dimensional regular grid. The idea of the algorithm is straightforward. The algorithm tracks critical points of the scalar function field to predict and determine the type of topology changes and locally adjust the rate of change of the scalar field to prevent disallowed changes. We consider three examples: Laplacian smoothing, sharpening, and anisotropic diffusion, and demonstrate that the algorithm makes it possible to control topology changes while retaining overall filter behavior and introducing relatively small errors.

Depending on problem semantics, the algorithm can either ensure complete topology preservation (e.g. for sharpening) or reduction in the number of topological features (for smoothing algorithms). The algorithm does not depend on dimension or structure of the grid in a fundamental way, and can be extended to three dimensions and unstructured grids.

2 Previous work

Analysis and simplification of the topology of vector fields and surfaces is a recurring topic in visualization, computational geometry and computer graphics. While many mesh simplification algorithms have allowed topology changes, in most cases these were unpredictable. One of the earliest examples of controlled topol-

[*]e-mail: gingold@mrl.nyu.edu
[†]e-mail: dzorin@mrl.nyu.edu

SPM 2006, Cardiff, Wales, United Kingdom, 06–08 June 2006.
© 2006 ACM 1-59593-358-1/06/0006 $5.00

ogy simplification is [He et al. 1996], in which filtering on volume rasters is used to simplify objects. Alpha shapes were used in subsequent work [El-Sana and Varshney 1997]. These approaches use a reasonable algorithmic definition of topology simplification but do not track feature changes and focus on individual surfaces or solids enclosed by surfaces. Other work focusing on surfaces includes [Guskov and Wood 2001] and [Wood et al. 2004]. Our work is more similar to analysis and simplification of the structure of height fields and vector fields, for which complete collections of contours and stream lines are considered. The foundation of a significant fraction of recent work in this area is Morse theory (e.g. [Milnor 1963]), which relates the topology of smooth manifolds to critical points of functions defined on these manifolds. [Helman and Hesselink 1991] applied critical point analysis to flow visualization. [de Leeuw and van Liere 1999] was one of the first examples of topology simplification for vector fields. The simplification of [de Leeuw and van Liere 1999] is discrete rather than continuous: whole regions were removed from the field. An alternative approach was proposed in [Tricoche et al. 2000], which merges critical points into higher-order points. [Tricoche 2002; Tricoche et al. 2001] describe how topology can be continuously simplified by removing pairs of critical points.

Two important approaches to analyzing topological structure of a scalar field are contour trees [Freeman and Morse 1967; Sircar and Cerbrian 1986; van Kreveld et al. 1997; Carr et al. 2000] and structures based on Morse-Smale complexes [Bajaj and Schikore 1998; Edelsbrunner et al. 2001; Edelsbrunner et al. 2003; Bremer et al. 2004].

The contour tree is a data structure that fully describes the topology of a scalar field, with contours passing through critical points as nodes. Contour trees have been extensively used for fast isosurface extraction and define a natural topological hierarchy. The 2D Morse-Smale complex has vertices at critical points which are connected by maximal descent paths: similar structures are also defined for an arbitrary number of dimensions. A topological hierarchy can also be defined using Morse-Smale complexes and feature *persistence*.

Both types of structures were used for topological simplification of scalar fields using associated hierarchies and different types of persistence functions, e.g. in recent papers [Bremer et al. 2004; Carr et al. 2004]. In both cases, the scalar field function values are updated to eliminate features locally. In [Bremer et al. 2004], smoothing is performed with constrained Morse-Smale complex boundaries, and the boundaries themselves are adjusted using smoothing.

While our algorithm can be used to construct topological hierarchies, this is not our primary goal. We aim to provide a tool that adds topology control to a variety of processing techniques for scalar data. Our primary concern is not visualization and analysis of a scalar field topology, rather, we aim to augment existing image processing tools with topological guarantees, while preserving the basic behavior of the tool.

The recent work [Sohn and Bajaj 2005] on time-varying contour topology deals with similar types of evolving scalar data. The goal of this work is accurate feature tracking in a given dataset, while our goal is to alter a time-dependent dataset to eliminate certain types of topological events.

The topological evolution resulting from filtering, smoothing in particular, is often considered in vision and medical imaging literature. The concept of *scale space* based on Laplacian smoothing (heat flow), proposed in [Witkin 1983], is used in a variety of applications, and one can consider similar types of constructions based on different flows, such as anisotropic diffusion [Perona and Malik 1990] or curvature flow. The topology of scale spaces was studied in [Damon 1995], from a mathematical point of view, and more recently in [Florack and Kuijper 2000].

3 Topological preliminaries

Our algorithm is based on the correspondence between topological features and pairs of critical points. *Topological events*, i.e. the changes in the topology of the sets of contours, are associated with changes in these pairs. For example, a new connected component appears if a pairs of critical points appear, and an existing component vanishes when two critical points merge.

In this section, we briefly review relevant definitions and facts from differentiable and discrete topology, which are used by our algorithm.

Smooth Morse theory. We restrict our attention to the case of functions defined on the plane. A function f is called a Morse function if it is at least twice differentiable, its values at critical points defined by $\nabla f = 0$ are distinct and its *Hessian*, i.e. the matrix of second derivatives, has nonzero determinant at critical points. Critical points with nondegenerate Hessian are called *simple*. If the Hessian vanishes at a critical point, it is called *complex*.

The Morse Lemma states that for a suitable choice of coordinates the function has the form $\pm x^2 \pm y^2$ in a neighborhood of any critical point. The number of minuses is the *index of the critical point*. Saddles have index 1, maxima have index 2, and minima have index 0. For functions defined on planar domains, it is convenient to add a point at infinity to the plane and assign a minimum to it with infinite negative value.

The indices of critical points x of functions defined on a sphere are known to satisfy

$$\sum_x (-1)^{\text{index(x)}} = 2 \tag{1}$$

$(-1)^{index(x)}$ is called the *topological charge* of a critical point.

Critical points can be used to describe the topological structure of the contour lines of the function f. The level set $f^{-1}(c)$ is a smooth curve unless c is a value at a critical point. Furthermore, if we consider an interval of values $[c_1, c_2]$ not containing critical points, the level sets for all $c \in [c_1, c_2]$ have the same topology. Thus, the critical points define all topological changes of level set curves.

The singular level sets corresponding to maxima and minima consist of isolated points and correspond to vanishing/appearing features if we regard the traversal of increasing values of $f(x)$ as advancing in time. Saddles correspond to the merging/splitting of features.

Discrete Morse theory. While one can define C^2 interpolants for functions especially on regular grids, studying critical points of such functions is difficult. It is preferable to generalize the notions of smooth Morse theory to piecewise-linear functions. We mostly follow [Edelsbrunner et al. 2001] in our definitions. A critical point of a piecewise linear function is always at a vertex. Its type can be inferred from comparing the values of the critical point with adjacent values. We consider *lower* and *upper* stars of a vertex. A lower star consists of all vertices w adjacent to v such that the function value $f(w) \leq f(v)$, the upper star consists of vertices w with $f(w) \geq f(v)$. (Figure 1, based on [Edelsbrunner et al. 2001]). Each star can be decomposed into continuous wedges. If one of the stars coincides with the entire neighborhood the point is a local minimum (upper star) or maximum (lower star). If each star has exactly one wedge, the point is considered regular. If the number of wedges in each star is $k+1$ for $k \geq 1$, then the star is a k-fold saddle (simple saddle for $k=1$). Unlike the smooth case, complex saddles are stable. The type of a vertex can be determined by the vertex *signature* i.e. a sequence of ones and zeros corresponding to the adjacent points, ordered counterclockwise, with one indicating that the value at the adjacent vertex is higher than the value at the center.

A discrete Morse function is any piecewise linear function for which the values at critical points are distinct. One can regard any function as a Morse function if ties are resolved using Simulation of Simplicity [Edelsbrunner and Mücke 1990]. As we only use value comparisons in our algorithm, it is sufficient to simply assign an ordering based on indices of vertices to break ties. However, it is necessary to use first-order perturbations to resolve ties for calculated event times (Section 4).

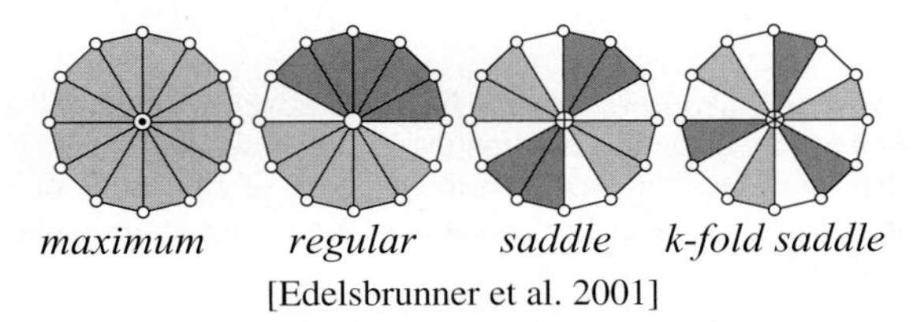

[Edelsbrunner et al. 2001]

Figure 1: Different vertex types.

Singularities of parametric families of functions. Smoothing a function using a flow equation, e.g. $\partial f/\partial t = \Delta f$, leads to a solution $f(x,t)$, which can be regarded as a one-parameter family of functions. While complex critical points of functions can be eliminated by small perturbations, this is no longer true for parametric families. To clarify this, consider any family of functions of one argument $f(x,t)$, such that $f(x,-1)$ has a maximum and a minimum and $f(x,1)$ has no extrema. In the beginning, the derivative of the function has two roots, and at $t=1$ it has no roots. Therefore, no matter what perturbation we use, there is a parameter value t_1 such that the derivative has exactly one root. One can easily see that this extremum cannot be generic: as the extrema merge, the Hessian is always positive at one and negative at the other. This implies that it is zero at the moment they merge.

By choosing a suitable coordinate system, a generic singularity with one parameter in one dimension can be reduced to the form

$$f(x,t) = x^3 + xt$$

Stable complex singularities arising in parametric families of functions are studied in singularity theory. It turns out that in two dimensions a single parameter singularity has a similar form

$$x^3 + xt + ay^2 \tag{2}$$

One can see (Figure 2) that it corresponds to two critical points (a saddle and a minimum) merging at a point for t increasing and a saddle-minimum pair appearing at a point for t decreasing. For $t=0$ the critical point is always complex.

Discrete case. Next we consider the discrete analogs of creation and annihilation events. Because of the presence of complex saddles, more event types are possible. In addition, because of the piecewise linear nature of the functions, the singularities do not move continuously but in discrete jumps from one vertex to another. So we introduce one more event type corresponding to singularities changing locations.

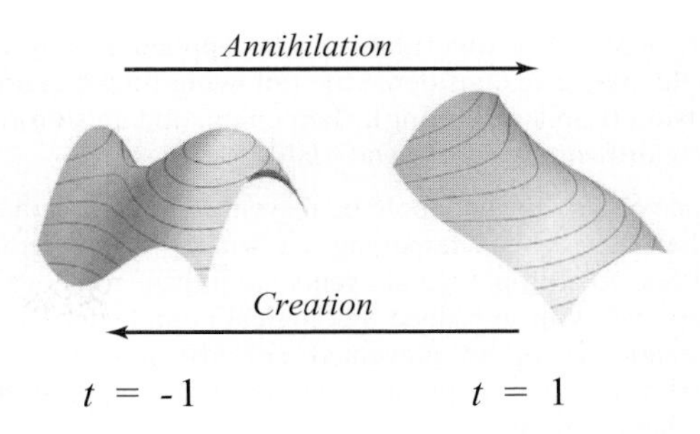

Figure 2: Annihilation and creation events.

We consider piecewise linear functions evolving piecewise linearly in time. The topological picture for such function changes discretely. Each change corresponds to a *value flip* at an edge (v,w), *i.e.* transition from configuration $f(v) > f(w)$ to the configuration $f(v) < f(w)$, with the relative order of all other adjacent functions remaining unchanged. Again, one can assume that two elementary flip events never coincide in time by simple tie-breaking.

Merge/Annihilation. We use the term merge events to denote any events which result in a critical point disappearing. Due to the presence of k-fold saddles, many variations of events are possible, with only the one involving a simple saddle and a maximum or minimum resulting in annihilation. Other types of events include saddle merges and a maximum or minimum absorbed by a saddle with a change in the number of folds. We consider all these events admissible.

Creation. Similarly, creation events are the events resulting in creation of a critical point, and can be of many types. Only one type (emergence of a saddle-minimum/maximum pair) results in critical points created from a regular point. We forbid all events that involve the creation of a maximum or a minimum. We allow saddle separation, as we view k-fold saddles as k-simple saddles merged together.

Exchange. Adjacent k-fold saddles may exchange folds; no critical points move in this case.

Move. This type of event corresponds to the situation when a critical point vanishes at one end of an edge with a point of the same type appearing at the other.

Non-event. Some flip events may result in no changes in the type of endpoints of the edge.

For each of our filter examples all event types are present for sufficiently complex images, unless topology control is applied.

4 Algorithm

The algorithm operates on *scalar values* $p(v)$ defined at vertices $v \in V$ of a mesh which evolve over time. We assume that there is a monotonic ordering of values at any fixed time, breaking ties in a uniform way using vertex indices. We use data defined on regular grids in our examples, but the algorithm can be used for arbitrary meshes.

In addition to the input data, the user defines the set of *disallowed* topological events.

The filter, which is separate from the topology control algorithm, for a given time step Δt and data $p^l(v)$ corresponding to time t_l produces *proposed values* $\bar{p}^{l+1}(v) = F(p^l(v), \Delta t)$ for the moment

$t^{l+1} = t^l + \Delta t$. The algorithm can be applied to any evolving scalar field. We have considered the following filter examples: linear diffusion (Laplace filtering), sharpening, and anisotropic edge-enhancing diffusion ([Perona and Malik 1990]).

The choice of disallowed topological events depends on the semantics of the filter. When sharpening one wishes to exaggerate existing features, so all topological events are prevented. For the other two filters, we with to reduce topological complexity by smoothing, so creation events are prevented. For diffusion, the updates are computed using either explicit or implicit time stepping, the latter allowing large time steps Δt.

A single step of the *outer loop* of the algorithm requests the proposed values $\bar{p}^{l+1}(v)$ from the filter and assumes linear evolution between $p^l(v)$ and $\bar{p}^{l+1}(v)$ for each value. The time of all possible topology events is computed, and the proposed values are adjusted to ensure that disallowed events do not occur. After adjustment, events need to be recalculated, and further adjustments may be necessary. The process is iterated until there are no disallowed events.

The algorithm may fail to produce progress, if for all points v computed update is below a threshold ε, which for any point v may be due to two reasons: either the local time step or $\bar{p}(v) - p(v)$ is too small, which means that the filter has converged to a limit value. The algorithm terminates if it either reached the target time, or failed to produce progress.

One iteration of the algorithm in detail. The algorithm uses two maps:

- *Critical point map* `status`(v) indicating if a vertex is a (discrete) maximum, minimum, saddle or regular point, and storing a critical point's ID. This map is initialized using the original data and incrementally updated at every step.
- *Value flip map* `flip`(v,w), where (v,w) is an edge, recording value flips observed at the current step, and the time for each flip (the timestamp).

Step 1: Obtain proposed values. Compute trial point positions $\bar{p}^{l+1}(v)$ using the uncontrolled filter.

$$\bar{p}^{l+1} = F(\Delta t, p^l(v)).$$

If no progress is made, i.e. the difference $|\bar{p}^{l+1} - p^l|$ does not exceed a user-defined threshold, the algorithm terminates.

Step 2: Identify and sort the flip events. We regard the evolution of the mesh as linear between p and $\bar{p}$. This guarantees that each edge may flip no more than once during a time step. For each edge (v,w) we determine whether or not it flips during the trial step and store the result in `flip`(v,w).

Then for all flip events we compute the exact time of the flip `flip.time`(v,w), using linearity and our tie-breaking scheme, and sort the flip edges according to this time.

Let a,b,c,d be four points, such that the pairs $(p(a),p(b))$ and $(p(c),p(d))$ change order. We assign unique infinitesimal perturbation scales $e(a),e(b),e(c),e(d)$ to each vertex, i.e. we regard the value at a vertex $v \in \{a,b,c,d\}$ as a polynomial $p(v)+\varepsilon e(v)$ in ε.

Let t be the the solution of $p(a)(1-t)+t\bar{p}(a) = p(b)(1-t)+t\bar{p}(b)$, and let t' be the solution of $p(c)(1-t)-t\bar{p}(c) = p(d)t-(1-t)\bar{p}(d)$. We assume that both solutions exist (otherwise there are no events that need to be ordered). Then

$$t = \frac{p(b)-p(a)+\varepsilon e(b)-\varepsilon e(a)}{(\bar{p}(a)-p(a))-(\bar{p}(b)-p(b))}$$

and

$$t' = \frac{p(d)-p(c)+\varepsilon e(d)-\varepsilon e(c)}{(\bar{p}(c)-p(c))-(\bar{p}(d)-p(d))}.$$

To compare t and t', we consider $f(\varepsilon) = t - t'$. If $f(0) > 0$ then no tie-breaking is necessary. However, it is possible that $f(0) = 0$. This is the case when using the perturbations is necessary: we break the tie by computing the first-order term of $f(\varepsilon)$, which determines the sign of $f(\varepsilon)$ for all sufficiently small $\varepsilon > 0$.

If $f(0) = 0$, then $f(\varepsilon)$ is given by

$$f(\varepsilon) = \varepsilon\left(\frac{e(b)-e(a)}{(\bar{p}(a)-p(a))-(\bar{p}(b)-p(b))} - \frac{e(d)-e(c)}{(\bar{p}(c)-p(c))-(\bar{p}(d)-p(d))}\right).$$

Step 3, Detect disallowed events. Next, we traverse the event list ordered by time, detecting creation events. For each flip event (v,w), we determine its type based on the changes of signatures of endpoints v and w in the critical point map for p^l and $\bar{p}^l$, as explained in Section 3.

If we find a disallowed event (v,w) at a time $t < t_{l+1}$ we set

$$\bar{p}(v) \leftarrow (\texttt{flip.time}(v,w) - \delta)(\bar{p}(v) - p(v)) + p(v)$$

and

$$\bar{p}(w) \leftarrow (\texttt{flip.time}(v,w) - \delta)(\bar{p}(w) - p(w)) + p(w)$$

and return to Step 2. In so modifying $\bar{p}$, we set the proposed values for v and w to δ before the flip along the line from $\bar{p}$ to p.

Thus the flip no longer occurs as the values of v and w evolve from p to $\bar{p}$. In other words,

$$sign(p(v) - p(w)) = sign(\bar{p}(v) - \bar{p}(w)).$$

This ensures that undesired topological events never occur.

If no creation events are found at the end of the event list traversal, then the inner loop is terminated, $p^{l+1}(v)$ is set to modified $\bar{p}^{l+1}(v)$ and the next increment is obtained from the filter.

The pseudocode for the algorithm is shown below.

repeat
    for all $v \in V$ **do**
        $\bar{p}(v) \leftarrow F(\Delta t, p(v))$
    end for
    repeat
        $\overline{\texttt{status}} \leftarrow \texttt{status}$
        `flip` ← all edge flips (v,w), sorted by time
        for all $f(v,w) \in$ `flip` **do**
            update $\overline{\texttt{status}}(v)$ and $\overline{\texttt{status}}(w)$
            if f is a disallowed event **then**
                modify $\bar{p}(v)$, $\bar{p}(w)$
                break
            end if
        end for
    until no undesired event is found
until no progress possible or target time reached

Depending on the type of the filter, the algorithm behaves in different ways: if locally the filter smoothes the values, it still may create features. However, these features are short-lived, so the values frozen by the algorithm are likely to be released quickly (see the discussion of the Laplacian smoothing example in Section 5). On the other hand, if the filter is locally enhancing and tends to create new features, the algorithm will prevent it from altering the image, which we consider the desirable behavior in such cases.

The return to Step 2 after modification of the proposed values is essential, as the value modification results in rearrangement of critical points. This results in a considerable increase in time vs. simple application of a filter for initial steps. However, following a sufficiently large number of smoothing steps, few critical points interact with each other in a given step, so fewer inner cycles are necessary for each time advance.

5 Results

We show the results of our algorithm for three different filters:

- Discrete Laplacian smoothing (diffusion):

$$p_L^{l+1}(v) = p^l(v) + \Delta t \sum_{edges(v,w)} (p^l(w) - p^l(v))$$

- Sharpening, for a fixed n,

$$p_S^{l+1}(v) = p^l(v) + \Delta t (p_L^n(v) - p^0(v))$$

- Discrete anisotropic diffusion ([Perona and Malik 1990]):

$$p_{AD}^{l+1}(v) = p^l(v) + \Delta t \sum_{edges(v,w)} \frac{p^l(w) - p^l(v)}{1 + \|p^l(w) - p^l(v)\|^2/k^2}$$

For diffusion, we have implemented both explicit and implicit time stepping, the linear systems solved using SuperLU [Demmel et al. 1999] in the latter case.

Note that for the sharpening filter for any time t is a linear interpolation between $p^0(v)$ and $p_L^n(v)$.

We use two artificial datasets for comparison: a ridge, showing the emergence of a high-persistence saddle for Laplacian smoothing without topology control, and a pure noise image, which almost entirely consists of critical points in the beginning.

We use several real datasets: the Puget Sound terrain map, a CT scan slice of a cow brain, a retina image, and a CT scan slice of a human torso. In the images showing critical points, crosses denote saddles, empty circles denote minima, and circles with dots denote maxima.

Figure 3 (right) compares the behavior of the Laplacian smoothing filter for artificial data with and without topology control. Note that while the topology change is prevented, there is little impact on the overall surface smoothness. Figure 4 shows a similar event in the Puget sound dataset.

Figure 3 (left) shows that even for very complex topologies (in the initial image almost every point is critical) the algorithm does not get stuck because of excessive numbers of frozen values.

The analysis of the topology of the scale space provides intuition into why this is the case: in [Damon 1995] it is shown that unlike the general case, there is an asymmetry between creation and annihilation events in the evolving data resulting from Laplace smoothing. In this case, the two types of events can be described by different normal forms and further analysis shows that newly created critical points have long expected lifetime only on ridges, i.e. on a small subset of the image. Thus, on average, the lifetime of created topological features is short, and values need to move only slightly slower to avoid feature creation entirely.

Next, we compare the results for three filters (Figure 7). For each filter we show five images: the original image, the filtered image without topology control, the filtered image with topology control, the relative magnitude of difference in each case, and the map of all created features without topology control. Note that most if not all created features are small, and hard to see in the image, and the filtering results in both cases are visually similar. We also observe that errors are small and localized for two filters (the maximal error is approx. 2% for Laplacian smoothing and 6% for anisotropic diffusion). While overall the changes introduced by topology control are small, for smoothing filters one can sometimes observe artifacts near values which are prevented from changing. In these areas the surface appears less smooth than in adjacent areas (for example, in the sharp ravines visible in the upper part of the terrain in Figure 7, left row). These artifacts exist for a relatively short time and are eliminated in subsequent filtering steps.

The errors are much higher and spread out for sharpening. This is not surprising as sharpening directly increases function value at locations with high-frequency detail, and our algorithm prevents the creation of some of these spurious features.

Figure 8 shows the numbers of critical points as a function of Laplacian smoothing iteration for several models. Note that the numbers oscillate for Laplacian smoothing but monotonically decrease with topology control. Significant oscillations early in the process of Laplacian smoothing are also present, although not visible because of the large total number of critical points. One can observe that while creation event are clearly a small fraction of the total number of events they have a visible impact on the topological evolution.

Figure 6 illustrates the algorithm performance (256x256 data, Laplacian smoothing, implicit time stepping). One can observe that a significant fraction of the time is spend on the first several steps, due to the need for a large number of inner loop iterations to resolve all topological events, mostly due to small scale noise. Although topology control increased the total time compared to uncontrolled Laplacian smoothing by a factor of 2.22, the first three steps accounted for 43.95% of the difference. Each uncontrolled smoothing step took, on average, 3.10 seconds; the data set required 128 total smoothing steps. (Results were generated on a 2.4 GHz Xeon.) For later steps, more than one inner loop iteration is rarely needed. If it is acceptable to pre-smooth the image without topology control or otherwise eliminate small-scale features e.g. by setting all values inside the critical contour bounding the feature to the contour value, one can drastically reduce the overhead of iterations in the first few steps. This is true as well for anisotropic diffusion. Sharpening, on the other hand, takes more time per step as virtual time increases. This is because we prevent any topological changes; the sharper the image becomes, the more undesired critical point changes sharpening attempts to effect.

Finally, we observe that by disallowing feature creation events we implicitly obtain a complete or partial topological hierarchy and a measure of feature persistence which can be used for topology visualization and simplification (cf. [Carr et al. 2004]).

Recall that each topological feature is associated with a pair of critical points (in 2D a saddle and a minimum or maximum). By observing annihilation events, we can establish a feature hierarchy, and use the critical point lifetime as feature persistence. This defines alternative feature persistence measures associated with different filter types. For example, anisotropic diffusion filters would give high persistence to features bounded by well-defined edges. Figure 9 compares the highest-persistence features given by the simplest persistence definition (value difference at the two critical points defining a feature) with the same number of high-persistence features given by anisotropic diffusion (specifically, stable features with infinite lifetime).

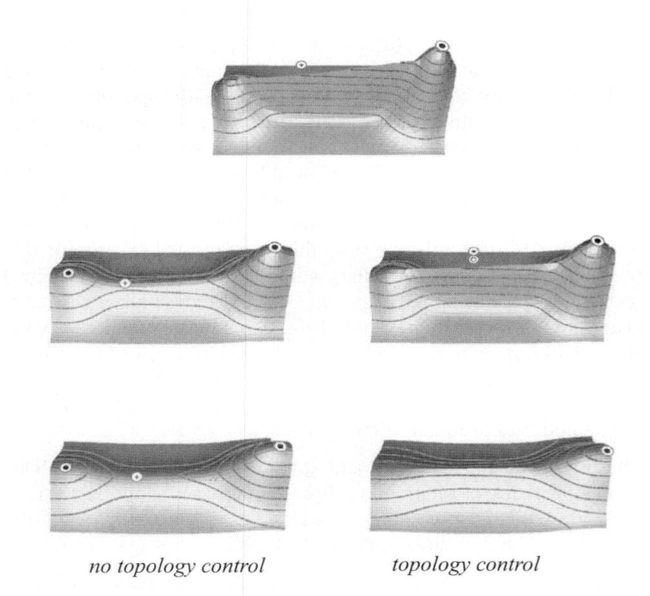

Figure 3: Comparison of our topology controlled smoothing algorithm with Laplacian smoothing for a simple artificial dataset.

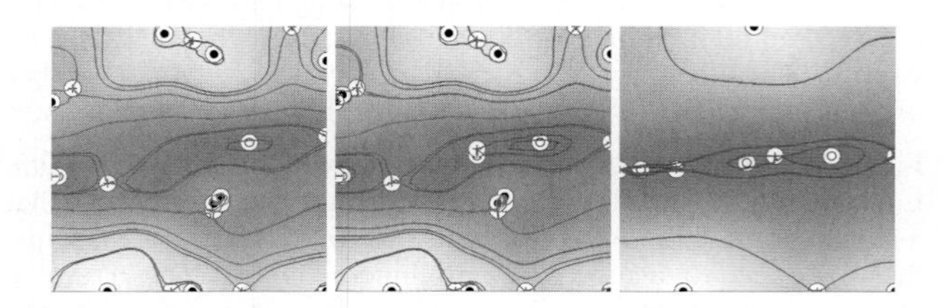

Figure 4: A creation event in the Puget Sound dataset (Laplacian smoothing).

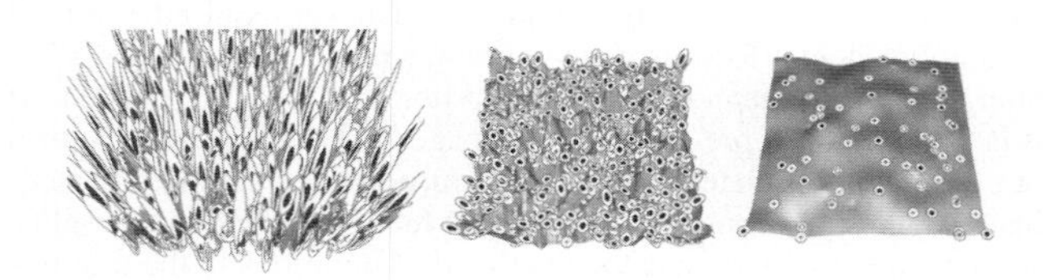

Figure 5: Topology controlled Laplacian smoothing for random initial data.

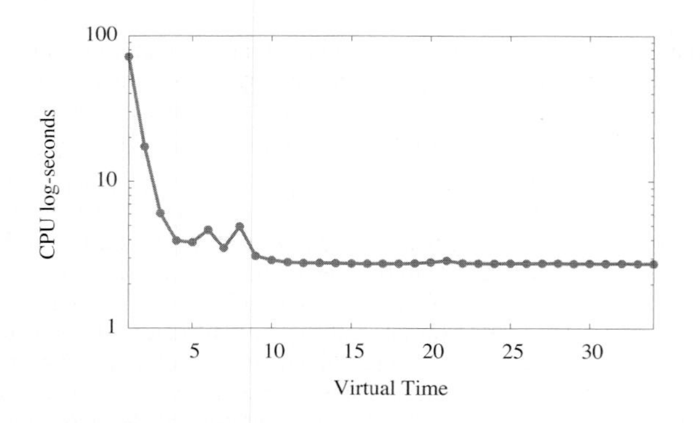

Figure 6: Algorithm performance: CPU time per virtual unit timestep, for implicit Laplacian smoothing, running on a Xeon 2.4 GHz.

6 Conclusions and future work

We have presented a simple algorithm that ensures that filtering results in topology preservation or monotonic topology simplification in the sense of the reduction of the number of critical points in a scalar field. We have demonstrated that for three filters and a number of test images, the constraints imposed by the topology control algorithm do not significantly affect the filtering process.

Clearly, our algorithm is a first step in this direction. While we have applied it to data sets defined on regular 2D meshes, there are no fundamental limitations on either mesh structure or the dimension of the problem. We plan to explore the behavior of the algorithm in 3D where creation events are more common and more types of topological events may occur. Specifically, critical points have indices 0 through 3; topological events occur between minima and index-1 saddles, index-1 and index-2 saddles, and maxima and index-2 saddles. It is these events that must be detected when an edge flips in a 3D mesh. The structure of the algorithm remains the same. It may, however, be necessary to find ways of improving the algorithm's efficiency.

One significant downside of approaches of this type is that artifacts are hard to predict. While we have observed very few, it would be desirable to have an algorithm which can alter the filter behavior in advance, spreading the modification necessary to prevent disallowed events to a larger number of points and reducing the necessary modification for each point.

Acknowledgements. We would like to thank Chris Wu, for his help with algorithm profiling and many useful suggestions; Chee Yap, Adrian Secord, and other NYU Computer Science colleagues for useful discussions.

References

ANDÚJAR, C., BRUNET, P., AND AYALA, D. 2002. Topology-reducing surface simplification using a discrete solid representation. *ACM Trans. Graph. 21*, 2, 88–105.

BAJAJ, C., AND SCHIKORE, D. 1998. Topology preserving data simplification with error bounds. *Journal on Computers and Graphics vol. 22*, 1, 3–12.

BREMER, P.-T., PASCUCCI, V., EDELSBRUNNER, H., AND HAMANN, B. 2004. A topological hierarchy for functions on triangulated surfaces. *IEEE Trans. Vis. Comput. Graphics 10*, 385–396.

CARR, H., SNOEYINK, J., AND AXEN, U. 2000. Computing contour trees in all dimensions. In *Symposium on Discrete Algorithms*, 918–926.

CARR, H., SNOEYINK, J., AND VAN DE PANNE, M. 2004. Simplifying flexible isosurfaces using local geometric measures. In *IEEE Visualization 2004*, 497–504.

CURLESS, B., AND LEVOY, M. 1996. A volumetric method for building complex models from range images. In *Proceedings of SIGGRAPH 96*, Computer Graphics Proceedings, Annual Conference Series, 303–312.

DAMON, J. 1995. Local Morse theory for solutions to the heat equation and Gaussian blurring. *J. Differential Equations 115*, 2, 368–401.

DE LEEUW, W., AND VAN LIERE, R. 1999. Collapsing flow topology using area metrics. In *VIS '99: Proceedings of the confer-*

ence on Visualization '99, IEEE Computer Society Press, Los Alamitos, CA, USA, 349–354.

DEMMEL, J. W., EISENSTAT, S. C., GILBERT, J. R., LI, X. S., AND LIU, J. W. H. 1999. A supernodal approach to sparse partial pivoting. *SIAM J. Matrix Analysis and Applications 20*, 3, 720–755.

DIEWALD, U., PREUER, T., AND RUMPF, M. 2000. Anisotropic diffusion in vector field visualization on Euclidean domains and surfaces. *IEEE Transactions on Visualization and Computer Graphics 6*, 2, 139–149.

EDELSBRUNNER, H., AND MÜCKE, E. 1990. Simulation of simplicity: a technique to cope with degenerate cases in geometric algorithms. *ACM Trans. Graph. 9*, 1, 66–104.

EDELSBRUNNER, H., LETSCHER, D., AND ZOMORODIAN, A. 2000. Topological persistence and simplification. In *Proc. 41st Ann. IEEE Sympos. Found Comput. Sci.*, 454–463.

EDELSBRUNNER, H., HARER, J., AND ZOMORODIAN, A. 2001. Hierarchical Morse complexes for piecewise linear 2-manifolds. In *Proc. 17th Ann. ACM Sympos. Comput. Geom.*, 70–79.

EDELSBRUNNER, H., HARER, J., NATARAJAN, V., AND PASCUCCI, V. 2003. Morse-Smale complexes for piecewise linear 3-manifolds. In *Proc. 19th Ann. Sympos. Comput. Geom.*, 361–370.

EL-SANA, J., AND VARSHNEY, A. 1997. Controlled simplification of genus for polygonal models. In *IEEE Visualization '97*, 403–412.

FLORACK, L., AND KUIJPER, A. 2000. The topological structure of scale-space images. *J. Math. Imaging Vis. 12*, 1, 65–79.

FREEMAN, H., AND MORSE, S. P. 1967. On searching a contour map for a given terrain profile. *Journal of the Franklin Institute 248*, 1–25.

GUSKOV, I., AND WOOD, Z. 2001. Topological noise removal. In *Graphics Interface 2001*, 19–26.

HAN, X., XU, C., AND PRINCE, J. 2001. A topology preserving deformable model using level sets. In *Computer Vision and Pattern Recognition*, vol. 2, 765–770.

HART, J. C. 1998. Morse theory for implicit surface modeling. In *Mathematical Visualization*, H.-C. Hege and K. Polthier, Eds. Springer-Verlag, Oct., 257–268.

HE, T., HONG, L., VARSHNEY, A., AND WANG, S. W. 1996. Controlled topology simplification. *IEEE Transactions on Visualization and Computer Graphics 2*, 2, 171–184.

HELMAN, J. L., AND HESSELINK, L. 1991. Visualizing vector field topology in fluid flows. *IEEE Comput. Graph. Appl. 11*, 3, 36–46.

MAHROUS, K., BENNETT, J., SCHEUERMANN, G., HAMANN, B., AND JOY, K. I. 2004. Topological segmentation in three-dimensional vector fields. *IEEE Transactions on Visualization and Computer Graphics 10*, 2, 198–205.

MILNOR, J. 1963. *Morse Theory*. Princeton Univ. Press, New Jersey.

NI, X., GARLAND, M., AND HART, J. C. 2004. Fair Morse functions for extracting the topological structure of a surface mesh. *ACM Transactions on Graphics 23*, 3 (Aug.), 613–622.

P.-T.BREMER, PASCUCCI, V., AND HAMANN, B. 2005. Maximizing adaptivity in hierarchical topological models. In *Proceedings of International Conference on Shape Modeling and Applications*, 298–307.

PERONA, P., AND MALIK, J. 1990. Scale-space and edge detection using anisotropic diffusion. *IEEE Trans. PAMI 12*, 7 (July), 629–639.

RANA, S., Ed. 2004. *Topological Data Structures for Surfaces: An Introduction to Geographical Information Science*. Wiley and Sons.

SIRCAR, J. K., AND CERBRIAN, J. A. 1986. Application of image processing techniques to the automated labelling of raster figitized contours. In *Int. Symp. on Spatial Data Handling*, 171–184.

SOHN, B., AND BAJAJ, C. 2005. Time-varying contour topology. *IEEE Transactions on Visualization and Computer Graphics 12*, 1, 14–25.

TONG, Y., LOMBEYDA, S., HIRANI, A. N., AND DESBRUN, M. 2003. Discrete multiscale vector field decomposition. *ACM Transactions on Graphics 22*, 3 (July), 445–452.

TRICOCHE, X., SCHEUERMANN, G., AND HAGEN, H. 2000. A topology simplification method for 2d vector fields. In *IEEE Visualization 2000*, 359–366.

TRICOCHE, X., SCHEUERMANN, G., AND HAGEN, H. 2001. Continuous topology simplification of planar vector fields. In *IEEE Visualization 2001*, 159–166.

TRICOCHE, X., WISCHGOLL, T., SCHEUERMANN, G., AND HAGEN, H. 2002. Topology tracking for the visualization of time-dependent two-dimensional flows. *Computers & Graphics 26*, 2 (Apr.), 249–257.

TRICOCHE, X. 2002. *Vector and Tensor Field Topology Simplification, Tracking, and Visualization*. PhD thesis, Universitat Kaiserslautern.

VAN KREVELD, M. J., VAN OOSTRUM, R., BAJAJ, C. L., PASCUCCI, V., AND SCHIKORE, D. 1997. Contour trees and small seed sets for isosurface traversal. In *Symposium on Computational Geometry*, 212–220.

WITKIN, A. P. 1983. Scale-space filtering. In *International Joint Conference on Artificial Intelligence*, 1019–1022.

WOOD, Z., HOPPE, H., DESBRUN, M., AND SCHRÖDER, P. 2004. Removing excess topology from isosurfaces. *ACM Transactions on Graphics 23*, 2 (Apr.), 190–208.

ZOMORODIAN, A. 2001. *Computing and Comprehending Topology: Persistence and Hierarchical Morse Complexes*. PhD thesis, University of Illinois at Urbana-Champaign.

ZOMORODIAN, A. 2005. *Topology for Computing*. Cambridge University Press, New York, NY.

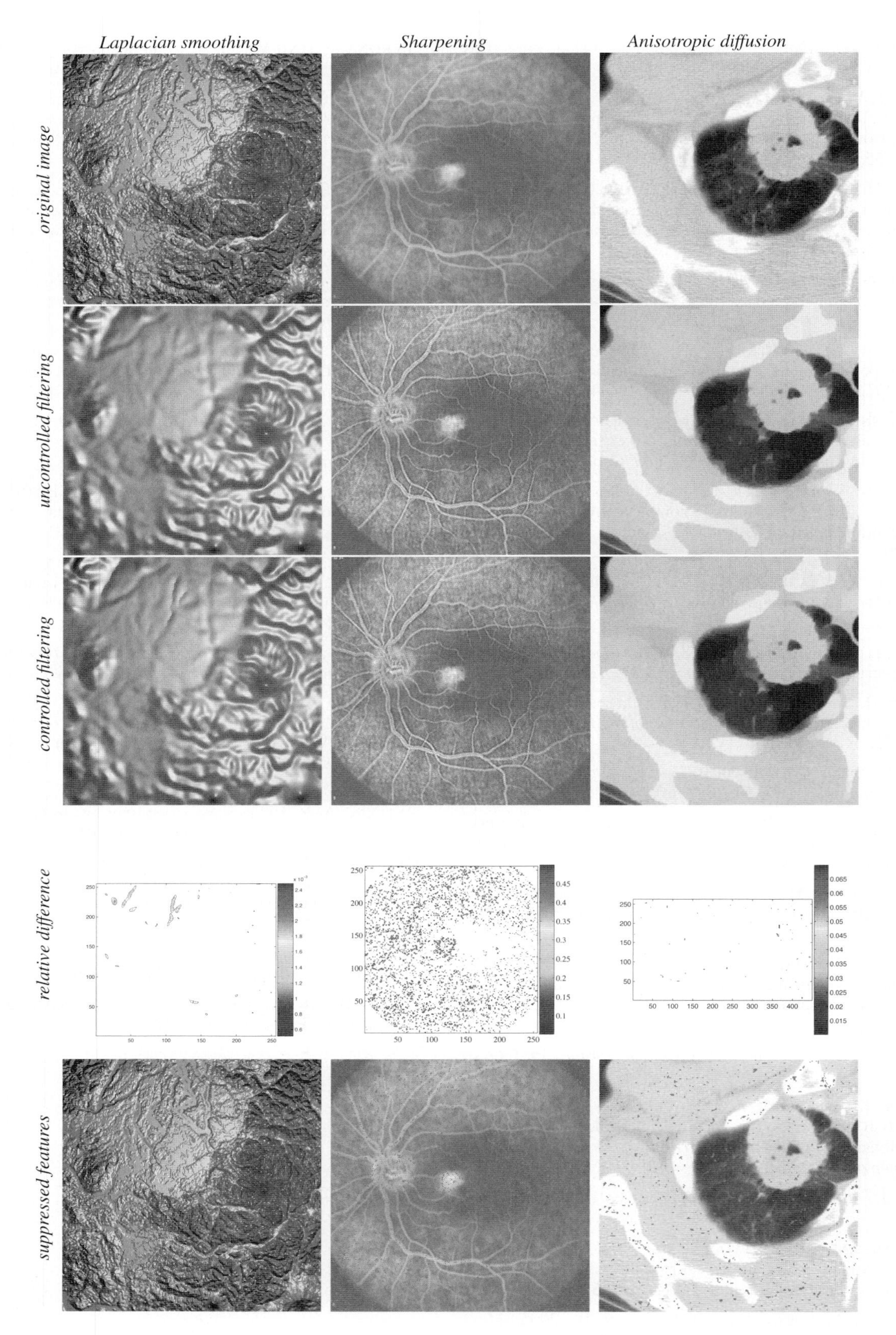

Figure 7: Filter examples for Laplacian (first column), sharpening (second column) and anisotropic diffusion (third column) filters. The images are: original, filtered without topology control, filtered with topology control, the difference between two filtered images, and the map of new features created without topology control (shown in red). For the anisotropic diffusion only a fragment of the complete image is shown, to make the small-scale features in the original more visible.

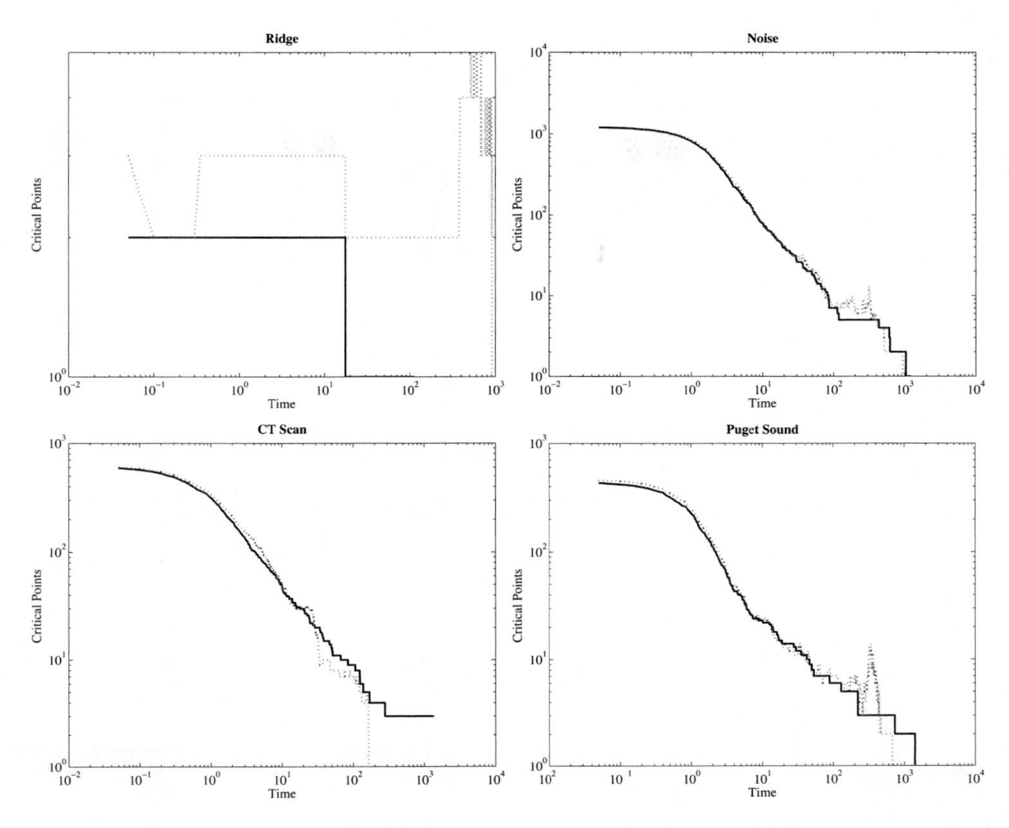

Figure 8: The number of critical points as functions of the iteration number for four datasets for Laplacian smoothing (red dotted line) and topology controlled Laplacian smoothing (blue solid line). Upper row: artifical datasets. Lower row: the cow brain CT scan and Puget Sound dataset.

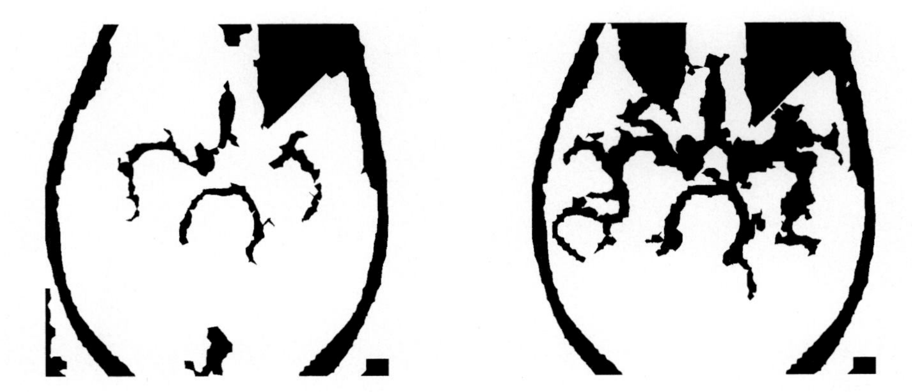

Figure 9: Several highest-persistence features, cow brain dataset. Left: persistence measured by difference of values of control points for a feature. Right: persistence measured by feature lifetime under anisotropic diffusion (in this case, stable features with infinite lifetime are shown).

Solving topological constraints for declarative families of objects

Hilderick A. van der Meiden* and Willem F. Bronsvoort†
Faculty of Electrical Engineering, Mathematics and Computer Science
Delft University of Technology
Mekelweg 4, NL-2628 CD Delft,The Netherlands

Abstract

Parametric and feature-based CAD models can be thought of to represent families of similar objects. In current modelling systems, however, model semantics is unclear and ambiguous in the context of families of objects.

We present the Declarative Family of Objects Model (DFOM), which enables to adequately specify and maintain family semantics. In this model, not only geometry, but also topology is specified declaratively, by means of constraints. A family of objects is modelled by a DFOM with multiple realisations. A member of the family is modelled by adding constraints, e.g. to set dimension variables, until a single realisation remains. The declarative approach guarantees that the realisation of a family member is also a realisation of the family.

The realisation of a family member is found by solving first the geometric constraints, then the topological constraints. From the geometric solution, a cellular model is constructed. Topological constraints indirectly specify which combinations of cellular model entities are allowed in the realisation. The system of topological constraints is translated into a boolean constraint satisfaction problem. The realisation is found by solving this problem. The feasibility of solving topological constraints has been investigated using an existing boolean satisfiability solver.

CR Categories: J.6 [Computer Applications]: Computer-Aided Engineering—Computer-Aided Design I.3.5 [Computer Graphics]: Computational Geometry and Object Modelling—Geometric algorithms, languages and systems F.2.2 [Analysis of Algorithms and Problem Complexity]: Non-numerical Algorithms and Problems—Geometrical problems and computations

Keywords: parametric and feature-based modelling, families of objects, declarative specification, topological constraints, satisfiability

1 Introduction

Parametric and feature-based modelling systems are used to create object models with a number of parameters. The set of objects that can be obtained by varying the parameters of a model, is often referred to as a family of objects, or family for short. Virtually all such modelling systems store a history of modelling operations on a B-rep. We refer to these systems as history-based modellers and to their models as history-based models. Typically, the procedure for modelling a family of objects, is to initially model a single object, called the prototype object or generic object. The history of modelling operations, used to construct the prototype object, is re-evaluated with different parameter values to generate other members of the family.

History-based models, even though they are the de-facto standard for commercial modelling systems, are not ideal for representing families of objects. Bidarra and Bronsvoort [2000a] identify six major problems with history-based models, of which the most relevant, in the context of families of objects, are the persistent naming problem, the feature ordering problem, and the inability to properly maintain feature semantics. The persistent naming problem basically is the problem of identifying topologically equivalent entities in different members of a family. This is a prerequisite for maintaining family semantics. Previous research on families of objects has focussed mainly on the persistent naming problem, e.g. [Rappoport 1997; Raghothama and Shapiro 1998; Bidarra et al. 2005b].

The feature ordering problem and maintenance of feature semantics have not received much attention. The order in which features are added to history-based models affects the resulting family of objects. This makes it difficult do design and edit family models. Also, the history-based modelling scheme does not have mechanisms for adequate specification and maintenance of semantics of features, and thus the resulting families have unclear semantics. In particular, topological properties cannot be adequately specified and maintained. These problems are discussed in more detail in Section 2.

We present a model for families of objects, the Declarative Family of Objects Model (DFOM). In this model, both geometry and topology are specified declaratively. Declarative specifications state properties of objects, typically by means of constraints, but not how to construct those objects. Procedural specifications, on the other hand, specify how to construct objects, but there is no guarantee that any property generically holds for those objects. For modelling families of objects, however, the ability to specify generic properties is essential.

Declarative specification of geometry using constraints is common practice in current modelling systems, but topology is practically always specified in a procedural way [Bettig et al. 2005]. Features in history-based systems correspond to set operations, or other operations that manipulate entities in a B-rep procedurally. These operations may change topological properties of the B-rep, even when this is not desired. In our new model, topological properties are specified by topological constraints that must hold for all members of a family. These constraints are imposed on topological entities that are either explicitly defined by features in the model, or implicitly by the interaction of features in the model. A topological constraint may be, for example, that a face of a feature must be on the boundary of the model, or that a feature may not be split into disjoint volumes. To find an explicit topology, the system of topological constraints is solved.

*e-mail:H.A.vanderMeiden@tudelft.nl
†e-mail:W.F.Bronsvoort@tudelft.nl

SPM 2006, Cardiff, Wales, United Kingdom, 06–08 June 2006.
© 2006 ACM 1-59593-358-1/06/0006 $5.00

The complete system of constraints to be solved thus consist of geometric and topological constraints. Although geometry and topology are closely tied, in our approach they can actually be treated separately. Geometric constraint solving is a well-developed field and is not further addressed here. For an overview of recent work we refer to [Sitharam et al. 2006]. Solving topological constraints on feature models, to the best of our knowledge, has not been addressed before.

In Section 2 we discuss previous work on families of objects. Section 3 is dedicated to the Semantic Feature Model, which is the basis for the DFOM. The DFOM itself is presented in Section 4. Solving topological constraints is discussed in Section 5, and results of solving experiments are presented in Section 6. Finally, we draw some conclusions in Section 7.

2 Previous work

Parametric and feature-based models can be thought of as dual–representation schemes [Shapiro and Vossler 1995], consisting of a parametric representation, e.g. a CSG representation, and a geometric representation, e.g. a B-rep. This view has led to considering two types of families: the parameter-space family and the representation-space family. The parameter-space family is the space of all parameter vectors that correspond to valid models. The representation-space family corresponds to the space of objects that can be obtained by operations on the geometric representation. The two families are interrelated by procedures and constraints that relate parameters to geometry, but this relation is not well understood.

A specific representation-space family is described by the concept of boundary representation deformation [Raghothama and Shapiro 1998]. Basically, a family of objects is defined by a prototype B-rep, and contains all objects that can be created by a continuous deformation of the prototype. The authors acknowledge that this definition of a family is too restrictive, because the boundary representation deformation cannot account for splitting and merging of entities. A more general definition is proposed in [Raghothama and Shapiro 2002]; a family may be considered equivalent to a category as defined by category theory (which allows to describe and compare the semantics of broad classes of mathematical objects, such as the category of sets and the category of topological spaces). Part families are defined as sub-categories of the category of cell-complexes, but an application of this theory to engineering practice is yet to be presented.

Suggested in [Rappoport 1997] are two possible models for families of objects: the classifying set model and the parametric set model. Here, a model is defined as a representation of a family of objects that we wish to query and perform operations on. The classifying set model implements a set-membership query, which, given a family model and an object model, answers whether the object is a member of the family. The parametric set model requires that member instances can be identified by a parameter vector. The parametric access query, given a family model and a parameter vector, returns a member model. A parametric set model for representing families of decomposed point sets is the Generic Geometric Complex (GGC) [Rappoport 1997]. The GGC represents families of Selective Geometric Complexes (SGCs). A SGC [Rossignac and O'Connor 1988] represents an object by carriers, which are n-dimensional algebraic or parametric geometries, and by entities, which are disjunct point sets obtained from intersections of carriers. The GGC represents a family of SGCs with the same carriers and a common subset of entities, marked as essential. Unfortunately, the model does not allow complex patterns of entities to be included

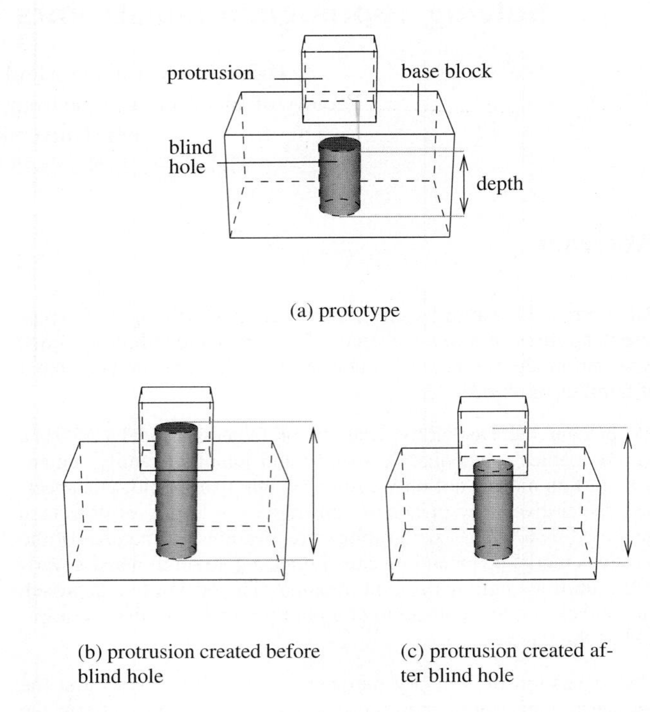

(a) prototype

(b) protrusion created before blind hole

(c) protrusion created after blind hole

Figure 1: Example of the feature ordering problem. When the depth of the blind hole feature in the prototype (a) is increased, either model (b) or model (c) may emerge, depending on the modelling history.

in or excluded from the family, so the class of families that can be modelled by the GGC is rather limited.

The models described above largely overcome the persistent naming problem, but other shortcomings of the history-based approach with respect to modelling families of objects, in particular the feature ordering problem and the problem of maintaining feature semantics, are not addressed.

The feature ordering problem manifests itself when instantiating new members of a family. Features, as they are implemented in history-based modelling systems, are essentially higher-level modelling operations, which are part of the modelling history. When modelling a prototype object, the order of modelling operations that seemed appropriate for that particular family member, may not yield the expected result when re-evaluating the history to create other family members. Consider, for example, Figure 1. A prototype object is first modelled, consisting of a base block, a block protrusion feature and a blind hole feature, as shown in Figure 1a. When instantiating a variant object where the depth of the blind hole is increased beyond the height of the base block, two results are possible, depending on the order in which the features were created in the prototype. If the block protrusion was created before the blind hole, the model shown in Figure 1b emerges. If, however, those features were created in the reverse order, the model shown in Figure 1c emerges. The implication of the feature ordering problem is that when modelling a family of objects, the order of operations must be taken into consideration, even though the effect may not be visible in the prototype. This complicates design and editing of family models, in particular models with many interacting features.

Maintaining feature semantics means that features must satisfy certain predefined properties, in particular, specific topological prop-

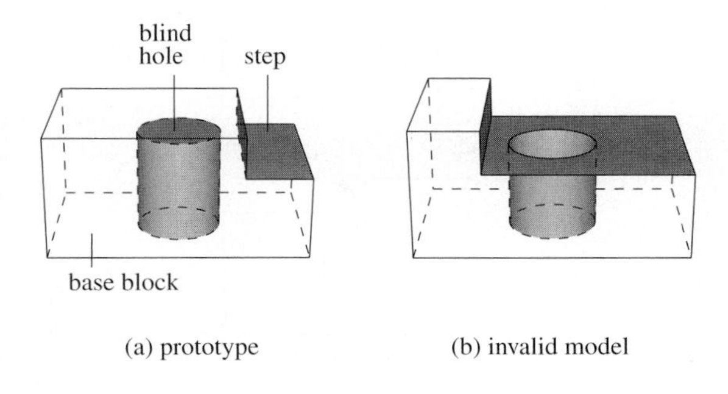

Figure 2: The semantics of a blind hole requires that the hole has a bottom, as in the prototype (a). When the step feature is changed as in model (b), the blind hole is changed into a through hole.

erties, in the context of any model. Due to interaction with other features, however, semantics of a feature may change. For example, Figure 2(a) shows a prototype object consisting of a base block, a blind hole feature and a step feature. The semantics of a blind hole requires that the hole has a bottom, i.e. that the hole does not cut entirely through the base block. When the step feature is changed as in Figure 2(b), the blind hole feature does cut through the model, thus the semantics of that feature has changed. If the semantics of one or more features changes, the model should be marked as invalid. History-based systems, however, only check the validity of a feature during instantiation of the feature in the model. If, due to interaction with other features, the semantics of a feature changes at later stages of the evaluation of the modelling history, this is not detected. In particular, topological properties of features cannot be adequately specified and maintained, because only the result of feature operations is stored in the B-rep, and insufficient feature information is available. Maintaining feature semantics is essential for families of objects, because the semantics must be the same for all members of a family. If feature semantics is not properly maintained, the family may contain invalid models.

By using a limited set of feature classes and strict adherence to proven modelling practice, undesirable situations as described above can sometimes be avoided. However, this practice has also made problems with history-based models more obscure and unpredictable.

An alternative to history-based models for defining families of objects is the Semantic Feature Model, which is defined only by a set of features and constraints, and keeps no record of modelling history. The work presented here is based on the Semantic Feature Modelling approach, which is elaborated in the following section.

3 The Semantic Feature Model

The Semantic Feature Model (SFM), introduced by Bidarra and Bronsvoort [2000a], is a declarative model, which allows feature semantics to be adequately specified and maintained. A SFM is defined by a set of features, instantiated from feature classes. A feature class consists of a shape part, a positioning part, a topological part, and an interface, as shown in Figure 3.

The interface of the feature class contains parameters that control the shape and position of the feature. Upon feature instantiation, the parameters must be assigned a value. Attach faces and positioning faces are set to refer to faces of other features in the model or to

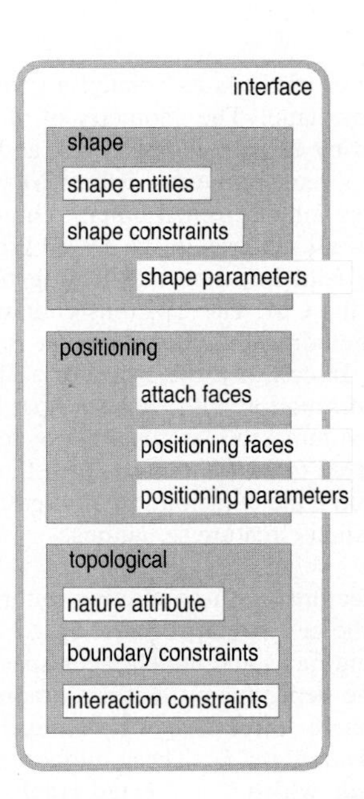

Figure 3: Elements of a feature class definition.

reference geometry. The feature shape is determined by solving geometric and algebraic constraints, specified in the feature class, on shape entities, attach faces and positioning faces. The topological part of a feature class consists of a nature attribute, boundary constraints and interaction constraints. The *nature* attribute can be *additive* or *subtractive*, indicating whether the feature adds or removes material to or from the object. A boundary constraint is associated with a feature face, and specifies that the face must be (partially) on the boundary of the model, or must not be on the boundary of the model. Interaction constraints are associated with a feature as a whole. Interactions with other features may create specific topological patterns, which can be disallowed by these constraints. Table 1 lists interactions commonly found in feature models [Bidarra and Bronsvoort 2000a].

The SFM relies on a geometric representation called the *cellular model* (CM), a type of cell-complex model that includes, in particular, volume cells [Bidarra et al. 2005a]. In the following, we will refer to cells of all dimensions as entities, and only to volume cells as cells. Entities are topological objects, i.e. vertices, edges and

Interaction type	Description
Splitting	Causes the boundary of a feature to be split into two or more disconnected subsets
Disconnection	Causes the volume of an additive feature (or part of it) to become disconnected from the model
Boundary clearance	Causes (partial) obstruction of a closure face of a subtractive feature
Volume clearance	Causes (partial) obstruction of the volume of a subtractive feature
Closure	Causes some subtractive feature volume to become a closed void inside the model
Absorption	Causes a feature to cease completely it's contribution to the model boundary

Table 1: Feature interaction types.

faces, and with each entity is associated a geometric description, e.g. a point, line or plane. The geometry of volume cells is completely determined by their bounding faces, and is not represented explicitly. An important property of the CM is that entities may not self-intersect or intersect each other. The CM is constructed from all entities of all features in the SFM. When feature entities intersect, they are split into non-intersecting new entities, which are then added to the CM. The CM thus contains the geometry of all the features, including the geometry that is not on the boundary of the model. In contrast, the B-rep of a history-based model loses geometric information with each set operation. Additionally, the CM, for each volume cell, stores a list of features that overlap with the cell, referred to as the *owner list* of the cell. This enables to correctly determine the contribution of overlapping features, per cell, and also to validate feature semantics.

For overlapping features with conflicting nature, it is determined per cell whether the cell contains material, by *dependency analysis*. The positioning faces and attach faces specified in the feature interface determine dependency relations among features. If feature f_1 refers to one or more faces of a feature f_2, then f_1 is said to be *directly dependent* on f_2. These relations are represented by a *dependency graph*, which is a directed graph, where every direct dependency of a feature f_1 on a feature f_2 is represented by an edge (f_1, f_2). A feature f_x is said to be dependent on a feature f_y if there is a path from f_x to f_y in the dependency graph. Feature precedence is a partial ordering derived from the feature dependency graph, as follows: if a feature f_x depends on a feature f_y, then f_x precedes f_y. For each cell in the CM, a precedence order is determined for the features in the owner list of the cell. The nature of the feature with the highest precedence determines whether the cell contains material. If the nature of the feature is additive, the cell contains material. If the nature of the feature is subtractive, the cell does not contain material.

When for all cells in the CM it has been determined whether it contains material, the validity of all features is checked by verifying that all boundary constraints and interaction constraints are satisfied. If any constraints are not satisfied, the model becomes invalid, and the user is guided through a recovery process, until validity has been restored.

A model for families of objects based on the SFM, is the Semantic Model Family [Bidarra and Bronsvoort 2000b]. A family consists of all models with the same features and constraints, but different parameter values. Boundary and interaction constraints guarantee that every member of the family has valid feature semantics. However, some families may be ambiguous, meaning that for some parameter values the resulting member is not well defined.

In most cases, a feature's contribution to the model is correctly determined by dependency analysis. Problems occur when there are features in the owner list of a cell that are independent and have conflicting natures. For example, Figure 4 shows the feature dependencies of the model in Figure 1. The block protrusion and blind hole features are both dependent on the base block, because they refer to is for positioning, but there are no dependencies between the two features. Thus no feature precedence can be determined, and again either Figure 1b or Figure 1c emerges, dependent on the order of feature creation, just like in history-based systems. Interestingly, the Semantic Feature Modelling approach can detect that the semantics of the blind hole feature in Figure 1c is incorrect, because the bottom of the blind hole in the cellular model does not correspond to the bottom of the blind hole in the feature definition (the boundary constraint on this feature face is not satisfied). However, this information is not used to determine the correct feature precedence order.

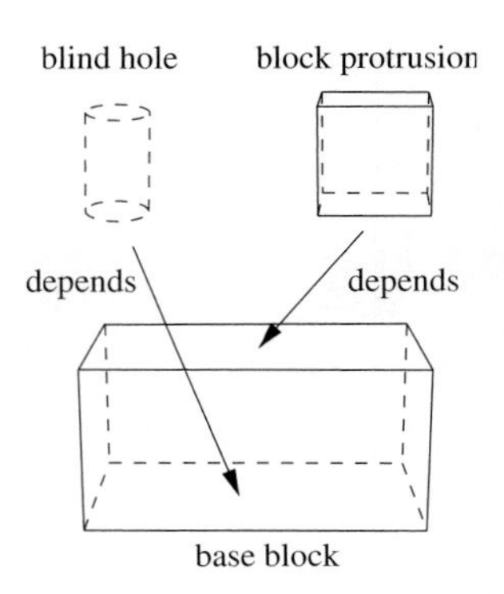

Figure 4: Feature dependencies for example Figure 1. The blind hole and the block protrusion depend on the base block, but are independent of each other.

The main shortcoming, more in general, of the SFM as a basis for defining families of objects, is that member models are determined by feature dependency analysis. This analysis cannot always unambiguously decide which cells should contain material. More importantly, the analysis is not directly related to the semantics of features as specified by topological constraints. Topological constraints are only checked after a model has been created, instead of being used to create a valid model. As a result, the model that satisfies the topological constraints is not always found. This problem is resolved with the model that is presented in the next section.

4 The Declarative Family of Objects Model

The Declarative Family of Objects Model (DFOM) is a generalisation of the Semantic Model Family. The DFOM is similarly defined by a set of features and constraints. Also, the geometric representation is a cellular model. However, whether the cells of the CM contain material is not determined by feature dependency analysis, but by solving topological constraints. Another difference with the Semantic Model Family is that the DFOM can represent both families of objects and single objects. Family members and subfamilies are modelled by adding constraints to a family model, e.g. to set dimension variables. Family membership can be determined by comparing the features and constraints of different DFOMs. Also, the definition of feature classes for the DFOM is more generic than for the Semantic Model Family. Altogether, the DFOM can represent a wider range of families, in an unambiguous way.

A feature class defines a canonical shape, constraints, and an interface. The canonical shape of a feature is represented by a canonical cellular model, referred to as the *feature CM*. The geometry of the entities in the feature CM is not completely determined. Only the type of the geometry is given, and the relations between the entities. For the volume cells, it is not yet specified whether they contain material. Geometric constraints are imposed on feature entities to determine the geometry and relative position and orientation of features. Topological constraints are imposed on feature entities, to specify topological properties of the feature that must be maintained. These constraints also determine the contribution of the feature to the model. Constraints may be part of a feature (feature constraints) or can be imposed between features (model constraints). The interface of a feature is the subset of the parameters of the features constraints that are made available to the user. Interface parameters may be numeric parameters, e.g. distances and angles, or geometric parameters, e.g. curves and surfaces. The latter can be used, in combination with geometric constraints, to position the feature relative to other features in the model.

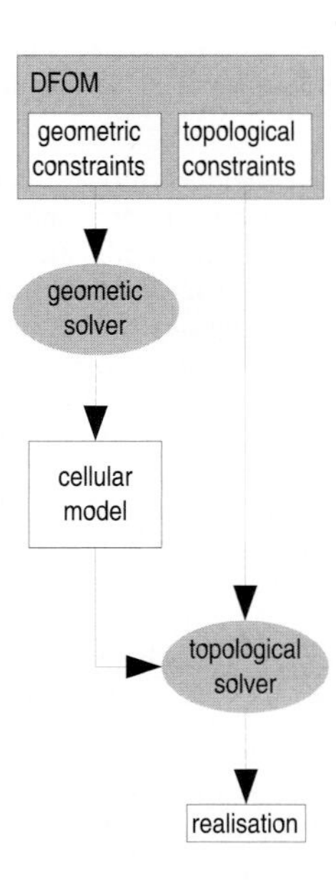

Figure 5: The interpretation of the DFOM consists of a geometric solving step, which yields a cellular model, and a topological solving step, which yields a realisation.

Feature classes no longer specify a nature attribute, as was the case for the SFM. Nature may, however, be implemented in the DFOM as a topological constraint that is imposed on feature cells. A constraint Nature(v, additive|subtractive) defines that the nature of the feature cell v is additive or subtractive respectively. If the same nature is specified for all the cells of a feature, the feature behaves similar to a feature in a SFM, but a DFOM feature may also have a mixed additive and subtractive nature. To further define feature semantics, boundary and interaction constraints can be used, but also other topological constraints, to define features with semantics that are not possible with the SFM.

A DFOM does not explicitly represent the geometry and topology of the model. Instead, a number of *realisations* is derived from a DFOM by a process called *interpretation*. A realisation is a cellular model with a value assigned to each cell, specifying whether the cell contains material. The interpretation of a DFOM involves two solving steps. First, the geometric constraints in the DFOM are solved. From the geometric solution, a *combined cellular model* is constructed, containing all geometry of all features. Then, the topological constraints in the DFOM are mapped to constraints on the combined cellular model. The resulting system of constraints is solved, and from the solution of this system, the realisation of the DFOM is constructed. This is illustrated in Figure 5.

Typically, if a DFOM represents a family of objects, the system of geometric constraints will be underconstrained, i.e. it has one or more degrees of freedom. This occurs when, for example, some dimensions of some features have not been specified. A system that is underconstrained, has an infinite number of solutions, and these cannot be represented explicitly. Implicitly, the model represents all the realisations that would have been obtained if we could generate all the geometric solutions and interpret them further. If the geometric system has a finite number of solutions, then all the solutions can be generated explicitly and interpreted further. Geometric systems can have a very large number of solutions, exponential to the number of constraints in the system. It is desirable that the number of geometric solutions is kept low. Therefore, feature classes should be defined in such a way that when they are instantiated in a model, there is only one geometric solution [van der Meiden and Bronsvoort 2005].

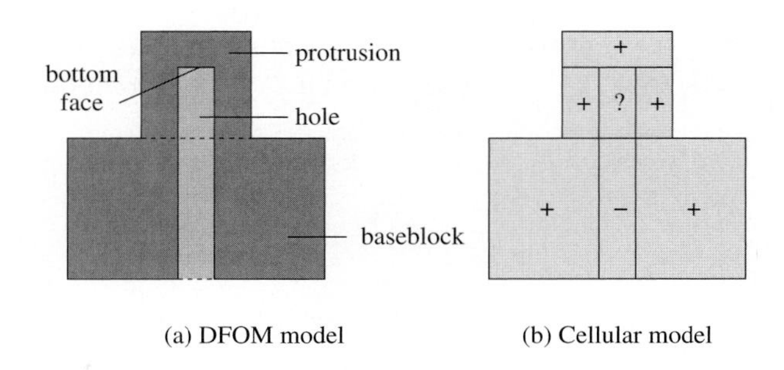

(a) DFOM model (b) Cellular model

Figure 6: A 2D model representing the feature ordering problem, corresponding to Figure 1. In the cellular model, the value of cells marked with a sign are determined by nature constraints. Cells marked with '+' contain material, cells marked with '-' do not contain material. The value of cells marked with '?' cannot be determined by nature constraints, and must therefore be determined by solving other topological constraints.

When the geometric constraint system has been solved, the geometry of all feature entities has been determined. The feature entities are then merged into the combined CM. The combined CM can be generated efficiently by adding feature entities one at a time [Bidarra et al. 2005a]. If the added entity intersects with other entities already in the combined CM, the intersecting entities are split into non-intersecting entities. For each entity in the combined CM, a list of references is kept to the original feature entities from which it was derived, referred to as the entity's *owner list*. This allows us to translate topological constraints on feature entities into constraints on entities of the combined CM. In the case of a nature constraint Nature(v, additive|subtractive), the translation is as follows. The set $C = \{c_1, ...c_k\}$ is the subset of the volume cells in the combined cellular model owned by v. For every volume cell c_i, determine the set F_i of features that contribute to this cell. The set D_i is the set of *dominant features*, i.e. the features in F_i on which no other features in F_i are dependent. The nature of these features is dominant over the nature of other features. Let the set N_i be the set of nature attribute values associated with the features in D_i. If the set N_i contains only additive, then the cell contains material. If the set N_i contains only subtractive, then the cell does not contain material. If the set contains no value or both values, then the cell's value is not determined by any nature constraint, but by other constraints, e.g. boundary and interaction constraints. The implementation of nature as a constraint in the DFOM thus resolves the ambiguity of feature dependency analysis of the SFM.

The model shown in Figure 6 represents a situation similar to Figure 1 in Section 2; a blind hole feature cuts though a base block, into a protrusion feature. The semantics of a blind hole requires that the bottom face of the hole is on the boundary of the model, expressed by a CompletelyOnBoundary constraint. Figure 6(b) shows that for one cell, the nature constraints cannot determine whether it should contain material. This cell corresponds to the intersection of the hole and the protrusion, which are independent and have a different nature. Therefore, the value of the cell can only be determined by solving other topological constraints. In this case, the cell should not contain material, because the bottom face of the hole must be on the boundary of the model.

In this example, and also in general, most cell values are determined by nature constraints. It therefore makes sense to first try to

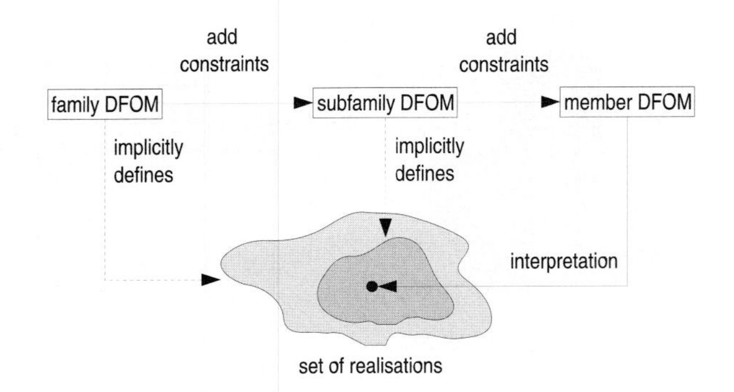

Figure 7: Specialisation and instantiation. By adding constraints to a DFOM, subfamilies and members are obtained. The realisations of a subfamily, and the realisation of a member, are in the set of realisations of the original family.

determine the values for cells carrying nature constraints. This can be done per cell, which is more efficient than considering all topological constraints and all cells at once. A relatively small number of cells is involved in complex feature interactions, such that their value cannot be determined by nature constraints. For those cells, the other topological constraints in the model are solved (see Section 5).

The interpretation of the DFOM, as described above, guarantees that realisations satisfy all constraints, and therefore the semantics specified by the model. A DFOM with more than one realisation thus represents a well-defined family of objects. However, because the set of realisations can be infinite, we cannot, in general, verify family membership by generating and comparing realisations. Therefore, family membership is defined differently, as follows.

A DFOM M represents a member of a family, represented by a DFOM F, if

- M contains the same set of features as F,
- M contains a superset of the constraints of F, and
- M has exactly one realisation.

In other words, members of a family are instantiated by adding more constraints, until the number of realisations is just one. Subfamilies are also defined in this way: they are represented by models with extra constraints and one or more realisations. Thus, adding constraints to a DFOM is equivalent to specialisation of a family. Instantiation is essentially the same as specialisation; members are subfamilies with just one realisation. Models with no realisations at all are invalid.

The relation between families of DFOMs and their realisations is illustrated in Figure 7. To better understand this relation, consider that a subfamily, defined by a superset of constraints, always has a smaller (or equal) number of realisations than the original family, because more constraints have to be satisfied. Also, the set of realisations is a subset of the set of realisations of the original family, because all the constraints of the original family DFOM have to be satisfied. A family member has only one realisation, which is thus guaranteed to be an element of the set of realisations of the family. On the other hand, we cannot guarantee that every realisation of a family DFOM has a corresponding member DFOM, because there may not be any combination of constraints that allows this realisation to be instantiated. Also, there may be DFOMs that are not members of a given family DFOM, because they contain different features, but whose realisations are in that family's set of realisations. It should thus be kept in mind that family membership in terms of DFOMs is not equivalent to set membership for realisations.

To verify DFOM family membership, we need to be able to compare sets of features and constraints, and to determine whether a model has zero, one or more realisations. The latter requirement implies that the geometric constraint solvers must be able to identify underconstrained and overconstrained situations, which is standard. In our implementation of topological constraint solving, this requirement is also met (see Section 5). Sets of features and constraints can easily be compared if features and constraints are uniquely identified by a name. This is only the case for models that were directly derived from a given family. In this case, verifying family membership is not interesting, because such models are by definition members of the family. To verify membership for models from another source, we can only consider the types of features and constraints, and the associations between them. In that case, the membership test is equivalent to graph matching.

Altogether, the DFOM described in this section allows families of objects to be specified with clear semantics. This semantics is declaratively specified by the user, thus families can be specified which include all, and only, the desired members. Family membership is well-defined, and can be verified procedurally.

5 Satisfying topological constraints

To find realisations of a DFOM, the topological constraints in the model must be solved. These constraints cannot be solved directly. Instead, from the topological constraints, a system of boolean constraints on the combined CM is derived. Realisations are found by solving the latter system of constraints.

Recall that a realisation is a set of assignments, specifying for each volume cell in the combined CM whether the cell contains material. All volume cells of the combined CM are thus mapped to boolean variables. Solutions of the boolean constraint system correspond to realisations of the DFOM, as follows. If a variable evaluates to `True`, the corresponding cell contains material; if it evaluates to `False`, the corresponding cell does not contain material. A boolean constraint is a predicate on a set of boolean variables, which may be expressed by a boolean function $satisfied(v_1, v_2, \ldots, v_k)$, where $v_1 \ldots v_k$ are boolean variables. If $satisfied$ evaluates to `True`, then the constraint is satisfied. If $satisfied$ evaluates to `False`, then the constraint is not satisfied.

Because realisations specify only whether volume cells contain material, they always represent proper manifold solids, with closed boundaries, and no dangling faces or edges. Constraints on lower dimensional entities, i.e. faces, edges and vertices, can therefore be expressed in terms of volume cells. Consider, for example, a constraint `CompletelyOnBoundary(f)`, meaning that a feature face f must be completely on the boundary of the model. To translate the constraint to boolean constraints, we need to determine first the set of faces $G = \{g_1, ..., g_k\}$ in the cellular model that overlap with f. All faces $g_i \in G$ must be on the boundary of the model. Each g_i has two adjacent volume cells, c_i^a and c_i^b. A face g_i is on the boundary if and only if exactly one adjacent volume cell contains material. Then the boolean constraint is:

$$\bigcap_{i=1}^{k}((c_i^a \cup c_i^b) \cap \neg(c_i^a \cap c_i^b))$$

The boolean constraint problem to be solved is the following: given a set of n boolean variables $V = \{v_1, v_2, \ldots, v_n\}$ and a set of m topological constraints $C = \{c_1, c_2, \ldots, c_m\}$, find an assignment $v_i := x_i$, where $x_i \in \{True, False\}$ for every $v_i \in V$, such that every constraint $c_j \in C$ is satisfied. This problem is known as the *boolean satisfiability problem*, or SAT problem, which is an NP-hard problem [Papadimitriou 1995]. The SAT problem has been well studied, and search algorithms exist that can find solutions efficiently for many instances. State-of-the-art solvers learn from earlier 'mistakes' and avoid search paths that are unlikely to lead to a solution. One such solving technique is *conflict-driven learning* [Silva and Sakalla 1996; Zhang et al. 2001], which has been implemented in the MINISAT solver [Een and Sörensson 2004] that was used in our experiments (see Section 6). This SAT solver, like most SAT solvers, is specialised to problems formulated as *conjunctive normal form* (CNF) clauses. A CNF clause is a set of literals, where a literal is a variable v_i or the negation of a variable v_i, where $v_i \in V$. A clause $(l_1, l_2, \ldots, l_j)$ is interpreted as $l_1 \cup l_2 \cup \ldots \cup l_j$. A set of clauses $\{c_1, c_2, \ldots, c_k\}$ is interpreted as $c_1 \cap c_2 \cap \ldots \cap c_k$.

Unfortunately, writing boolean constraints as CNF clauses can, result in an exponential number of clauses [Papadimitriou 1995]. Even if a constraint has an efficient formulation in CNF, it may be very difficult to find it. Some topological constraints can perhaps be formulated in CNF manually, but this is cumbersome for all but the most trivial constraints. It is, however, feasible to formulate most topological constraints using only the common boolean operators, AND, OR, and NOT, as done for the `CompletelyOnBoundary` constraint in the example above. Such boolean expressions are more easily converted to CNF clauses than general boolean constraints, using a method called *operator expansion*. This method replaces each subexpression with a new variable. The meaning of the operators between subexpressions, is expressed by CNF clauses on the new variables. For example, the expression $a \cup (b \cap c)$ is expanded as follows. The subexpression $b \cap c$ is substituted by a new variable x. Now the expression becomes $a \cup x$, which is already a CNF expression. We also represent the expression $x = b \cap c$ in CNF, which becomes:

$$(b \cup c \cup \bar{x}) \cap (b \cup \bar{c} \cup \bar{x}) \cap (\bar{b} \cup c \cup \bar{x}) \cap (\bar{b} \cup \bar{c} \cup x)$$

The complete expression thus corresponds to the following set of clauses:

$$\{(a,x), (b,c,\bar{x}), (b,\bar{c},\bar{x}), (\bar{b},c,\bar{x}), (\bar{b},\bar{c},x)\}$$

Operator expansion introduces a large number of new variables in the problem, one for each operator in the original boolean expression. However, writing expressions in CNF form without introducing new variables may again result in an exponential number of clauses. Finding the optimal formulation for any expression is not a trivial task; in fact, it is as difficult as solving the expression. We believe, however, that CNF clauses generated by operator expansion are not more difficult to solve than expressions created by other rewriting methods. This is supported by the experiments described in the next section.

6 Results

To gain insight in the feasibility of solving topological constraints, we have experimented with solving idealised models. The models are 2D, and consist of orthogonal rectangular features. They do not represent realistic models, but are intended to generate topological constraint systems of a complexity that can be expected from realistic models. The entities of the cellular model are also orthogonal rectangular cells. The cells in the cellular model are non-overlapping, thus overlapping features result in split cells. Associated with each cell are one or more features, and a boolean variable. Topological constraints are formulated as boolean expressions (using AND, OR and NOT operators) on boolean variables associated with cells in the combined CM. These are converted into CNF clauses via the operator expansion method described in Section 5. The system of CNF clauses is then solved with the MINISAT solver [Een and Sörensson 2004].

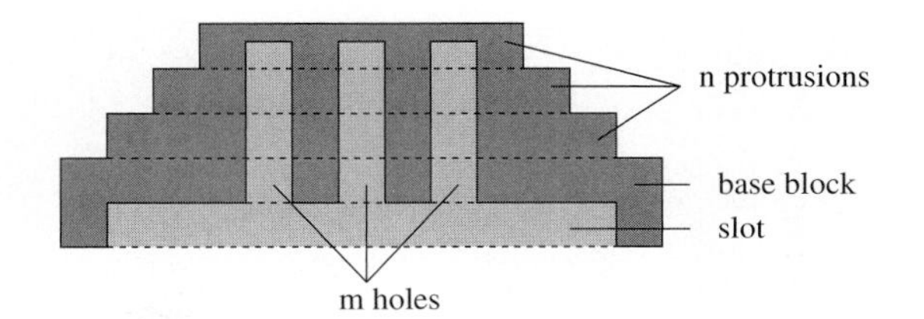

Figure 8: 2D model with $n = 3$ protrusion features and $m = 3$ hole features.

$m \times n$	# cells	# clauses	time (msec)
10x10	259	1457	3.9
16x16	601	3467	8.1
21x21	996	5802	11.9
25x25	1384	8102	17.4
28x28	1717	10079	23.1
31x31	2086	12272	28.7
34x34	2491	14681	39.4
37x37	2932	17306	41.3
39x39	3246	19176	50.7
41x41	3576	21142	52.6
44x44	4101	24271	64.1
46x46	4471	26477	82.1
48x48	4857	28779	77.5
49x49	5056	29966	80.6
51x51	5466	32412	96.4
53x53	5892	34954	91.0
55x55	6334	37592	122.0
56x56	6561	38947	117.3
58x58	7027	41729	151.4
59x59	7266	43156	144.0

Table 2: Solving statistics for the model in Figure 8.

The first model consists of a variable number of horizontal protrusions, intersected by a variable number of vertical blind holes, as shown in Figure 8. This model has been solved with different parameter values to determine the solving times for increasing numbers of feature interactions. The hole features and the protrusion features are all dependent on a base block, but are pairwise independent of each other. Thus, the values of the intersection cells of the holes and the protrusions cannot be determined by feature dependency analysis. Instead, each hole has a topological constraint that it may not be split, i.e. that there must be a passage from the top of the hole to the bottom of the hole. The model has exactly one solution for any given number of holes and protrusions.

Table 2 shows, for different model sizes, the number of cells in the cellular model, the number of clauses in the constraint system, and the solving time in milliseconds. The size of the model is expressed as $m \times n$, where m is the number of blind hole features and n is the number of protrusion features. The number of cells in the model is roughly quadratic to the number of features in the model, because the model is designed to have a large number of feature interactions. The number of clauses is linear with respect to the number of cells in the model. The solving time in the table is the sampled mean over 50 experiments. The standard deviation for the solving times is approximately 10% of the mean, probably due to some random choices made in the MINISAT solver [Een and Sörensson 2004].

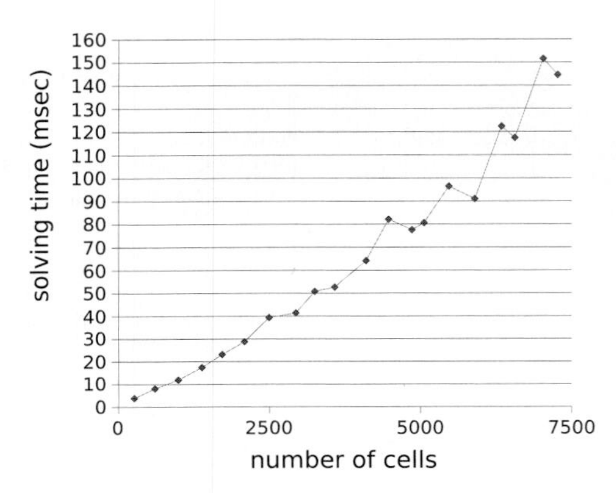

Figure 9: Plot of solving times against the number of cells for the model in Figure 8.

Figure 9 shows a plot of the solving times against the number of cells in the model. The latter quantity is representative for the complexity of the model, because it depends on the number of feature interactions. The plot is quite jagged, due to the large standard deviation, but seems to suggests that solving time grows slightly faster than linear with respect to the number of cells in the model. From other solving statistics (not shown), it is clear that the problem is solved mostly by propagation, indicating that the complexity of the problem is low [Silva and Sakalla 1996]. We have also run the experiment on a more complex model, shown in Figure 10. A plot of the solving times against the number of cells for this model is shown in Figure 11. This plot also shows a slightly more than linear increase of the solving time with the number of cells.

It should be noted that, in these models, the number of cells grows quadratically with the number of features, and the number of clauses quickly gets very large. This is because the example models were designed to maximise the number of feature interactions. Yet solving times are relatively small, and do not seem to grow alarmingly with the number of cells. In realistic models, the number of cells per feature will likely be much smaller, and even better solving times can be expected.

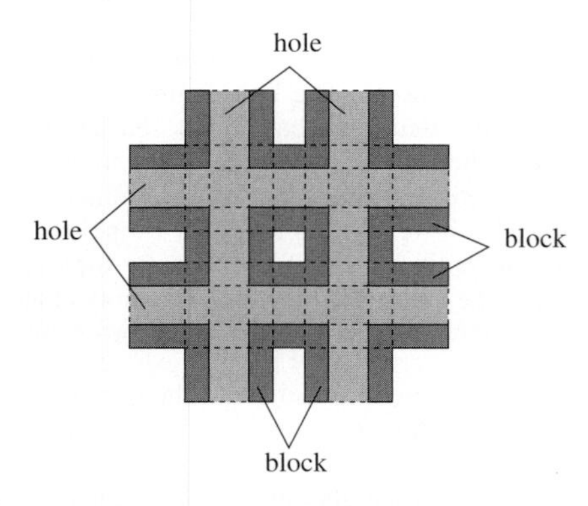

Figure 10: 2D model with $n \times n$ intersecting blocks and through holes, shown for $n = 2$.

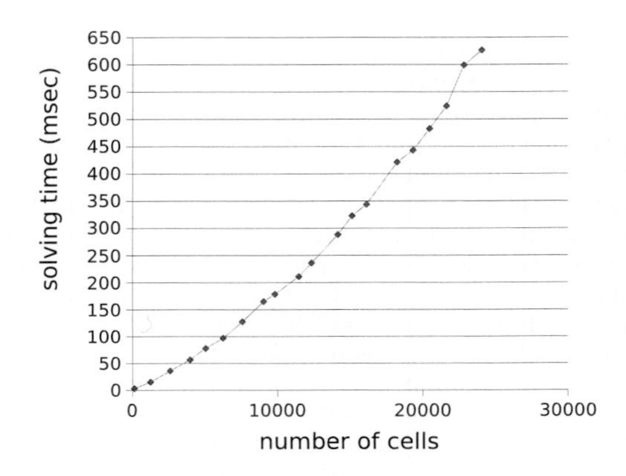

Figure 11: Plot of solving times against the number of cells for the model in Figure 10.

7 Conclusions

The Declarative Family of Objects Model (DFOM), presented here, allows families of objects to be specified by constraints on geometry and topology. It does not have the problems associated with history-based models. In particular, it correctly maintains feature semantics, and is not hampered by the feature ordering problem. Feature and constraint definitions of the Semantic Feature Model have been re-implemented in the DFOM. The ambiguity of feature dependency analysis is overcome by solving topological constraints.

Solving systems of topological constraints is equivalent to the satisfiability problem, which is NP-hard. Our experiments with the MINISAT solver, however, show that solving topological constraints is feasible, even for large models with many interacting features. The CNF problems derived from such models, although not hard to solve for the MINISAT solver, are very large. Generating these problems is costly. An algorithm that efficiently solves topological constraints or boolean expressions directly would be very useful. The MINISAT solver can be extended with new types of constraints, but the effectiveness of the solver's optimisation strategy for such constraints is unknown.

When an invalid model is constructed, i.e. a model that has no realisations, the user should be given meaningful feedback that can help restore validity, e.g. which constraints are involved, and which parts of the model are underconstrained or overconstrained. Currently, the solver is not able to give such feedback.

The DFOM allows a large variety of new features and constraints to be defined, which results in the required flexibility to adequately define the semantics of a family of objects. It therefore seems a good basis for better ways to model families of objects.

Acknowledgements

H.A. van der Meiden's work is supported by the Netherlands Organisation for Scientific Research (NWO).

References

BETTIG, B., BAPAT, V., AND BHARADWAJ, B. 2005. Limitations of parametric operators for supporting systematic design. In *CDROM Proceedings DECT-2005, ASME International Design Engineering Technical Conferences, September 24-28, Long Beach, California, USA*, ASME.

BIDARRA, R., AND BRONSVOORT, W. F. 2000. Semantic feature modelling. *Computer-Aided Design 32*, 3, 201–225.

BIDARRA, R., AND BRONSVOORT, W. F. 2000. On families of objects and their semantics. In *Proceedings of Geometric Modeling and Processing 2000, April 10-12, Hong Kong, China*, IEEE Computer Society, 101–111.

BIDARRA, R., MADEIRA, J., NEELS, W., AND BRONSVOORT, W. F. 2005. Efficiency of boundary evaluation for a cellular model. *Computer-Aided Design 37*, 12, 1266–1284.

BIDARRA, R., NYIRENDA, P. J., AND BRONSVOORT, W. F. 2005. A feature-based solution to the persistent naming problem. *Computer-Aided Design and Applications 2*, 1-4, 517–526.

EEN, N., AND SÖRENSSON, N. 2004. An extensible SAT solver. In *Theory and Applications of Satisfiability Testing, 6th International Conference, SAT 2003, Santa Margherita Ligure, Italy, May 5-8, 2003, Selected Revised Papers*, Springer Verlag, E. Giunchiglia and A. Tacchella, Eds., vol. 2919 of *Lecture Notes in Computer Science*.

PAPADIMITRIOU, C. H. 1995. *Computational Complexity*. Addison-Wesley.

RAGHOTHAMA, S., AND SHAPIRO, V. 1998. Boundary representation deformation in parametric solid modeling. *ACM Transactions on Graphics 17*, 4, 259–286.

RAGHOTHAMA, S., AND SHAPIRO, V. 2002. Topological framework for part families. *Journal of Computing and Information Science in Engineering 2*, 4, 246–255.

RAPPOPORT, A. 1997. The Generic Geometric Complex (GGC): a modeling scheme for families of decomposed pointsets. In *Proceedings Solid Modeling '97, Fourth ACM symposium on Solid Modeling and Applications, May 14-16, Atlanta, Georgia, USA*, ACM Press, C. M. Hoffmann and W. F. Bronsvoort, Eds., 19–30.

ROSSIGNAC, J. R., AND O'CONNOR, M. A. 1988. SGC: a dimension-independent model for pointsets with internal structures and incomplete boundaries. In *Proceedings of the 1988 IFIP/NSF Workshop on Geometric Modeling, Renselaerville, New York, USA*, M. Wozny, J. Turner, and K. Preiss, Eds., 145–180.

SHAPIRO, V., AND VOSSLER, D. L. 1995. What is a parametric family of solids? In *Proceedings of the Third ACM/IEEE Symposium on Solid Modeling and Applications, May 17-19, Salt Lake City, Utah, USA*, ACM Press, C. M. Hoffmann and J. R. Rossignac, Eds., 43–54.

SILVA, J. P., AND SAKALLA, K. A. 1996. Grasp - a new search algorithm for satisfiability. In *Proceedings of ICCAD 1996, IEEE/ACM International Conference on Computer-Aided Design, November 10-14, San Jose, California, USA*, IEEE Computer Society Press, 220–227.

SITHARAM, M., OUNG, J.-J., ZHOU, Y., AND ABREE, A. 2006. Geometric constraints within feature hierarchies. *Computer-Aided Design 38*, 1, 22–38.

VAN DER MEIDEN, H. A., AND BRONSVOORT, W. F. 2005. An efficient method to determine the intended solution for a system of geometric constraints. *International Journal of Computational Geometry and Applications 15*, 3, 279–298.

ZHANG, L., MADIGAN, C. F., MOSKEWICZ, M. W., AND MALIC, S. 2001. Efficient conflict driven learning in a boolean satisfiability solver. In *Proceedings of ICCAD 2001, IEEE/ACM International Conference on Computer-Aided Design, November 4-8, San Jose, California, USA*, IEEE Computer Society Press, 279–285.

Geometric and Physical Modelling in Medical Image Processing: Methods, Applications and Examples

Jürgen Weese

Geometrical and physical modelling is of increasing importance in medical image processing. Geometric models are typically used in segmentation and registration methods to support image visualization and the extraction of measurements from images. Increasingly more, image derived geometrical models are the input for a more refined analysis using physical models. An example is the analysis of blood flow patterns in the heart or in aneurysms with CFD simulations.

The lecture starts by discussing geometric modeling of organs for model-based segmentation. This comprises the representation of simple structures such as bones as well as complex organs such as the heart by suitable surface meshes. Furthermore, modeling of the shape variability between different individuals is considered and the value for achieving an accurate segmentation is discussed. Finally, an outlook and examples are provided, regarding physical modeling in upcoming medical applications.

Jürgen Weese received a degree and a PhD in physics from the University of Freiburg, Germany, in 1990 and 1993, respectively. From 1993 to 1994, he was with the Freiburger Material Research Center as Head of the Scientific Information Processing group. He joined the Philips Research Laboratories Hamburg in the end of 1994 and became Senior Scientist in 2000. In 2001 he was appointed Principal Scientist and moved to the Philips Research Laboratories Aachen, Germany. Since he joined Philips Research, his research work focuses on medical image processing, image guided surgery and interventional X-ray applications. Currently, he is coordinator of several projects and activities in this area, (co-) author of about 90 scientific publications in conference proceedings and regular journals, and (co-) inventor of more than 50 patents or patent applications.

SPM 2006, Cardiff, Wales, United Kingdom, 06–08 June 2006.
© 2006 ACM 1-59593-358-1/06/0006 $5.00

Real-time Haptic Incision Simulation using FEM-based Discontinuous Free Form Deformation

Guy Sela[1,*] Jacob Subag[1,†] Alex Lindblad[2,3,‡] Dan Albocher[1,§] Sagi Schein[1,¶] Gershon Elber[1,||]

Abstract

Computer-aided surgical simulation is a topic of increasingly extensive research. Computer graphics, geometric modeling and finite-element analysis all play major roles in these simulations. Furthermore, real-time response, interactivity and accuracy are crucial components in any such simulation system. A major effort has been invested in recent years to find ways to improve the performance, accuracy and realism of existing systems.

In this paper, we extend the work of [Sela et al. 2004], in which we used Discontinuous Free Form Deformations (DFFD) to artificially simulate real-time surgical operations. The presented scheme now uses accurate data from a Finite-Element Model (FEM), which simulates the motion response of the tissue around the scalpel, during incision. The data is then encoded once into the DFFD, representing the simulation over time. In real-time, The DFFD is applied to the vertices of the surface mesh at the actual incision location and time. The presented scheme encapsulates and takes advantage of both the speed of the DFFD application, and the accuracy of a FEM. In addition, the presented system uses a haptic force feedback device in order to improve realism and ease of use.

CR Categories: I.3.5 [Computer Graphics]: Computational Geometry and Object Modeling—Physically based modeling; I.6.8 [Simulation and Modeling]: Types of Simulation—Visual

Keywords: Free-Form Deformation, Finite Element Model, Surgical Simulation

1 Introduction

Today, surgical simulators constitute an active research subject. Surgical simulators allow physicians to practice and hone their skills inside a virtual environment before entering the operating room. Such pre-operative training procedures have been shown to significantly improve the results of actual procedures [Seymour et al. 2002]. This is especially true with the recent increase in the use of endoscopic and laparoscopic procedures.

In order to maximize the potential gain in such virtual-reality training, a surgical simulation system should replicate the surgical environment as closely as possible in terms of look and feel. Conveying a realistic impression is difficult. Because of the complexity of such a task, it is best grasped when broken into smaller undertakings. One of the most important roles of any surgical simulator is to realistically animate - in real-time - the way tissue (skin and flesh or internal organs, etc.) behaves under cutting operations. A virtual cutting simulator should supply the following basic capabilities. First, it should have some mechanism for real time collision detection. Such a mechanism should control the location, direction and orientation of a virtual scalpel and constantly test for intersections with the model. Second, a cutting module should implement geometric operations that would progressively cut through the model, modifying its topology and constructing new geometry (the geometry of the cut) as needed and over time. Third, the cut model should reflect the physical behavior as accurately as possible, mainly presenting tissue behavior over time. Another important detail not to be overlooked is the user interface. A haptic force feedback device is invaluable in providing realistic interaction behavior, both from the visual and the palpable point of view.

When dealing with surface meshes, the actual task of cutting the tissue can be divided into two sub-tasks. First, there is the surface modeling task, in which the model surface should be split along the route of the scalpel as it advances. Second, the geometry around the cut should change, reflecting the shape and orientation of the cutting tool and the internal strain and stress properties of the tissue. In this work, we propose a framework that performs these two tasks. The framework is based upon an augmented variant of Free Form Deformation (FFD) [Sederberg and Parry 1986], which allows discontinuities and openings to be created in geometric models. The Discontinuous FFD (DFFD) [Schein and Elber 2005] is continuous everywhere except at the incision, and hence it has the ability to continuously deform the geometry around the cut. Moreover, we incorporate previously simulated results, using a Finite-Element Model, into the deformation function in order to make the behavior of the cut as realistic as possible.

Because FEM simulations are difficult to compute in real-time, an alternative approach could apply physical simulations to a low resolution representation of the model and encode it into the DFFD during the interaction, only to be immediately applied to the fine resolution representation of the geometry. This alternative approach would, of course, entail a much higher processing overhead, as the FEM simulation will need to be executed during run time, but on the other hand it will allow for more adaptable results than the first approach. In this work, we will concentrate on the first approach, in which the DFFDs are constructed off-line.

The proposed FEM-DFFD synergy is of low real-time computational complexity while retaining reasonable accuracy. Consequently, the algorithm is capable of handling complex geometric models at interactive rates. The FEM calculations are conducted

*guysela@cs.technion.ac.il
†jsubag@cs.technion.ac.il
‡alind@washington.edu
§sdannya@cs.technion.ac.il
¶sagi.schein@hp.com
||gershon@cs.technion.ac.il
[1]Department of Computer Science, Technion - Israel Institute of Technology, Haifa 32000, Israel.
[2]Human Interface Technology Lab, University of Washington, PO Box 352142 Seattle, WA 98195-2142
[3]Department of Civil and Environmental Engineering, University of Washington, PO Box 352700, Seattle, WA 98195-2700.

SPM 2006, Cardiff, Wales, United Kingdom, 06–08 June 2006.
© 2006 ACM 1-59593-358-1/06/0006 $5.00

once as a preprocessing stage using a straight-line scalpel path, whereas the deformation is applied to a limited local set of mesh vertices at every time step, mapping the straight path to a deformed path following the virtual scalpel.

The rest of this work is organized as follows. In Section 2, we give an overview of the previous work on the problems of cutting through and deforming geometric models. In Section 3, we describe the proposed cutting simulation approach; Section 4 presents a few examples and finally, we conclude in Section 5.

2 Related Work

Throughout the years, the problems of cutting through geometric models and deforming 3D models, for general as well as for medical purposes, have been tackled from many directions. In this section, and due to space constraints, we only consider a small subset of the relevant work. In Section 2.1, we look at work dealing with cutting through polygonal or tetrahedral meshes and in Section 2.2, we consider results related to the incorporation of FEM simulation results into medical simulations.

2.1 Mesh and Surface Cutting

Earlier work on mesh-cutting dealt mainly with surface-based meshes. Bruyns and Senger [Bruyns and Senger 2001] suggested a method of cutting polygonal meshes interactively without any post-processing, simply by splitting the affected polygons into several new polygons. A different approach proposed by Neinhuys and Van der Stappen [Nienhuys and van der Stappen 2004] suggested combining a local Delauny-based triangulation step as part of the mesh-cutting process. Edge-flip operations are used on the faces affected by the cutting operation in order to eliminate triangles with large circumferences. This method can be applied to both 2D and 3D surface meshes. A different approach, by Ellens and Cohen [Ellens and Cohen 1995], directly incorporated arbitrary-shaped cuts into tensor product B-spline surfaces. This approach has the advantage of operating over an inherently smooth surface, but requires modifications to the standard definition of trimmed B-spline surfaces.

Other works considered volumetric data models, mostly in the form of tetrahedral meshes. Using volumetric data models is beneficial as it can represent both the outer surface of the model as well as its inner parts. Ganovelli and O'Sullivan [Ganovelli and O'Sullivan 2001] proposed cutting tetrahedral meshes while re-meshing the tetrahedra around the cut to achieve the required level of smoothness. Since such splitting operations could degrade the quality of the mesh, they suggested using edge-collapse operations in order to remove low-quality tetrahedra from the mesh. Bielser et al. [Bielser et al. 1999] described cutting through tetrahedral meshes based on the observation that, topologically, there are only five distinct ways to cut a tetrahedron. Once the system detects a collision between a tetrahedron and the cutting scalpel, the case is mapped to one of the five available cutting configurations. The scheme uses a generic subdivision which replaces every original tetrahedron to be split with 17 new tetrahedra. This results in the introduction of many additional tetrahedra into the model and degrades the performance of the system over time. To circumvent this problem, Neinhuys and Van der Stappen [Nienhuys and van der Stappen 2001] proposed locally aligning the edges of the triangular faces around the cut to the route of the virtual scalpel. The movements of the scalpel inside a triangle are recorded and the vertices adjacent to the motion-curve are snapped onto it. Then, the triangles are separated along these aligned edges. Another approach, proposed by Forest et al. [Forest et al. 2002], treats cutting through tetrahedral meshes as a material removal problem. In [Forest et al. 2002], tetrahedra are removed from the mesh when hit by the pointing device. This trivially conserves the three-manifoldness of the tetrahedral mesh but results in a loss of mass of the volumetric model. Since the fineness of the cut is tightly coupled to the fineness of the model, this could result in sharp edges around the incision, something that is infrequently found when cutting human tissue.

2.2 Finite Element Deformation

Bro-Nielsen [Bro-Nielsen 1998] employed a linear elastic material model in order to gain speed at the expense of accuracy. The problem was that linear elastic models are only sufficient when dealing with small deformations. In another effort, by Mor and Kanade [Mor and Kanade 2000], model deformation was achieved by employing linear FEM over the cut model. Another problem that was tackled in [Mor and Kanade 2000] is the introduction of progressive cutting to prevent delays during the cutting procedure. Neinhuys and Van der Stappen [Nienhuys and van der Stappen 2000] tried to combine a FEM simulation by using an iterative solution to the set of equations with a conjugate gradient method. This method allows for alterations of the mesh topology at run time, as there is no preprocessing required. Vigneron et al. [Vigneron et al. 2004] take advantage of the XFEM method used in fracture mechanics to model cuts and resections in human tissue. XFEM does not require continuity within the mesh element and is thus useful for modeling cracks and cuts. Nonetheless, this can not be done in real time, as large systems of equations must still be solved during the actual simulation. Berkley et al. [Berkley et al. 2004] showed how constraints can be used to simulate suturing and support general real-time displacement-based interaction with finite element models. Wu and Heng [Wu and Pheng-Ann 2004] described a GPU-assisted system that uses coarse models to support limited real-time interaction.

Solving a large set of linear equations takes time. Even the use of iterative solvers is time consuming and rarely yields interactive frame rates. In our work, we incorporate the data from an off-line FEM simulation, trying to overcome the problems of solving these systems of equations at run time.

3 The Algorithm

The proposed algorithm operates in four phases:

- A first preprocessing stage. A FEM is created, simulating the cutting operation over time and in a canonical setup. Further, the locations of the individual elements are recorded at every time step. This stage is described in detail in Section 3.1.
- A second preprocessing stage, in which the data from the FEM simulation is encoded into a DFFD deformation model over time. This deformation is described in Section 3.2, and its encoding is discussed in Section 3.3.
- The real-time cutting stage. While the user is moving the scalpel along the skin, the skin polygons of the mesh are split in order to represent the cut. This step is presented in Section 3.4.
- The deformation stage. Following the cutting operation, the deformation is applied to the polygons around the cut, and

over time, splitting the cut open using the aforementioned deformation function. This stage is detailed in Section 3.5.

In addition, the specific support and integration of the haptic device into this simulation environment is described in Section 3.6.

3.1 Finite Element Model of Skin and Tissue

The finite element formulation used in this research has two separate element types: one for the skin, and one for the soft tissue below it. The skin element is a two-dimensional surface element that is bonded to the soft tissue. It has no stiffness in the transverse direction and is used to introduce tensile skin pre-stresses into the model. The soft tissue element is a volumetric element that smears the subdermal muscles, fat, tendons, etc. into one material representing the bulk behavior of the inhomogeneous continuum. This allows for a more accurate modeling of the skin and subcutaneous tissue, which inherently have very different material properties.

The finite element model is a three-dimensional volume that represents the patch of tissue being explicitly modelled. This volume is discretized with triangular constant strain elements and compatible prismatic wedge elements for the skin and soft tissue respectively. Along the sub-dermal boundaries that connect this volume to the rest of the body, distributed springs simulate the displacement-traction boundary conditions.

The cut path is defined before meshing occurs, and is used to define the shape of the mesh. Elements on the boundaries are aligned with the cut path and duplicate nodes are created along it. The duplicate nodes represent topologically distinct points on either side of the cut path. Without any additional constraints, the model represents the behavior of a tissue patch in which a cut has been introduced. Initial closure of the cut is imposed through the use of displacement constraints, as seen in Figures 1 and 2. These are algebraic constraints enforcing the condition that corresponding points on either side of the cut have similar displacements. As the virtual scalpel traverses the cut path, these constraints are relaxed, allowing the cut to open.

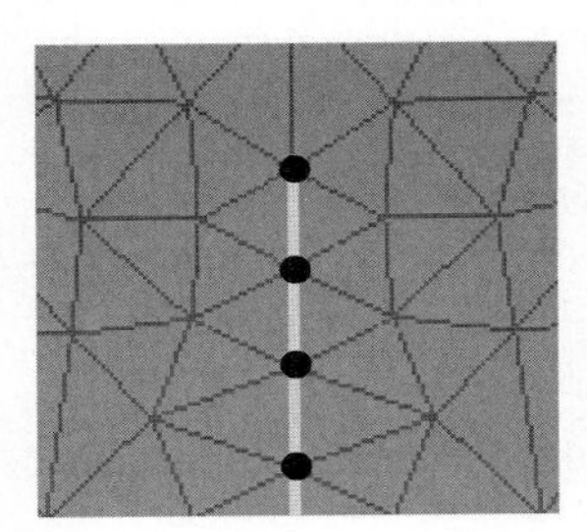

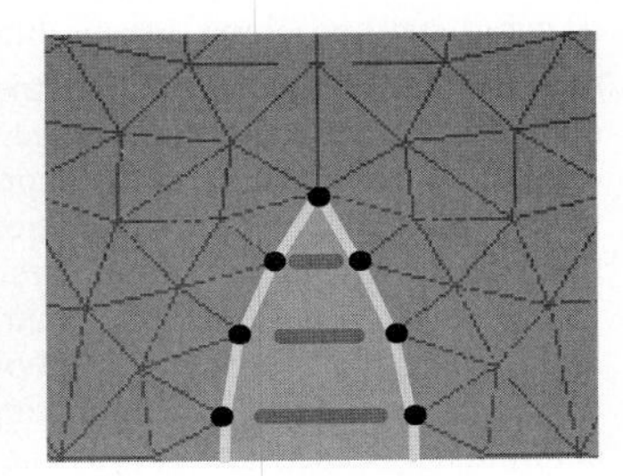

Figure 1: Physical and conceptual representations of the cut while the displacement constraints are active. Yellow (light) lines indicate locations of the cut whereas red (dark) bars indicate the actual displacement constraints.

3.1.1 Governing Equations

In general, the finite element equilibrium equations are written as $\mathbf{Ku} = \mathbf{f}$ where $\mathbf{K}$ is the stiffness matrix assembled from the stiffness matrices of the skin, soft tissue, and spring boundary elements, $\mathbf{u}$ is the vector of nodal displacements and $\mathbf{f}$ is the vector of externally applied loads. By adding to this system the constraints which model tissue separation at the cut path, the original system of linear equations is augmented to form the following 2×2 block system:

$$\begin{bmatrix} \mathbf{K} & \mathbf{C}^T \\ \mathbf{C} & \mathbf{0} \end{bmatrix} \begin{Bmatrix} \mathbf{u} \\ \mathbf{v} \end{Bmatrix} = \begin{Bmatrix} \mathbf{f} \\ \mathbf{0} \end{Bmatrix}. \tag{1}$$

$\mathbf{C}\,\mathbf{u} = \mathbf{0}$ are constraints for enforcing closure of the cut. Each equation expresses the fact that two corresponding points on either side of the cut have the same displacements, in a given direction, when the cut is closed at that point. Algebraically, this may be written for two points A and B as:

$$\sum \phi_i(A)u_i - \sum \phi_i(B)u_i = 0, \tag{2}$$

where $\phi_i(X)$ is the value of the i-th finite element nodal shape function at location X, and the sum is over all nodal degrees of freedom in a given direction. The values $\phi_i(X)$ are the entries of the coefficient matrix $\mathbf{C}$. Given the compact support of finite element shape functions, $\mathbf{C}$ is very sparse. The vector $\mathbf{v}$ is a vector of Lagrange multipliers representing generalized forces necessary for enforcing the constraints (closing the cut).

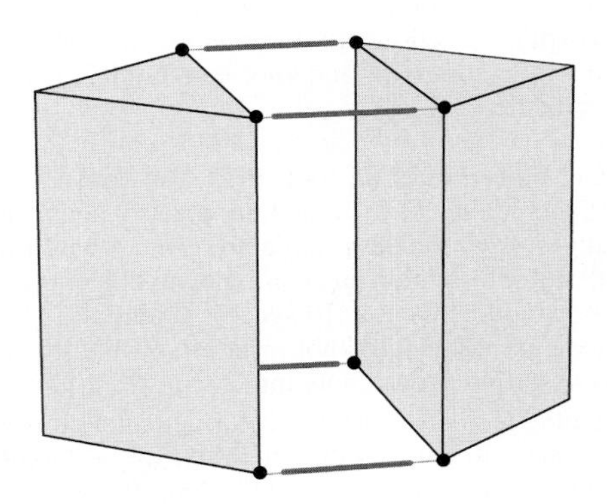

Figure 2: Conceptual image depicting the constraints in three dimensions. These are two elements that surround the cut, where the gap between them is part of the cut, but does not represent its full depth; which is typically modeled using multiple finite element layers. The linkages connecting the two elements represent the rigid displacement constraints that restrict the x,y and z displacements of the two nodes to be the same. Each set of linear prism elements are connected by four nodal constraints.

3.1.2 Solution Methods

Performing a block Gauss elimination on Equation (1) gives the following two equations,

$$\mathbf{CK}^{-1}\mathbf{C}^T\mathbf{v} = \mathbf{CK}^{-1}\mathbf{f} \tag{3}$$

$$\mathbf{Ku} = \mathbf{f} - \mathbf{C}^T\mathbf{v}, \tag{4}$$

where the solution of Equation (3) provides the values of the Lagrange multipliers and the solution of Equation (4) provides the displacement field. To advance the simulation as the tissue is cut, one set of displacement constraints, pertaining to the x, y, and z components of two attached nodes, is removed from $\mathbf{C}$ and Equations (3) and (4) are solved. This process is repeated until the cut is completed.

3.2 The Discontinuous Deformation Model

DFFD extends FFD to support the mapping of continuous domains into discontinuous ranges. Traditionally, FFD defines a mapping from $D \subset \mathbb{R}^3 \Rightarrow \mathbb{R}^3$, warping a region of $\mathbb{R}^3$ into another region of $\mathbb{R}^3$. When such mapping is applied to an embedded geometric object, $\mathcal{O}$, it also deforms its shape to follow the prescribed space warping operation.

FFDs may be defined using arbitrary functions $F : \mathbb{R}^3 \Rightarrow \mathbb{R}^3$. However, due to their robustness and controllability, trivariate tensor product B-spline functions are usually used. Such functions are defined as,

$$F(u,v,w) = \sum_{i=0}^{l}\sum_{j=0}^{m}\sum_{k=0}^{n} P_{ijk}B_i^o(u)B_j^o(v)B_k^o(w), \quad (5)$$
$$(u,v,w) \in [U_{min},U_{max}) \times [V_{min},V_{max}) \times [W_{min},W_{max}),$$

where P_{ijk} are the control points and $B_i^o(u)$ are the univariate basis functions of order o, in all three directions. As a result of using trivariate B-spline functions for deformation, a wealth of modeling tools, such as degree-raising and knot-insertion[Cohen et al. 2001], is made available.

As with any other modeling tool, FFD also possess several limitations. Specifically, FFD cannot support mapping into discontinuous ranges. Nevertheless, there are cases where the ability to model both the deformation and tearing of an object inside a unified framework, would be useful. To specify potential C^{-1} discontinuity into F, we use a standard knot-insertion procedure [Cohen et al. 1980]. By inserting order knots into $t = t_0 \in \{u,v,w\}$ parametric axis, a potential C^{-1} discontinuity is formed along this iso-surface. For brevity and without loss of generality, we will assume henceforth that knot insertion always occur along the v parametric axis at $v = v_0$ and that are no existing knots at $v = v_0$. As a result of inserting order knots into $v = v_0$, new control points are formed that interpolate the iso-surface $F(u, v = v_0, w)$. By manipulating the control points that are on or near iso-surface $F(u, v = v_0, w)$, a rich family of shapes can be modeled. In particular, for a virtual incision application the shapes of an arbitrary scalpel can now be realized.

FFD is commonly used to deform polygonal models. In such cases, the deformation is usually approximated by applying the deformation function, F, to the vertices of the model. However, in the case of DFFD, polygons that cross the discontinuity would not be mapped properly. To ameliorate this problem, the DFFD algorithm would split crossing polygons such that the edges of the polygon are clipped against the plane $v = v_0$. As a result of this clipping operation, new, non-crossing polygons replace the old crossing polygon. The splitting operation occurs in the parametric domain of F, hence the edges of a crossing polygon are always clipped against an axis-aligned plane, making the clipping operation much simpler.

One direct result of the above split operation is that closed models become open. Since for some applications this is an undesirable consequence, the DFFD algorithm should also supply means to seam the cut. For stability, vertices near the cut are translated in v by $\pm\varepsilon$ such that each is mapped to either side of the discontinuity. This operation only guarantees a C^0 continuity between the added geometry and the original one, at the incision location. In the next section, we will show how the results of an FEM simulation from Section 3.1 could be incorporated into the process of modeling the DFFD function.

3.3 Representing the DFFD Over Time

The result of the FEM simulation is a set of points $\{P_i^j\}$ where P_i^j is the location of point i, at frame j. Our interest lies in the volumetrically minimal subset that contains the points surrounding the cut from the first point at which the scalpel breached the skin, advancing along the cut, and until they reach a steady state, when the cut is fully open. We approximate the points' movement over time using a 4-variate (u,v,w,t) smooth B-spline function, which will represent the DFFD over time.

In our implementation, we used 4th order (cubic) B-splines and uniform open-end knot vectors, with the exception of the v-axis knot vector in which we added a discontinuity at the middle of the domain, simulating the cut (which is along the u-axis). This choice of orders, number of control points and knot vectors affects the level of accuracy. For example, increasing the number of control points (i.e., the degrees of freedom) would provide a DFFD that more accurately describes the results of the FEM simulation. A Least Squares (LS) fit problem could be defined which is linear in the 4-variate's control points' coordinates and corresponds to the following set of constraints:

$$\forall i,j \sum_{k,l,m,n} Q_{k,l,m,n}B_k(u_i^j)B_l(v_i^j)B_m(w_i^j)B_n(t_i^j) = P_i^j,$$

where $Q_{k,l,m,n}$ are the 4-variate's control points of indices k,l,m,n, $B_\alpha(\beta)$ are the α'th B-Spline basis functions of the selected orders and knot sequences evaluated at parameter value $\beta \in \{u_i^j, v_i^j, w_i^j, t_i^j\}$ of P_i^j.

The (u_i^j, v_i^j, w_i^j) parametric values of points $\{P_i^j\}$ were taken from the initial frame (i.e. the parametric values of point P_i^j are set to be the Euclidean coordinates of P_i^0), and the t parametric value is set to be j.

This technique's most significant drawback is evident in the case of regions in the deformed space that have too few sampled points and are thus underdetermined. For such regions, the LS algorithm results in control points being reduced to the zero point. In order to avoid such problems we modified the equation system's right-hand side to be $P_i^j - P_i^0$. Thus, the resulting control points' coordinates are actually the difference between the coordinates of the desired control points and the control points coordinates' values for an identity DFFD with identical orders, knot vectors and domain, meaning that the desired result is $I + Q$ where Q is the least squares result and I holds the control points coordinates values for an identity DFFD as specified above, i.e., $I(u,v,w,t) = (u,v,w,t)$. The resulting DFFD, and following the off-line FEM simulation, continuously describes the way the volume of the canonical tissue surrounding the cut deforms over time. An example is depicted in Figure 3.

3.4 Splitting the Geometry

DFFDs can be applied to surface meshes, volumetric data sets and parametric models. For the sake of efficiency, we focus our work on triangular surface meshes. When cutting through a triangular mesh, the most basic operation is triangle splitting. Any triangle that has vertices on both sides of the cut must be split before the DFFD mapping is applied to it. Moreover, vertices on the cut line must be treated with care, as will be discussed shortly. This splitting operation is conducted incrementally as the virtual scalpel advances, one triangle at a time, following the path of the cut, along the skin surface.

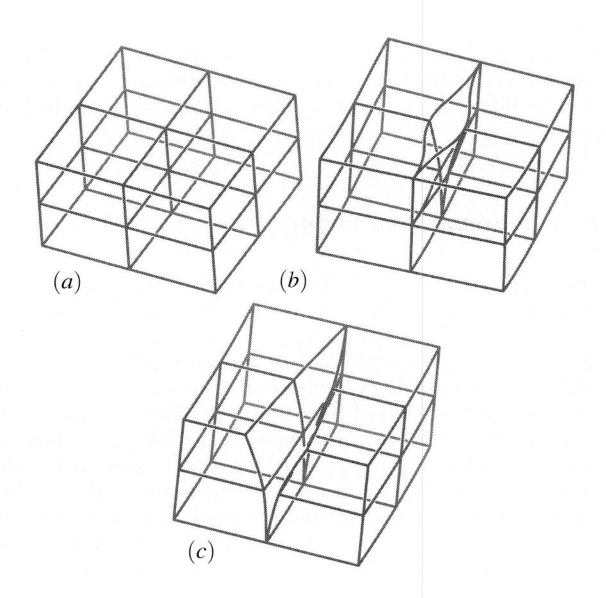

Figure 3: Sampling of the 4-variate DFFD at $t = 0$ (a), $t = 0.2$ (b) and at $t = 0.8$ (c).

The following notations will be used: Let $C(s)$ be an arc-length parametric representation of the cut line following the path of the virtual scalpel. Similarly, $N(s)$ defines the orientation of the virtual scalpel at every time t.

Usually, splitting a triangular face creates three new triangles; one triangle on one side of the cut, and two on the other side (see Figure 4). These three new triangles replace the original face, which is then purged. These original triangles have both entry and exit cut locations, found by calculating the closest point on a segment (the entry or exit edge) to $C(s)$. In addition, we examine how close this linear approximation of the cut line is. If the straight line between the entry and exit locations is sufficiently close to $C(s)$, that triangle is split as just described. Otherwise, the triangle must be subdivided recursively into smaller triangles before the split can take place. Figures 5 shows an example of this more complex case.

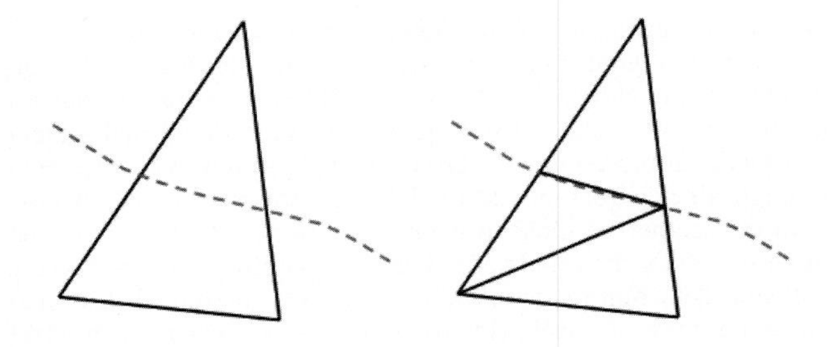

Figure 4: Triangle split, normal operation. The dashed line indicates the cut path.

If the cut passes very close to a vertex, we handle things a bit differently, as the splitting method that was proposed above will create one very small triangle and one very long and thin triangle. In this case, we move the vertex at hand onto the cut, duplicate it, and use one copy on each side of the cut, see Figure 6.

While we only process the skin surface, we also seek to model the deepness of the cut, and model this new geometry on the fly. In order to do this, the entry and exit locations are duplicated a number of times and moved in the direction of $-N(s)$ into the body. Triangles are then used to tessellate these interior vertices, all the way to the bottom of the cut. The resulting geometry can be seen in Figure 7.

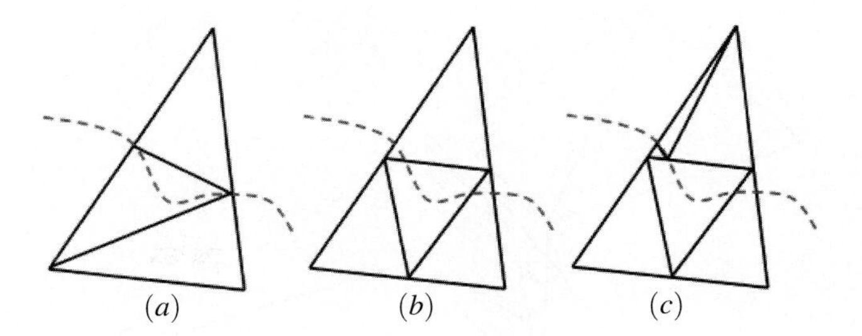

Figure 5: When the straight line between the entry and exit points is not close enough to the cut path, the triangle is subdivided and the algorithm continues. (a) shows the naive split, (b) shows the subdivision, and (c) shows the first child triangle after the split.

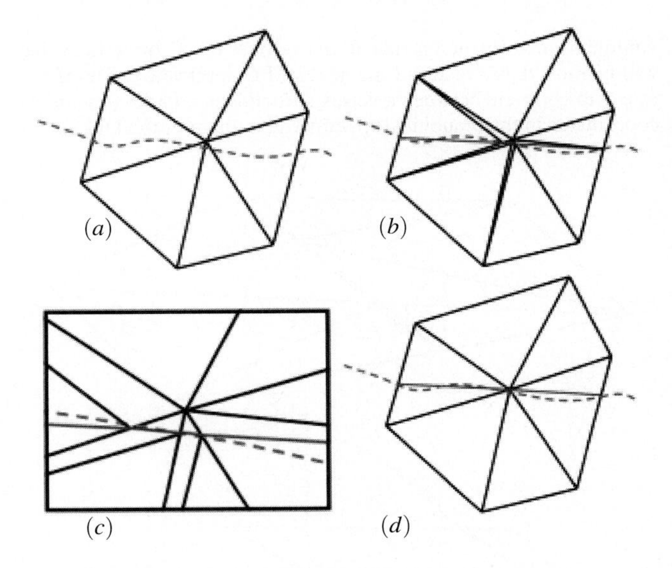

Figure 6: When the scalpel path lies very close to a vertex, the geometry is split in a different manner. The dashed line indicates the scalpel path. (a) shows the initial configuration. (b) shows the splitting of the neighborhood triangles according to the basic algorithm, (c) is an enlargement of the center vertex area. (d) shows our modified solution for the problem, which moves the original vertex to the path and splits only two triangles.

3.5 Time Dependant Deformation of the Geometry

The DFFD is applied at regular time intervals to a list, $\mathscr{L}$, of active vertices around the cut. The active vertices in $\mathscr{L}$ are vertices of the original mesh near the cut, in addition to the vertices created when the triangles around the cut were split. A vertex is entered into $\mathscr{L}$ when the scalpel passes near it (in the case of original vertices) or when it is created (for the new ones), and removed from $\mathscr{L}$ when it is no longer affected by the DFFD application (as is detailed later on). At every time step, $\mathscr{L}$ is updated, and the scalpel path ($C(s)$) and scalpel orientation curve ($N(s)$), are recreated. From these curves and the width of the cut, we reconstruct the incision parametrization volume (IPV), see Figure 8. The IPV is actually the

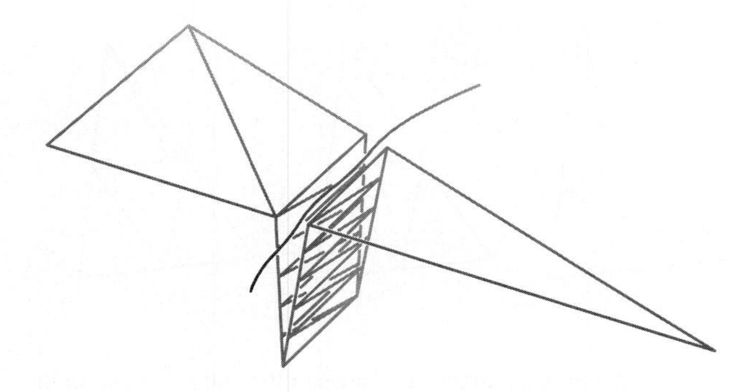

Figure 7: The deepness of the cut polygons modelled and added to the mesh.

volume inside of which reside all the vertices of $\mathscr{L}$, or vertices that will be moved as a result of the next DFFD application. The IPV's role is to correlate between a vertex's Euclidean coordinates and its coordinates in the (canonical) parametric domain of the DFFD.

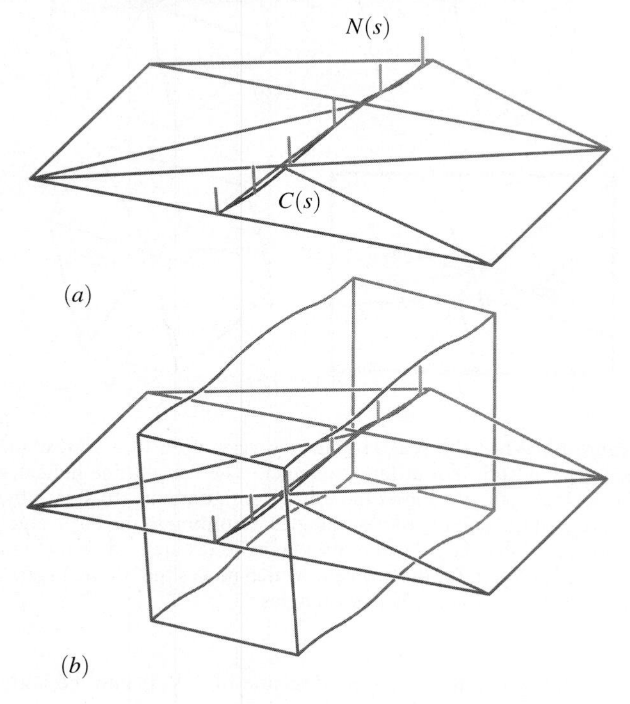

Figure 8: IPV construction. The volume is constructed along the curve, appearing with respective scalpel orientations at regular intervals, as seen in (a). The volume is seen in (b).

The vertices in $\mathscr{L}$ are mapped through the IPV, creating the u,v,w and t coordinates for the DFFD. For every vertex V_i in $\mathscr{L}$, let $v_s = \arg\min_s \|C(s) - V_i\|$. Then, u and w are set by their distance to $C(v_s)$ along $N(v_s)$ and $T(v_s) \times N(s)$, respectively, where $T(s) = C'(s)$. The t parameter is, again, the time in which the vertex was entered into $\mathscr{L}$. Since only parameter values of new vertices, that enter $\mathscr{L}$, need to be recomputed for every DFFD iteration, the procedure is quite efficient.

Hence, we have $(u',v',w')=DFFD(u,v,w,t)$ $= DFFD(IPV^{-1}(x,y,z),t)$. We then apply the IPV to these (u',v',w') parametric coordinates in order to find the new Euclidean location of the vertex.

A vertex $V_i \in \mathscr{L}$, inside the volume surrounding the cut, will move an amount that is the sum of all the small deformations assigned to it in all the iterations of all DFFDs, while V_i is in the incision volume.

3.6 Force Feedback Support

In order to provide a more realistic use for the proposed approach, we enabled a SensAble(tm) PHANToM Desktop(tm) haptic device [SensAble Technologies, Inc.] to work with the system. see Figure 9. The device consists of a pen-sized handle connected to a robotic arm with flexible engine-enforced joints. These allow the arm to move in 6 degrees of freedom (DOF): three spatial coordinates and pitch, roll and yaw. The engines at the arm joints allow the device to apply force to the holder of the handle, providing it with force feedback ability. This pen-sized handle was used as scalpel in our simulation, so when adjusting the haptic resistance correctly, the combination of the visual and haptic user interface provided us with a convincing look and feel of an actual incision process.

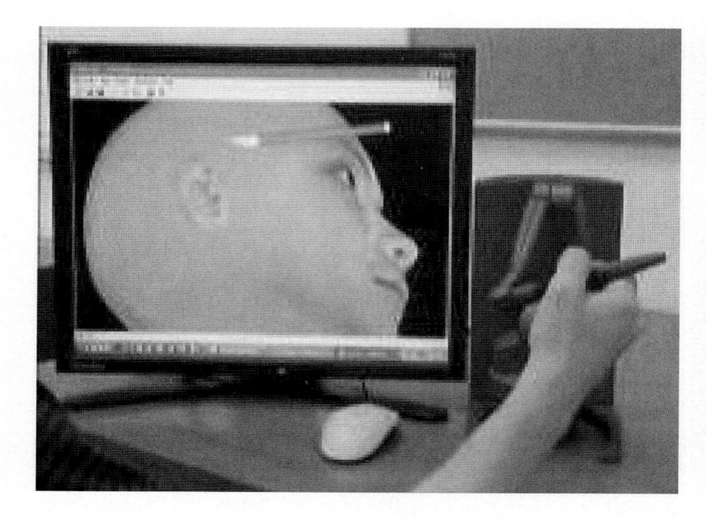

Figure 9: PHANToM Haptic Device in operation.

Due to the sensitivity of the human tactile system, a force feedback device requires update rates of 1kHz in order to provide high definition simulations (see [Tan et al. 1994]). Therefore, our algorithm must handle collision queries between the virtual scalpel and model at these rates. A brute force algorithm, which iterated through all polygons of the model and performed collision tests with the scalpel, resulted in poor frame rates with models of the order of tens of thousands of polygons. Therefore, to speed up the collision detection process, we preprocessed the model data, and created a uniform voxel grid around the model, where each voxel holds a list of polygons that intersect the voxel. This limited our collision queries to only the polygons in the voxels intersecting the scalpel, and effectively reduced our calculations to a few hundred line-polygon intersection tests at most, for every collision query between the virtual scalpel and the model.

Our force feedback model has two stages: a pre-puncture stage and a post-puncture stage. When light forces are applied to the skin, a force model is used to simulate the scalpel touching the skin, without penetrating it [Terzopoulos et al. 1987]. In this model, the force is a function of the depth to which the scalpel is pushed in the direction normal to the skin. When the scalpel pierces the skin, this force is replaced by a viscous drag force (according to Stokes' model) $-bT(t)$, [Terzopoulos et al. 1987] (where b is an approximated constant of viscosity extracted from experiments and $T(t)$ is the the speed vector of the scalpel's cutting path). This simulates the movement of the scalpel while cutting through skin or flesh. Constants were chosen empirically, to provide realistic touch and cut force feedback. Eventually, the 6DOF control over the scalpel, along with the force feedback feature, provided us with intuitive control over the scalpel, and raised the simulation to a higher level of realism.

4 Results

Figures 10, 11 and 12 show snapshots from a few incision simulations generated with our implementation. In Figure 10, an incision simulating a brow-lift is shown, Figure 11 shows a typical incision made in a face-lift operation and Figure 12 shows a side view of an incision being made across the bridge of the nose, displaying a side view of the cut.
The simulations were performed on a P4-2.8GHz desktop computer with 1GB of RAM. The original head model consists of 12108 polygons. The cuts shown in Figures 10, 11 and 12 added 453, 1443 and 1082 polygons, respectively, including the polygons representing the deepness of the cut.

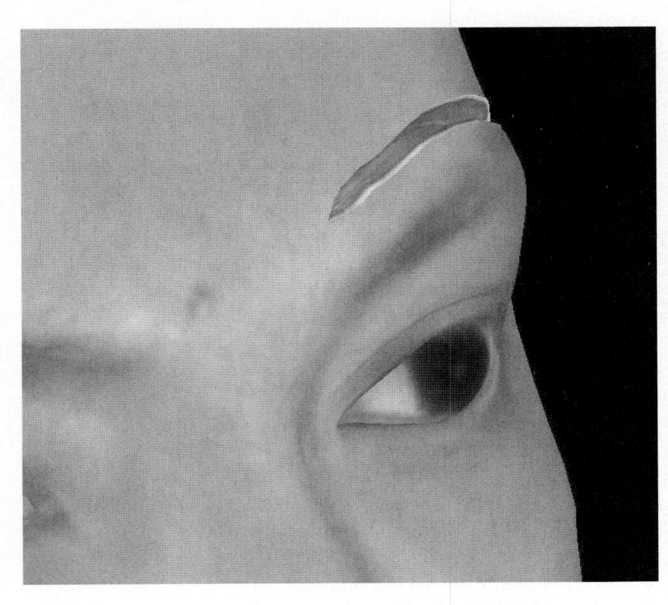

Figure 10: A brow-lift cut simulation.

Figures 10,11,12 and 13 are all from a real-time film recording of the actual use of the force feed-back scalpel. The full incision video and other results are available at:
http://www.cs.technion.ac.il/~guysela/incision.htm

5 Conclusions and Future Work

We have presented an enhancement to our previous work [Sela et al. 2004] in which we proposed a method to perform real-time incision simulation using 3D DFFDs. In this enhancement, we detail how to compute an incision simulation using an off-line FEM simulation and incorporate this simulation's results into a 4D DFFD representation over time. Finally, we demonstrated how this 4D DFFD can

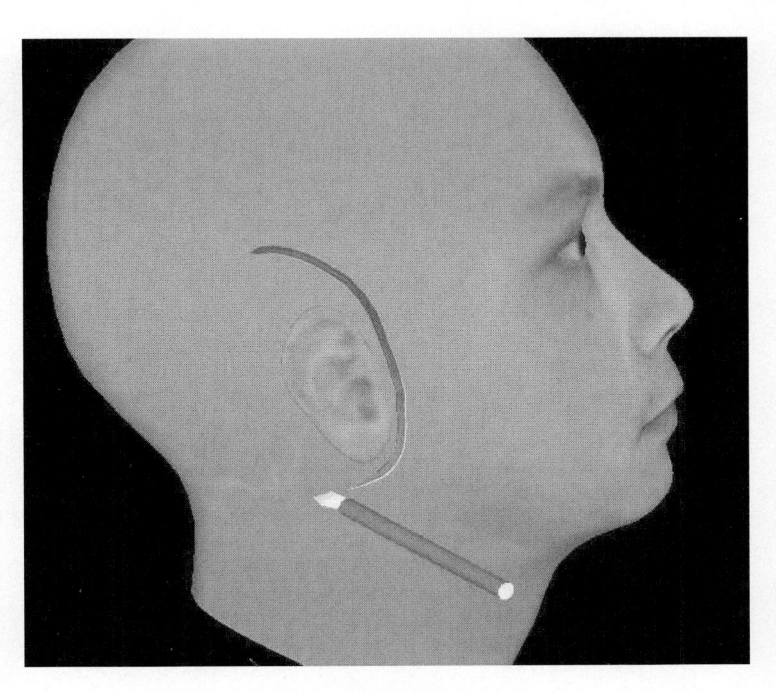

Figure 11: A face-lift cut simulation.

be used to simulate real-time incisions on a 3D model using a haptic device. This method is modular and, therefore, flexible enough to accommodate different methods of incision simulation as well as different model representations. It is also computationally simple enough to yield real-time frame rates on desktop computers.

Future extensions may include the usage of more than one simulation results, i.e., using different FEM simulations and corresponding DFFDs for different types of incisions, handling incisions with variable width and depth, depending on material properties, and the application of the method to volumetric data. Another possible extension could be the exploration of different surface mesh cutting techniques. Our system uses a recursive, non-adaptive scheme, and an adaptive one could reduce the number of polygons added to the model at the expense of additional computation.

Furthermore, the presented framework can be extended to model different cut shapes, representing bent scalpels, and altogether different deformations of the model such as bend, protrude, twist, etc., functioning as a cutting modeling tool.

Another potential use of the technology presented in this paper is to convey the results of large scale finite element models to scientists and engineers. Frequently numerical models in many important application domains such as structural mechanics simulations, wave propagation in heterogeneous media, weather prediction, etc. generate multiple, large-scale 3D data sets. The multiple data sets arise from simulations that are run with different parameters, discontinuities, boundary conditions, interfaces, etc, in order to fully quantify a phenomenon or evaluate a design. An important task for the scientist/engineer user is to browse the massive data generated from these simulations with the goal of finding critical patterns and features and assessing changes in the model response due to changes in loading, size of cracks, contact regions and a myriad of application-specific model parameters. The task is tedious and time consuming and in the current state of the art often involves either a round-trip call to the underlying finite element simulation, or manual access and parsing of data files from a separate storage subsystem. This prevents real-time, free style, exploration of the model response. Further, and perhaps more importantly, this *staccato* mode of ex-

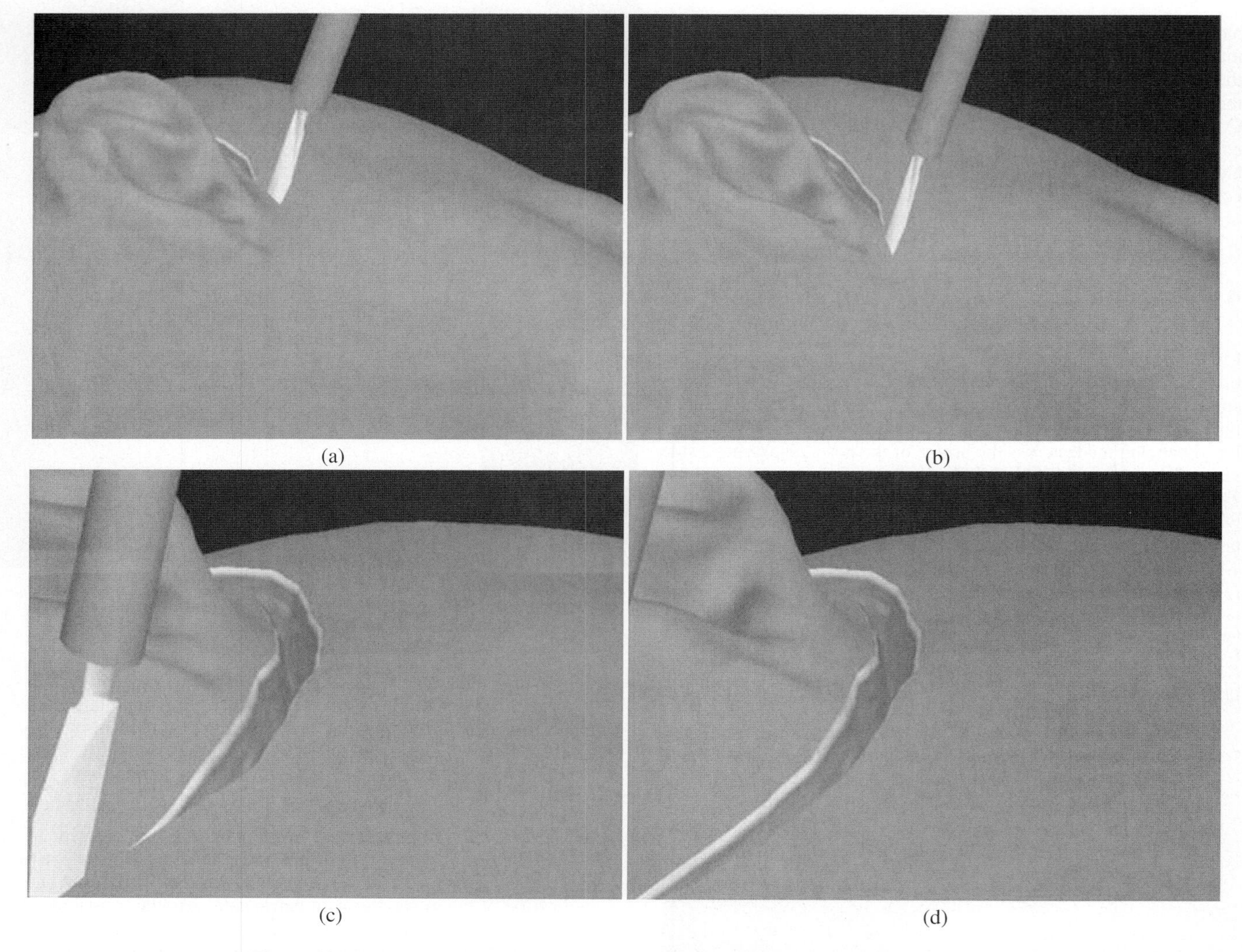

Figure 13: Four snapshots of the interactive incision process of the face-lift simulation.

ploring response data is, by its nature, a hinderance for understanding how the response is affected by changes in model structure, topology, boundary conditions and other parameters. Discontinuous FFD presents a practical mechanism for storing finite element data sets with discontinuities and conveying them effectively to end users. We have shown that by using DFFD, even data sets with spatial discontinuities can be encoded and then replayed, rendered and manipulated at interactive speeds. This allows users to explore the data produced by the numerical simulations and quickly find the specific spatial regions, combination of parameters, or critical lengths of contacts/discontuities, etc. that are of most interest. This information can then be used to access portions of the underlying raw data for additional, more detailed analysis. DFFD provides the ability to essentially summarize and capture large amounts of finite element data sets, while allowing users to visualize and browse warped data, possibly with discontinuities and then "zoom in" to access specific data items at interactive rates. This is a key supporting technology for the development of scientific computing applications.

One drawback of the system is the inability to deal with more complex situations, such as large lacerations, tissue removal or the creation of a skin flap. Another limitation is due to the fact that our FE model assumes that the entire simulated volume is comprised of tissue without the presence of, say, bones. As the framework performs the FEM calculation only once, it is not possible to encode such information into the FFD since the location and orientation of a bone is unknown during preprocessing. One more noteworthy point is the fact that although we used a multivariate B-Spline function for encoding the FEM data, it is by no means the only method. Any other representation that captures the simulation data over time would suffice as long as it can be retrieved quickly enough when required in real-time for fast calculation of the vertices' new locations.

6 Acknowledgments

This research was support in part by the Israeli Ministry of Science Grant No. 01-01-01509, in part by European FP6 NoE grant 506766 (AIM@SHAPE), and partially by the fund for promotion of research at the Technion, Haifa, Israel.

The third author would like to thank the Department of Energy's Computational Science Graduate Fellowship, and the Krell Institute, for funding a portion of this research.

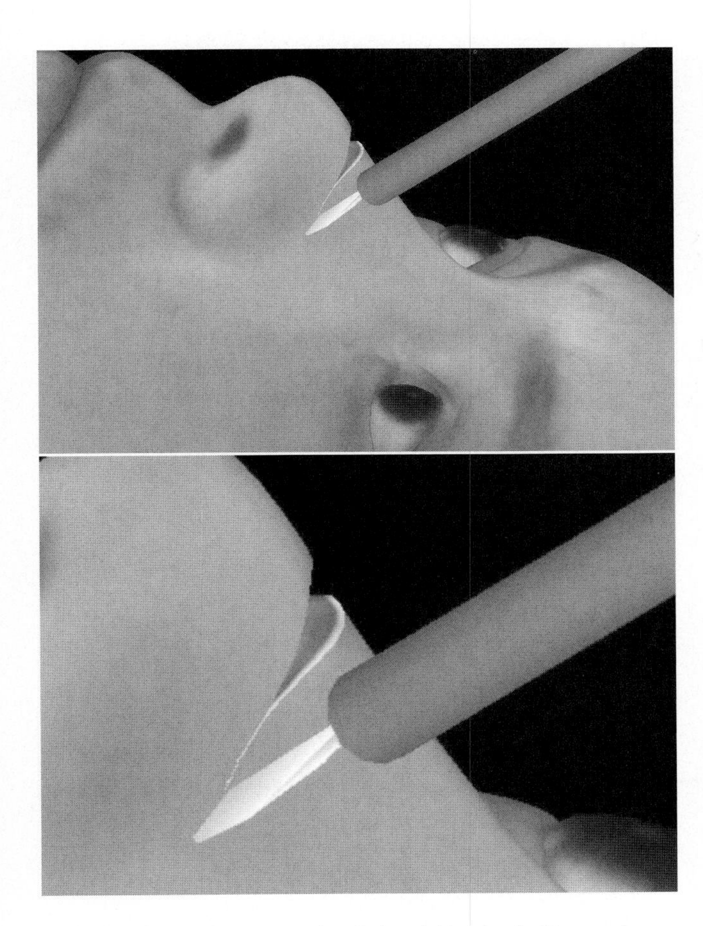

Figure 12: Rhinoplasty cut simulation (side view). Bottom image presents a zoom of the relevant area of the top image.

The face model was retrieved from the 3D-cafe web site, www.3dcafe.com. All the illustrations were prepared with the aid of the Irit solid modeling system, www.cs.technion.ac.il/~irit.

We would like to thank Jihad El-Sana who allowed us to use his PHANToM(tm) device for the duration of this project.

References

BERKLEY, J., TURKIYYAH, G., BERG, D., GANTER, M. A., AND WEGHORST, S. 2004. Real-time finite element modeling for surgery simulation: An application to virtual suturing. *IEEE Trans. Vis. Comput. Graph. 10*, 3, 314–325.

BIELSER, D., AND GROSS, M. H. 2000. Interactive simulation of surgical cuts. In *Proceedings of Pacific Graphics 2000*, I. C. S. Press, Ed., 116 –125.

BIELSER, D., MAIWALD, V. A., AND GROSS, M. H. 1999. Interactive cuts through 3-dimensional soft tissue. In *Computer Graphics Forum (Eurographics '99)*, P. Brunet and R. Scopigno, Eds., vol. 18(3), 31–38.

BRO-NIELSEN, M., AND COTIN, S. 1996. Real-time volumetric deformable models for surgery simulation using finite elements and condensation. *Computer Graphics Forum 15*, 3, 57–66.

BRO-NIELSEN, M. 1998. Finite element modeling in surgery simulation. *Proceedings of the IEEE 86*, 3, 490–503.

BRUYNS, C., AND SENGER, S. 2001. Interactive cutting of 3d surface meshes. *Computers & Graphics 25*, 4, 635–642.

CAMARA, O., DELSO, G., BLOCH, I., AND FOEHRENBACH, H. 2002. Elastic thoracic registration with anatomical multiresolution. In *International Journal of Pattern Recognition and Artificial Intelligence*, World Scientific. Special Issue on Correspondence and Registration Techniques.

COHEN, E., LYCHE, T., AND RIESENFELD, R. 1980. Discrete bsplines and subdivision techniques in computer aided geometric design and computer graphics. *Computer Graphics and Image Processing 14*, 87–111.

COHEN, E., RIESENFELD, R. F., AND ELBER, G. 2001. *Geometric modeling with splines: an introduction*. A. K. Peters, Ltd., Natick, MA, USA.

DELINGETTE, H. 1998. Towards realistic soft tissue modeling in medical simulation. *Proceedings of the IEEE : Special Issue on Surgery Simulation* (Apr.), 512–523.

ELLENS, M. S., AND COHEN, E. 1995. An approach to c^{-1} and c^0 feature lines. *Mathematical Methods for Curves and Surfaces*, 121–132.

EUGENE, M., TARDY, J., THOMAS, J. R., AND BROWN, R. 1995. *Aesthetic Facial Surgery*. Mosby, January.

FOREST, C., DELINGETTE, H., AND AYACHE, N. 2002. Cutting simulation of manifold volumetric meshes. In *Medical Image Computing and Computer-Assisted Intervention (MICCAI)*, Springer, Tokyo, Japan, T. Dohi and R. Kikins, Eds., vol. 2489, 235–244.

GANOVELLI, F., AND O'SULLIVAN, C. 2001. Animating cuts with on-the-fly re-meshing. In *Eurographics 2001 - short presentations*, 243–247.

GANOVELLI, F., CIGNONI, P., MONTANI, C., AND SCOPIGNO, R. 2000. A multiresolution model for soft objects supporting interactive cuts and lacerations. ??–??

MASUTANI, Y., AND KIMURA, F. 2001. Modally controlled free form deformation for non-rigid registration in image-guided liver surgery. In *Proceedings Medical Image Computing and Computer-Assisted Intervention (MICCAI)*, Springer, Utrecht, NL, W. J. Niessen and M. A. Viergever, Eds., vol. 2208 of *Lecture Notes in Computer Science*, 1275–1278.

MOR, A., AND KANADE, T. 2000. Modifying soft tissue models: Progressive cutting with minimal new element creation. In *Medical Image Computing and Computer-Assisted Intervention (MICCAI).*, Springer-Verlag, vol. 1935, 598–607.

NIENHUYS, H.-W., AND VAN DER STAPPEN, A. F. 2000. Combining finite element deformation with cutting for surgery simulations. In *EuroGraphics Short Presentations*, A. de Sousa and J. Torres, Eds., 43–52.

NIENHUYS, H.-W., AND VAN DER STAPPEN, A. F. 2001. A surgery simulation supporting cuts and finite element deformation. In *Medical Image Computing and Computer-Assisted Intervention*, Springer-Verlag, Utrecht, The Netherlands, W. J. Niessen and M. A. Viergever, Eds., vol. 2208 of *Lecture Notes in Computer Science*, 153–160.

NIENHUYS, H.-W., AND VAN DER STAPPEN, A. F. 2004. A delaunay approach to interactive cutting in triangulated surfaces.

In *Fifth International Workshop on Algorithmic Foundations of Robotics*, Springer-Verlag Berlin Heidelberg, Nice, France, J.-D. Boissonnat, J. Burdick, K. Goldberg, and S. Hutchinson, Eds., vol. 7 of *Springer Tracts in Advanced Robotics*, 113–129.

SCHEIN, S., AND ELBER, G. 2005. Discontinous free-form deformation. In *The 12th Pacific Conference on Graphics and Applications (PG)*, 227–236.

SCHNABEL, J. A., RUECKERT, D., QUIST, M., BLACKALL, J. M., CASTELLANO-SMITH, A. D., HARTKENS, T., PENNEY, G. P., HALL, W. A., LIU, H., TRUWIT, C. L., GERRITSEN, F. A., HILL, D. L. G., , AND HAWKES, D. J. 2001. A generic framework for non-rigid registration based on nonuniform multi-level free-form deformations. In *Proceedings in Medical Image Computing and Computer-Assisted Intervention (MICCAI)*, W. J. Niessen and M. A. Viergever, Eds., vol. 2208, 573–581.

SEDERBERG, T. W., AND PARRY, S. R. 1986. Free-form deformation of solid geometric models. *Computer Graphics 20* (Aug), 151–160.

SELA, G., SCHEIN, S., AND ELBER, G. 2004. Real-time incision simulation using discontinuous free form deformation. In *Medical Simulation: International Symposium, ISMS 2004, Cambridge,MA, USA, June 17-18, 2004*, Springer, S. Cotin and D. N. Metaxas, Eds., vol. 3078 of *Lecture Notes in Computer Science*, 114–123.

SENSABLE TECHNOLOGIES, INC. PHANTOM haptic device. `http://www.sensable.com`

SEYMOUR, N. E., GALLAGHER, A. G., ROMAN, S. A., O'BRIEN, M. K., BANSAL, V. K., ANDERSEN, D. K., AND SATAVA, R. M. 2002. Virtual reality training improves operating room performance: Results of a randomized, double-blinded study. *Annals of Surgery 236*, 4, 458–464.

SHI-MIN, H., HUI, Z., CHIEW-LAN, T., AND JIA-GUANG, S. 2001. Direct manipulation of ffd: Efficient explicit solutions and decomposible multiple point constraints. *The Visual Computer 17*, 6, 370–379.

TAN, H., SRINIVASAN, M., EBERMAN, B., AND CHANG, B., 1994. Human factors for the design of force-reflecting haptic interfaces.

TERZOPOULOS, D., PLATT, J., BARR, A., AND FLEISCHER, K. 1987. Elastically deformable models. *Computer Graphics (Proc. SIGGRAPH'87) 21*, 4, 205–214.

VIGNERON, L. M., VERLY, J. G., AND WARFIELD, S. K. 2004. Modelling surgical cuts, retractions, and resections via extended finite element method. In *Proceedings of the 7th Int. Conf on Medical Image Computing and Computer-Assisted Intervention - MICCAI 2004 (2)*, Springer Verlag, Saint-Malo, France, C. Barillot, D. Haynor, and P. Hellier, Eds., vol. 3217 of *LNCS*, 311–318.

WU, W., AND PHENG-ANN, H. 2004. A hybrid condensed finite element model with gpu acceleration for interactive 3d soft tissue cutting. *Computer Animation and Virtual Worlds (CAVW) Journal 15*, 3-4, 219–227.

Conformal Virtual Colon Flattening

Wei Hong Xianfeng Gu Feng Qiu Miao Jin Arie Kaufman

Center for Visual Computing (CVC) and Department of Computer Science *
Stony Brook University
Stony Brook, NY 11794-4400, USA

(a) Holomorphic 1-form

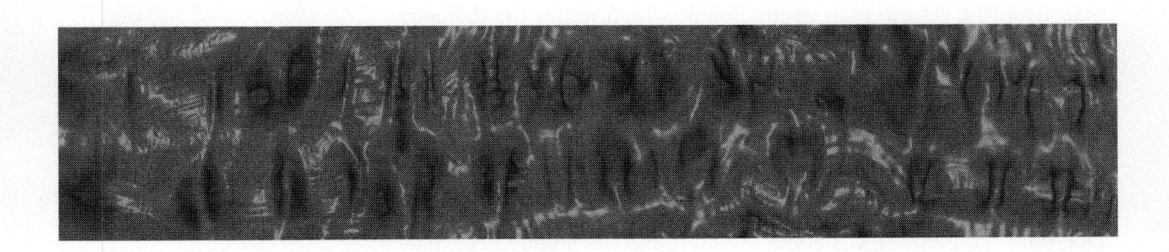

(b) Conformal virtual colon flattening

Figure 1: Conformal virtual colon flattening: (a) Illustrates the holomorphic one-form on the colon surface by texture-mapping a checker board image. (b) Exhibits the conformal flattening induced by (a).

Abstract

We present an efficient colon flattening algorithm using a conformal structure, which is angle-preserving and minimizes the global distortion. Moreover, our algorithm is general as it can handle high genus surfaces. First, the colon wall is segmented and extracted from the CT data set of the abdomen. The topology noise (i.e., minute handle) is located and removed automatically. The holomorphic 1-form, a pair of orthogonal vector fields, is then computed on the 3D colon surface mesh using the conjugate gradient method. The colon surface is cut along a vertical trajectory traced using the holomorphic 1-form. Consequently, the 3D colon surface is conformally mapped to a 2D rectangle. The flattened 2D mesh is then rendered using a direct volume rendering method accelerated with the GPU. Our algorithm is tested with a number of CT data sets of real pathological cases, and gives consistent results. We demonstrate that the shape of the polyps is well preserved on the flattened colon images, which provides an efficient way to enhance the navigation of a virtual colonoscopy system.

Keywords: Conformal Mapping, Direct Volume Rendering, Virtual Colonoscopy

*Email: {weihong|gu|qfeng|mjin|ari}@cs.sunysb.edu

SPM 2006, Cardiff, Wales, United Kingdom, 06–08 June 2006.
© 2006 ACM 1-59593-358-1/06/0006 $5.00

1 Introduction

Virtual colonoscopy uses computed tomographic (CT) images of a patient's abdomen and a virtual fly-through visualization system [Hong et al. 1997] that allows the physician to navigate within a 3D model of the colon searching for polyps, the precursors of cancer. Virtual colonoscopy has been successfully demonstrated to be more convenient and efficient than the traditional optical colonoscopy. However, because of the length of the colon, inspecting the entire colon wall is time consuming and prone to errors. Moreover, polyps behind folds may be hidden, which results in incomplete examinations.

Virtual dissection is an efficient visualization technique for polyp detection, in which the entire inner surface of the colon is displayed as a single 2D image. The straightforward method [Balogh et al. 2002; Wang and Vannier 1995] starts with uniformly resampling the colonic central path. At each sampling point, a cross section orthogonal to the path is computed. The central path is straightened and the cross sections are unfolded and remapped into a new 3D volume. The isosurface is then extracted and rendered. In this method, nearby cross sections may overlap at high curvature regions. As a consequence, a polyp might appear twice or be missed completely in the flattened image. Balogh et al. [2002] have presented an iterative method to correct cross sections, using two consecutive ones at a time. Wang et al. [1998; 1999] have used electrical field lines generated by a local charged path to generate curved cross sections instead of planar sections. If the complete path is charged, then the cross sections tend to diverge, avoiding overlaps. However, due to the expensive computation of the global charge, the authors only locally charge the path, which cannot guarantee that the curved cross sections do not intersect each other any more.

Paik et al. [2000] have used cartographic projections to project the whole solid angle of the camera. This approach samples the solid angle of the camera and maps it onto a cylinder, which is mapped finally to the image. However, this method causes distortions in shape. Bartrolí et al. [2001b] have proposed a method to move a camera along the central path of the colon. For each camera position a small cylinder tangent to the path is defined. Rays starting at the cylinder axis and being orthogonal to the cylinder surface are traced. The cylinder is then opened and mapped to a 2D image. The result is a video where each frame shows the projection of a small part of the inner surface of the colon onto the cylinder. This avoids the appearance of double polyps since intersections can only appear between different frames. However, this approach does not provide a complete overview of the colon. They have presented a new two step technique to deal with double appearance of the polyps and nonuniform sampling problems [Bartrolí et al. 2001a]. First, curved rays are cast along the negative gradient of the distance map from the central path of the colon, which return the distance between the camera and the intersection points on the colon surface. Then, the height field is unfolded and the nonlinear 2D scaling is applied to achieve area preservation. However, it is important to this method that the central path is smooth and has as many linear segments as possible.

Haker et al. [2000] have proposed a method based on the discretization of the Laplace-Beltrami operator to flatten the colon surface onto the plane in a manner which preserves angles. The flattened colon surface is colored according to its mean curvature. A morphological method is used to remove minute handles resulting from the segmentation algorithm, because their algorithm requires the input surface to be a topologically open-ended cylinder. However, the color-coded mean curvature of the extracted surface is not efficient for polyp identification, and it requires a highly accurate and smooth surface mesh to achieve a good mean-curvature calculation. Furthermore, our method maps the colon surface to a planar rectangle, while their method maps the colon surface to a planar parallelogram.

We propose a novel method for colon flattening by computing the conformal structure of the surface, represented as a set of holomorphic 1-form basis. It has the following advantages: (1) The algorithm is rigorous and theoretically solid, which is based on the Riemann surface theory and differential geometry; (2) It is general, so it can handle high genus surfaces; (3) The global distortion from the colon surface to the parametric rectangle is minimized, which is measured by harmonic energy; (4) It is angle preserving, so the shape of the polyps is preserved; (5) The topology noise is removed automatically by our shortest loop algorithm. Combined with the direct volume rendering method, the flattened 2D colon image provides an efficient way to enhance virtual colonoscopy systems.

The remainder of this paper is organized as follows. The shortest loop algorithm for topological denoising is presented in Section 2. The algorithm to flatten the colon surface with conformal mapping is discussed in Section 3. The direct volume rendering algorithm for the flattened colon surface is described in Section 4. The implementation and experimental results are reported in Section 5. In Section 6, concluding remarks are drawn, and the future work of this subject is summarized.

2 Topological Denoising

The colon surfaces reconstructed from a CT data set usually have complicated topologies caused by the noise and inaccuracy of the reconstruction methods. In general several spurious handles will be introduced to a surface. This topological noise complicates our flattening algorithm, and introduces large distortions.

It is challenging to locate these handles and remove them using special "topology surgery". El-sana and Varshney [1997] have proposed a topology controlled simplification method for polygonal models. Tiny tunnels are identified by rolling a sphere with small radius over the object. Guskov and Wood [2001] have presented a local wave front traversal algorithm to discover the local topologies of the mesh and identify features such as small tunnels. The mesh is then cut and sealed along non-separating cuts, reducing the topological complexity of the mesh. These methods are efficient for tiny handle identification. However, we find that handles are not tiny in our colon data sets as shown in Figure 2. Our approach is different in that it identifies handles by locating the shortest loop for each homotopy class.

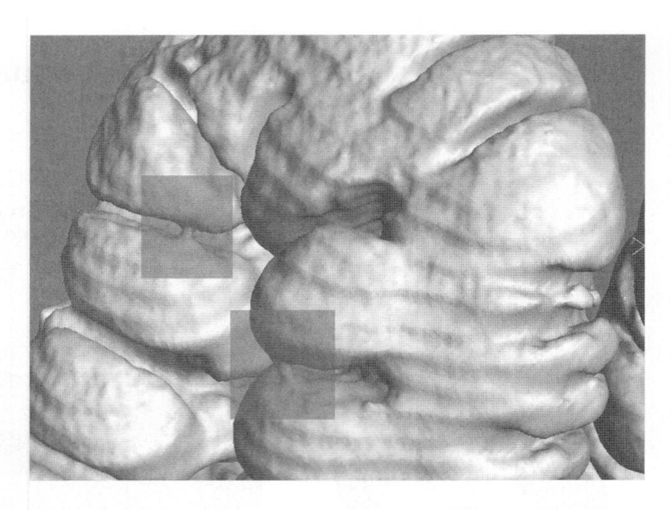

Figure 2: A zoom-in view of a colon surface with two handles.

2.1 Handle Identification

Intuitively, the topology of a closed oriented surface is determined by its number of handles (genus). Two closed curves are *homotopic* if they can deform to each other on the surface. Homotopic equivalence classes form the so-called *homotopy* group, which has finite generators, i.e. *homotopy basis*. Each handle corresponds to two generators. A handle can be removed by cutting the handle along one of its generators, and filling the resulting holes as shown in Figure 3.

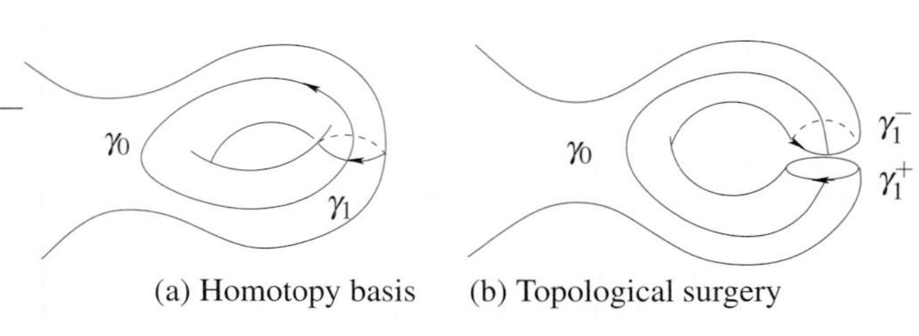

(a) Homotopy basis (b) Topological surgery

Figure 3: Homotopy basis and topological surgery.

In order to remove a handle, it is highly desirable to locate the shortest loop. It is natural to compute the shortest loop using *universal covering space*. Suppose that $\bar{M}$ and M are two surfaces, then $(\bar{M}, \pi)$ is said to be a covering space of M if $\pi : \bar{M} \rightarrow M$ is a surjective continuous map with every $p \in M$ having an open neighborhood U such that every connected component $\pi^{-1}(U)$ is mapped homeomorphically onto U by π. If $\bar{M}$ is simply connected, then it is said to be a *universal covering space* of M. A simply connected region $\tilde{M} \subset \bar{M}$ is called a *fundamental domain*, if the restriction of π on $\tilde{M}$ is bijective. Intuitively, one can slice M along some curve set (cut graph) to obtain a topological disk (a fundamental domain), and glue fundamental domains coherently to form the universal covering space. For any point $p \in M$, its preimages are the discrete set $\pi^{-1}(p) = \{\bar{p}_0, \bar{p}_1, \bar{p}_2, \bar{p}_3 \cdots\} \subset \bar{M}$. If $\bar{\gamma}_k$ is a curve connecting $\bar{p}_0$ and $\bar{p}_k$ in the universal covering space $\bar{M}$, then $\gamma_k = \pi(\bar{\gamma}_k)$ is a closed loop on M. By going through all end points $\bar{p}_k$, γ_k goes through all

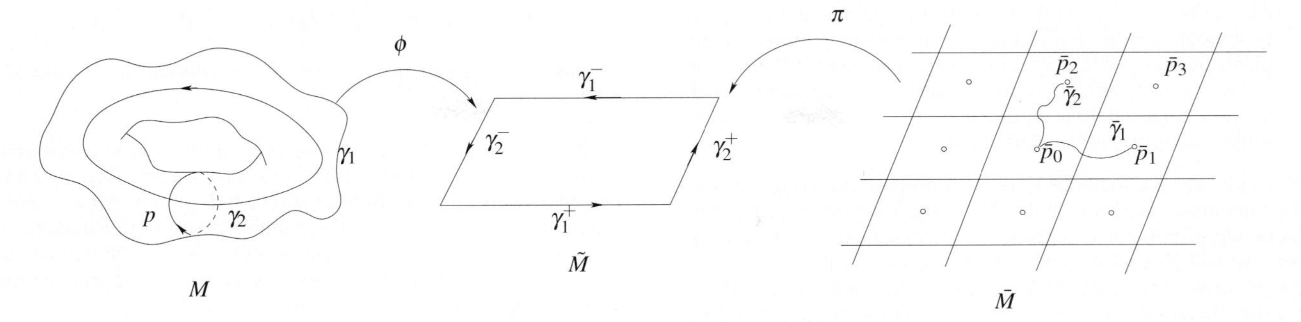

Figure 4: Topology concepts: Two curves γ_1, γ_2 on a surface M form a *cut graph*. M is sliced open along the cut graph to become a *fundamental domain* $\tilde{M}$, γ_i is mapped to γ_i^+ and γ_i^-. By gluing many copies of $\tilde{M}$ such that γ_i^+ is glued with γ_i^{-1}, the *universal covering space* $\bar{M}$ can be obtained. $\pi : \bar{M} \to M$ is the projection map. Any point p on M has a discrete preimage set $\pi^{-1}(p) = \{\bar{p}_0, \bar{p}_1, \bar{p}_2, \cdots\}$. Any closed curves through p on M are lifted as curve segments connecting two points in $\pi^{-1}(p)$, e.g., γ_1 is lifted as $\bar{\gamma}_1$, γ_2 is lifted as $\bar{\gamma}_2$. The shortest loops on M correspond to the shortest path on $\bar{M}$.

homotopy classes. In order to find the shortest loop γ_k, we can find the shortest path $\bar{\gamma}_k$ in the universal covering space instead. Figure 4 demonstrates the concepts of fundamental domain and universal covering space using a genus one surface. It illustrates the idea of lifting a loop to a path and converting the shortest loop problem to the shortest path problem. In computational topology, the algorithms to compute cut graph, homotopy basis, and fundamental domain are well developed [Dey and Schipper 1995; Lazarus et al. 2001; Gu et al. 2002; Erickson and Har-Peled 2003].

However, it has been proven by Erickson and Har-Peled [2003] that this problem is NP-hard. Erickson and Whittlesey [2005] have presented a greedy algorithm to compute the shortest system of loops in $O(nlogn)$ time. However, each loop may not be the shortest loop in its *homotopy* group. Given a system of loops, Colin de Verdière and Lazarus [2005] have proposed a method to compute the shortest simple loop homotopic to a given simple loop (a loop without self intersection). However, the shortest loop within each homotopy group may not be simple. In our case, the surfaces extracted from the segmented colon data sets usually only have a small number of handles as shown in Figure 2. In order to compute the shortest loop, we first simplify the mesh while preserving the topology of the finest mesh. A finite portion of the universal covering space is constructed using the coarsest mesh. The shortest loop is computed in the universal covering space and lifted back to the finest mesh, which approximates the shortest loop on the finest mesh. Our experiments show that this algorithm is manageable in our case.

2.2 Denoising Algorithm

The main procedure of our denoising algorithm is described as follows:

1. Compute the cut graph and homotopy basis.
2. Simplify the cut graph, then slice along the simplified cut graph to form the fundamental domain.
3. Glue finite copies of the fundamental domain coherently to construct a finite portion of the universal covering space.
4. Compute the shortest loop by finding the shortest path in the universal covering space.

After the shortest loop is obtained, we slice the mesh along the loop and fill the holes to remove a handle. This procedure is repeated until all handles are removed.

3 Conformal Flattening

In our method, the colon surface is conformally mapped to a planar rectangle. Conformal maps are extremely valuable for real applications because of their special properties:

- Conformal maps are *angle preserving* (*local shape preserving*). Because analytic functions are angle preserving, therefore by definition, conformal maps preserve angles. For example, any two intersecting curves γ_1 and γ_2 are mapped to $f(\gamma_1)$ on M_2 and $f(\gamma_2)$ by a conformal map f, then the intersection angle between γ_1 and γ_2 equals to the intersection angle between $f(\gamma_1)$ and $f(\gamma_2)$. Physicians usually identify polyps based on the shape of polyps. Polyps can still be identified based on their shape in the flattened colon image because of local shape preserving.
- Conformal maps minimize elastic energy (*harmonic energy*). One can treat M_1 as a rubber surface, the mapping to another surface will introduce stretching distortion and generate the elastic energy. It has been proven [Jost 2002] that conformal maps minimize the harmonic energy. It is highly desirable in practice to find the best match between two surfaces which minimize the distortion.
- Conformal maps are *intrinsic*. Conformal maps are determined by the metric, not the embedding. For example, one can change a surface by rotation, translation, folding, or bending without stretching, and the conformal parameterization is invariant. This is valuable for surface registration purpose.
- Conformal maps are *stable* and easy to compute. Computing conformal maps is equivalent to solving an elliptic geometric PDE [Schoen and Yau 1997], which are stable and insensitive to the noise and the resolution of the data. If two surfaces are similar to each other, then the corresponding conformal maps are similar also.
- Conformal parameterization simplifies geometric processing from 3D to 2D. By parameterizing a surface, we map it to

the planar domain with local shape preservation. Some of the 3D geometric features are carried by the mapping with high fidelity. For example, Figure 9 illustrates the polyp on a colon surface both in 3D and in the conformal parameter domain. It is obvious that the shape of the polyp is well preserved on the plane. It is easier to process in the planar domain than in the 3D domain. Furthermore, many differential operators (such as the Laplace-Beltramin operator) are in the simplest form under conformal parameterization.

Recently, several methods have been proposed to compute conformal mapping. Desbrun et al. [2002] and Lévy et al. [2002] developed algorithms for computing conformal maps of topological disks. Gu and Yau introduced a method for computing conformal maps of surfaces with arbitrary topologies [Gu and Yao 2003]. In our work, the colon surface has complicated topologies, therefore we adapted this method.

In the following sections, we first briefly introduce the major concepts and theorems used in our colon flattening algorithms. For general background material on Riemann surface theory, we refer to [Jost 2002]. A thorough discussion about surface parameterization methods can be found in [Floater and Hormann 2005]. Then, the detail of the flattening algorithm will be presented.

3.1 Riemann Surface Theory

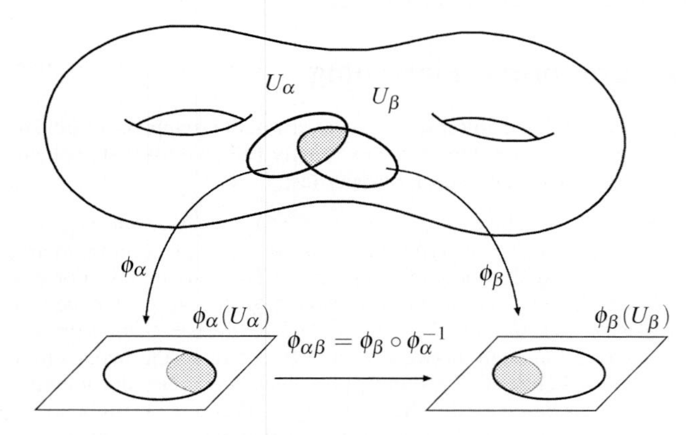

Figure 5: Riemann surface: The manifold is covered by a set of charts (U_α, ϕ_α), where $\phi_\alpha : U_\alpha \to \mathbf{R}^2$. If two charts (U_α, ϕ_α) and (U_β, ϕ_β) overlap, the transition function $\phi_{\alpha\beta} : \mathbf{R}^2 \to \mathbf{R}^2$ is defined as $\phi_{\alpha\beta} = \phi_\beta \circ \phi_\alpha^{-1}$. If all transition functions are analytic, then the manifold is a Riemann surface. The atlas $\{(U_\alpha, \phi_\alpha)\}$ is a conformal structure.

A manifold can be treated as a set of open sets in $\mathbf{R}^2$ glued coherently.

Definition 3.1 *A 2-dimensional manifold is a connected Hausdorff space M for which every point has a neighborhood U that is homeomorphic to an open set V of $\mathbf{R}^2$. Such a homeomorphism $\phi : U \to V$ is called a coordinate chart. An atlas is a family of charts $\{(U_\alpha, \phi_\alpha)\}$, where U_α constitutes an open covering of M.*

Definition 3.2 (Analytic Function) *A complex function $f : \mathbf{C} \to \mathbf{C}, (x,y) \to (u,v)$ is analytic (holomorphic), if it satisfies the following Riemann-Cauchy equation*

$$\frac{\partial u}{\partial x} = \frac{\partial v}{\partial y}, \frac{\partial u}{\partial y} = -\frac{\partial v}{\partial x}.$$

A conformal atlas is an atlas with special transition functions.

Definition 3.3 (Riemann Surface) *Suppose M is a 2-dimensional manifold with an atlas $\{(U_\alpha, \phi_\alpha)\}$. If all chart transition functions*

$$\phi_{\alpha\beta} := \phi_\beta \circ \phi_\alpha^{-1} : \phi_\alpha(U_\alpha \bigcap U_\beta) \to \phi_\beta(U_\alpha \bigcap U_\beta)$$

are analytic, then the atlas is called a conformal atlas, and M is called a Riemann surface.

Two conformal atlases are *compatible* if their union is still a conformal atlas. All the compatible conformal atlases form a *conformal structure* of the manifold as shown in Figure 5. All oriented 2-dimensional manifolds with Riemannian metrics are Riemann surfaces and have conformal structures [Jost 2002], such that on each chart (U_α, ϕ_α) with local parameter (u,v), the metric can be represented as $ds^2 = \lambda(u,v)(du^2 + dv^2)$.

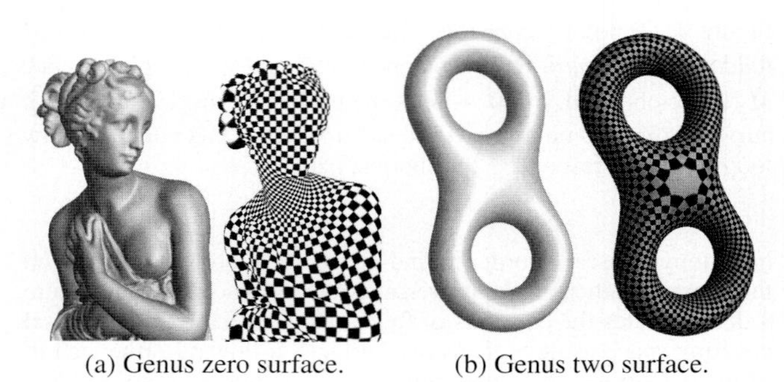

(a) Genus zero surface. (b) Genus two surface.

Figure 6: Holomorphic 1-form examples for genus zero and genus two surfaces.

3.1.1 Holomorphic 1-form

In order to flatten the surface, we need special differential forms defined on the conformal structure.

Definition 3.4 (Holomorphic 1-form) *Given a Riemann surface M with a conformal structure $\mathscr{A}$, a holomorphic 1-form ω is a complex differential form, such that on each local chart $(U, \phi) \in \mathscr{A}$,*

$$\omega = f(z)dz,$$

where $f(z)$ is an analytic function, $z = u + iv$ is the local parameter in the complex form.

The holomorphic 1-forms of a closed genus g surface form a g complex dimensional linear space, denoted as $\Omega(M)$. It is noted that a genus zero surface has no holomorphic 1-forms. A conformal atlas can be constructed by using a basis of $\Omega(M)$. Considering its geometric intuition, a holomorphic 1-form can be visualized as two vector fields $\omega = (\omega_x, \omega_y)$, such that the curls of ω_x and ω_y equal zero. Furthermore, one can rotate ω_x about the normal by a right angle to arrive at ω_y,

$$\nabla \times \omega_x = 0, \nabla \times \omega_y = 0, \omega_y = n \times \omega_x.$$

3.1.2 Conformal Parameterization

Suppose $\{\omega_1, \omega_2, \cdots, \omega_g\}$ is a basis for $\Omega(M)$, where g is genus of M. We can find a collection of open disks $U_\alpha \subset M$, such that U_α form an open covering of M, $M \subset \cup U_\alpha$. We define $\phi_\alpha^k : U_\alpha \to \mathbf{C}$ using the following formula, first we fix a base point $p \in U_\alpha$, for any point $q \in U_\alpha$,

$$\phi_\alpha^k(q) = \int_\gamma \omega_k,$$

where the path $\gamma : [0,1] \to U_\alpha$ is an arbitrary curve connecting p and q and inside $U_\alpha, \gamma \subset U_\alpha, \gamma(0) = p, \gamma(1) = q$. It can be verified that we can select a $\phi_\alpha^k, k = 1,2,\cdots,g$, such that ϕ_α^k is a bijection, we simply denote it as ϕ_α. Then the atlas $\{(U_\alpha, \phi_\alpha^k)\}$ is a conformal atlas.

For a genus one closed surface M, given a holomorphic 1-form $\omega \in \Omega(M)$, we can find 2 special curves $\Gamma = \gamma_1 \cup \gamma_2$, such that $\tilde{M} = M/\Gamma$ is a topological disk. Furthermore, on each open set U_α, if the curve $\int_{\gamma_1} \omega$ is a horizontal line in the parameter plane, then γ_1 is a *horizontal trajectory*. In the current work, we choose γ_2 such that $\int_{\gamma_2} \omega$ is a vertical line in the parameter plane, namely, γ_2 is a vertical trajectory. Γ is called a *cut graph*.

Then, by integrating ω on $\tilde{M}$, $\tilde{M}$ is conformally mapped to a parallelogram, as shown in Figure 4. Figure 6 illustrates holomorphic 1-forms on surfaces. The texture coordinates are obtained by integrating the 1-form on the surface.

3.1.3 Conformal Maps

Suppose M_1 is a Riemann surface with a conformal atlas $\{(U_\alpha, \phi_\alpha)\}$, and M_2 is another Riemann surface with conformal atlas $\{(V_\beta, \tau_\beta)\}$.

Definition 3.5 (Conformal Map) *A map $f : M_1 \to M_2$ is a* conformal map, *if its restriction on any local charts (U_α, ϕ_α) and (V_β, τ_β),*

$$f_\alpha^\beta := \tau_\beta \circ f \circ \phi_\alpha^{-1} : \phi_\alpha(U_\alpha) \to \tau_\beta(V_\beta)$$

is analytic.

3.2 Flattening Algorithm

The concepts of Riemann surface and conformal map are defined using continuous mathematics. Computing conformal parameterization is equivalent to solving an elliptic partial differential equation on surfaces.

Unfortunately, in reality, all surfaces are represented by discrete piecewise linear meshes, which are not differentiable in general. Fortunately, the solution to the elliptic PDE can be approximated accurately by piecewise linear functions using the finite element method [Reddy 2004]. The convergence and accuracy have been thoroughly analyzed in finite element field.

Therefore, our algorithm is mainly based on the finite element method. The key step is to use piecewise linear functions defined on edges to approximate differential forms. Furthermore, the forms minimize the harmonic energy, and the existence and the uniqueness are guaranteed by Hodge theory [Schoen and Yau 1997].

In the following discussion, we assume the surfaces are represented by meshes using the halfedge data structure. We use f to denote a face, e for a halfedge, e^- for the dual halfedge of e, M for mesh, $\tilde{M}$ for the fundamental domain of M.

3.2.1 Double Covering

In our case, after the topological noise removal, the surface is a closed genus zero surface. Because the genus zero surface has no holomorphic 1-form, a *double covering* method is used to construct a genus one surface. Two holes are first punched on the input surface. Then, a mesh M with two boundaries is obtained. The algorithm to construct a closed genus one mesh is described as follows:

1. Make a copy of mesh M, denoted as M', such that M' has all vertices in M, if $[v_0, v_1, v_2]$ is a face in M, then $[v_1, v_0, v_2]$ is a face of M'.
2. Glue M and M' along their boundaries, if a halfedge $[v_0, v_1]$ is on the boundary of M $[v_0, v_1] \in \partial M$, then $[v_1, v_0]$ is on the boundary of M'. Glue $[v_0, v_1]$ with $[v_1, v_0]$.

The resulting mesh is closed and symmetric, with two layers coincided. It is noted that a general genus one surface can be conformally mapped to a planar parallelogram, but not a rectangle. In our case, the genus one surface is obtained by the double covering method. The Riemann metric defined on the double covered surface is symmetric. Each boundary where we glue two surfaces is mapped to a straight line. Thus, the denoised genus zero colon surface can be conformally mapped to a rectangle.

3.2.2 Computing Harmonic and Holomorphic 1-form

After getting the homotopy basis $\{\gamma_1, \gamma_2, \cdots, \gamma_{2g}\}$, it is easy to compute the holomorphic 1-form basis.

1. Select γ_k, compute $\omega_k : K_1 \to \mathbf{R}$, form the boundary condition:

   $$\sum_{e \in \gamma_i} \omega_k(e) = \delta_i^k, \omega_k(\partial f) = 0, \forall f \in K_2, \quad (1)$$

   where

   $$\delta_i^k = \begin{cases} 1 & : \ i = k \\ 0 & : \ i \neq k \end{cases}$$

   K_1 is the edge set of M and K_2 is the face set of M.

2. Under the above linear constraints, compute ω_k minimizing the quadratic energy,

   $$E(\omega_k) = \sum_{e \in K_1} k_e \omega_k^2(e), \quad (2)$$

   using linear constrained least square method, where k_e is the weight associated with each edge, suppose the angles in the adjacent faces against edge e are α, β, then $k_e = \frac{1}{2}(\cot\alpha + \cot\beta)$ [Pinkall and Polthier 1993]. Solving this equation is equivalent to solving Riemann-Cauchy equation using finite element method. In our implementation, we preprocess the colon mesh to remove noises and improve the quality of triangulations, such that the number of the obtuse triangles are minimized. As a result, the Laplacian matrix is positive and definite.

3. On face $[v_0, v_1, v_2]$, its normal n is computed first, and a unique vector v in the same plane of v_0, v_1, v_2 is obtained by solving the following equations:

   $$\begin{cases} < v_1 - v_0, v > & = & \omega_k([v_1, v_0]) \\ < v_2 - v_1, v > & = & \omega_k([v_2, v_1]) \\ < n, v > & = & 0 \end{cases} \quad (3)$$

   Rotate v about n a right angle, $v^* = n \times v$, then define

   $$\omega_k^*([v_i, v_j]) := < v_j - v_i, v^* > .$$

The harmonic 1-form basis is represented by $\{\omega_1, \omega_2, \cdots, \omega_{2g}\}$, and the holomorphic 1-form basis is given by $\{\omega_1 + i\omega_1^*, \omega_2 + i\omega_2^*, \cdots, \omega_{2g} + i\omega_{2g}^*\}$.

3.2.3 Conformal Parameterization

Suppose we have selected a holomorphic 1-form $\omega : K_1 \to \mathbf{C}$, then we define a map $\phi : \tilde{M} \to \mathbf{C}$ by integration. The algorithm to trace the horizontal trajectory and the vertical trajectory on $\phi(\tilde{M})$ is as follows:

1. Pick one vertex $p \in \tilde{M}$ as the base vertex.

2. For any vertex $q \in \tilde{M}$, find the shortest path $\gamma \in \tilde{M}$ connecting p to q.

3. Map q to the complex plane by

$$\phi(q) = \sum_{e \in \gamma} \omega(e).$$

4. Pick a vertex $p \in M$, trace the horizontal line γ on the plane region $\phi(\tilde{M})$ through $\phi(p)$ as shown in Figure 7. If γ hits the boundary of $\phi(\tilde{M})$ at the point $\phi(q)$, q must be in the cut graph Γ, then there are two points q^+, q^- on the boundary of $\tilde{M}$, $\partial\tilde{M}$. Assume γ hits $\phi(q^+)$, then we continue to trace the horizontal line started from $\phi(q^-)$, until we return to the starting point $\phi(p)$. The horizontal trajectory is $\phi^{-1}(\gamma)$.

5. Trace the vertical trajectory similar to step 4.

6. The new cut graph $\tilde{\Gamma}$ is the union of the horizontal and vertical trajectories. Cut the surface along $\tilde{\Gamma}$ to get $\tilde{M}'$, and compute $\tilde{\phi}$. Then $\tilde{\phi}(\tilde{M}')$ is a rectangle, $\tilde{\phi}$ is a conformal map.

4 Direct Volume Rendering

The result of the flattening algorithm is a triangulated rectangle where the polyps are also flattened. The rendering of the flattened colon image is crucial for the detection of polyps. Haker et al. [2000] use color-coded mean curvature to visualize the flattened colon surface. Although it can show the geometry information of the 3D colon surface, it is still unnatural for the physicians to detect the polyps. The shape of the polyps is a good clue for polyp detection. In this section, we describe a direct volume rendering method to render the flattened colon image. Each pixel of the flattened image is shaded using a fragment program executed on the GPU, which allows the physician to move and zoom a viewing window to inspect the entire flattened inner colon surface. The idea of our rendering algorithm is to map each pixel of the flattened image back to the 3D colon surface, i.e., the volume space. The pixel is shaded using a volumetric ray-casting algorithm in the volume space.

4.1 Camera Registration

In order to perform the ray-casting algorithm, the ray direction needs to be determined for each vertex of the 3D colon surface first. A number of cameras are uniformly placed on the central path of the colon. The ray direction of a vertex is then determined by the nearest camera to that vertex.

Our camera registration algorithm starts with approximating the central path with a B-spline and resampling it into uniform intervals. Each sampling point represents a camera. Each vertex is then

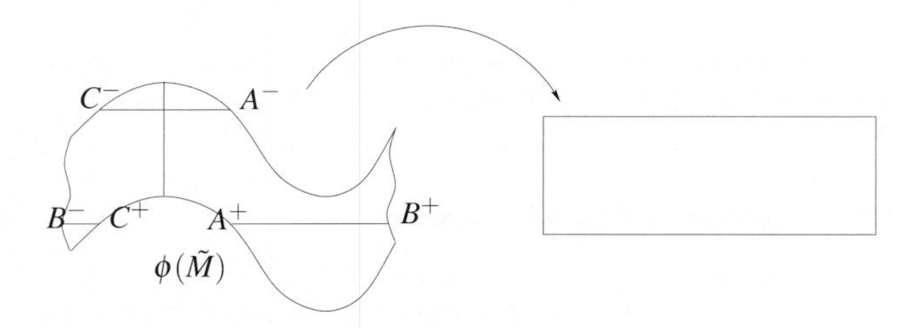

Figure 7: Tracing horizontal trajectory.

Figure 8: The colon is divided into a number of segments.

registered with a sampling point on the central path. The registration procedure is implemented efficiently by first dividing the 3D colon surface and central path into N segments. The registration is then performed between the correspondent segments of colon and the central path. The division of the 3D colon is done by classifying the vertices of the flattened 2D mesh into N uniform segments based on their height. As a consequence, the vertices of the 3D colon mesh are also divided into N segments, as shown in Figure 8. We then trace $N-1$ horizontal lines on the flattened 2D mesh, which uniformly divide the 2D mesh into N segments. Each traced horizontal line corresponds to a cross contour on the 3D mesh. In fact, we do not need to really trace the horizontal lines. For each horizontal line, we only need to compute the intersection points of the horizontal line and edges intersecting with it. For each intersection point, the corresponding 3D vertex of the 3D colon mesh is then interpolated. The centroid of these interpolated 3D vertices is computed and registered with a sampling point of the central path. Therefore, the central path is also divided into N segments, and each segment of the 3D colon mesh corresponds to a segment of the central path. Although the division of the 3D colon surface and the central path is not uniform as that of the 2D mesh, it does not affect the accuracy of the camera registration.

For each vertex of a colon surface segment, we find its nearest sampling point in its corresponding central path segment and the neighboring two segments. This algorithm is efficient because for each vertex, the comparison is performed only with a small number of sampling points on the central path. For each vertex, we only record the B-spline index of the sampling points, instead of its 3D coordinates.

4.2 Volumetric Ray-Casting

To generate a high-quality image of the flattened colon, only coloring the vertices of the polygonal mesh and applying linear interpolation is not sufficient. We need to determine the color for each pixel of the 2D image. This can be performed efficiently using a fragment program on the GPU. For each vertex of the flattened polygonal mesh, we pass its corresponding 3D coordinates and camera index through texture coordinates to the fragment program. When the flattened polygonal mesh is rendered, each pixel of the flattened image will obtain its barycentric interpolated 3D coordinates and camera index. Its 3D position may not be exactly on the colon surface, but very close to the colon surface. Because we use a direct volume rendering method to determine the color for the pixel, it does not affect the image quality. We use the interpolated camera index to look up its corespondent sampling point on the central path. Then, the ray direction is determined and volumetric ray casting algorithm is performed using an opaque transfer function. By this method, we can determine the color for each pixel on the flattened image to generate a high-quality image.

Since our flattened image is colored per-pixel, we can provide the physician with a high-quality zoom-in view of a suspicious area on the flattened image in real-time. Because each vertex is registered with a sampling point on the central path, the flattened colon image can be easily correlated with the navigation of a virtual colonoscopy system. The correlated 3D view of the suspicious area can be also shown simultaneously.

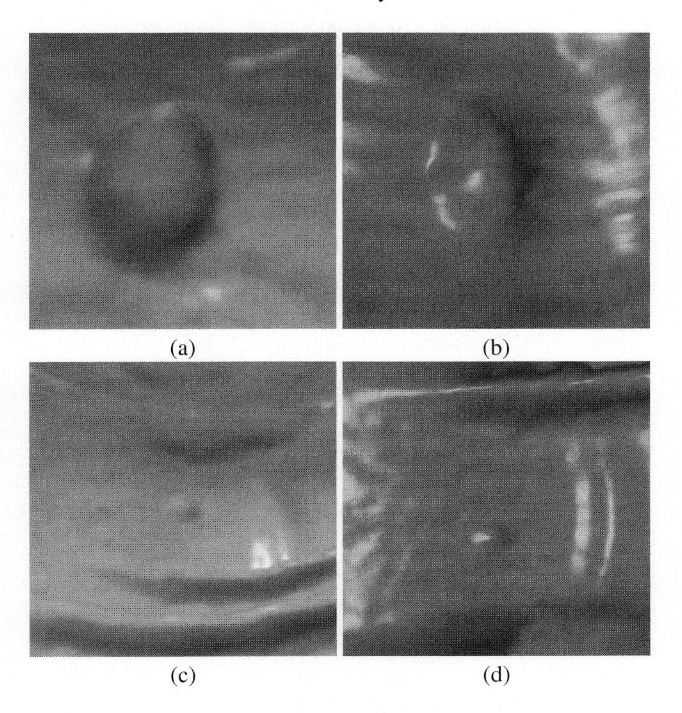

Figure 9: (a) A close view of a polyp rendered with volumetric ray casting; (b) A view generated from the flattened colon image showing the same polyp. (c) A view containing a small polyp generated from the navigation of a virtual colonoscopy system; (d) A view of the same smaller polyp generated from the flattened colon image.

5 Implementation and Results

We have implemented our conformal flattening and rendering algorithm in C/C++. All of the experiments are performed on a uniprocessor 3.0 GHz Pentium IV PC running Windows XP, with 2G RAM and NVIDIA Geforce 6800GT graphics board. A large number of colon CT data sets have been used to test our algorithms. All the data sets have a large number of slices (> 350), and the resolution of each slice is 512×512. They all exhibit similar results.

5.1 Preprocessing

Before our colon flattening algorithm can be applied, we need to perform the following tasks to extract the colon surface from the CT data set. First, a segmentation algorithm [Lakare et al. 2000] is applied, and a binary mask is generated, which labels the voxels belonging to the colon interior and the colon wall. This algorithm ensures a fast and accurate segmentation and electronic cleansing with the ability to consider the partial volume effect. Second, the rendering algorithm involves the central path of the colon. The central path is automatically extracted from the CT data set based on an accurate DFB-distance field with the exact Euclidian values [Wan et al. 2002]. The path is then approximated by a B-spline curve. Finally, given the binary mask and the CT data set, an enhanced dual contouring method [Zhang et al. 2004] is used to extract the simplified colon surface while preserving the finest resolution isosurface topology. Since our algorithm can deal with small handles, we do not need to remove these handles in the preprocessing step. All the algorithms used in the preprocessing step are robust and efficient, and can be executed in seconds on the PC platform.

5.2 Discussion

One of the CT colon data sets that we use has the resolution of $512 \times 512 \times 460$ and contains two polyps. The diameter of one polyp is 9 mm, and the diameter of the smaller polyp is 4 mm. Five handles are removed from the colon surface automatically, before applying the flattening algorithm. A flattened image of the whole colon using our rendering algorithm is shown in Figure 10. The resolution of the flattened image is 196×4000, and it is shown in three separate images. The total rendering time for these images is about 300ms. The larger polyp surrounded by a circle can be inspected from Figure 10(a). However, the smaller polyp located in Figure 10(c) is hard to recognize. Therefore, in a real medical application, the resolution should be at least four times higher than the one we used in this paper. In fact, we do not need to pre-compute such a high resolution flattened image. Our rendering algorithm with the acceleration of the commodity graphics hardware can provide a real-time high-quality zoom-in function, which allows the physician to interactively inspect the entire flattened colon at various resolutions.

In Figure 9(a), we show the larger polyp rendered using the volumetric ray-casting algorithm by positioning a camera in front of the polyp. In Figure 9(b), we show a zoom-in view generated by our flattening rendering algorithm showing the same polyp. We can clearly see that our flattening algorithm well preserves the shape of the polyp. In Figures 9(c) and 9(d), we show a small polyp in a 3D fly-through image and a zoom-in flattened image. The smaller polyp can be clearly recognized in the zoom-in image.

The whole process of our algorithm can be completed in about 10 minutes. Most steps of our algorithm are done within seconds or minutes. The most time consuming part of our algorithm is computing the harmonic holomorphic 1-form using the conjugate gradient method, which takes about several minutes. The good news is that the conjugate gradient method can be accelerated with the GPU [Bolz et al. 2003].

6 Conclusions

We have presented an efficient colon flattening algorithm using a conformal structure. Our algorithm is general for all high genus surfaces, and does not require the input surface to be a topological cylinder. The topology noise (i.e., minute handle) is removed automatically by our shortest loop algorithm. We have proven that our algorithm is angle preserving and the global distortion is minimal. The shape of the polyps on the flattened colon image is well preserved, and can be easily identified by a physician. The flattened colon image is rendered with a direct volume rendering method accelerated with commodity graphics hardware. We demonstrate that the conformal colon flattening image cooperates well with a fly-through virtual colonoscopy system.

We have some on-going research work in this area. We are experimenting with clustering algorithms and pattern recognizing techniques to detect polyps automatically on the flattened colon image. Since our flattening algorithm is not limited to a genus one surface, we are in the process of applying our algorithm to other human organs, such as the heart.

Acknowledgement

This work has been partially supported by NSF grant CCR-0306438, NSF grant DMS-0528363, NSF CAREER grant CCF-0448339, and NIH grants CA082402 and CA110186. One of the authors owns a fraction of Viatronix shares. The CT colon data sets are courtesy of Stony Brook University Hospital.

References

BALOGH, E., SORANTIN, E., NYUL, L. G., PALAGYI, K., KUBA, A., WERKGARTNER, G., AND SPULLER, E. 2002. Colon unraveling based on electronic field: Recent progress and future work. *Proceedings SPIE 4681*, 713–721.

BARTROLÍ, A. V., WEGENKITTL, R., KÖNIG, A., AND GRÖLLER, E. 2001. Nonlinear virtual colon unfolding. *IEEE Visualization*, 411–418.

BARTROLÍ, A. V., WEGENKITTL, R., KÖNIG, A., GRÖLLER, E., AND SORANTIN, E. 2001. Virtual colon flattening. *VisSym Joint Eurographics - IEEE TCVG Symposium on Visualization*, 127–136.

BOLZ, J., FARMER, I., GRINSPUN, E., AND SCHRÖDER, P. 2003. Sparse matrix solvers on the gpu: Conjugate gradients and multigrid. *ACM Transactions on Graphics 22*, 3, 917–924.

CORMEN, T. H., LEISERSON, C. E., RIVEST, R. L., AND STEIN, C. 2001. *Introduction to Algorithms, Second Edition*. The MIT Press.

DESBRUN, M., MEYER, M., AND ALLIEZ, P. 2002. Intrinsic parameterization of triangle meshes. *Eurographics (Computer Graphics Forum)*, 209–218.

DEY, T. K., AND SCHIPPER, H. 1995. A new technique to compute polygonal schema for 2-manifolds with application to null-homotopy detection. *Discrete and Computational Geometry 14*, 93–110.

EI-SANA, J., AND VARSHNEY, A. 1997. Controlled simplification of genus for polygonal models. *IEEE Visualization*, 403–412.

ÉRIC COLIN DE VERDIÈRE, AND LAZARUS, F. 2005. Optimal system of loops on an orientable surface. *Discrete and Computational Geometry 33*, 3, 507–534.

ERICKSON, J., AND HAR-PELED, S. 2003. Optimally cutting a surface into a disk. *ACM Symposium on Computational Geometry*, 215–228.

ERICKSON, J., AND WHITTLESEY, K. 2005. Greedy optimal homotopy and homology generators. *ACM-SIAM Symposium on Discrete Algorithms*, 1038–1046.

FLOATER, M. S., AND HORMANN, K. 2005. Surface parameterization: a tutorial and survey. *Advances in Multiresolution for Geometric Modelling*, 157–186.

GU, X., AND YAO, S.-T. 2003. Global conformal surface parameterization. *ACM Symposium on Geometry Processing*, 127–137.

GU, X., GORTLER, S. J., AND HOPPE, H. 2002. Geometry images. *ACM Transactions on Graphics 21*, 3, 355–361.

GUSKOV, I., AND WOOD, Z. 2001. Topological noise removal. *Graphics Interface*, 19–26.

HAKER, S., ANGENENT, S., TANNENBAUM, A., AND KIKINIS, R. 2000. Nondistorting flattening maps and the 3d visualization of colon ct images. *IEEE Transactions on Medical Imaging 19*, 7, 665–670.

HONG, L., MURAKI, S., KAUFMAN, A., BARTZ, D., AND HE, T. 1997. Virtual voyage: Interactive navigation in the human colon. *Proceedings of ACM SIGGRAPH*, 27–34.

JOST, J. 2002. *Compact Riemann Surfaces*. Springer.

LAKARE, S., WAN, M., SATO, M., AND KAUFMAN, A. 2000. 3d digital cleansing using segmentation rays. *IEEE Visualization*, 37–44.

LAZARUS, F., POCCHIOLA, M., VEGTER, G., AND VERROUST, A. 2001. Computing a canonical polygonal schema of an orientable triangulated surface. *ACM Symposium on Computational Geometry*, 80–89.

LÉVY, B., PETITJEAN, S., RAY, N., AND MAILLOT, J. 2002. Least squares conformal maps for automatic texture atlas generation. *ACM Transactions on Graphics 21*, 3, 362–371.

MASSEY, W. S. 1990. *Algebraic Topology: An Introduction*. Springer.

PAIK, D. S., BEAULIEU, C. F., JEFFREY, R. B. J., KARADI, C. A., AND NAPEL, S. 2000. Visualization modes for ct colonography using cylindrical and planar map projections. *Journal of Computer Assisted Tomography 24*, 179–188.

PINKALL, U., AND POLTHIER, K. 1993. Computing discrete minimal surfaces and their conjugates. *Experimental Mathematics 2*, 1, 15–36.

REDDY, J. N. 2004. *An Introduction to Nonlinear Finite Element Analysis*. Oxford University Press.

SCHOEN, R., AND YAU, S.-T. 1997. *Lectures on Harmonic Maps*. International Press.

WAN, M., LIANG, Z., KE, Q., HONG, L., BITTER, I., AND KAUFMAN, A. 2002. Automatic centerline extraction for virtual colonoscopy. *IEEE Transactions on Medical Imaging 21*, 1450–1460.

WANG, G., AND VANNIER, M. W. 1995. Gi tract unraveling by spiral ct. *Proceedings SPIE 2434*, 307–315.

WANG, G., MCFARLAND, E. G., BROWN, B. P., AND VANNIER, M. W. 1998. Gi tract unraveling with curved cross section. *IEEE Transactions on Medical Imaging 17*, 318–322.

WANG, G., DAVE, S. B., BROWN, B. P., ZHANG, Z., MCFARLAND, E. G., HALLER, J. W., AND VANNIER, M. W. 1999. Colon unraveling based on electronic field: Recent progress and future work. *Proceedings SPIE 3660*, 125–132.

ZHANG, N., HONG, W., AND KAUFMAN, A. 2004. Dual contouring with topolgy-preserving simplification using enhanced cell representation. *IEEE Visualization*, 505–512.

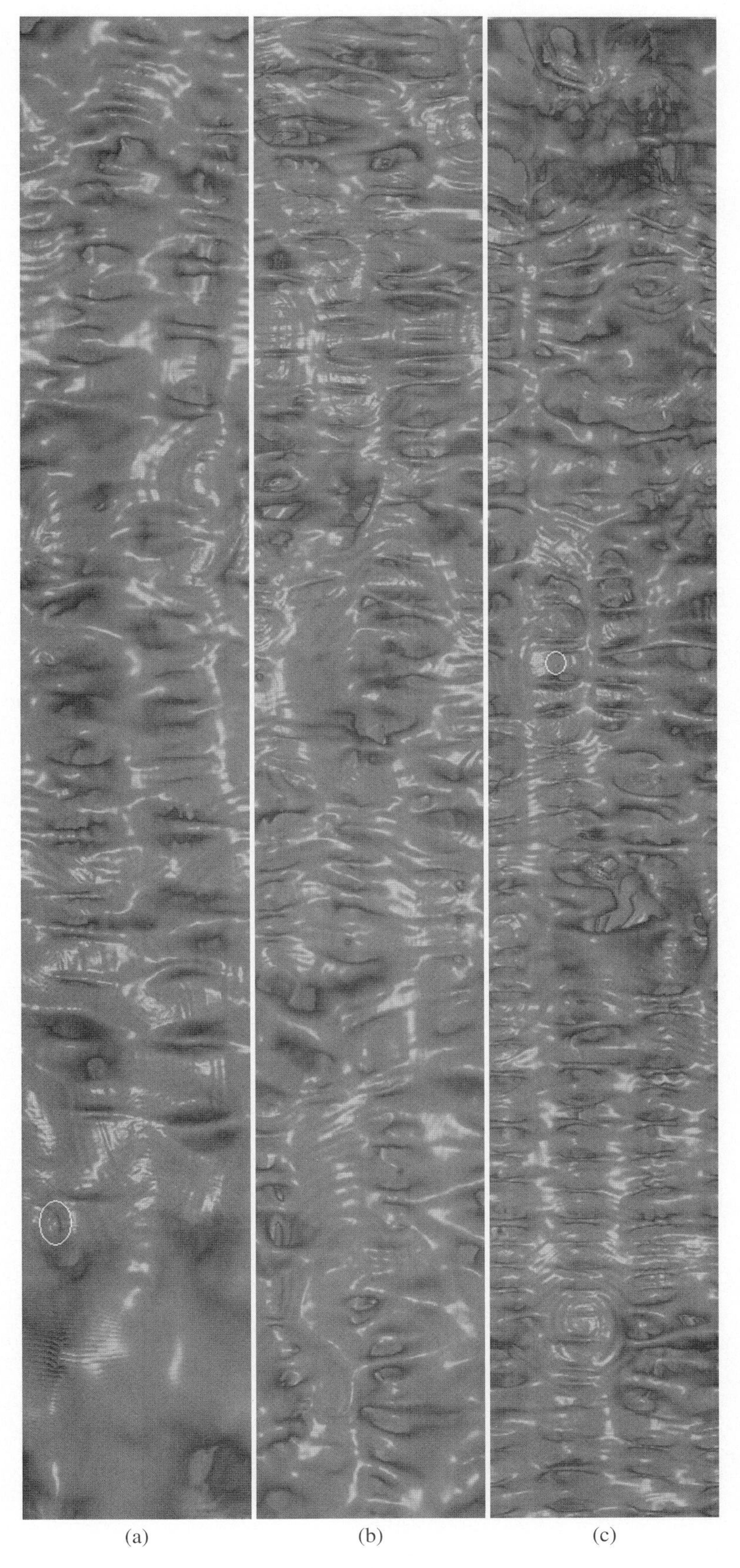

Figure 10: A flattened image for a whole colon data set is shown in three images. The bottom of image (a) is the rectum of the colon, and the top of image (c) is the cecum of the colon.

Interactive simulation of one-dimensional flexible parts

Mireille Grégoire*
DaimlerChrysler Research and Technology

Elmar Schömer†
Johannes Gutenberg Universität Mainz, Germany

Abstract

Computer simulations play an ever growing role for the development of automotive products. Assembly simulation, as well as many other processes, are used systematically even before the first physical prototype of a vehicle is built in order to check whether particular components can be assembled easily or whether another part is in the way. Usually, this kind of simulation is limited to rigid bodies. However, a vehicle contains a multitude of flexible parts of various types: cables, hoses, carpets, seat surfaces, insulations, weatherstrips... Since most of the problems using these simulations concern one-dimensional components and since an intuitive tool for cable routing is still needed, we have chosen to concentrate on this category, which includes cables, hoses and wiring harnesses.

This paper presents an interactive, real-time, numerically stable and physically accurate simulation tool for one-dimensional components. The modeling of bending and torsion follows the Cosserat model and is implemented with a generalized spring-mass system with a mixed coordinate system which features usual space coordinates for the positions of the points and quaternions for the orientation of the segments joining them. This structure allows us to formulate the springs based on the coordinates that are most appropriate for each type of interaction and leads to a banded system that is then solved iteratively with an energy minimizing algorithm.

CR Categories: I.6.5 [Simulation and Modeling]: Model Development

Keywords: Cable simulation, Cosserat model, Modelling of torsion

1 Introduction

The automotive industry aims to reduce development costs and time while meeting the demand for quality and for an increasing model range. In order to meet this challenge, more and more work is done digitally. Styling reviews, Digital Mock Ups and assembly simulations are used at an early stage in the development process. Thus, potential problems can be detected and solved much earlier, long before the first physical prototype is built. Assembly simulation is one of the applications used in the construction design process: the virtual prototype is tested for feasibility and ease of assembly, which can be optimized. Like in other Virtual Reality applications, the physical behavior of the components needs to be simulated: a collision detection system and a contact simulation that impede the interpenetration of the objects and allow them to slide one on the other are required. Our work will be based on the Virtual Reality software *veo*, developed by DaimlerChrysler Research and Technologies, which already has a real-time collision detection and interactive contact simulation [Buck and Schömer 1998] as well as a real-time multibody dynamics [Sauer and Schömer 1998]. A realistic way of dealing with flexible parts -which are currently treated as rigid bodies by the simulation - is however not yet included. This implies that deformations that can occur in the physical world cannot be simulated, which limits the possibilities of the tool. A typical example of this would be an assembly simulation in which a cable should be pushed slightly aside to permit the mounting of another part. Studies from the business units show that most of the problems concerning flexible parts are encountered with cables, wire harnesses and hoses. These parts are also concerned with another particular application: the routing. Wire harnesses are on the one hand growing more and more complex as the use of electrical and electronic components in vehicles grows; on the other hand they often need to be modified to accommodate for changes in surrounding parts or for optimization. To meet the interests of the business units, we have chosen to first concentrate on simulating these parts since they cover the most urgent need. For these bodies, one of their dimension (the length) is much bigger than the other two and their centerline contains most of the information needed to represent them. This one-dimensional nature leads to simplifications in the simulation compared to other objects such as flexible surfaces or bodies.

1.1 Previous work

Several approaches are used for simulating flexible bodies, and one-dimensional ones in particular. The spectrum of solution ranges from a purely graphical representation of oscillations without any physics and with very low computational requirements [Barzel 1997] to a complete finite element simulation of cloth to predict its mechanical properties [Finckh et al. 2004]. Hergenröther [Hergenroether and Daehne 2000] models a cable as a chain of cylinder segments connected by ball joints. The chain has at first two segments and is iteratively refined, doubling the number of segments at each iteration, thus giving an inexact but fast dynamic simulation and a more exact, slower static one. Loock [Loock and Schömer 2001] implements a cable as a spring-mass system with torsion springs for the bending forces that are proportional to the bending angle.

There are few approaches that deal with the torsion. This is unfortunate, since, as shown for example in [Goss et al. 2005], it plays a crucial part in the deformation of a cable. One of the most interesting approaches taking the torsion into account is the one from Pai [Pai 2002]. He uses a Cosserat model as a base for the simulation of suture strands during laparoscopic surgery. These strands are objects visually well approximated by a curve but nevertheless present three-dimensional body properties like twisting. In the typical use case in surgery simulation, the position and direction of the strand are defined at one end (corresponding to the end fixed in human tissues) and the forces and moments are defined at the other end, corresponding to the needle haptic device. The ordinary differential equation resulting from the Cosserat model is integrated in two passes to calculate all the variables.

*e-mail: mireille.gregoire@daimlerchrysler.com

†e-mail:schoemer@informatik.uni-mainz.de

SPM 2006, Cardiff, Wales, United Kingdom, 06–08 June 2006.
© 2006 ACM 1-59593-358-1/06/0006 $5.00

1.2 Specific contributions

Being aware of the importance of torsion for the deformations of a cable, it became one of our main interests. After several attempts, we have finally chosen a generalized spring-mass system for modeling the cables. This system uses a mixed coordinate system that contains at the same time ordinary position coordinates and quaternions representing the orientation of segments on the basis of which the torsion is easy to calculate. For enforcing constraints like the conservation of length, we introduce an integral force analog to the action of a proportional-integral controller.

2 The Cosserat model for rods

The Cosserat model for rod-like solids (with one dimension - the length - much greater than the other two cross-section dimensions) is a model from continuum mechanics. It models such a three-dimensional body as a one-dimensional one while taking into account the properties of the cross-section. A rod is represented by its centerline (a curve in the usual three-dimensional space) associated to frames (whose vectors are the so-called directors) which represent material orientation and deformation. Such a model is well suited for real-time applications since it has a smaller number of variables (compared to finite elements models for example) and can nevertheless take into account a great number of physical properties. Large deformations are neither a problem since all the properties of the system can be defined relatively to the local frames.

2.1 Description

We consider only unstretchable and unshearable bodies (which represent the vast majority of objects of interest in the context of automotive construction). We furthermore assume that the cross-section is homogeneous and undeformable, and that the mechanical properties are constant along the length of the cable. The general Cosserat model without the above mentioned restrictions is explained in [Antman 1995] or in [Rubin 2000]. A cable of length L is parameterized by its arc length s. It is described by a function associating to each point of the centerline of a reference configuration (for example a state without tensions like a straight line without torsion) a vector $\mathbf{r}(s)$ describing the position of the point of the centerline and a directors frame, $(\mathbf{d_1}(s), \mathbf{d_2}(s), \mathbf{d_3}(s))$ representing material directions. Under the above mentioned restrictions the directors frame is a right-handed orthonormal basis and a member of the special orthogonal group $SO(3)$ (the group of rotations of $\mathbb{R}^3$).

$$\begin{array}{rcl} [0,L] & \rightarrow & \mathbb{R}^3 \times \mathrm{SO}(3) \\ s & \mapsto & (\mathbf{r}, (\mathbf{d_1}, \mathbf{d_2}, \mathbf{d_3})) \end{array}$$

The basis of directors is adapted to the curve: the third director $\mathbf{d}_3$ points in the tangent direction of the curve. The vectors $\mathbf{d}_1$ and $\mathbf{d}_2$ show the position of two material lines in the cross section of the rod. The evolution of the basis $(\mathbf{d}_k)_{1 \leq k \leq 3}$ along the curve is represented by the Darboux vector ω. Similar to the angular velocity vector (replacing the time derivative with a derivative along the arc length), this vector has the following property:

$$\frac{\mathrm{d}\mathbf{d}_k}{\mathrm{d}s} = \omega \times \mathbf{d}_k \text{ for } k = 1, 2, 3$$

where $\times$ represents the cross-product.

2.2 Forces and torques

At the point of arc length s_0, the rod has a tension $\mathbf{n}(s_0)$ and an inner torque $\mathbf{m}(s_0)$. The rod is also submitted to external distributed forces like the weight or the contact forces, with a linear density f. The static equilibrium leads to

$$\left\{ \begin{array}{rcll} \mathbf{0} & = & \frac{\mathrm{d}\mathbf{n}}{\mathrm{d}s}(s_0) & + \quad \mathbf{f}(s_0) \\ \mathbf{0} & = & \frac{\mathrm{d}\mathbf{m}}{\mathrm{d}s}(s_0) & + \quad \frac{\mathrm{d}\mathbf{r}}{\mathrm{d}s}(s_0) \times \mathbf{n}(s_0) \end{array} \right.$$

2.3 Material properties

We know from continuum mechanics that the torque at a point of a rod is

$$\mathbf{m}(s) = B_\tau \tau \,\mathbf{T} + B_\kappa \kappa \,\mathbf{B}$$

with κ the curvature, τ the torsion, $\mathbf{T} = \mathbf{d_3}$ the tangent, $\mathbf{B}$ the binormal. The coefficients B_κ and B_τ are defined in a similar way to the moments of inertia and depend on the material properties (Young's modulus E and shear modulus G) and the geometry of the cross section. For a circular and homogeneous cross section with radius R, we have

- bending stiffness:

$$B_\kappa = EI = \iint_{crosssection} E x^2 \, dA = E \frac{\pi R^4}{4}$$

- torsional stiffness:

$$B_\tau = GJ = \iint_{crosssection} G r^2 \, dA = G \frac{\pi R^4}{2}$$

where x and r are the distances respectively to the bending axis (in the cross-section, passing by the centerline) and to the torsion axis (the tangent to the centerline). In the case of a non-homogeneous cross section, the stiffnesses can be calculated by considering E and G as a function of the positions. In particular, for a hollow body like a hose, the result can be expressed with the help of the inner radius R_{inner}: $B_\kappa = E \frac{\pi(R^4 - R_{inner}^4)}{4}$

2.4 Equations

The combination of these equations results in the following differential equations:

$$\left\{ \begin{array}{rcl} \frac{\mathrm{d}\mathbf{n}}{\mathrm{d}s} & = & -\mathbf{f} \\ \frac{\mathrm{d}\mathbf{m}}{\mathrm{d}s} & = & -\mathbf{d}_3 \times n \\ \frac{\mathrm{d}\mathbf{d}_k}{\mathrm{d}s} & = & u \times \mathbf{d}_k \\ \omega & = & \kappa_1 \mathbf{d}_1 + \kappa_2 \mathbf{d}_2 + \tau \mathbf{d}_3 \\ \mathbf{m} & = & B_\kappa(\kappa_1 \mathbf{d}_1 + \kappa_2 \mathbf{d}_2) + B_\tau \tau \mathbf{d}_3 \end{array} \right.$$

These equations have the remarkable property that the torsion is constant over the length of the rod.

$$\tau = \tau_0$$

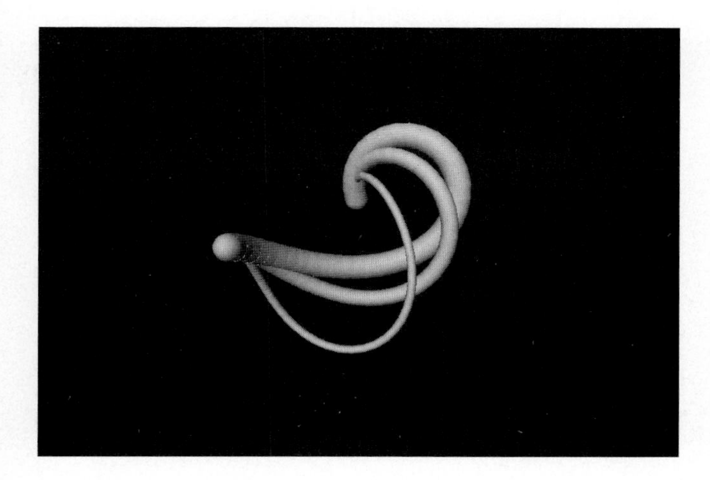

Figure 1: Influence of the radius : the different radii lead to different stiffnesses B_κ and B_τ. The influence on the form of the cable can be important. A change in material parameters would produce similar results.

2.5 Relation with the Frenet frame

Let us consider the Frenet frame (tangent $\mathbf{T}$, normal vector $\mathbf{N}$ and binormal vector $\mathbf{B}$) of the centerline. At any given point of the curve, the directors and the Frenet frames have at least a common vector, the tangent ($\mathbf{d}_3 = \mathbf{T}$). Therefore a rotation around $\mathbf{d}_3$ by an angle θ exists which transforms $(\mathbf{N}, \mathbf{B}, \mathbf{T})$ in $(\mathbf{d}_1, \mathbf{d}_2, \mathbf{d}_3)$. This angle θ shows the position of the material lines relatively to the Frenet frames: it is a "pure material torsion". The Cosserat torsion is then the Frenet torsion τ_f (geometrical torsion of the centerline) augmented with this material torsion.

$$\tau = \tau_f + \frac{d\theta}{ds}$$

The Darboux vector has the following expression:

$$\omega = \kappa_1 \mathbf{d}_1 + \kappa_2 \mathbf{d}_2 + \tau \mathbf{d}_3$$

where κ_1 and κ_2 are the components of the curvature on $\mathbf{d}_1$ and $\mathbf{d}_2$:

$$\kappa \mathbf{B} = \kappa_1 \mathbf{d}_1 + \kappa_2 \mathbf{d}_2$$

3 First attempts

3.1 Shooting method for ODE

The Cosserat model gives us a set of ordinary differential equations where the tensions and inner moments are coupled to the positions and orientations. In order to solve this, we need to know either the forces, moments, position and orientation at one end of the cable, or a combination of them distributed over the two ends. Since it is much easier for the user in the absence of force feedback to specify the position and orientation at one point, for example with a spacemouse, we wanted to have boundary conditions of geometrical type (position and orientation) at both ends of the cable. We therefore have to solve the following problem: knowing the positions and the directors frames at the two endpoints, we are looking for the deformation of the cable under the condition that the length of the cable should remain constant. This is a so-called two-points-boundary-value problem where the boundary conditions are spread on the two endpoints of the integration interval.

The solution of Pai [Pai 2002] is implemented for geometrical boundary conditions at one end of the strand and for dynamical boundary conditions at the other end, corresponding to the use of a haptic device. This particular configuration allows to solve the problem with two passes: the first one to calculate the forces and torques, and the second one in the opposite direction for the geometrical variables.

Unfortunately, an attempt to adapt this method to the desired type of boundary conditions was not successful. The usual method for solving such problems are so-called "shooting methods": there are two known and incomplete sets of boundary conditions, one at $s = 0$ and the other one at $s = L$. The method consists in completing the set at $s = 0$, integrating the differential equation with these initial conditions, and considering the obtained final conditions as a function of the initial ones, looking iteratively (for example with a Newton algorithm) for the appropriate initial conditions. Unfortunately, the initial conditions are difficult to determine, and the search does not always converge and sometimes aborts, for example when the curvature radius at a point is of the same order of magnitude as the integration step length. Furthermore, such an approach makes it difficult to integrate external forces such as contact forces.

3.2 Simple spring-mass-model

In order to avoid these difficulties, we experimented with a different approach using a spring-mass system. The rod is modeled as a sequence of mass points (lying on the centerline of the cable) which are connected with different kinds of springs: linear springs for the length conservation and torsion springs for the bending. The knowledge of the centerline leaves one degree of freedom unspecified, namely the material rotation around the centerline.

Since the Cosserat theory also considers the material direction, we need a new variable at each point to represent it, for example θ, as the angle between the Frenet and the directors frames. The global torsion is the sum of both the Frenet torsion - calculated from the coordinates of the mass points - and the pure material torsion - calculated as the derivative of the fourth coordinate θ-. The energy is then $E = \frac{1}{2}(B_\kappa \kappa^2 + B_\tau \tau^2)$ and the forces are calculated as the negative of the gradient of the energy.

We studied two ways of computing the Frenet torsion: one using the binormal vector and one using a function of the derivatives of the coordinates. For the first one, since $\mathbf{B}$ is a unit vector, the change of $\mathbf{B}$ is a rotation and the torsion is - in the discretized system - proportional to the angle between the two binormal vectors at the points i and $i+1$:

$$\tau_f = \frac{\angle(\mathbf{B}_i, \mathbf{B}_{i+1})}{|s_{i+1} - s_i|}$$

where s_i is the arc length at point i. Since the plane defined by the three points $i-1$, i, $i+1$ contains the tangent $\mathbf{T}$ and the normal $\mathbf{N}$, the binormal at point i is orthogonal to both $\mathbf{u_{i-1}}$ and $\mathbf{u_i}$ and is calculated as:

$$\mathbf{B}_i = \frac{\mathbf{u_{i-1}} \times \mathbf{u_i}}{\|\mathbf{u_{i-1}} \times \mathbf{u_i}\|}$$

with $\mathbf{u_i}$ the unit vector between $\mathbf{r}(s_i)$ and $\mathbf{r}(s_{i+1})$:

$$\mathbf{u_i} = \frac{\mathbf{r}(s_{i+1}) - \mathbf{r}(s_i)}{\|\mathbf{r}(s_{i+1}) - \mathbf{r}(s_i)\|}$$

This scheme is - in particular in the case of a small curvature - extremely sensitive to noise in the position of the mass points. When for example $\mathbf{u_{i-1}}$ and $\mathbf{u_i}$ are fixed, and the angle between $\mathbf{u_i}$ and $\mathbf{u_{i+1}}$ is small, a very small change in $\mathbf{u_{i+1}}$ could mean a huge difference for τ_f, just like at the Earth poles a small change in position can mean a large change in longitude. Accordingly, the forces in case of a small curvature are huge, which leads to numerical instability and impedes the convergence.

Inflexion points represent another problem: the binormals are not defined at such points and undergo a discontinuity (opposite directions). In a configuration with (almost) inflexion points, the direction of the binormal can change from one iteration to another, which is naturally impractical. An easy solution is to precalculate the binormal before each iteration and, when necessary, to multiply it by -1 in order to insure $\mathbf{B}_i^T \mathbf{B}_{i+1} \geq 0$, and to keep these places in memory to take into account the possible sign changes for the subsequent calculations.

Curvature and torsion can also be calculated as functions of the derivatives of $\mathbf{r}(s)$. After discretization and simplification, the expressions for the curvature and the torsion become

$$\kappa = \frac{1}{2L} \|\mathbf{u_i} \times (\mathbf{u_{i+1}} - \mathbf{u_{i-1}})\|$$

$$\tau_f = \frac{1}{L^3 \kappa^2} \mathbf{u_i}^T (\mathbf{u_{i+1}} \times \mathbf{u_{i-1}})$$

with L the length of a segment. This scheme is more stable and less noise-sensitive as the previous one, but it is still insufficient. The torsion remains in both cases the problem. Consequently, another way of calculating the torsion is needed.

4 Implementation

4.1 Mixed coordinates

We are looking for a coordinate system in which the torsion is easily expressed and not too noise-sensitive. The torsion depends on the difference of orientation along the tangent to the centerline of two segments. This orientation can be described as a rotation of $SO(3)$ from a reference orientation. From the several representations of rotations of $SO(3)$, we have chosen unit quaternions. (Quaternions are 4-tuples that can be seen as a generalization of complex numbers. They will be described in more detail in section 5.) Many properties speak in favor of unit quaternions: they only have 4 coordinates and one constraint (they must have a unit length), which is an advantage compared to rotation matrices (with 9 components and 6 constraints). Furthermore, the rotation of a vector is easily expressed and the composition of two rotations can also be easily calculated as the product of the two corresponding quaternions. This group structure is an advantage for the simulation. Quaternions also lack singular points and gimbal lock, contrarily to Euler or Cardan angles. So we used a system with seven coordinates for each point: the three usual space coordinates and a quaternion which represents the orientation of a segment between two points.

4.2 Notation

The following notations will be used in the rest of the paper:

n is the number of discretization points of the cable. The cable begins and ends with a discretization point and the centerline is linearly interpolated between two points. $\mathbf{X} \in \mathbb{R}^{7n-4}$ is the vector of all coordinates of the cable and therefore of all scalar unknowns. $X_i \in \mathbb{R}$ is the i-th component of $\mathbf{X}$. $\mathbf{x_i} \in \mathbb{R}^3 = (X_{7i+1}, X_{7i+2}, X_{7i+3})$ is the position of the point i. $\mathfrak{q}_i \in \mathbb{R}^4 = (X_{7i+4}, X_{7i+5}, X_{7i+6}, X_{7i+7})$ is a quaternion representing the orientation of the segment between the points i and $i+1$.

For representing the orientations, we can arbitrarily choose an orientation reference - in our case a right-handed orthonormal basis whose third vector is in the direction $\mathbf{ref} = (0,1,0)$. This corresponds to the default orientation of the axis of cylinders in OpenInventor/OpenGL, which will simplify the graphical representation afterwards.

Additionally, $\mathbf{F} \in \mathbb{R}^{7n-4}$ is the vector of all forces; $F_i \in \mathbb{R}$ is the i-th component of $\mathbf{F}$. L_{ref} is the reference length for each segment. It is equal to the total length of the cable divided by the number of segments $(n-1)$. $L_i = \sqrt{(\mathbf{x_{i+1}} - \mathbf{x_i})^2}$ is the actual length of segment i.

5 Forces

The interactions taken into account are bending and torsion, weight, stretch forces (unextensibility), normalization of quaternions and coherence between the quaternions and the space coordinates. Handles (fixed points) can be defined optionally: each one defines the additional constraint that the cable should pass by a particular point with a particular orientation. By default, the two endpoints are considered to be fixed in position and orientation: their coordinates are excluded from the solver range.

The use of space coordinates and quaternions allows to apply the forces to the most appropriate kind of coordinates:

- On the space coordinates:
  - conservation of length
  - handles (position)
  - weight
- On the quaternions:
  - bending and torsion forces
  - normalization of the quaternions
  - handles (orientation)
- On both the quaternions and the position:
  - coherence between position and quaternions

The main disadvantage of this method is the increased number of variables. But since every type of interaction is relatively easy to calculate, while at the same time the system matrix is strongly banded, the system as a whole is easy to solve.

5.1 Derivation of the forces

For each kind of interaction, we first define an energy function and then derive the forces from this function, following the use in mechanics:

$$F_i = -\frac{\partial E}{\partial X_i}$$

We also define the symmetric Hessian matrix $\mathbf{H} \in \mathbb{R}^{7n-4 \times 7n-4}$, such that

$$H_{i,j} = -\frac{\partial^2 E}{\partial X_i \partial X_j} = \frac{\partial F_i}{\partial X_j}$$

Since each type of interaction can be decomposed as a sum of interactions between either two points, two quaternions or two points and a quaternion, we calculate the energy, the forces and the Hessian as a sum over element groups. In order to enhance the calculation speed, it is important to calculate only once the partial terms that appear several times in the three functions.

For a global position $\mathbf{X}$, we can formulate the energy as the sum of the energies over all points and segments and over all interactions.

$$\begin{aligned} E &= \sum_{Interactions} \sum_{i} E_{Interaction,i} \\ &= \sum_{i=1}^{n-1} E_{Length,i} + \sum_{i=1}^{n-1} E_{QuatNorm,i} \\ &+ \sum_{i=1}^{n} E_{Weight,i} + \sum_{i=1}^{n-1} E_{Coh,i} \\ &+ \sum_{i=1}^{n-2} E_{Bending,i} + \sum_{i=1}^{n-2} E_{Torsion,i} \end{aligned}$$

The forces and the Hessian can be calculated in a similar way. In the following, $k_{Interaction}$ is the constant relative to the interaction $Interaction$ that will be used for defining the relative weights of the different interactions.

5.2 Stretch forces and conservation of the length

We first use strong linear springs for the length conservation. The energy is defined classically as

$$E_{Length,\,P,\,i} = \frac{1}{2} k_{Length} (L_i - L_{ref})^2$$

where k_{Length} is the constant of the spring. If we consider only this spring, the minimum of the energy is clearly at $L_i = L_{ref}$. But the points i and $i+1$ are also submitted to other interactions that result in a disturbing force that stretches (or compresses) the spring. Since the spring can only exert a force when it is not at equilibrium, it alone cannot enforce exactly the constraint of a constant length: we use an additional force to achieve this. If we look at control theory, a spring alone is a proportional controller: the actuating variable (the force) is proportional (with the proportionality constant k_{Length}) to the difference between the desired value (L_{ref}) and the actual one (L_i). The forces for segment i between point i and $i+1$ due to the stretch forces are

$$\mathbf{f}_{P,\,i} = -\mathbf{f}_{P,i+1} = k_{Length}(L_i - L_{ref})\mathbf{u_i}$$

We want to use the equivalent of a PI-(proportional-integral) controller, which has the property of being able to enforce a constraint exactly so that the steady-state error is null. The PI-controller combines a proportional part (the spring) with an integral one. Its actuating variable is proportional to the integral over the time of the difference between the desired and actual value. When the error is null, the proportional force is also null, but the integral one counteracts the perturbations. The proportional force is nonetheless essential to the stability of the system. As a consequence of the presence of the integral force, the proportional constant can be reduced (in comparison with a spring-only system), which makes the system less stiff as a whole. To implement this, we have added constant forces $f_{I,\,i}\mathbf{u_i}$ on point i and $-f_{I,\,i}\mathbf{u_i}$ on point $i+1$. This force must have the same direction as the proportional one. The total force on point i due to the stretch of segment i is then

$$\mathbf{f}_i = (f_{I,\,i} + k_{Length}(L_i - L_{ref}))\mathbf{u_i}$$

This corresponds to an energy of

$$E_{Length,\,i} = \frac{1}{2} k_{Length}(L_i - L_{ref})^2 + f_{I,\,i}(L_i - L_{ref})$$

We have now $n-1$ additional parameters $f_{I,\,i}$. For solving the system, we use an iterative approach: we first solve the system holding the $f_{I,\,i}$ constant, then adjust their values in function of the result and solve anew the system until an equilibrium is reached (The constraints are enforced within an ε tolerance that should be chosen slightly bigger than the numerical precision of the system solver).

For updating the integral forces, we adjust each one individually: the new force is the old one augmented with the difference of lengths multiplied by a constant

$$f_{I,\,i,\,new} = f_{I,\,i,\,old} + k_{I,\,Length}(L_i - L)$$

This constant $k_{I,\,Length}$ should not have a value too big compared to k_{Length} for ensuring stability.

5.3 Weight

We consider the cable as a chain of mass points with equal masses m. The total mass of the cable is $n\,m$. The energy is $E_{Weight,i} = -m\mathbf{g}^T\mathbf{x_i}$ where $\mathbf{g} \in \mathbb{R}^3$ is the acceleration of gravity.

5.4 Coherence between quaternions and positions

It is important to ensure that the bending and torsion forces are correctly coupled to the positions. The unit vector $\mathbf{u}_i = \frac{\mathbf{x_{i+1}} - \mathbf{x_i}}{\|\mathbf{x_{i+1}} - \mathbf{x_i}\|} \in \mathbb{R}^3$ in the direction between the two points i and $i+1$ should be equal the reference direction **ref** rotated by the quaternion $\mathfrak{q}_\mathbf{i}$. Let us consider a rotation of an angle θ around an axis $\mathbf{v} \in \mathbb{R}^3$ (with $\|\mathbf{v}\| = 1$). The corresponding quaternion is then

$$\mathfrak{q} = (q_0, q_1, q_2, q_3) = (q_0, \mathbf{q}) = (\cos\frac{\theta}{2}, \sin\frac{\theta}{2}\mathbf{v})$$

The image $\mathbf{b} \in \mathbb{R}^3$ of a vector $\mathbf{a} \in \mathbb{R}^3$ by a rotation represented by a quaternion $\mathfrak{q}$ can be calculated by: $(0, \mathbf{b}) = \mathfrak{q} \cdot (0, \mathbf{a}) \cdot \overline{\mathfrak{q}}$ where $\overline{\mathfrak{q}} = (q_0, -\mathbf{q})$ is the conjugated of $\mathfrak{q}$ and where the quaternion product

$$\mathfrak{p} \cdot \mathfrak{q} = (p_0 q_0 - \mathbf{p}^T\mathbf{q}, p_0\mathbf{q} + q_0\mathbf{p} + \mathbf{p} \times \mathbf{q})$$

is used. For a constant $\mathbf{ref} = (0, 1, 0)$, the image can be directly calculated as

$$Rot(\mathbf{ref}) = (2(q_1q_2 - q_0q_3), q_0^2 - q_1^2 + q_2^2 - q_3^2, 2(q_2q_3 - q_0q_1)).$$

The energy is defined as

$$E_{Coh,i} = k_{Coh}\|(0, \mathbf{u_i}) - \mathfrak{q}_\mathbf{i} \cdot (0, \mathbf{ref}) \cdot \overline{\mathfrak{q}_\mathbf{i}}\|^2$$

In the same manner as for the length conservation, we introduce an integral force for each of the three components of the coherence between quaternions and positions.

5.5 Quaternion norm

Quaternions represent a pure rotation only when they have a unit norm, otherwise a scaling effect is introduced. The energy is simply defined as

$$E_{QuatNorm,i} = k_{QuatNorm}(\|\mathfrak{q}_\mathrm{i}\| - 1)^2$$

for $i \in 1...n-1$ with $\|\mathfrak{q}\| = \sqrt{q_0^2 + q_1^2 + q_2^2 + q_3^2}$. This term is easy to enforce and does not pose any difficulty.

5.6 Bending and torsion

The bending and torsion forces are calculated jointly. They are determined by the two consecutive quaternions $\mathfrak{q}_\mathbf{i}$ and $\mathfrak{q}_\mathbf{i+1}$. (Observe that the previous formulations needed three segments and that the Frenet torsion is not needed anymore.) The relative rotation from segment i to segment $i+1$ is represented by

$$\mathfrak{q}_{\mathbf{i}\to\mathbf{i+1}} = \mathfrak{q}_{\mathbf{i+1}}\overline{\mathfrak{q}_\mathbf{i}}.$$

(The inverse of a quaternion $\mathfrak{q}$ is $\frac{\overline{\mathfrak{q}}}{\|\mathfrak{q}\|}$ and for unitary quaternions simply $\overline{\mathfrak{q}}$.). This rotation of an angle θ around an axis $\mathbf{v}$ is nothing else but the integral of the Darboux vector over the length of a segment:

$$\theta \mathbf{v} = \int_{s_i}^{s_{i+1}} \omega(s)\,\mathrm{d}s.$$

We also have

$$\mathfrak{q}_{\mathbf{i}\to\mathbf{i+1}} = (\cos\frac{\theta}{2}, \sin\frac{\theta}{2}\mathbf{v}).$$

Since $\frac{\theta}{2}$ can be expected to be small, it follows the approximation

$$\omega = \frac{2}{L}\mathfrak{q}_{\mathbf{i}\to\mathbf{i+1}}.$$

On the other hand, the general properties of the Darboux vector give

$$\omega = \kappa\mathbf{B} + \tau\mathbf{T}.$$

The curvature κ and the torsion τ can then be calculated. The tangent vector $\mathbf{T}$ is defined (symmetrically in $\mathbf{u_i}$ and $\mathbf{u_{i-1}}$) as

$$\mathbf{T}_i = \frac{\mathbf{u_{i-1}} + \mathbf{u_i}}{\|\mathbf{u_{i-1}} + \mathbf{u_i}\|}.$$

A decomposition of ω on the basis formed by $\mathbf{T}$ and $\mathbf{B}$ (which is defined as the direction of the residual component $\omega - \tau\mathbf{T} = \kappa\mathbf{B}$) gives us:

$$\tau = \omega \cdot \mathbf{T}$$

and

$$\kappa = \|\omega - \tau\mathbf{T}\|.$$

The energy is

$$E_{BendingTorsion,i} = E_{Bending,i} + E_{torsion,i} = \frac{1}{2}B_\kappa\kappa^2 + \frac{1}{2}B_\tau\tau^2$$

A possibility would be to use only the quaternions as coordinates and calculate iteratively the positions. The disadvantage is that the system matrix would be a full matrix (the position of point i depends on all the quaternions between 1 and $i-1$), which makes the calculation much slower. External forces like contact forces are also difficult to implement.

5.7 Handles

Each handle (fixed point) is attached either to a point or to a segment. The two possibilities are offered for an easy interface with the graphical representation: the user can attach it either to a sphere or a cylinder. The handle can fix either only the position or both the position and orientation at a point. In the case of a sphere, the orientation is taken as the (normed) mean value of the two quaternions surrounding it; in the case of a cylinder, the position is the mean value of the positions of the two surrounding points. The energy, if

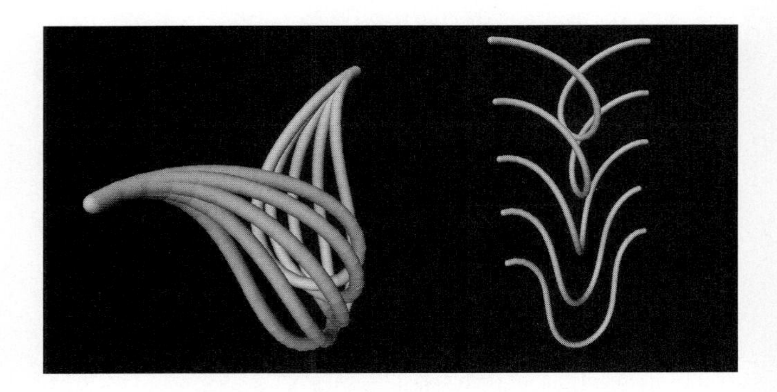

Figure 2: Influence of the torsion : the same cable is submitted to different values of the torsion by rotating one of its extremities

a handle is associated with the point $\mathbf{x_i}$ and with the quaternion $\mathfrak{q}_\mathbf{i}$, is

$$E_{HandleSphere,i} = k_{Handle}((\mathbf{x_i} - \mathbf{x_{Handle}})^2 + (\frac{\mathfrak{q}_\mathbf{i-1} + \mathfrak{q}_\mathbf{i}}{\|\mathfrak{q}_\mathbf{i-1} + \mathfrak{q}_\mathbf{i}\|} - \mathfrak{q}_\mathbf{Handle})^2)$$

or

$$E_{HandleCylinder,i} = k_{Handle}(\frac{\mathbf{x_i} - \mathbf{x_{i+1}}}{2} - \mathbf{x_{Handle}})^2 + (\mathfrak{q}_\mathbf{i} - \mathfrak{q}_\mathbf{Handle})^2)$$

5.8 Contact forces

When a collision is detected (the mass point i has penetrated inside of another body), we apply a spring and an integral force at this point to keep it on the surface. Let $\mathbf{y} \in \mathbb{R}^3$ be the orthogonal projection of $\mathbf{x_i}$ on the surface and $\mathbf{n} \in \mathbb{R}^3$ the normal at $\mathbf{y}$, oriented towards the outside and normed. The penetration distance is then $d = (\mathbf{y} - \mathbf{x_i})^T\mathbf{n}$ and the force is

$$\mathbf{f_{Collision,\,i}} = k_{Collision}\,d\mathbf{n} + f_{I,\,Collision,\,i}\mathbf{n}$$

and the energy

$$E_{Collision,\,i} = \frac{1}{2}k_{Collision}\,d^2 + f_{I,\,Collision,\,i}\,d$$

When the penetration distance of a point is 0 (within the tolerance), if its integral force $f_{I,\,Collision,\,i}$ is negative, it means that the point needs to be attracted towards the object in order to be on the surface: in this case,it is not a collision point anymore and the spring and the integral forces are removed.

$f_{I,Collision,i}$ can be updated with two different methods: an individual or a global one. The first method is similar to the one we used for the length:

$$f_{I,i,new} = f_{I,i,old} + k_{I,Collision}d$$

The second method uses the derivative of the forces as a predicator for the behavior of the system upon a small change of the values of the forces. The Taylor series for the forces is $\mathbf{F}(\mathbf{X}+\mathbf{h}) = \mathbf{F}(\mathbf{X}) + \mathbf{H}\mathbf{h}$ with $\mathbf{h} \in \mathbb{R}^{7n-4}$ the difference between the old and the new position. Since the old position is an equilibrium, $\mathbf{F}(\mathbf{X}) = \mathbf{0}$. In order to fulfill the non-penetration constraint, we need that

$$\mathbf{h}^T\tilde{\mathbf{n}} = d$$

with $\tilde{\mathbf{n}} \in \mathbb{R}^{7n-4}$ the "direction" of the constraint: $\tilde{\mathbf{n}}_k = 0$ except for $\tilde{\mathbf{n}}_{7i+1}$, $\tilde{\mathbf{n}}_{7i+2}$, $\tilde{\mathbf{n}}_{7i+3}$ that take the values of $\mathbf{n}$. Additionally, the new force is $\mathbf{F}(\mathbf{X}+\mathbf{h}) = \Delta f_I\tilde{\mathbf{n}}$ (the variation of the integral force

$\Delta f_I = f_{I,mew} - f_{I,old}$ is not included in the Hessian). Combining the two equations leads to

$$\Delta f_I \tilde{\mathbf{n}} = \mathbf{Hh}$$

and further to

$$\Delta f_I = \frac{d}{\tilde{\mathbf{n}}^T \mathbf{H}^{-1} \tilde{\mathbf{n}}}$$

In the case of several contact points, the new force is $\mathbf{F}(\mathbf{X}+\mathbf{h}) = \sum_{i \in Contactpoints} \Delta f_{I,i} \tilde{\mathbf{n}}_\mathbf{i}$ and the condition to fulfill $\tilde{\mathbf{n}}_\mathbf{i}^T \mathbf{h} = d_i$. The combination of these relations results in a matrix equation:

$$\mathbf{M}(\Delta f_{I,1}, ..., \Delta f_{I,c})^T = (d_1, ..., d_c)^T$$

with c the number of contact points and $\mathbf{M} \in \mathbb{R}^{c \times c}$ a matrix such that $\mathbf{M}_{i,j} = \tilde{\mathbf{n}}_\mathbf{i}^T \mathbf{H}^{-1} \tilde{\mathbf{n}}_\mathbf{j}$. Although we have to solve the above system, this method converges experimentally much faster than the other: it only needs one or two iterations to find the correct values for the f_I. It cannot be used as such in the case of the length conservation or of the coherence between positions and quaternions since the direction $\tilde{\mathbf{n}}$ would not be constant.

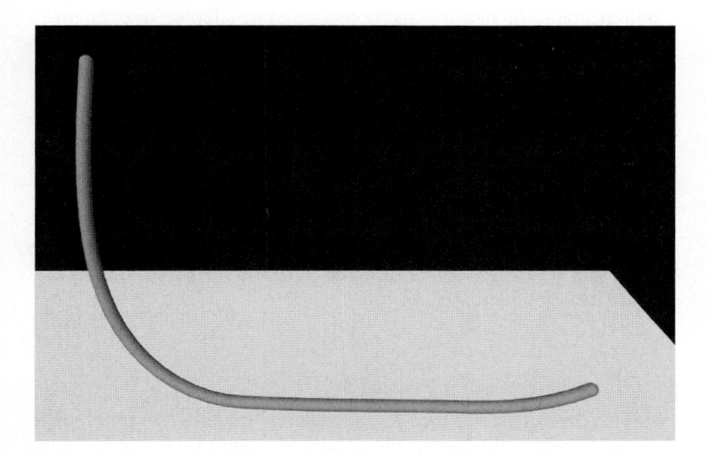

Figure 3: Deformation of a cable due to the contact to a plane

6 Solver

6.1 General principles

Since the cables and hoses do not have a high dynamic range, considering a static solution at each time step is sufficient for most of the applications we are concerned with, like wire routing or assembly simulation. Dynamic phenomena like fast oscillations are excluded by the quasi-static system: the system is after each time step at equilibrium. Numerical oscillations, which are often a problem for stiff spring-mass systems, are also excluded by the absence of speed as variables. We use an energy minimizing algorithm for solving the system. Note that the system could be easily modified to become a dynamic one, introducing the velocities as supplementary variables and solving the system for example with the algorithm proposed by Baraff [Baraff 1996]. It would be necessary to add damping forces for each kind of interaction, following the formulation of Baraff: if the constraint is $C(\mathbf{X}) = 0$, the energy is $E = \frac{1}{2} k C(\mathbf{X})^T C(\mathbf{X})$, the forces are of the form $-k \frac{\partial C(\mathbf{X})}{\partial \mathbf{X}} C(\mathbf{X})$ and the damping forces of the form $-k \frac{\partial C(\mathbf{X})}{\partial \mathbf{X}} \dot{C}(\mathbf{X})$.

If the norm of the forces vector $\mathbf{F}$ is null (in practice small enough $\|\mathbf{F}\| < \varepsilon$ in order to account for numerical error), the system is at equilibrium and we have the solution we were looking for. If not, we minimize the energy until we find an equilibrium. The basic hypothesis is that the new solution (for slightly modified conditions: a point has been moved between the two frames for example) should be near to the old one, and the old solution vector $\mathbf{X_{old}}$ can thus be used as a good starting point for the search of the new solution. Our algorithm is iterative. In each loop, a sequence of different algorithms is used until a smaller value for the energy is found. If a particular algorithm gives a solution, $\mathbf{X_{old}}$ is replaced by the new solution and a new loop begins until the equilibrium is reached. If it does not find a better solution, the next algorithm is used. When the difference either in position $\|\mathbf{X_{new}} - \mathbf{X_{old}}\|$ or in energy $\|E_{new} - E_{old}\|$ is smaller than a predetermined value, the loop is stopped. The different algorithms that we use are in order: Newton's Method applied to the forces, non-linear Conjugate Gradient Method, Linear Conjugate Gradient Method, Steepest Descent Method and if all else fails, a linear search along the forces. It is important to note that in the immense majority of the cases, only the first or the first two are used; the other algorithms serve as a security for particular stiff cases or for tuning the different constants for the interactions. This structure allows us to have at the same time a fast response in usual cases while retaining the robustness necessary to deal with stiff cases.

6.2 Individual algorithms

The first algorithm is the Newton Method applied to the forces. It is particularly efficient near the equilibrium. However, it is well-known that it only converges if the starting solution is near enough to the equilibrium. If the energy in a point $\mathbf{X}+\mathbf{h}$ is approximated by the Taylor series in $\mathbf{X}$, $E_{approx} = E(\mathbf{X}) - \mathbf{F}^T \mathbf{h} - \frac{1}{2} \mathbf{h}^T \mathbf{Hh}$. This approximation is minimum when its gradient is null, i.e. for $\mathbf{F} + \mathbf{Hh} = 0$ which leads to $\mathbf{h} = -\mathbf{H}^{-1}\mathbf{F}$ and $\mathbf{X_{new}} = \mathbf{X_{old}} - \mathbf{H}^{-1}\mathbf{F}$. For inverting the Hessian, which is a strongly banded matrix, we use a simple Gauss algorithm slightly modified for taking into account all the zero-elements of the matrix.

The conjugate gradient method uses a family of $\mathbf{H}$-conjugated vectors (two vectors $\mathbf{u}$ and $\mathbf{v}$ are $\mathbf{H}$-conjugated if $\mathbf{u}^T \mathbf{Hv} = 0$ with $\mathbf{H}$ a symmetric positive definite matrix) to look iteratively for the solution to the system $\mathbf{Hh} + \mathbf{F} = 0$. In the non-linear method, the matrix $\mathbf{H}$ is recalculated during the process, while the linear method keeps it constant.

The steepest descent and the line search use the forces as a search direction (The forces are the opposite of the gradient, and thus indicate the direction of the steepest descent). Both methods look for a coefficient α such that $\mathbf{X_{new}} = \mathbf{X_{old}} + \alpha \mathbf{F}$ has a minimum energy. In the case of the steepest descent, the Hessian is used to approximate $E(\alpha) = E(\mathbf{X_{new}}(\alpha))$ by a parabola and finding its minimum, which is found for $\alpha = -\frac{\mathbf{F}^T \mathbf{F}}{\mathbf{F}^T \mathbf{HF}}$. The linear search uses decreasing powers of 2 ($\alpha = \frac{1}{2^k}$) until it finds a solution with a lower energy.

7 Integration in a Virtual Reality environment

The simulation is integrated in the Virtual Reality software *veo* of DaimlerChrysler Research and Technology. It provides the whole environment for the simulation, such as graphics, scenes and objects handling... The cable is represented as a sequence of spheres and cylinders. The spheres are set at the discretization points. The

Number of Points n	Mean Calculation Time in ms
10	6,25
20	12,95
30	19,49
40	21,43
50	25,55
60	27,11
80	29,89
100	38,86
120	45,94
140	54,97
160	60,92
180	80,04
200	77,09
250	104,45
300	128,31

Table 1: Influence of the number of points on calculation time. The calculation time is proportional to the number of discretization points. The banded structure of the Hessian allow a good scalability of the system. Performance of simulation on a Pentium M 1.5GHz.

cylinders are set along the segments joining two consecutive points. The cylinders and spheres can be moved (in the geometrical limits permitted by the simulation) easily by selecting them and moving them with the spacemouse. At each frame, the simulation checks whether the eventually selected objects in the scene correspond to a part of a cable. If yes, its new position and orientation are taken into account, the simulation calculates the new solution and the new positions and orientations of all the spheres and cylinders are passed to the Virtual Reality software for representation.

A seamless integration in the processes already in place is necessary for the users. The data from the construction - usually in CATIA - are tessellated and converted automatically to the OpenInventor format, used for the Virtual Reality application. We also have an interface that allows to convert all or part of a tessellated wire harness or hose to a flexible one having the same characteristics (like length, radius, position...). A partial conversion is useful since some of the construction parts represent a huge harness extended in the whole vehicle. The connectivity information of the tessellation triangles is parsed to create a graph representing the interconnections of the vertices. Geometrical considerations allow to find cross-sections on the surface at a given point that we use for cutting the tree at the appropriate vertices and for calculating the coordinates of points on the centerline, from which we construct the new flexible object. A copy of the tessellation with the remaining points is also created for the graphical representation. A complete new cable can also be created between two points of the scene.

A graphical user interface is available: it groups the commands that are useful for the simulation. The simulation can be turned on or off, handles can be added or removed at any selected point or can be set equidistant automatically along the cable, geometrical parameters like the radius or the length and material parameters like the Young Module and the Poisson coefficient can be changed interactively, the new centerline of the cables can be exported to CATIA via the VDA-FS (*Verband der Automobileindustrie - Flaechen-Schnittstelle*)format and can then be used as a draft for the construction. Our tool has already been used and tested by pilot users who have reacted positively to the possibilities offered to them. It is used in various contexts: on the one hand during assembly simulation and the other hand for cable routing. In assembly simulation, flexible parts like cables or hoses are often in the way for mounting another part of the vehicle. The idea is to make the problematic part flexible (which is represented as a rigid body), push it apart out

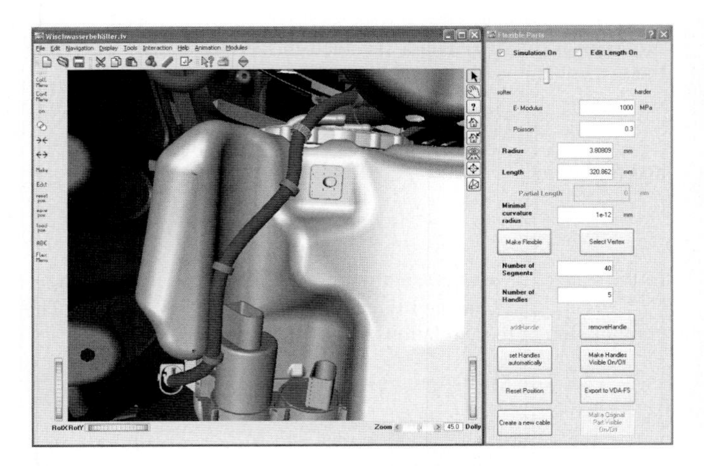

Figure 4: The Graphical User Interface

of the way as it could be done with the physical parts and continue with the assembly simulation. For wire routing, a new path for the cable needs to be defined and constructed or adapted to changes in the construction of other parts. The length of the cable may need to be adapted to the new form. We have two special modes for this, changing the handling of the stretch forces. In both cases, the integral forces relative to the length are removed. In the first mode, the user chooses two points on the cable between which the segments can be elongated either by pulling them or by typing the new length in a text field in the GUI. The reference length is then replaced either by the actual length of the segment or by the value of the text input. The length can then be changed smoothly from one iteration to another until the user is satisfied with the result. In the second mode, the stretch energy is set to

$$E_{Length,i} = \frac{k_{Length}}{2} L_i^2$$

with a very soft value for k_{Length}. The cable tends to stretch or to become shorter, finding automatically an optimal length for given endpoints and handles, thus allowing the user to route intuitively without worrying about the length. Playing with the ratio of k_{Length} and the bending and torsion stiffnesses make the form of the cable vary form "rather straight" to "rather bend".

8 Conclusion and future work

In this paper, we presented a virtual environment suitable for the simulation of cables. Our approach modeled the cable with an extended spring-mass system that was solved with an energy minimizing algorithm. The cable was modeled using the Cosserat theory, taking into account the conservation of length, the weight, the bending and the torsion. In order to easily formulate the bending and torsion interaction, we used a mixed coordinate system where each mass point had three space coordinates and the orientation of each segment was represented by a normalized quaternion. Each type of interaction was then calculated on the base of the coordinate type that is best suited: on the one hand, the length conservation and the contact forces with the positions, on the other hand the bending and torsion with the quaternions. Additionally, forces for the coherence between positions and quaternions and for the normalization of quaternions were employed.

For constraints that must be exactly enforced, like the length conservation, the coherence between positions and quaternions or the collision, we added an integral force at each point in the same direction as the proportional spring force. The system was solved for a constant value of those forces which were then updated as a function of the result. This mechanism allowed softer springs (and thus a less stiff system) while at the same time exactly enforcing the constraints.

Future work will include the extension of the functionalities of the user interface to better meet user needs, a better embedding of the collision response and an extension of the tool to also simulate other kinds of flexible parts.

References

ANTMAN, S. S. 1995. *Nonlinear Problems of Elasticity*, vol. 107 of *Applied Mathematical Sciences*. Springer Verlag.

BARAFF, D. 1996. Linear-time dynamics using Lagrange multipliers. *Computer Graphics 30*, Annual Conference Series, 137–146.

BARZEL, R. 1997. Faking dynamics of ropes and springs. *IEEE Comput. Graph. Appl. 17*, 3, 31–39.

BUCK, M., AND SCHÖMER, E. 1998. Interactive rigid body manipulation with obstacle contacts. *Journal of Visualization and Computer Animation 9*, 243–257.

FINCKH, H., STEGMAIER, T., AND PLANCK, P. H., 2004. Numerische Simulation der mechanischen Eigenschaften textiler Flächengebilde - Gewebeherstellung.

GOSS, V. G. A., VAN DER HEIJDEN, G. H. M., THOMPSON, J. M. T., AND NEUKIRCH, S., 2005. Experiments on snap buckling, hysteresis and loop formation in twisted rods.

HERGENROETHER, E., AND DAEHNE, P., 2000. Real-time virtual cables based on kinematic simulation.

LOOCK, A., AND SCHÖMER, E. 2001. A virtual environment for interactive assembly simulation: From rigid bodies to deformable cables. In 5th *World Multiconference on Systemics, Cybernetics and Informatics (SCI'01)*, vol. 3, 325–332.

PAI, D. K. 2002. Strands: Interactive simulation of thin solids using cosserat models. *Comput. Graph. Forum 21*, 3, 347–352.

RUBIN, M. B. 2000. *Cosserat Theories: Shells, Rods and Points*. Kluwer Academic Publ., Dordrecht,.

SAUER, J., AND SCHÖMER, E. 1998. A constraint-based approach to rigid body dynamics for virtual reality applications. In *Proc. ACM Symposium on Virtual Reality Software and Technology*, 153–161.

Computing Surface Hyperbolic Structure and Real Projective Structure

Miao Jin *
Stony Brook University

Feng-Luo†
Rutgers University

Xianfeng Gu‡
Stony Brook University

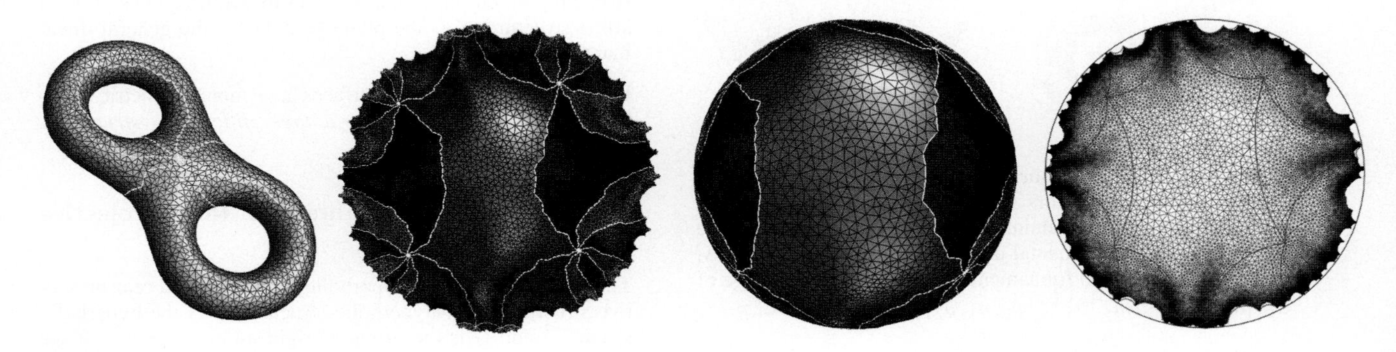

Figure 1: **The hyperbolic structure and real projective structure of a genus two closed surface. From left to right, the original surface with a set of canonical fundamental group basis; the isometric embedding of its universal covering space in the Poincaré disk with the hyperbolic uniformization metric; the isometric embedding in the Klein disk which induces the real projective structure; the isometric embedding in the Poincaré disk, where the boundaries of fundamental domains are straightened to hyperbolic lines.**

Abstract

Geometric structures are natural structures of surfaces, which enable different geometries to be defined on the surfaces. Algorithms designed for planar domains based on a specific geometry can be systematically generalized to surface domains via the corresponding geometric structure. For example, polar form splines with planar domains are based on affine invariants. Polar form splines can be generalized to manifold splines on the surfaces which admit affine structures and are equipped with affine geometries.

Surfaces with negative Euler characteristic numbers admit hyperbolic structures and allow hyperbolic geometry. All surfaces admit real projective structures and are equipped with real projective geometry. Because of their general existence, both hyperbolic structures and real projective structures have the potential to replace the role of affine structures in defining manifold splines.

This paper introduces theoretically rigorous and practically simple algorithms to compute hyperbolic structures and real projective structures for general surfaces. The method is based on a novel geometric tool - *discrete variational Ricci flow*. Any metric surface admits a special uniformization metric, which is conformal to its original metric and induces constant curvature. Ricci flow is an efficient method to calculate the uniformization metric, which determines the hyperbolic structure and real projective structure.

The algorithms have been verified on real surfaces scanned from sculptures. The method is efficient and robust in practice. To the best of our knowledge, this is the first work of introducing algorithms based on Ricci flow to compute hyperbolic structure and real projective structure.

More importantly, this work introduces the framework of general geometric structures, which enable different geometries to be defined on manifolds and lay down the theoretical foundation for many important applications in geometric modeling.

CR Categories: I.3.5 [Computer Graphics]: Computational Geometry and Object Modeling—Geometric algorithms;

Keywords: Geometric Structures, Hyperbolic Structure, Real Projective Structure, Hyperbolic geometry, Real Projective Geometry, Affine Geometry, Ricci Flow, Riemann Uniformization

1 Introduction

According to Felix Klein's Erlanger program, a geometry is the study of properties of a space X invariant under a group G of transformations of X. For example, planar Euclidean geometry is the geometry of 2-dimensional Euclidean space $\mathbb{R}^2$ invariant under rigid motions (translations, rotations). The central invariant is the distance between any two points. Planar affine geometry studies the invariants of the plane under affine transformations (non singular linear maps), where the invariants are parallelism, barycentric coordinates. Real projective geometry on real projective space $\mathbb{RP}^2$ studies the invariants under projective transformation (linear rational maps), where the cross ratio is the central invariant.

*e-mail: mjin@cs.sunysb.edu
†e-mail:fluo@math.rutgers.edu
‡e-mail:gu@cs.sunysb.edu

SPM 2006, Cardiff, Wales, United Kingdom, 06–08 June 2006.
© 2006 ACM 1-59593-358-1/06/0006 $5.00

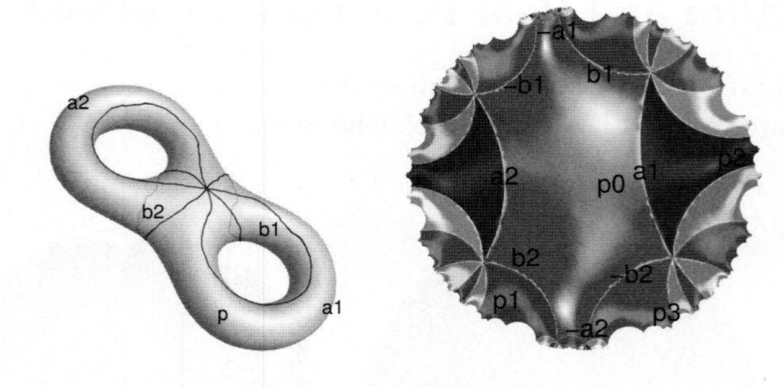

Figure 2: A genus two surface with a set of canonical fundamental group generators $\{a_1, b_1, a_2, b_2\}$ is on the left. A finite portion of its universal covering space is shown on the right. Different fundamental domains are colored differently. The boundary of each fundamental domain is the preimage of $a_1 b_1 a_1^{-1} b_1^{-1} a_2 b_2 a_2^{-1} b_2^{-1}$. $\{p_0, p_1, p_2\}$ are preimages of p.

Most algorithms in geometric modeling, computational geometry and computer graphics are constructed on planar spaces, in which different algorithms are based on different geometries. For example, planar Delaunay triangulation uses Euclidean geometry, and the distances among points play the central role; Splines with planar domains based on polar forms use affine geometry, and barycentric coordinates play central roles. The fundamental task of geometric modeling is to study shapes, therefore it is highly desirable to find a systematic way to generalize the conventional mature planar constructions onto the surfaces. Hence, we need a solid theoretical tool to define different geometries on surfaces.

Geometric structures are natural surface structures, which enable different geometries to be defined on surfaces coherently and allow general planar algorithmic constructions to be generalized onto surfaces directly.

1.1 Geometric Structures

Surfaces are *manifolds*, where no global coordinates exist in general. Instead, a manifold M is covered by a set of open sets $\{U_\alpha\}$. Each U_α can be parameterized by a local coordinate system and a map $\phi_\alpha : U_\alpha \to \mathbb{R}^2$ maps U_α to its parameter domain. (U_α, ϕ_α) is a local chart for the manifold M. A particular point p may be covered by two local coordinate systems (U_α, ϕ_α) and (U_β, ϕ_β). The transformation of the local coordinates of p in (U_α, ϕ_α) to those in (U_β, ϕ_β) is formulated as the chart transition map $\phi_{\alpha\beta} = \phi_\beta \circ \phi_\alpha^{-1}$. All the charts form an *atlas* $\{(U_\alpha, \phi_\alpha)\}$.

If all chart transition maps are rigid motions on $\mathbb{R}^2$, then we can discuss the concepts of angle, distance, and parallelism on the surface locally. These geometric measurements can be calculated on one chart, but the results are independent of the choice of the chart. Namely, we can define Euclidean geometry on the surface. Similarly, if all transition maps are affine, then we can define parallelism and midpoint of a line segment on the surface. If all transition maps belong to a particular transformation group of $\mathbb{R}^2$, we can define the corresponding geometry on the surface. We say a surface M has a (X, G) *structure*, where X is a topological space, G is a subgroup of the transformation group of X, if M has an atlas $\{(U_\alpha, \phi_\alpha)\}$, such that the parameter domain $\phi_\alpha(U_\alpha) \subset X$ is in space X, and the transition maps $\phi_{\alpha\beta} \in G$ are in G.

Surfaces have rich (X, G) geometric structures. A genus zero surface has a spherical structure, where X is the unit sphere $\mathbb{S}^2$ and G is the rotation group. A genus one surface has an affine structure, which plays vital roles in manifold splines. For an affine structure, X is the plane $\mathbb{R}$ and G is the general linear maps $GL(\mathbb{R}, 2)$.

Besides (X, G) structures, surfaces have more geometric structures, such as *topological structures*, *differential structures*, and *conformal structures*.

1.2 Hyperbolic Structure and Real Projective Structure

This paper focuses on hyperbolic structures and real projective structures. For hyperbolic structure, X is the hyperbolic space $\mathbb{H}^2$ and G is the group of rigid motions on $\mathbb{H}^2$. If we use Poincaré model for $\mathbb{H}^2$, then G is the Möbius transformation group, (complex linear rational maps). For real projective structure, X is $\mathbb{RP}^2$ and G is the real projective maps $PGL(\mathbb{R}, 3)$, (real linear rational maps). Compared with the affine structure, whose transition maps are linear, hyperbolic and real projective structures have more complicate transition maps. But compared with other structures, such as differential and conformal structures, they have much simpler transition maps.

The existence of different (X, G) structures are determined by the topology of the surface. For example, a surface has an affine structure if and only if its Euler characteristic number is zero [Benzècri 1959; J.W.Milnor 1958]. Conventional polar form splines are based on affine geometry, therefore, it is impossible to generalize conventional splines on the surface without extraordinary points. Fortunately, all surfaces have real projective structures, and if a spline scheme is based on cross ratio, it can be generalized on surfaces directly.

Real projective structures are closely related to hyperbolic structures. Hyperbolic structures induce real projective structures in a natural way.

1.3 Previous Works

Geometric structures have been implicitly and explicitly applied in geometric modeling, computer graphics and medical imaging. For the genus zero case, the spherical structure was studied for texture mapping in [Gotsman et al. 2003; Praun and Hoppe 2003] and for conformal brain mappings in [Gu et al. 2004; Haker et al. 2000] . Algorithms for computing conformal structures were introduced in [Gu and Yau 2003; Jin et al. 2004], and the method is based on computing holomorphic differentials on surfaces.

Hyperbolic structure was applied in [Ferguson et al. 1992] for topological design of surfaces, where the high genus surfaces were represented as quotient spaces of the Poincaré disk over Fuchsian group actions. In [Grimm and Hughes 2003], Grimm and Hughes defined parameterizations for high genus surfaces and constructed functions on them. Wallner and Pottmann introduced the concept of spline orbifold in [Wallner and Pottmann 1997], which defined splines on three canonical parameter domains, the sphere, the plane and the Poincaré. The key difference between these works and our current one is that our method computes the hyperbolic metric which is conformal to the original metric on the surface, but their works only consider the topology and ignore the geometry of the surface. For many real applications, such as texture mapping, shape analysis and spline constructions, conformality between the original and the final metrics is highly desirable.

Manifold splines based on polar forms are introduced in [Gu et al. 2005]. They demonstrated the equivalence between manifold splines and the affine structure and gave a systematic way to generalize splines defined on planar domains to manifold domains. They also constructed affine structures using conformal structures for surfaces with arbitrary topologies.

Recently, [Goldman 2003] examines some possible alternative mathematical foundations for Computer Graphics, such as Grassmann spaces and tensors. General geometric structures on manifolds contribute to the theoretical foundations for graphics and geometric modeling.

Ricci flow on surfaces was first introduced by Hamilton [Hamilton 1988]. Theoretical results of combinatorial Ricci flow have been summarized in [Chow and Luo 2003]. Conventional Ricci flow can be formulated as the gradient descent method for optimizing a special energy form, and the deficiency of its speed makes Ricci flow impractical. In our current work, we improved the theoretical results in [Chow and Luo 2003] by considering surface Riemannian metric induced from $\mathbb{R}^3$ instead of from the combinational structure. We replaced the gradient descent method with Newton's method to speedup Ricci flow completion by tens of times. We named this novel algorithm the *discrete variational Ricci flow*. A practical system for computing hyperbolic and real projective structures for real surfaces has been developed based on discrete variational Ricci flow.

Circle packing was first introduced by Thurston in the seventies in [Thurston 1976], with an algorithm. A practical software system for circle packing can be found in [Stephenson 2005]. The hyperbolic metrics computed in their system are not conformal to the original metrics. Recently, circle packing has been generalized to circle patterns [Bobenko and Schroder 2005] [Bobenko and Springborn 2004] and used for surface parameterization in [Kharevych et al. 2005], which focuses on Euclidean geometry. Circle packing, circle pattern and discrete Ricci flow can be unified using the derivative cosine law [Luo and Gu 2006].

Our current work is based on a novel theoretical tool - discrete variational Ricci flow and focuses on hyperbolic structure and real projective structure instead of Euclidean structure. Furthermore the hyperbolic metrics computed using our method are conformal to the original metrics. The conformal hyperbolic metrics convey much geometric information of the surfaces, which are valuable for the purposes of shape analysis.

1.4 Contribution

This paper introduces general geometric structures, such as the affine structure, the hyperbolic structure, and the real projective structure. Geometric structures allow different geometries to be defined on manifolds directly and lay down the theoretical foundations for graphics and geometric modeling.

Compared to other structures, the hyperbolic structure and real projective structure have not been fully studied. This paper aims to introduce a novel practical algorithm to compute hyperbolic structure and real projective structure. The algorithm is based on a recently developed theoretical tool in differential geometry field- Ricci flow. To the best of our knowledge, this paper is the first to numerically compute hyperbolic structure using Ricci flow, and also the first one to introduce a practical method to compute real projective structures for arbitrary closed surfaces. Therefore, the major contributions of this paper are:

- Introduce a novel theoretical framework : Geometric Structures, which enable algorithms defined on Euclidean domains to be systematically generalized to manifold domains.
- Introduce a novel geometric tool : discrete Variational Ricci flow.
- Practical, efficient algorithm for computing hyperbolic structures for surfaces of genus at least two. The hyperbolic metric is conformal to the original metric on the surface.
- Practical and efficient algorithm to compute Real projective structures for arbitrary surfaces.

2 Theoretical Background

The algorithms require some basic concepts from algebraic topology, differential geometry and Riemann surface. We assume the readers are familiar with these subjects. In this section, we briefly review the basic concepts necessary for our discussion. For detailed explanation, we refer readers to [Thurston 1997].

2.1 (X,G) structure for Manifolds

Given a manifold S, suppose (U_α,ϕ_α) and (U_β,ϕ_β) are two charts, $U_\alpha \cap U_\beta \neq \Phi$, then the chart transition is defined as

$$\phi_{\alpha\beta} : \phi_\alpha(U_\alpha \cap U_\beta) \to \phi_\beta(U_\alpha \cap U_\beta).$$

In what follows X will be a space with a geometry on it and G is the group of transformations of X which preserves this geometry. An atlas $\{U_\alpha,\phi_\alpha\}$ on a manifold is called *(X,G) atlas* if all chart transitions are in group G and the parameter domains $\phi_\alpha(U_\alpha)$ are in X, $\phi_\alpha(U_\alpha) \subset X$.

Two (X,G) atlases are *compatible*, if their union is still an (X,G) atlas. Compatible relation is an equivalent relation of atlases, and each equivalent class is called an (X,G) *structure*.

Any surface with negative Euler number has a hyperbolic structure $(\mathbb{H}^2, PGL(2,\mathbb{C}))$. Suppose $\{(U_\alpha,\phi_\alpha)\}$ is a hyperbolic atlas, then $\phi_\alpha(U_\alpha)$ is in the hyperbolic space and the transition maps are hyperbolic rigid motions. If we use Poincaré model, $\phi_{\alpha\beta}$ is a Möbius transformation.

Similarly, for a real projective atlas, the parameter domains are subsets of real projective space $\mathbb{RP}^2$, and the transition maps are real projective transformations $PGL(3,\mathbb{R})$ (linear rational functions).

2.2 Riemann Surface Uniformization

Any orientated metric surface M has a special $(\mathbb{C},\Omega)$structure, where Ω represents the holomorphic functions on the complex plane. Namely, the chart transition maps are conformal maps. This structure is called the *conformal structure*. A surface with a conformal structure is called a *Riemann surface*.

Suppose a surface M is embedded in the Euclidean space $\mathbb{R}^3$, then it has the Riemannian metric induced by the Euclidean metric, denoted as g_0. A *conformal class of metrics* on M refers to a family of Riemannian metrics $\{e^{2\phi}g_0\}$ and the so called *conformal factor* $e^{2\phi}$ represents the area stretching under the metric $e^{2\phi}g_0$.

The Riemann Uniformization theorem claims that any Riemann surface M with metric g_0 admits a uniformization metric g, such that g and g_0 are conformal to each other $g=e^{2\phi}g_0$ and (M,g) has constant Gaussian curvature on the interior points and zero geodesic curvature on the boundary points. The constant Gaussian curvature is one of the three $\{-1,0,1\}$. Surfaces with positive Euler numbers have spherical uniformization metrics with $+1$ curvature; surfaces with zero Euler number have flat uniformization metrics with 0 curvature; surfaces with negative Euler numbers have hyperbolic uniformization metrics with -1 curvature. The uniformization metrics induce the spherical, Euclidean and hyperbolic structures on surfaces respectively.

2.3 Ricci flow

Surface Ricci flow is first introduced by Hamilton in [Hamilton 1988]. The main idea is to conformally deform the Riemannian metric of the surface driven by its curvature, then the metric will flow to the uniformization metric eventually.

Suppose M is a surface with the metric tensor $g=(g_{ij})$, and K is the current Gaussian curvature, then Ricci flow is defined as

$$\frac{\partial g_{ij}}{\partial t} = -2Kg_{ij}.$$

It is proven that the Ricci flow with normalized total surface area will flow the metric such that the Gaussian curvature on the surface is constant, namely the Gaussian curvature function $\lim_{t\to\infty}K(t,p)$ converges to a constant function. It is also proven that the convergence rate of Ricci flow is exponential, namely, for any point p on the surface,

$$|K(t,p)-K(\infty,p)| < c_1 e^{-c_2 t},$$

where c_1, c_2 are two constants determined by the surface itself.

2.4 Fundamental Group and Universal Covering Space

If two closed curves on surface M can deform to each other without leaving the surface, they are homotopic to each other. The closed loops on the surface can be classified by homotopic classes. Two closed curves sharing common points can be concatenated to form another loop. This operation defines the multiplication of homotopic classes naturally. Therefore, all the base pointed homotopy classes form the so called *the first fundamental group* of M, and is denoted as $\pi_1(M)$. For genus g closed surface, the fundamental group has $2g$ generators. A set of fundamental group basis $\{a_1,b_1,a_2,b_2,\cdots,a_g,b_g\}$ is canonical, if a_i,b_i have one geometric intersection, but a_i,a_j have zero geometric intersection, and a_i,b_j have zero geometric intersection. (The definition of geometric intersection number between two simple curves a,b is the $\min\{|x\cap y|$: x homotopic to a, y homotopic to b$\}$.)

Suppose that $\bar{M}$ and M are surfaces. Then $(\bar{M},\pi)$ is said to be a covering space of M if π is a surjective continuous map with every $p\in M$ having an open neighborhood U such that every connected component of $\pi^{-1}(U)$ is mapped homeomorphically onto U by π. Furthermore, if $\bar{M}$ is simply connected, $(\bar{M},\pi)$ is the universal covering space of M.

A transformation of the universal covering space $\phi:\bar{M}\to\bar{M}$ is a *deck transformation*, if $\pi=\pi\circ\phi$. All deck transformations form a group G, the so called *Fuchsian group* of M, which is isomorphic to the fundamental group of M in the following way. Suppose $p\in M$ is an arbitrary point on M, its preimages are $\pi^{-1}(p)=\{p_0,p_1,p_2,\cdots,p_n,\cdots\}$ on $\bar{M}$. Suppose a deck transformation $\phi\in G$ maps p_0 to p_k. Then we can draw a curve segment on the universal covering space $\gamma:[0,1]\to\bar{M}$ connecting p_0,p_k and its projection $\pi(\gamma)$ is a loop on M, the homotopy class of γ is determined by p_0,p_k only and independent of the choice of γ. Then we get a bijective map from deck transformations to the first fundamental group of M.

A *fundamental domain* F is a subset of $\bar{M}$, such that the universal covering space is the union of conjugates of F, $S=\bigcup_{g\in G} gF$, and any two conjugates have no interior point in common. Given a canonical fundamental group generators $\{a_i,b_i\}$, where $i=1,2\cdots,g$, we can slice M along the curves and get a fundamental domain with boundary

$$\{a_1b_1a_1^{-1}b_1^{-1}a_2b_2a_2^{-1}b_2^{-1}\cdots a_gb_ga_g^{-1}b_g^{-1}\}.$$

Suppose a surface M has genus at least two with a hyperbolic metric, its universal covering space $\bar{M}$ can be isometrically embedded in the hyperbolic space. The embedding produces a tessellation of the hyperbolic space, and each tile is a fundamental domain. The deck translations which map fundamental domains to fundamental domains are rigid motions in the hyperbolic space. Figure 2 illustrates the fundamental group generators of a genus two surface and its universal covering space.

2.5 Hyperbolic Space Models

There are two common models for hyperbolic geometry.

2.5.1 Poincaré Model

The Poincaré model is a unit disk $\mathbb{D}^2$ in the complex plane with the Riemannian metric $ds^2=\frac{4dzd\bar{z}}{(1-z\bar{z})^2}$.

The geodesics are circular arcs perpendicular to the boundary of the unit disk $\partial\mathbb{D}^2$. The isometric transformations in this model is the so called Möbius transformation with the form

$$\phi(z)=e^{i\theta}\frac{z-z_0}{1-\bar{z}_0 z}, z,z_0\in\mathbb{C}, \theta\in[0,2\pi).$$

The above Möbius transformation maps z_0 to the center of the disk, and rotate the whole disk by angle θ. Hyperbolic circles are also Euclidean circles.

Suppose M is a closed surface with genus $g > 1$, then M has a hyperbolic uniformization metric. The universal covering space $\bar{M}$ with this metric is isometric to $\mathbb{H}^2$. Each fundamental domain is a polygon with $4g$ sides and each side is a hyperbolic geodesic. This formes a tessellation of $\mathbb{H}^2$. The deck transformations of $\bar{M}$ are Möbius transformation on $\mathbb{H}^2$.

The Poincaré model is a conformal model, whereas the Klein model is a real projective model.

2.5.2 Klein Model

The Klein model is another model of hyperbolic space defined also on the unit disk $\mathbb{D}^2$. Any geodesic in the Klein model is a chord of the unit circle of the boundary of $\mathbb{D}^2$. The map from the Poincaré model to the Klein model is $\beta : \mathbb{H}^2 \to \mathbb{D}^2$,

$$\beta(z) = \frac{2z}{1+\bar{z}z}, \beta^{-1}(z) = \frac{1-\sqrt{1-\bar{z}z}}{\bar{z}z}z. \quad (1)$$

Any Möbius transformation in the Poincaré model ϕ becomes a real projective transformation in the Klein model $\beta \circ \phi \circ \beta^{-1}$.

3 Hyperbolic Discrete Variational Ricci flow

This section explains the algorithms to compute the hyperbolic uniformization metric of a surface with negative Euler number. In practice, all surfaces are represented as triangular meshes.

A triangular mesh is a two dimensional simplicial complex, denoted by $M = (V, E, F)$, where V is the set of all vertices, E is the set of all non-oriented edges and F is the set of all faces. We use $v_i, i = 1, \cdots, n$ to denote its vertices, e_{ij} to denote an oriented edge from v_i to v_j, and f_{ijk} to denote an oriented face with vertices v_i, v_j, v_k which are ordered counter-clockwisely such that the face normals toward outside.

In reality, surfaces are approximated by meshes. The concepts of conformal maps, hyperbolic metrics and Ricci flow are also translated from the smooth surface category to the discrete mesh category. We show the Riemannian metrics of meshes induced by Delaunay triangulations converge to the metric on the smooth surface in [Dai et al. 2006]. The convergence of conformal structures is proved in [Luo 2006]. The convergence of hyperbolic structure remains open.

3.1 Circle Packing Metric

The following key observation plays vital role for systematically translating smooth Ricci flow to discrete Ricci flow. A conformal mapping between two smooth surfaces maps infinitesimal circles to other infinitesimal circles, changing the radii of the infinitesimal circles, but preserving the intersection angles among them.

In order to translate conformal mappings from the smooth surface category to the discrete mesh category, Thurston defined *circle packing* as the following,

1. Change infinitesimal circles to circles with finite radii.
2. Each circle is centered at a vertex like a cone, and the radius is denoted as γ_i for vertex v_i.
3. An edge has two vertices. The two circles intersect each other with an intersection angle, and the angle is denoted as Φ_{ij} for edge e_{ij}, called the *weight*.

Therefore a mesh with circle packing (M, Γ, Φ), where M represents the triangulation (connectivity), $\Gamma = \{\gamma_i, v_i \in V\}$ are the vertex radii and $\Phi = \{\Phi_{ij}, e_{ij} \in E\}$ are the angles associated with each edge. A *discrete conformal mapping* $\tau : (M, \Gamma, \Phi) \to (M, \bar{\Gamma}, \Phi)$ solely changes the vertex radii Γ, but preserves the intersection angles Φ.

In reality, a discrete conformal mapping can approximate a smooth conformal mapping with arbitrary accuracy. If we keep subdividing the mesh and construct refiner and refiner circle packing, the discrete conformal mappings will converge to the smooth conformal mapping. For a rigorous proof, we refer the readers to [Rodin and Sullivan 1987].

A circle packing (M, Φ, Γ) uniquely determines a so called *circle packing metric*. The length l_{ij} associated with the edge e_{ij} is computed using the cosine law,

$$l_{ij} = \sqrt{\gamma_i^2 + \gamma_j^2 + 2\gamma_i\gamma_j \cos\Phi_{ij}}. \quad (2)$$

In practice, the mesh is embedded in $\mathbb{R}^3$, the mesh has an induced Euclidean metric, the Euclidean length of edge e_{ij} is denoted as $\bar{l}_{ij}$. We select an appropriate circle packing metric (Γ, Φ), such that it approximates the Euclidean metric as close as possible,

$$\min_{\Gamma,\Phi} \sum_{e_{ij} \in M} |l_{ij} - \bar{l}_{ij}|^2.$$

Given a circle packing metric on a mesh, such that the edge lengths for each face f_{ijk} satisfy the triangle inequality $l_{ij} + l_{jk} > l_{ki}$, then the face can be realized in the hyperbolic space. We denote the angle at vertex v_i as θ_i^{jk} and compute it by the hyperbolic cosine law:

$$\cos\theta_i^{jk} = \frac{\cosh l_{ij} \cosh l_{ik} - \cosh l_{jk}}{\sinh l_{ij} \sinh l_{ik}}.$$

The discrete Gaussian curvature K_i at an interior vertex v_i is defined as

$$K_i = 2\pi - \sum_{f_{ijk} \in F} \theta_i^{jk}, \qquad v_i \notin \partial M, \quad (3)$$

while the discrete Gaussian curvature for a boundary vertex v_i is defined as

$$K_i = \pi - \sum_{f_{ijk} \in F} \theta_i^{jk}, \qquad v_i \in \partial M. \quad (4)$$

3.2 Hyperbolic Ricci Flow

Given a mesh with circle packing metric (M, Γ, Φ), the discrete hyperbolic Ricci flow is defined as

$$\frac{\partial \gamma_i}{\partial t} = -\sinh\gamma_i K_i \quad (5)$$

A solution to equation 5 exists and is *convergent*

$$\lim_{t \to \infty} K_i(t) = 0.$$

A convergent solution *converges exponentially* if there are positive constants c_1, c_2, so that for all time $t \geq 0$

$$|K_i(t) - K_i(\infty)| \leq c_1 e^{-c_2 t}, |\gamma_i(t) - \gamma_i(\infty)| \leq c_1 e^{-c_2 t}.$$

In theory, the discrete Ricci Flow is guaranteed to be exponentially convergent [Chow and Luo 2003].

3.3 Hyperbolic Discrete Variational Ricci flow Using Newton's Method

Discrete hyperbolic Ricci Flow is the solution to an energy optimization problem, namely, it is the negative gradient flow of some convex energy, which was first introduced by [de Verdiére 1991] for tangential circle packing, and [Chow and Luo 2003] generalizes it to all intersection cases. Therefore we can use Newton's method to further improve the convergence speed.

Let $u_i = \ln \tanh \frac{\gamma_i}{2}$, under this change of variable, the Ricci Flow in Equation 5 takes the following form: $\frac{du_i}{dt} = -K_i$,

We can define an energy form

$$f(\mathbf{u}) = \int_{\mathbf{u}_0}^{\mathbf{u}} \sum_{i=1}^{n} K_i du_i,$$

where $\mathbf{u} = (u_1, u_2, \cdots, u_n)$, $\mathbf{u}_0$ is $(0, 0, \cdots, 0)$. Thus $\frac{\partial f}{\partial u_i} = K_i$, that is, the Ricci Flow 5 is the negative gradient flow of the energy $f(\mathbf{u})$.

The Hessian matrix of the energy f is

$$\frac{\partial^2 f}{\partial u_i \partial u_j} = \frac{\partial K_i}{\partial u_j} = \frac{\partial K_i}{\partial r_j} \sinh r_j,$$

From the definition of curvature K_i, hyperbolic cosine law and the definition of circle packing metric, the above formulae has a closed form. The Hessian matrix can be easily verified to be positive definite. As f is strictly convex, it therefore has a unique global minimum, so Newton's method can be used to stably locate this minimum.

4 Computing Hyperbolic Structures

After computing the hyperbolic uniformization metrics described in the last section 3, we focus on computing hyperbolic structures of general surfaces in the current section.

For any surface with boundaries, we can convert it to a closed symmetric surface using the double covering method [Gu and Yau 2003]. First we make two copies of the surface, then we reverse the orientation of one of them and glue the two copies along their corresponding boundaries. The double covered surface admits a uniformization metric conformal to its original metric and the original boundary curves become geodesics under this metric. The real projective structures of the original surface and its double covering can be induced by this uniformization metric. Therefore, in the following discussion, we always assume the surfaces are closed unless otherwise noted.

4.1 Algorithm Pipeline

In order to compute the hyperbolic structure, we develop the following algorithms. Suppose M is a surface with genus g greater than one with hyperbolic uniformization metric, which is computed in the previous section, the major steps for computing a hyperbolic atlas of M are

1. Compute the fundamental domain of the surface M; construct a finite portion of its universal covering space $(\bar{M}, \pi)$, π is the projection map; construct a canonical fundamental group generators $\Gamma = \{a_1, b_1, \cdots, a_g, b_g\}$.
2. Isometrically embed the universal covering space (or a fundamental domain) onto the Poincaré disk.
3. Compute the Fuchsian group generators (deck transformation group of $\bar{M}$) corresponding to a_i, b_i, denoted them as $\{\phi_1, \phi_2, \cdots, \phi_{2g}\}$, where ϕ_i's are Möbius transformation on the Poincaré disk.
4. Construct a hyperbolic atlas

The algorithms in the first step, such as computing the fundamental domain, universal covering space and canonical homology basis, have been studied in computational topology and computer graphics literature [Colin de Verdière and Lazarus 2002] [Erickson and Whittlesey 2005]. We adopted the methods introduced in [Carner et al. 2005]. The following discussion will explain the other steps in detail.

4.2 Hyperbolic Embedding

In the second step, we need to isometrically embed the universal covering space $\bar{M}$ in the Poincaré disk. Let the isometric embedding be

$$\phi : \bar{M} \to \mathbb{H}^2. \tag{6}$$

First, we select a face f_{012} from $\bar{M}$ arbitrarily. Suppose three edge lengths are $\{l_{01}, l_{12}, l_{20}\}$, the corner angles are $\{\theta_0^{12}, \theta_1^{20}, \theta_2^{01}\}$ under the hyperbolic uniformization metric. We can simply embed the triangle as

$$\phi(v_0) = 0, \phi(v_1) = \frac{e^{l_{01}} - 1}{e^{l_{01}} + 1}, \phi(v_2) = \frac{e^{l_{02}} - 1}{e^{l_{02}} + 1} e^{i\theta_0^{12}}.$$

Second, we can embed all the faces which share an edge with the first embedded face. Suppose a face f_{ijk} is adjacent to the first face, vertices v_i, v_j have been embedded. A hyperbolic circle is denoted as $(\mathbf{c}, r)$, where $\mathbf{c}$ is the center, r is the radius. Then $\phi(v_k)$ should be one of the two intersection points of $(\phi(v_i), l_{ik})$ and $(\phi(v_j), l_{jk})$. Also, the orientation of $\phi(v_i), \phi(v_j), \phi(v_k)$ should be counter-clock-wise. In Poincaré model, a hyperbolic circle $(\mathbf{c}, r)$ coincides with an Euclidean circle $(\mathbf{C}, R)$,

$$\mathbf{C} = \frac{2 - 2\mu^2}{1 - \mu^2 |\mathbf{c}|^2} \mathbf{c}, R^2 = |\mathbf{C}|^2 - \frac{|\mathbf{c}|^2 - \mu^2}{1 - \mu^2 |\mathbf{c}|^2},$$

where $\mu = \frac{e^r - 1}{e^r + 1}$. The intersection points between two hyperbolic circles can be found by intersecting the corresponding Euclidean circles. The orientation of triangles can also be determined using Euclidean geometry on the Poincaré disk.

Then, we can repeatedly embed faces which share edges with embedded faces in the same manner, until we embed enough portions of the whole of $\bar{M}$ to the Poincaré disk.

4.3 Fuchsian Group Generator

Let $\pi(M)$ be the fundamental group of M, with a set of canonical basis $\Gamma = \{a_1, b_1, \cdots, a_g, b_g\}$, while all curves only share one base point as shown in the left frame of figure 2. For each closed curve $\gamma : [0,1] \to M$ on M, $\gamma(0) = \gamma(1)$, we use $[\gamma]$ to denote its homotopy class.

Let $(\bar{M}, \pi)$ be the universal covering space embedded in the Poincaré disk as shown in the right frame of figure 2. Let $\bar{p}, \bar{q} \in \bar{M}$ be two points, such that $\pi(\bar{p}) = \pi(\bar{q})$. Suppose $\bar{\gamma}$: $[0,1] \to \bar{M}$ is a curve, such that $\bar{\gamma}(0) = \bar{p}, \bar{\gamma}(1) = \bar{q}$, then $\pi(\bar{\gamma})$ is a closed loop on M. Its homotopy class is $[\pi(\bar{\gamma})]$. It can be verified that $[\pi(\bar{\gamma})]$ is determined by $\bar{p}$ and $\bar{q}$, independent of the choice of $\bar{\gamma}$, therefore, we use $[\bar{p}, \bar{q}]$ to denote the homotopy class of $[\pi(\bar{\gamma})]$.

Given a base point on the surface $p \in M$, its preimages are $\pi^{-1}(p) = \{\bar{p}_0, \bar{p}_1, \bar{p}_2, \cdots, \} \subset \bar{M}$. Then $[\bar{p}_0, \bar{p}_k], k = 0, 1, 2, \cdots$ will traverse the whole fundamental group $\pi(M)$. For each k, there exists a unique deck transformation $\tau_k : \bar{M} \to \bar{M}$, such that $\tau_k(\bar{p}_0) = \tau_k(\bar{p}_k)$, furthermore, τ_k is a Möbius transformation. The group formed by $\{\tau_k\}$ is called the *Fuchsian group* of M, and denoted as $F(M)$. Then we construct a map η between the fundamental group $\pi(M)$ and the Fuchsian group $F(M)$,

$$\eta : \pi(M) \to F(M), [\bar{p}_0, \bar{p}_k] \to \tau_k.$$

It can be verified that the map is an isomorphism.

The goal of the algorithm in this subsection is to compute the isomorphism η. It is sufficient to find the Möbius transformations corresponding to the curves in Γ.

First we choose two distinct points p and q on the surface, $p, q \in M$, such that p, q are not on any curves in Γ, $p, q \notin \Gamma$. The preimages of Γ $\pi(\Gamma)$ partition the covering space $\bar{M}$ to fundamental domains, $\{\Sigma_0, \Sigma_1, \Sigma_2, \cdots\}$. The preimages of p and q are

$$\pi^{-1}(p) = \{\bar{p}_k\}, \pi^{-1}(q) = \{\bar{q}_k\}, k = 0, 1, 2, \cdots,$$

both $\bar{p}_k$ and $\bar{q}_k$ are interior points of Σ_k.

Suppose Σ_0 and Σ_1 are two fundamental domains on $\bar{M}$ adjacent to each other, $\pi(\Sigma_0 \cap \Sigma_1) = a_2$. There exists a unique Möbius transformation ψ_1, such that $\psi_1(\bar{p}_0) = \bar{p}_1, \psi_1(\bar{q}_0) = \bar{q}_1$. In order to compute ψ_1, we construct a Möbius transformation μ_0 to map $\bar{p}_0$ to the origin, $\bar{q}_0$ to a real positive number, and μ_1 to map $\bar{p}_1$ to the origin and $\bar{q}_1$ to the same positive number, then

$$\psi_1 = \mu_1 \circ \mu_0^{-1}.$$

Therefore the problem is deduced to find a Möbius transformation μ, such that for a given pair of complex numbers inside the Poincaré disk $z_0, z_1 \in \mathbb{H}^2$,

$$\mu(z_0) = 0, arg(\mu(z_1)) = 0.$$

μ can be constructed straightforwardly,

$$\mu = e^{-i\theta_0} \frac{z - z_0}{1 - \bar{z}_0 z}, \theta_0 = \arg \frac{z_1 - z_0}{1 - \bar{z}_0 z_1}.$$

In this way, by choosing different Σ_k's adjacent to Σ_0, we can compute all the Möbius transformations ψ_k which are the generators of the Fuchsian group of M. The isomorphism $\eta : \pi(M) \to F(M)$ can be constructed in the following way:

$$\eta(a_i) = \tau_k, if[\pi(\Sigma_0 \cap \Sigma_k)] = [b_i^{-1}],$$

and

$$\eta(b_i) = \tau_k, if[\pi(\Sigma_0 \cap \Sigma_k)] = [a_i].$$

4.4 Hyperbolic Atlas

First we construct a family of open sets $\{U_\alpha\}$, such that the union of the open sets covers the surface M, $M \subset \bigcup U_\alpha$. Then we locate one preimage of each U_α in the universal covering space $\bar{M}$, $\pi^{-1}(U_\alpha)$. The embedding of the preimage $\pi^{-1}(U_\alpha)$ in the Poincaré disk gives the local coordinates of U_α, namely

$$\phi_\alpha := \phi \circ \pi^{-1},$$

where ϕ is the embedding map defined in equation 6. If one point $p \in M$ on the surface M is covered by two charts (U_α, ϕ_α) and (U_β, ϕ_β), suppose $\bar{p}_\alpha \in \pi^{-1}(U_\alpha)$ and $\bar{p}_\beta \in \pi^{-1}(U_\beta)$, and the homotopy class $[\bar{p}_\alpha, \bar{p}_\beta] = [\gamma_1 \gamma_2 \cdots \gamma_n]$, then the chart transition map $\phi_{\alpha\beta}$ has the form

$$\phi_{\alpha\beta} = \eta(\gamma_n) \circ \eta(\gamma_{n-1}) \cdots \eta(\gamma_1),$$

where $\gamma_k, k = 1, 2, \cdots, n$ are fundamental group generators, $\gamma_k \in \Gamma$. $\phi_{\alpha\beta}$ is Möbius transformation on the Poincaré disk. Therefore $\{(U_\alpha, \phi_\alpha)\}$ is a hyperbolic atlas.

5 Computing Real Projective Structure

Any oriented metric surface has a real projective structure. Real projective structures are induced by spherical structures, Euclidean structures and hyperbolic structures. This section introduces the algorithms to compute real projective atlases for surfaces with different topologies.

5.1 Genus zero surfaces

Closed genus zero surfaces have spherical structures. The universal covering space of a closed genus zero surface is itself. Therefore, the surface can be conformally mapped to the unit sphere.

A method based on heat flow to construct conformal maps between a closed genus zero surface to the unit sphere $\mathbb{S}^2$ is introduced in [Gu et al. 2004]. The spherical uniformization metrics are induced by these conformal maps.

We set six tangent planes at the intersection points between the unit sphere and the axes, then project the sphere onto these tangent planes using central projection. This procedure produces the real projective atlas for the surface.

Figure 10 demonstrates a real projective atlas for a closed genus zero surface, Michelangelo's David head model.

5.2 Genus one surfaces

Any genus one oriented surface admits a flat uniformization metric, such that the Gaussian curvature is zero everywhere. Its universal covering space can be embedded in the plane. Each fundamental domain is a parallelogram, and the deck transformations are translations on the plane.

The flat uniformization metric on a genus one closed surface can be induced by the holomorphic 1-forms on it. A holomorphic 1-form can be treated as a pair of vector fields with zero divergence and circulation, and are orthogonal to each other. The algorithms for computing holomorphic 1-forms are introduced in [Gu and Yau 2003; Gu et al. 2005]. By integrating a holomorphic 1-form, its universal covering space can be conformally mapped to the plane. This induces an affine atlas for the surface. This construction method is similar to the algorithms in section 4. Affine structure can be treated as a special case of real projective structure.

5.3 High genus surfaces

For a closed surface M with genus $g > 1$, from its hyperbolic structure we can deduce its real projective atlas. But the real projective structure can not deduce the hyperbolic structure. Suppose $\{(U_\alpha, \phi_\alpha)\}$ is a hyperbolic atlas of M, a real projective atlas $\{(U_\alpha, \tau_\alpha)\}$ can be straightforwardly constructed. Let

$$\tau_\alpha = \beta \circ \phi_\alpha,$$

and

$$\tau_{\alpha\beta} = \beta \circ \phi_{\alpha\beta} \circ \beta^{-1}.$$

where β is the map from the Poincaré model to the Klein model as defined in equation 1. Suppose $\phi_{\alpha\beta}$ has the form

$$\phi_\alpha = e^{i\theta}\frac{z - z_0}{1 - \bar{z}_0 z},$$

where $z_0 = x_0 + iy_0$. We use homogenous coordinates (xw, yw, w) to parameterize the points (x,y) on the Klein model, then the transition map $\tau_{\alpha\beta}$ has the following form $\tau_{\alpha\beta} = \frac{1}{\lambda}OT$, where $\lambda = x_0^2 + y_0^2 - 1$, O is the rotation matrix

$$O = \begin{pmatrix} \cos\theta & -\sin\theta & 0 \\ \sin\theta & \cos\theta & 0 \\ 0 & 0 & 1 \end{pmatrix}$$

and T is

$$T = \begin{pmatrix} 1 + x_0^2 - y_0^2 & 2x_0y_0 & -2x_0 \\ 2x_0y_0 & 1 - x_0^2 + y_0^2 & -2y_0 \\ 2x_0 & 2y_0 & -1 - x_0^2 - y_0^2 \end{pmatrix}.$$

6 Implementation

The algorithms are implemented using $c++$ on the windows platform. All the models are represented as triangular meshes and some of them are scanned from real sculptures. In order to manipulate the meshes efficiently, we use a common mesh library in our implementation: the OpenMesh [Sovakar and Kobbelt 2004] library, because it is flexible, efficient and versatile. In the following we demonstrate some of our experimental results.

Our experiments demonstrate that the hyperbolic discrete variational Ricci flow algorithms are efficient and Robust. Figure 9 shows the the performance of Ricci flow on eight model. The upper blue curve is using hyperbolic Ricci flow with gradient descendent method, compared to the lower red curve with Newton's method. It is obvious that Newton's method converges much faster.

Fuchsian group generators of eight model on Poincaré Model.

	$e^{i\theta}$	z_0
a_1	$-0.631374 + i0.775478$	$+0.730593 + i0.574094$
b_1	$+0.035487 - i0.999370$	$+0.185274 - i0.945890$
a_2	$-0.473156 + i0.880978$	$-0.798610 - i0.411091$
b_2	$-0.044416 - i0.999013$	$+0.035502 + i0.964858$

Fuchsian group generators of knotty model on Poincaré Model.

	$e^{i\theta}$	z_0
a_1	$-0.736009 + i0.676972$	$+0.844441 - i0.402575$
b_1	$+0.227801 - i0.973708$	$-0.851267 - i0.510906$
a_2	$-0.521197 + i0.853436$	$-0.528917 + i0.702150$
b_2	$-0.090798 - i0.995869$	$+0.973285 + i0.206533$

Figure 10 illustrates a genus zero closed surface case. First we conformally map the surface to a unit sphere, and induce the real projective atlas from its spherical image.

Figure 3 illustrates a genus one closed surface. We compute a holomorphic 1-form on the surface, which induces an affine structure. Figure (b) demonstrates one fundamental domain, (c) is a finite portion of its universal covering space with nine fundamental domains.

Figure 1 shows the process of computing hyperbolic atlas and real projective atlas for a genus two closed surface. Figure (a) is the original surface with 7000 faces, (b) is the isometric embedding of its universal covering space in the Poincaré disk with its hyperbolic uniformization metric, (c) shows the result after straightening the boundaries of fundamental domains to hyperbolic lines, (d) is the embedding in Klein model, which induces a real projective atlas. The most time consuming part is to compute its hyperbolic uniformization metric using discrete Ricci flow, which took 100 seconds on a 1.7G CPU plus 1M RAM laptop. So the total time for the whole process is no more than 2 minutes.

Figures 4,7,8 illustrate the same process for different genus two closed surfaces. Figure 5 demonstrates the isometric embedding of the universal covering space of a genus three surface, the scanned sculpture surface, on the Poincaré model and Klein model. Figure 6 is the same process for a genus four surface, a scanned Greek model. These results show our method is capable to handle surfaces with complicated geometries and topologies.

7 Conclusion

In this paper, we introduce a general framework to define different geometries on surfaces via geometric structures. Algorithms designed for Euclidean space can be systematically generalized to surface domains directly.

In order to compute hyperbolic structures and real projective structures on general surfaces, a theoretically rigorous and

Fuchsian group generators of eight model on Klein Model.

$$a_1 \begin{pmatrix} 10.3242 & 8.39204 & -13.2671 \\ -2.9578 & -1.08348 & 2.98706 \\ -10.6929 & -8.4024 & 13.6359 \end{pmatrix} \quad b_1 \begin{pmatrix} 4.86605 & -26.0235 & -26.4556 \\ 2.14144 & -5.86616 & -6.16422 \\ -5.22151 & 26.6577 & 27.1826 \end{pmatrix}$$

$$a_2 \begin{pmatrix} 6.59028 & 4.02982 & 7.65972 \\ -5.0888 & -1.69285 & -5.26893 \\ 8.26606 & 4.25502 & 9.35056 \end{pmatrix} \quad b_2 \begin{pmatrix} -0.963558 & -28.3933 & 28.392 \\ 1.08105 & 2.27398 & -2.31077 \\ -1.04743 & -28.4667 & 28.5035 \end{pmatrix}$$

Fuchsian group generators of Knotty model on Klein Model.

$$a_1 \begin{pmatrix} 5.45671 & -1.57356 & -5.59033 \\ -12.4179 & 6.33332 & 13.9038 \\ -13.527 & 6.4488 & 15.0189 \end{pmatrix} \quad b_1 \begin{pmatrix} -82.4315 & -50.31 & -96.5662 \\ 85.687 & 50.6147 & 99.5144 \\ 118.896 & 71.3578 & 138.669 \end{pmatrix}$$

$$a_2 \begin{pmatrix} -0.985119 & 2.85311 & -2.84793 \\ -4.65847 & 5.57247 & -7.19401 \\ 4.6553 & -6.18002 & 7.80157 \end{pmatrix} \quad b_2 \begin{pmatrix} -22.6068 & -5.81234 & 23.3206 \\ 192.165 & 40.6573 & -196.417 \\ -193.488 & -41.0585 & 197.799 \end{pmatrix}$$

practically simple algorithm is explained. Also a novel and powerful geometric tool called variational Ricci flow is introduced, which is efficient, robust and easy to implement.

In the future, we will focus on designing novel spline schemes which are based on hyperbolic structures or real projective structures, and constructing manifold splines without extraordinary points. We also prove whether the discrete Ricci Flow will converge , as one refines the meshes approximating a smooth surface, toward the smooth Ricci Flow.

Acknowledgement

We thank Stanford university and Cindy Grimm for the surface models.

This work was partially supported by the NSF CAREER Award CCF-0448339 and NSF DMS-0528363 to X. Gu.

References

BENZÈCRI, J. P. 1959. Varits localement affines. *Sem. Topologie et Gom. Diff., Ch. Ehresmann 7*, 229–332.

BOBENKO, A., AND SCHRODER, P. 2005. Discrete willmore flow. In *Eurographics Symposium on Geometry Processing*.

BOBENKO, A. I., AND SPRINGBORN, B. A. 2004. Variational principles for circle patterns and koebe's theorem. *Transactions of the American Mathematical Society 356*, 659.

CARNER, C., JIN, M., GU, X., AND QIN, H. 2005. Topology-driven surface mappings with robust feature alignment. In *IEEE Visualization*, 543–550.

CHOW, B., AND LUO, F. 2003. Combinatorial ricci flows on surfaces. *Journal Differential Geometry 63*, 1, 97–129.

COLIN DE VERDIÈRE, É., AND LAZARUS, F. 2002. Optimal system of loops on an orientable surface. In *Proceedings of the 43rd Annual IEEE Symposium on Foundations of Computer Science*, 627–636.

DAI, J., LUO, W., YAU, S.-T., AND GU, X. 2006. Geometric accuracy analysis for discrete surface approximation. In *Geometric Modeling and Processing*. submitted.

DE VERDIÉRE, C. 1991. Yves un principe variationnel pour les empilements de cercles. (french) "a variational principle for circle packings". *Invent. Math. 104 3*, 655–669.

ERICKSON, J. G., AND WHITTLESEY, K. 2005. Greedy optimal homotopy and homology generators. In *Proc. 16th Symp. Discrete Algorithms*, ACM and SIAM, 1038–1046.

FERGUSON, H., ROCKWOOD, A. P., AND COX, J. 1992. Topological design of sculptured surfaces. In *SIGGRAPH*, 149–156.

GOLDMAN, R. 2003. Computer graphics in its fifth decade: Ferment at the foundations. In *11th Pacific Conference on Computer Graphics and Applications (PG'03)*, 4–21.

GOTSMAN, C., GU, X., AND SHEFFER, A. 2003. Fundamentals of spherical parameterization for 3d meshes. *ACM Trans. Graph. 22*, 3, 358–363.

GRIMM, C., AND HUGHES, J. F. 2003. Parameterizing n-holed tori. In *IMA Conference on the Mathematics of Surfaces*, 14–29.

GU, X., AND YAU, S.-T. 2003. Global conformal parameterization. In *Symposium on Geometry Processing*, 127–137.

GU, X., WANG, Y., CHAN, T. F., THOMPSON, P. M., AND YAU, S.-T. 2004. Genus zero surface conformal mapping and its application to brain surface mapping. *IEEE Trans. Med. Imaging 23*, 8, 949–958.

GU, X., HE, Y., AND QIN, H. 2005. Manifold splines. In *Symposium on Solid and Physical Modeling*, 27–38.

HAKER, S., ANGENENT, S., TANNENBAUM, A., KIKINIS, R., SAPIRO, G., AND HALLE, M. 2000. Conformal surface parameterization for texture mapping. *IEEE Trans. Vis. Comput. Graph. 6*, 2, 181–189.

HAMILTON, R. S. 1988. The ricci flow on surfaces. *Mathematics and general relativity 71*, 237–262.

JIN, M., WANG, Y., YAU, S.-T., AND GU, X. 2004. Optimal global conformal surface parameterization. In *IEEE Visualization*, 267–274.

J.W.MILNOR. 1958. On the existence of a connection with curvature zero. *Comm. Math. Helv. 32*, 215–223.

KHAREVYCH, L., SPRINGBORN, B., AND SCHRODER, P. 2005. Discrete conformal mappings via circle patterns.

LUO, W., AND GU, X. 2006. Discrete curvature flow and derivative cosine law. In *preprint*.

LUO, W. 2006. Error estimates for discrete harmonic 1-forms over riemann surfaces. In *Communications in Analysis and Geometry*. submitted.

MUNKRES, J. 1984. *Elements of Algebraic Topology*. Addison-Wesley Co.

PETERSEN, P. 1997. *Riemannian Geometry*. Springer Verlag.

PRAUN, E., AND HOPPE, H. 2003. Spherical parametrization and remeshing. *ACM Trans. Graph. 22*, 3, 340–349.

RATCLIFFE, J. G. 1994. *Foundations of Hyperbolic Manifolds*. Springer Verlag.

RODIN, B., AND SULLIVAN, D. 1987. The convergence of circle packings to the riemann mapping. *Journal of Differential Geometry 26*, 2, 349–360.

SOVAKAR, A., AND KOBBELT, L. 2004. Api design for adaptive subdivision schemes. *Computers & Graphics 28*, 1, 67–72.

STEPHENSON, K. 2005. *Introduction To Circle Packing*. Cambridge University Press.

THURSTON, W. 1976. *Geometry and Topology of 3-manifolds*. Princeton lecture notes.

THURSTON, W. P. 1997. *Three-Dimensional Geometry and Topology*. Princeton University Presss.

WALLNER, J., AND POTTMANN, H. 1997. Spline orbifolds. *Curves and Surfaces with Applications in CAGD*, 445–464.

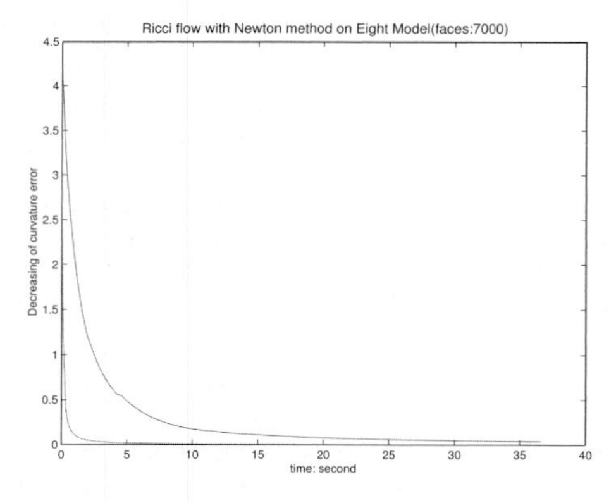

Figure 9: **Comparison of performance of Ricci flow with Newton's method (lower red curve) over the gradient descent method (upper blue curve). Both converge exponentially.**

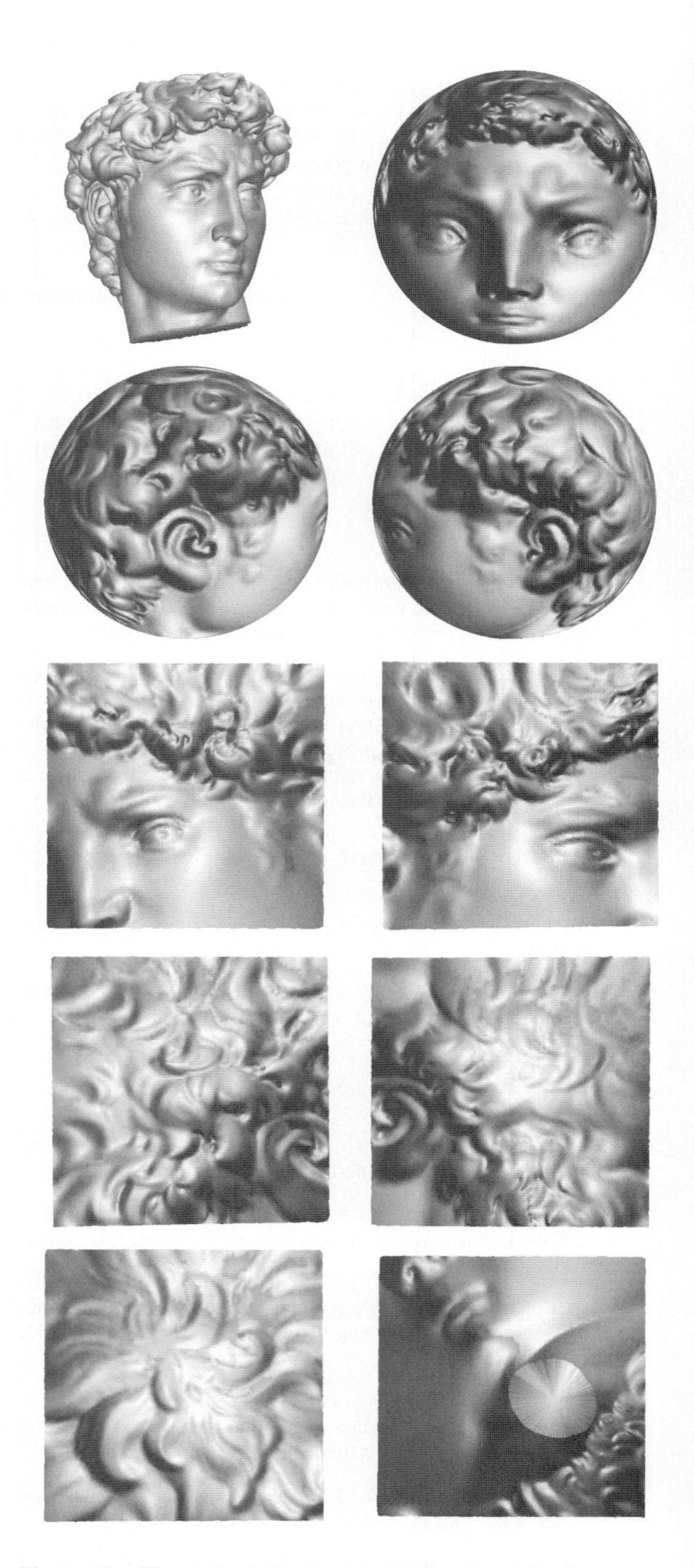

Figure 10: **The spherical structure and real projective structure of a genus zero closed surface.**

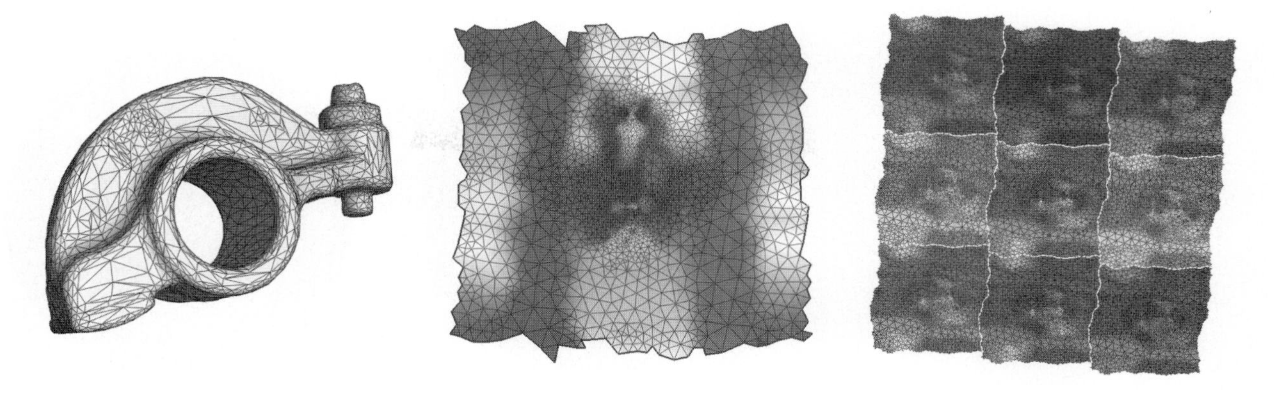

Figure 3: **The Euclidean structure of a genus one closed surface: (a) the original surface (b) one fundamental domain (c) part of its universal covering space**

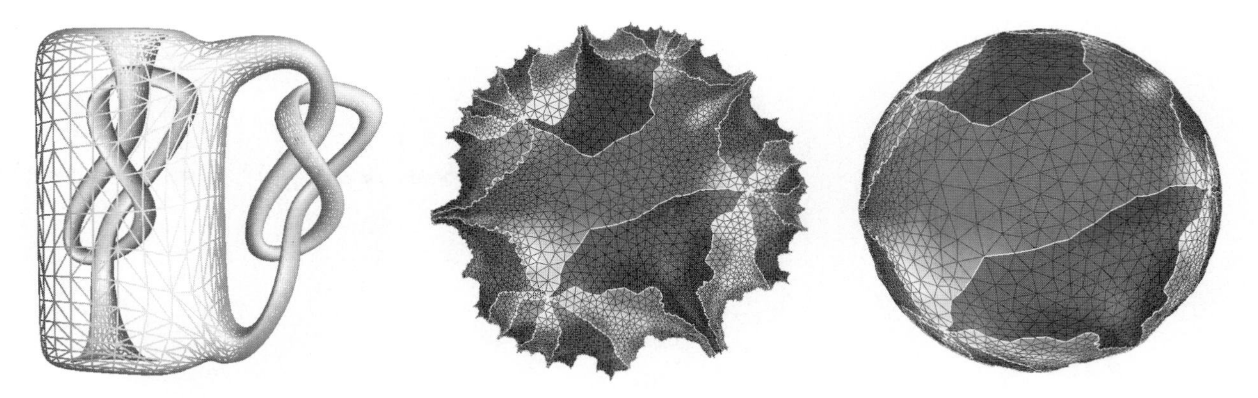

Figure 4: **The hyperbolic structure of a genus two closed surface: (a) the original surface (b) the isometric embedding of its universal covering space in the Poincaré disk (c) the isometric embedding in the Klein disk which induces real projective structures**

Figure 5: **The hyperbolic structure of a genus three closed surface: (a) the original surface (b) the isometric embedding of its universal covering space in the Poincaré disk (c) the isometric embedding in the Klein disk which induces real projective structures**

Figure 6: **The hyperbolic structure of a genus four closed surface: (a) the original surface (b) the isometric embedding of its universal covering space in the Poincaré disk (c) the isometric embedding in the Klein disk which induces real projective structures**

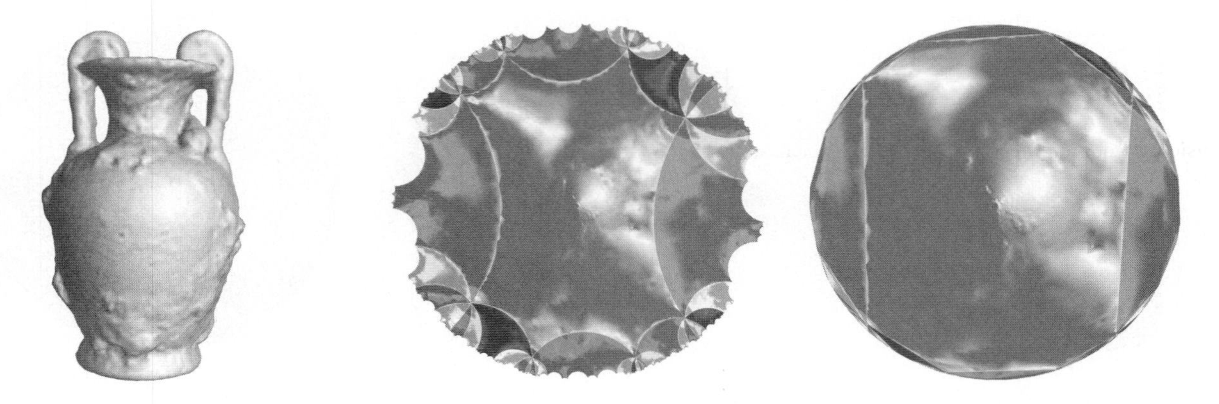

Figure 7: **The hyperbolic structure of a genus two closed surface: (a) the original surface (b) the isometric embedding of its universal covering space in the Poincaré disk (c) the isometric embedding in the Klein disk which induces real projective structures. The boundaries of fundamental domains are straightened to hyperbolic lines.**

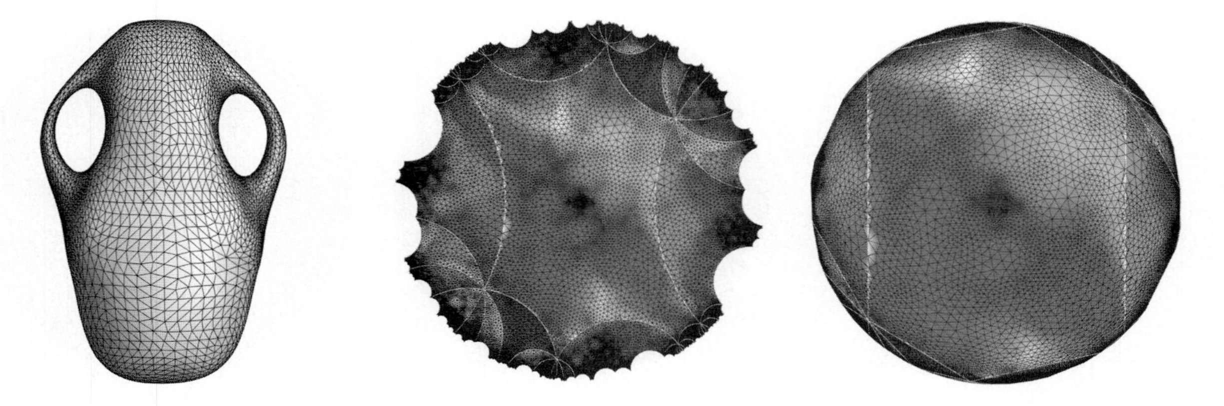

Figure 8: **The hyperbolic structure of a genus two closed surface: (a) the original surface (b) the isometric embedding of its universal covering space in the Poincaré disk (c) the isometric embedding in the Klein disk which induces real projective structures.The boundaries of fundamental domains are straightened to hyperbolic lines.**

Automated Mixed Dimensional Modelling for the Finite Element Analysis of Swept and Revolved CAD Features

T T Robinson[a], C G Armstrong[a], G McSparron[b], A Quenardel[c], H Ou[a] & R M McKeag[b]
[a]School of Mechanical & Aerospace Engineering,
[b]School of Electronics, Electrical Engineering & Computer Science,
The Queen's University of Belfast, UK
[c]Snecma

Abstract

Thin-walled aerospace structures can be idealised as dimensionally reduced shell models. These models can be analysed in a fraction of the time required for a full 3D model yet still provide remarkably accurate results. The disadvantages of this approach are the time taken to derive the idealised model, though this is offset by the ease and rapidity of design optimisation with respect to parameters such as shell thickness, and the fact that the stresses in the local 3D details can not be resolved.

A process for automatically creating a mixed dimensional idealisation of a component from its CAD model is outlined in this paper. It utilises information contained in the CAD feature tree to locate the sketches associated with suitable features in the model. Suitable features are those created by carrying out dimensional addition operations on 2D sketches, in particular sweeping the sketch along a line to create an extruded solid, or revolving the sketch around an axis to create an axisymetric solid. Geometric proximity information provided by the 2D Medial Axis Transform is used to determine slender regions in the sketch suitable for dimensional reduction. The slender regions in the sketch are used to create sheet bodies representing the thin regions of the component, into which local 3D solid models of complex details are embedded. Analyses of the resulting models provide accurate results in a fraction of the run time required for the 3D model analysis.

Also discussed is a web service implementation of the process which automatically dimensionally reduces 2D planar sketches in the STEP format.

1 Introduction

During preliminary design of complex, thin-walled aerospace structures, finite element analysis (FEA) is carried out on idealised models comprised of lower dimensional elements such as beams, shells or plates. This computationally efficient analysis enables rapid optimisation of the global behaviour of the structure. Dimensional reduction is a popular simplification technique that involves the idealisation of simple regions in a model for which the behaviour can be predicted using engineering principles. Thin sheets of material, which have large lateral dimensions relative to their thickness, are approximated by their mid surface with a thickness attribute. Slender solids have one large dimension relative to the other two, and are approximated by a line along their neutral axis with cross sectional attributes [Donaghy et al. 2000]. The combination of elements of different dimension in one analysis model is referred to as mixed dimensional modelling.

Current aerospace practice makes extensive use of idealised stiffened shell models, in which 1D stiffeners reinforce 2D manifold shell models. These can determine natural frequencies with sufficient accuracy in run times which are 2-3 orders of magnitude smaller than the equivalent analysis on a detailed solid model. However generating the idealisation from a 3D solid Digital Mock Up is time consuming and involves a skilled analyst using engineering judgement to determine the most suitable idealisation of the component, and then creating the equivalent analysis model. This significantly reduces the advantages offered by the approach.

The inclusion of idealisation and dimensional reduction tools is becoming more common in commercial finite element analysis pre-processors. Many pre-processors [Altair, MSC, PTC (ProE), Unigraphics, Ideas] offer "mid surface extraction", allowing the user to extract the mid surface of a part, or from between two opposite faces. Some of the more advanced tools also have the ability to extract a surface from a group of user defined surfaces.

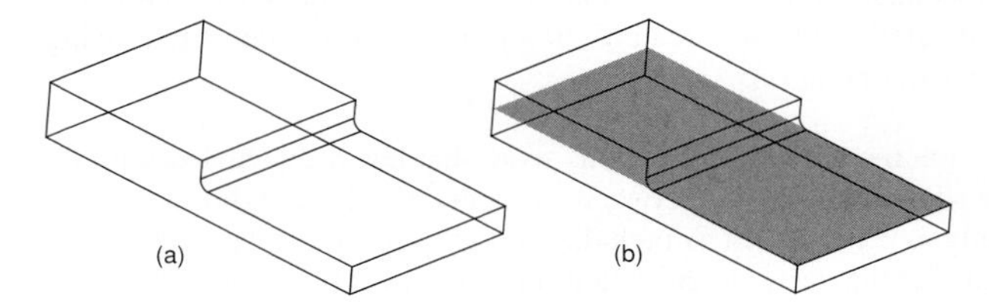

Figure 1 Part body and part body with idealisation

Typically the idealisations offered by such programs are poor, and do not accurately represent the geometry they are idealising. Figure 1(a) shows a part body and (b) the same part body with the mid surface calculated by a commercial CAD package in grey. Clearly the calculated mid surface does not accurately represent the part, and it is not obvious what the mid surface should be in the vicinity of the change in thickness. Problems also exist for "T" shaped sections, where the faces from which the mid surface are to be extracted are ambiguous. The result on sections such as this can be an incorrect or incomplete idealisation. Another problem is that the idealisation tools cannot idealise regions of a part, and will reduce all of a component where perhaps only some areas are suitable for reduction. In Figure 1 a full 3D analysis

SPM 2006, Cardiff, Wales, United Kingdom, 06–08 June 2006.
© 2006 ACM 1-59593-358-1/06/0006 $5.00

would be required to capture the stress concentration at the change in section.

The aim of this research is to automate the preparation of simplified models from the CAD preliminary design information, allowing the advantages of appropriate idealisation to be realised without the expense of preparing the simplified models manually.

Aerospace CAD models are often comprised of swept and revolved features, generated by carrying out dimensional addition operations on planar sketches. This paper details an implementation that idealises the sketches used to generate these features, located by searching the CAD model feature tree for suitable feature types. The procedures outlined have been implemented in CATIA V5, but a similar approach is available for most commercial CAD systems.

In section 2 a summary is given of the Medial Axis Transform (MAT) [Blum 1967; Blum 1973], which identifies properties of the object shape that are useful for idealisation purposes.

Section 3 details how geometric proximity information about 2D profiles provided by the 2D MAT can be used to determine which regions in the profile are suitable for dimensional reduction based on their aspect ratio. The identified regions are replaced by their mid line, a dimensional reduction from 2D to 1D.

Section 4 details the process of locating the appropriate features for dimensional reduction, and extracting the associated sketch in the feature tree of a CAD model.

Section 5 details how the dimensionally reduced sketch is used to create an idealised model of the 3D component. When the subsequent sweep or revolve operations are applied to the dimensionally reduced sketch (as opposed to the original), the resulting mixed-dimensional geometry is an idealised representation of the component, subdivided into slender and non-slender parts. The slender regions are modelled as sheets, and the non-slender parts are represented as detailed local solid models embedded within them. Material in the CAD model which is not associated with the feature being reduced is identified using Boolean operations.

Section 6 summarises previous work that has been carried out into the accurate coupling between elements of different dimension for analysis. This work is included to highlight the state of the art in terms of the accurate continuation of stress contours across mixed dimensional interfaces. There is also an example showing why mixed dimensional interfaces have to be displaced away from stress concentration features.

In section 7 an explanation is given of how the dimensional reduction of a 2D planar sketch using the procedures detailed in this paper has been implemented as a web service, based on models in the STEP format. This allows remote access to the idealisation functionality without the local installation of software, which is often difficult in large multi-partner aerospace consortia.

Section 8 compares the results of an analysis of a mixed dimensional model with the results of the analysis of a 3D solid model and an idealised model produced using the current industry best practice.

2 The Medial Axis Transform

The Medial Axis Transform (MAT) is a skeleton-like representation of geometric shape. The Medial Axis is created by tracing out the centre of the maximal inscribed disc as it rolls around the interior of a surface (Figure 2) or of the maximal inscribed sphere as it rolls around the interior of a solid. This plus the radius function forms the Medial Axis Transform, which is a complete, unambiguous representation of the original shape.

2.1 2D MAT

A medial edge occurs where the inscribed disc is in contact with the surface boundary in two places. The radius of the maximal inscribed disc is recorded along the length of the medial edge.

Where medial edges meet, medial vertices occur. These represent the position of the centroid of the maximal inscribed disc when the set of boundary elements the disc is in contact with changes. Medial vertices also occur where a medial edge terminates with zero radius at a convex corner vertex, or where the inscribed circle is in curvature contact with the object boundary [Ramanathan and Gurumoorthy 2003; Hanniel et al. 2005].

The geometric information supplied by the 2D MAT has been utilised by many researchers for a range of purposes. [Gursoy et al. 1991] demonstrated how the MAT can be used to decompose a shape into simpler regions. [Ang et al. 2002; Tam et al. 1991] detail how the information can be used to generate an adaptive mesh for a surface. Similar to the requirements of this work [Armstrong et al. 1998; Donaghy et al. 2000; Suresh 2003] make use of the 2D MAT to determine suitable regions in a model for dimensional reduction.

The TranscenData implementation of the MAT [TranscenData], commercially available within the CADfix software, was used for the work outlined in this paper. It records as attributes for each medial edge the radius of the largest, smallest and average maximal inscribed disc.

2.2 3D MAT

The 3D MAT consists of medial surfaces as well as edges and vertices. A medial surface exists where the inscribed sphere is in contact with the boundary faces of the solid in two places, a medial edge where it is in contact with the faces in three places, and a medial vertex where it is in contact in four or more places. As for the 2D case, a medial vertex also occurs where a medial edge is terminated at a corner vertex, or where finite contact occurs between the inscribed sphere and the boundary of the body. Medial edges and vertices can also be caused by curvature contact.

Although much research has been done into the generation of the 3D MAT [Lee et al. 1994; Lee et al. 1997; Sherbrooke et al. 1996] a robust, commercially available implementation is currently not available. Like the 2D MAT, the 3D implementation has been widely used for shape interrogation and simplification [Sheehy et al. 1996], and adaptive meshing [Gursoy et al. 1991].

3 Dimensional reduction of 2D surfaces

Dimensional reduction is when regions of simple geometry are approximated by elements of lower dimension with the reduced dimension recorded as an attribute. In analysis terms this reduces the size of the model and therefore the expense of the analysis.

For a 2D planar surface a dimensional reduction is achieved by slender regions being approximated by their mid lines (1D elements), with the thickness of the region applied as an attribute.

The dimensional reduction of 2D surfaces in this paper is facilitated by geometric proximity information provided by the TranscenData implementation of the 2D MAT, and uses a procedure similar to that of [Donaghy et al. 2000].

The procedures outlined below have been incorporated into software built on top of the CADfix implementation of the 2D MAT.

The steps used in the software are:

1. Input profile/sketch
2. Generate MAT (Figure 2)
3. Identify medial edges representing slender regions (Figure 2)
4. Extend chunky regions one medial axis diameter into the slender regions (Figure 5)
5. Partition model into slender and chunky regions (Figure 6)
6. Replace slender regions with the equivalent midline (Figure 7)

3.1 Identification of slender regions

The MAT provides the geometric information to identify the medial edges representing slender regions.

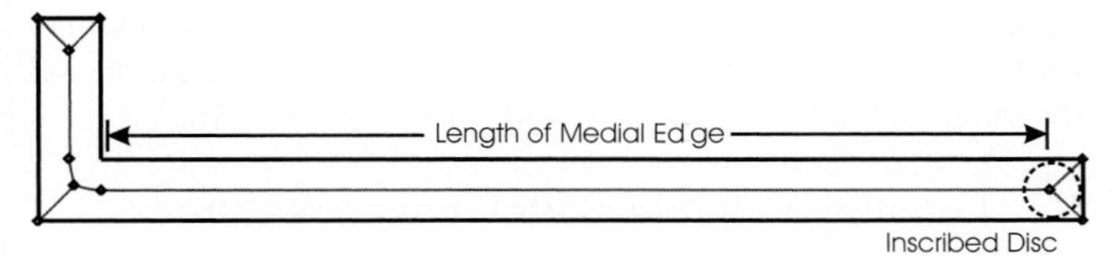

Figure 2. Profile, MAT and measurements used to determine aspect ratio.

3.1.1 Aspect Ratio

The geometric proximity information provided by the MAT is used to determine where the slender regions in the profile occur. A slender region is long relative to its thickness. The measure of aspect ratio (AR) is used to determine if a region is slender. The aspect ratio of a medial edge is calculated using the length of the edge, and the diameter of the largest maximal inscribed disc [Equation 1].

$$AR = \frac{\text{Length of medial edge}}{\text{Diameter of largest maximal inscribed disc}} \quad [1]$$

The diameter of the largest maximal inscribed disc that traces out the medial edge under investigation is used in the calculation of aspect ratio to ensure a conservative estimate of slenderness.

A region is considered slender if it has an aspect ratio greater than a critical value (Critical Aspect Ratio, CAR). Figure 3 shows the effect of varying the CAR on the mixed dimensional model produced. As the CAR is increased the amount of dimensional reduction is reduced. It should be noted that reasonable analysis results can be obtained with a CAR of around 2.

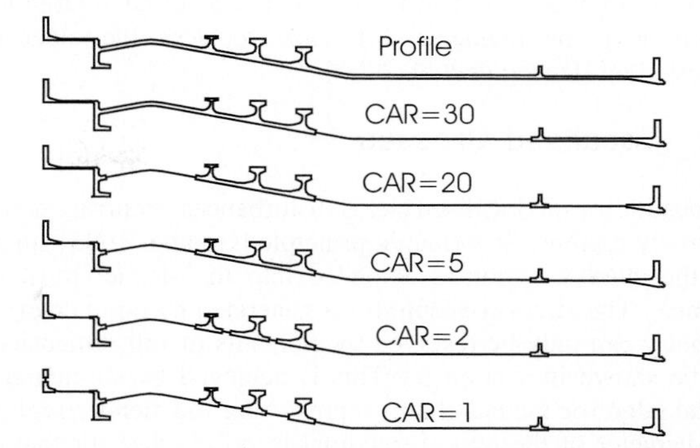

Figure 3. Models produced by varying critical aspect ratio.

3.1.2 Difference in diameter

When the models considered are of clean geometry the sole consideration of aspect ratio as a measure of slenderness is appropriate. However, it is common for models of even simple geometry to contain complex topology, which complicates the interpretation of the MAT information. An example of this is shown in Figure 4, where the top and bottom edges bounding the horizontal slender region are each comprised of several segments. As a result, the slender region is associated with 5 medial edges (ME1 to ME5), all of which are quite short. None are slender based on the aspect ratio computation of [Equation 1].

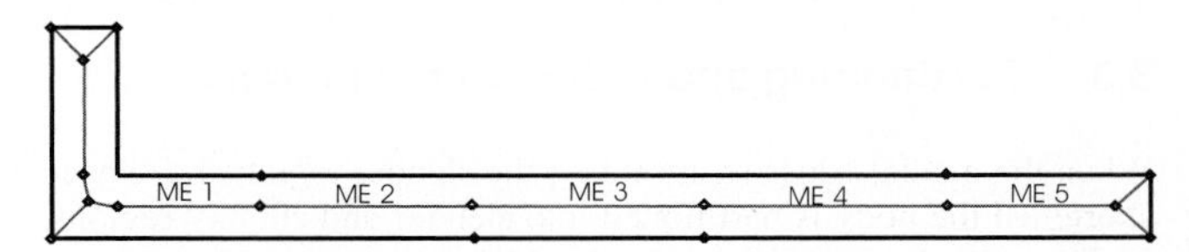

Figure 4. Complex topology producing a number of medial edges.

If a region is represented by more than one medial edge, some or all of which have an aspect ratio below the critical value, it is necessary to group them together to determine whether the region they represent is slender, and if so to dimensionally reduce it. A measure is required to determine whether shorter medial edges should be considered as part of a group (or chain) of medial lines.

Two adjacent medial lines are to be considered as representing the same region if they have similar medial diameters. The approach adopted in this paper considers the difference in diameter of the maximal inscribed discs that created each medial edge, and compares it with that of the neighbouring medial edge. If the difference is small (< 5%) they are considered to represent the same region. This check is continued for adjacent medial edges in both directions until one is located at each end that does not meet the condition or a branch point (where more than two medial edges meet) is met. A chain of medial edges created in this way is considered to represent one region. The aspect ratio of the chain is determined using [Equation 1], but instead of using the length and diameter of one medial edge, the combined length of the medial edges, and the largest maximum disc diameter for the entire chain is used. If the calculated aspect ratio exceeds the critical value the region is considered slender.

[Lin 2005] uses a similar measure to determine where potential problem areas for extrusion die manufacture occur.

The diameter difference measure also ensures the region is not overly tapered, since the beam, shell and plate theories used for the idealisation assume that the elements do not have large taper. In this way the diameter difference replaces the taper ratio suggested by [Donaghy et al. 2000].

3.2 Localised stresses

To account for the localised stress disturbances occurring adjacent to chunky regions, St Venant's principle [Goodier 1933] suggests that the chunky region is extended into the slender by a short distance. The stress concentrations generated by rapid changes in geometry can only be resolved by elements of full dimension, as will be shown in section 6. This is achieved by shortening the medial edge (or medial chain) representing the slender region by one diameter of the largest maximal inscribed disc for that edge (or chain) at each end (Figure 5). This reduction in length affects the aspect ratio of the region. To account for this [Equation 1] is replaced with [Equation 2].

$$AR = \frac{\text{Length of medial edge}}{\text{Diameter of largest maximal inscribed disc}} - 2 \quad [2]$$

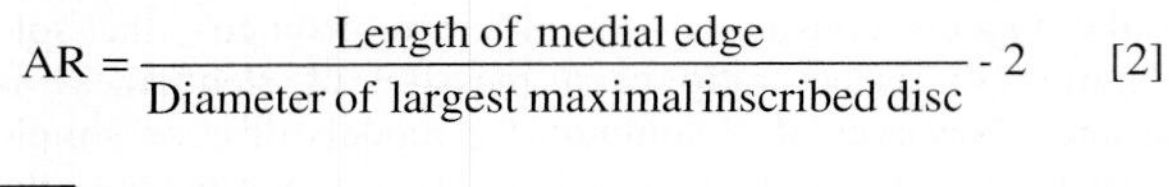

Figure 5. Shorten slender regions by one disc diameter.

3.3 Partitioning and reducing the model

When the medial edges representing the slender regions have been shortened the body is partitioned into slender and chunky regions. In Figure 6, A represents the only region that has an aspect ratio above a given critical value and is considered slender.

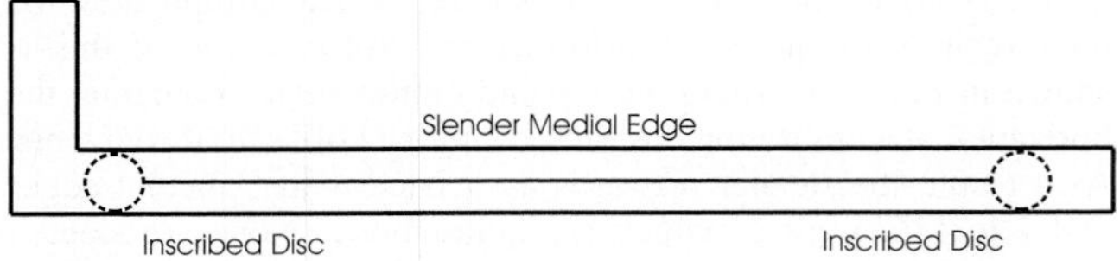

Figure 6. The partitioned profile.

The surface representing the slender region is removed, and replaced with its mid line, which takes the geometry of the associated medial edge. The result is shown in Figure 7.

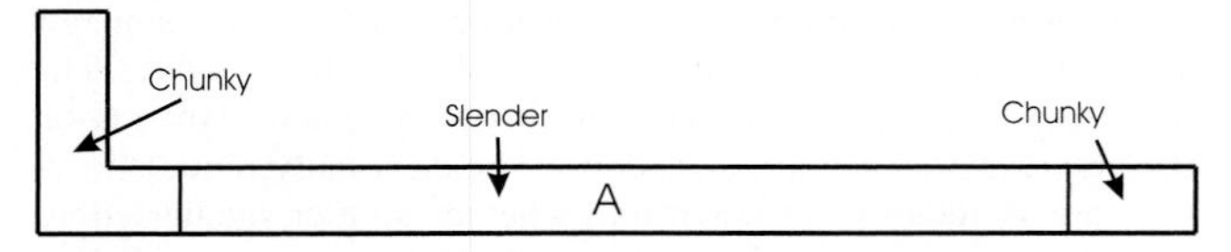

Figure 7. The mixed dimensional result.

3.4 Ignoring perturbations

A perturbation is a single short medial edge caused by a small local feature in the geometry, an example of which is shown in Figure 8(a).

The region within the touching points of the MAT vertex where the perturbation meets the other medial edges (shaded in gray in Figure 8(a)) would typically be considered to represent a chunky region. The branch point would be offset by one inscribed disc diameter in each direction along the medial edges (as shown in Figure 8(b).) and the surface partitioned. Providing the regions outside the partition have an aspect ratio greater than the critical aspect ratio they can be dimensionally reduced (Figure 8(c)).

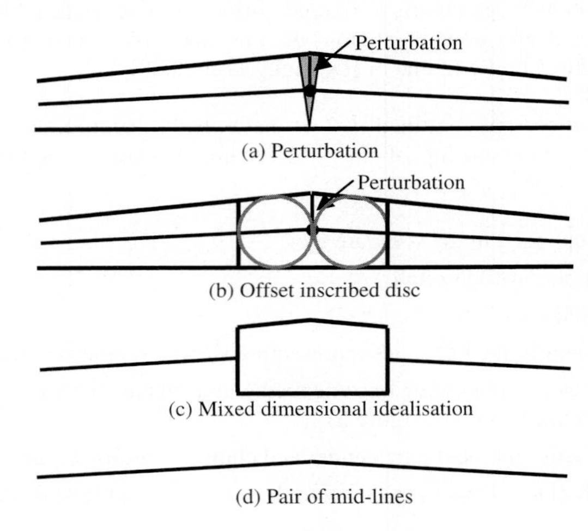

Figure 8. Perturbation.

Typically, if the feature is small relative to the thickness of the part it has negligible effect on the global model. Within the idealisation procedures medial edges shorter than the maximal diameter of the inscribed disc of the adjacent medial edges are ignored. This means that areas that would be slender without the small feature are modelled as slender. The result for the model shown in Figure 8(a) is a pair of mid-lines (Figure 8(d)).

4 Locating suitable sketch based features

When using modern feature based CAD systems, it is common practice to 3D model components by sweeping a 2D sketch along a line to create an extruded solid, or revolving a sketch around an axis to create an axisymmetric solid. Carrying out these operations on the 2D/1D mixed dimensional profiles created using the procedures described in section 3 results in a mixed dimensional idealisation of the original CAD feature.

The first step is to locate all of the suitable features in the model by using a scripting interface to the CAD package. Visual Basic (VB) scripting in CATIA V5 allows each feature in the model to be visited, and checked whether it is a suitable feature type (A "Pad" or a "Shaft" feature). This is achieved in CATIA V5 using the "TypeName(Feature)". Some sketch based features remove material from the model (e.g. a pocket feature) and are not suitable for reduction.

It is also necessary to record the parameter information about the feature so that the appropriate operation can be used on the mixed dimensional profile. This information includes the starting and ending lengths for a sweep, or angles for a revolve.

Figure 9 shows a CAD model and its feature tree, in which the feature "Shaft.1", and its associated sketch (Sketch.1) are visible.

The sketch can be extracted, converted to a planar surface, and exported to the dimensional reduction software using scripting.

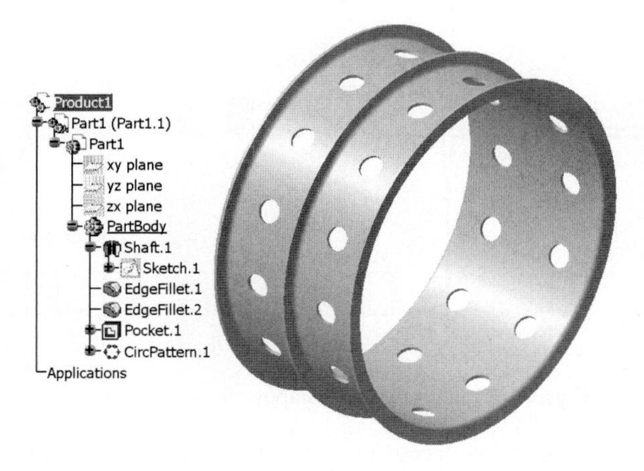

Figure 9. CAD model and feature tree.

5 Modelling with the dimensionally reduced surface

Once dimensionally reduced the resulting mixed dimensional surface is imported back into the CAD environment, and is used instead of the sketch in the feature operation.

5.1 Creating the idealised feature

When a dimensionally reduced profile (Figure 10) is imported into the CAD package, the same dimensional addition operation is used as on the original sketch. This may not be the same feature type as was used for the original sketch as unfortunately most CAD packages have different operations for solid and surface bodies.

Figure 10. The mixed dimensional surface.

In CATIA V5 any operation resulting in a solid, like the revolution of a surface representing the chunky region is a part design feature (like the shaft feature used in the example). Lines representing the slender regions must be revolved or extruded using a surface design tool, (such as the revolute function), as the result is a surface. Figure 11 shows the result of revolving the mixed dimensional profile from Figure 10. In the feature tree three shaft features are visible, representing the solid parts of the model, shown in dark shading in Figure 11. The two revolute features, created by revolving the mid lines, are the grey regions of the model.

5.2 Adding additional features

The procedures above approximate only the feature the sketch is associated with (Shaft.1 in the CAD model feature tree shown in Figure 9). Other features in the model (EdgeFillet.1, EdgeFillet.2, Pocket.1 and CircPattern.1 in Figure 9) have to be incorporated into the mixed dimensional representation of the sketch based feature to create an idealisation of the overall component. In this paper the "feature model" is defined as the geometric model created by suppressing a given set of features (everything after Shaft.1 in Figure 9).

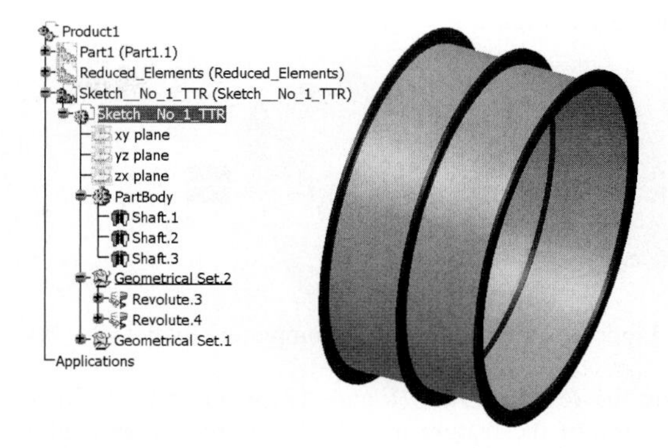

Figure 11. The mixed dimensional idealisation (C).

After creating a mixed dimensional model of the feature there are two possible sources of error in the representation:

1. Material not approximated by the mixed dimensional model of the feature.
2. Material modelled by the mixed dimensional model of the feature, which should not be.

The first group refers to material that is added to the model after the original feature was created from the sketch, but which does not create thin sheets of material suitable for dimensional reduction. An example would be edge fillets applied to the interior corners (EdgeFillet.1 in Figure 9), or bosses added to a component.

The second group refers to material that is removed from the original feature at a later stage. Examples would be the pocket features (Pocket.1 in Figure 9) used to create the holes in the component shown in Figure 9, or fillets on exterior corners (EdgeFillet.2 in Figure 9).

A Boolean symmetric difference between the feature model and the component model (with all features unsupressed) can be used to identify the material in both categories. The three entities used in the description of the Boolean operations are the component model (A), the feature model (B), and the mixed dimensional idealisation of the feature model shown in Figure 11, (C).

Figure 12 shows the result of subtracting the feature model from the component model (A – B). The result is the internal fillets that are added to the interior corners of the stiffener, which is the material added by EdgeFillet.1 in Figure 9.

Figure 12. Component model – Feature model (A – B).

Figure 13 shows the result of subtracting the component model from the feature model (B – A). The result is the material removed by the holes and the external fillets added at each end of the component, or Pocket.1, CircPattern.1 and EdgeFillet.2 respectively in Figure 9.

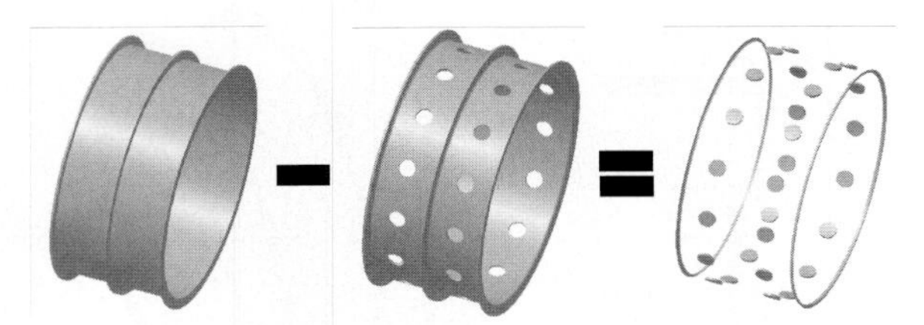

Figure 13. Feature model - Component model (B – A).

Adding the result of A-B (Figure 12) to the mixed dimensional idealisation of the feature model, C (Figure 11), and subtracting the result of B-A (Figure 13) produces a mixed dimensional idealisation of the component (Figure 14), where all material in the model is considered.

$$\text{Idealisation} = C + (A - B) - (B - A)$$

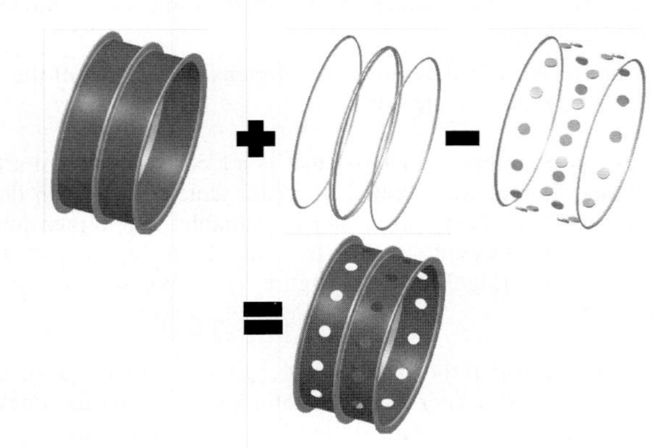

Figure 14. Add and subtract extra material (C + (A-B) – (B-A)). Sheet bodies are in dark shading, solids are light.

The operation depicted in Figure 14 may in some instances result in sliver sheet and solid features. Sliver solids should be added to the adjacent solid chunky parts. Sliver sheets should be thickened and added to the adjacent chunky parts.

5.3 Managing interfaces between idealised features.

All of the material in the model is represented in the result of the operation in Figure 14, but it is possible not all of the material is properly connected in the idealisation. Figure 15 shows the cross section through a component, the features used to create it, their idealisations, and the different interfaces that occur.

The component in Figure 15 is comprised of four features (labelled A-D), two of which (A and C) are suitable for idealisation because they have slender regions. The structural representation of the component is shown, with the same four features labelled a to d. The different interfaces that occur between the idealisations of the features are also shown. There are three different categories of interface:

1. Solid - solid (between a and d)
2. Sheet - sheet (between a and c)
3. Solid - sheet (between a and b)

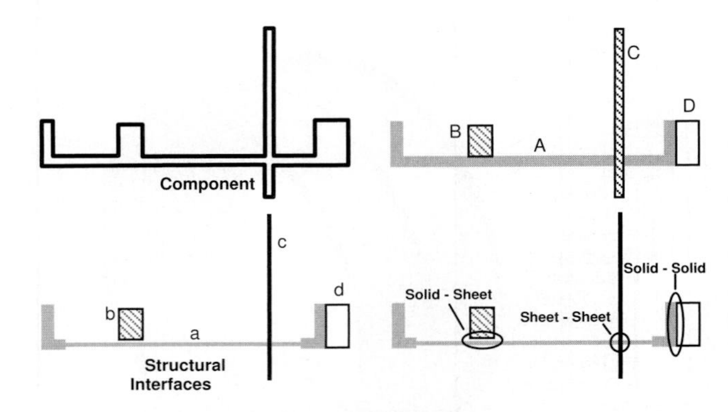

Figure 15. Interfaces in a component idealisation.

The solid - solid interface requires no further operation, as at the common boundary between the two there is automatically full contact making it suitable for analysis.

The sheet – sheet interface is modelled in reduced dimensional elements after the idealisation. If accurate local stresses are required, the region where slender features intersect cannot be modelled using reduced dimensional elements and must be modelled in full dimension. The procedure for increasing the dimension in such regions is shown in Figure 16.

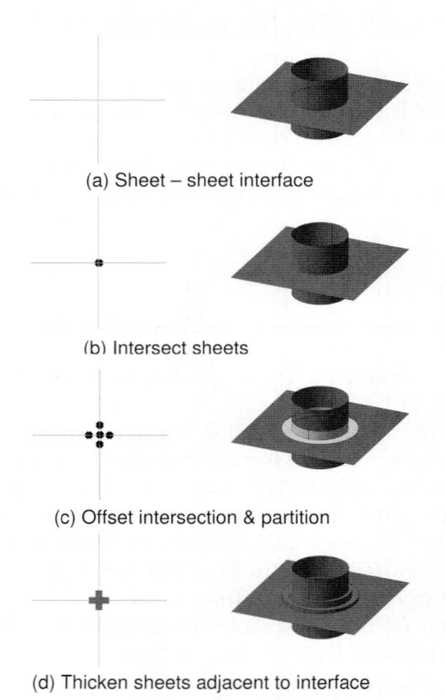

Figure 16. Procedure to model sheet - sheet interface.

Figure 16(a) shows the reduced dimension element – reduced dimension element interface in both 2D and 3D. The first step in creating a full dimensional interface is to intersect the two components (Figure 16(b)), which creates a point in the 2D scenario, and a line in the 3D. The point or line created by intersecting the elements should then be offset by an amount equal to 1.5 sheet thicknesses in each direction (Figure 16(c)).

This is analogous to the offset of the chunky region in section 3.2. In the area adjacent to the intersection, shown in white in Figure 16(c), thickening operations are applied to create a full 3D model of the intersection, shown in Figure 16(d). It is in this solid region that the localised stress is modelled during the analysis.

The solid – sheet interface occurs in regions where material is added to a slender feature later in the feature tree. When the slender feature is idealised, the solid body representing the later feature is no longer in contact with the sheet as is shown between a and b in Figure 15. Clearly this is not suitable and the material in the adjacent sheet should be considered as part of the chunky solid rather than as a thin sheet. The procedures to achieve this are shown in Figure 17.

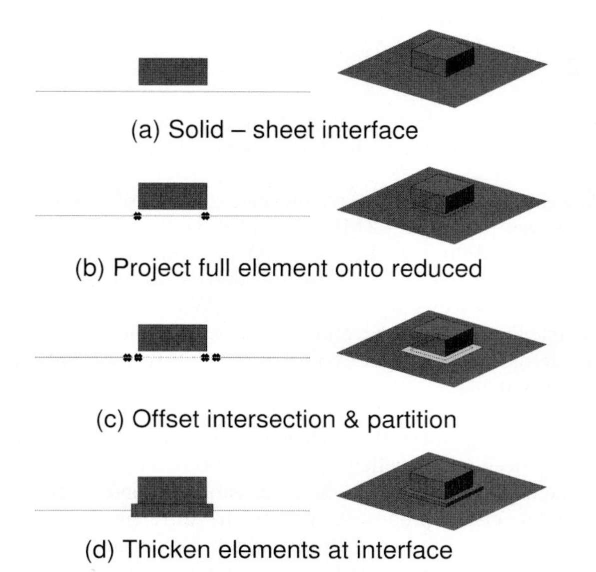

(a) Solid – sheet interface
(b) Project full element onto reduced
(c) Offset intersection & partition
(d) Thicken elements at interface

Figure 17. Solid - sheet interface.

Figure 17(a) shows the reduced dimension element and full dimension element not in contact in a 2D and 3D situation. The first step in the integration of the two is to project the boundary of the face of the solid nearest to the reduced dimension element onto the reduced dimension element. This creates points in the 2D scenario and lines in the 3D. These lines are then offset by an amount equal to one thickness in the same direction as the outward facing normal of the faces adjacent to the face that was projected onto the reduced element. The offset is to allow for localised stresses that occur adjacent to the chunky region. The point or lines created by the offset are used to partition the reduced dimension elements. The partitioned region adjacent to the solid part, shown in white in Figure 17(c), is thickened by half the thickness in each direction. The thickened interface is automatically united with the solid feature, shown in Figure 17(d).

5.4 Implementation difficulties using CATIA V5 scripting

The above procedures have been automated using CATIA V5 scripting with the exception of the two operations detailed below. The implementation of both of these operations would be trivial using a low level modelling kernel, but not in a commercial CAD system.

Unfortunately it is not possible to carry out Boolean operations between solid and sheet bodies in CATIA V5, through either the scripting interface or graphical user interface. Figure 18 shows that the Boolean subtraction of a solid cylinder from a sheet body will result in the sheet body without the cylinder removed. The result should be a sheet with a hole in it.

The procedures discussed in section 5.3 for the sheet - sheet interface, and the solid - sheet interface can be automated for CATIA V5 using VB scripting, with the exception of the projection of the full dimension element onto the reduced element shown in Figure 17(b). This cannot be achieved because of the lack of topological information for solid bodies in the scripting interface. VB scripting of CATIA V5 does not allow the identification and selection of individual faces for a solid body, and as such it is impossible to identify which face is to be projected onto the reduced dimensional element.

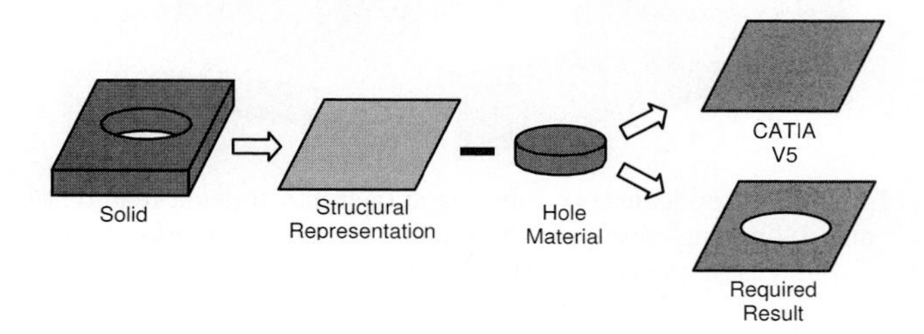

Figure 18. Problems with the Boolean Subtraction.

The prototype implementation required interactive picking of the relevant faces to illustrate the proof of concept studies described here.

6 Mixed dimensional coupling

To allow accurate finite element analysis of the mixed dimensional idealisation, the solid elements representing the chunky regions need to be embedded into the global shell models of the component. The accurate modelling of stress at mixed dimensional interfaces is not achieved successfully by many of the coupling techniques that are available in commercial CAD systems. This section reviews successful research previously carried out in this area.

The majority of finite element pre-processors have inbuilt coupling techniques that are used to join regions of different dimension together for analysis. The problem with many of these techniques is that spurious stresses are generated at the interface. The majority of these techniques are derived by considering the displacements of the nodes on either side of the mixed dimensional interface.

The accurate coupling of regions of different dimension has been investigated by [McCune et al. 2000; Monaghan et al. 2000; Shim et al. 2002]. In dimensionally reduced regions it is assumed that the stress through the thickness of the region adheres to the conventional strength of materials theories i.e. that in-plane loading and/or plate bending causes stress which varies linearly through the thickness, whilst out-of-plane shear forces cause shear stress which varies parabolically through the thickness. It is the ability to determine the distribution of stress through the thickness of the plate that allows a more accurate coupling technique to be derived. St Venant's Principle [Goodier 1933] suggests that the stress in a body will assume this distribution away from a detail feature. Figure 19 shows a component with transverse shear loading applied at one end. Figure 19 (a and c) shows that the stress in the body is not linear through the thickness of the mixed dimensional interface adjacent to the fillet radius. Figure 19 (b and d) shows how the stress in the body is linear through the thickness of the mixed dimensional interface at one thickness away from the fillet.

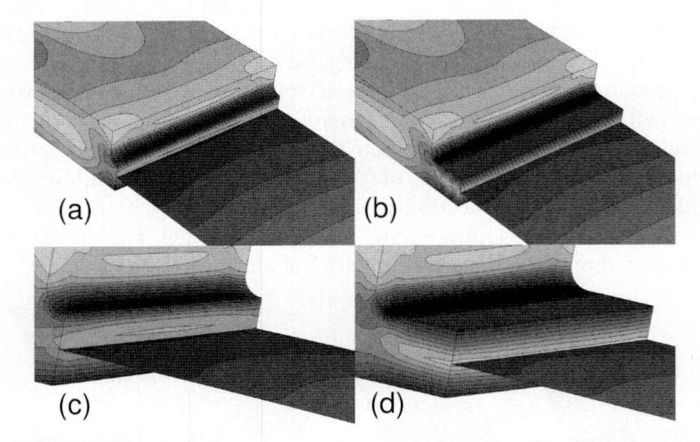

Figure 19. Application of coupling at (a and c) at detail feature, (b and d) one thickness away from detail feature. Contours shown are of Von Mises equivalent stress.

McCune introduced the procedures for 3D-2D coupling [McCune et al. 2000], Monaghan extended it to 3D-1D coupling [Monaghan et al. 2000] and Shim automated the procedure for laminates of multiple materials [Shim et al. 2002]. The underlying principle of the techniques is to equate the work done on the boundary of the reduced dimensional side of the mixed dimensional interface to the work done by the stresses on the full dimensional side. The stresses on the 3D side of the interface can be written in terms of forces and moments on the reduced side. This allows the forces and moments to be eliminated, resulting in multipoint constraint equations (MPCs) relating the displacements and rotations of the reduced dimensional side to the displacements on the 3D side. In practical terms this is achieved by writing equations for each node on the reduced dimensional side of the element, relating each degree of freedom for that particular node to the degrees of freedom of the 3D nodes adjacent to it.

7 Dimensional reduction web service

In order to allow the dimensional reduction software to be tested, and to allow its inclusion in automated work flows, a method of accessing the tool over the internet was developed. This brought several advantages over distributing the software to end user partners as development continued:

- Proprietary software was being used which not all partners had access to.
- Partners did not all work on the same platform.
- Partners did not have to install any software on their machines.
- Changes could be tested immediately by partners.
- The latest version of the software was always available.

7.1 Web page

Initial deployment was via a system in which a user could access the software via a web browser. A web page [QUB] exists containing a form which allows a user to enter the location of a STEP file on their system and an e-mail address to receive the reduced file. On submission the file and the e-mail address are uploaded to the web server. A PHP script on the server causes the dimensional reduction software to process the STEP file and generates a web page informing the user that the process is underway and that results will be e-mailed to the provided address.

When a STEP file has been reduced successfully an email is sent to the provided address containing the reduced STEP file as an attachment, otherwise an email is sent detailing the error.

This was an acceptable method of testing the software, but was limited to having a user interact with the web page.

7.2 Web services

To expand upon the internet distribution it was decided to create a web service interface to the dimensional reduction software.

A web service is a method of sharing web based applications using web standards. XML (Extended Markup Language) [W3 (1)] is used to tag the data, SOAP (Simple Object Access Protocol) [W3 (2)] is used for transferring the data and WSDL (Web Services Description Language) [W3 (3)] is used to define the interface to the service in a machine readable way. All three are defined by W3, the World Wide Web Consortium and are open standards.

The advantage over the web page based approach is that the service need not be called by human interaction; it can be part of an automated procedure. It is possible to have a number of different client programs, written in different languages or operating on different platforms which can all access the same service. It is also possible to create a GUI for the web service on any target platform.

7.2.1 Service Implementation

Web services may be implemented in any language that supports the WSDL and SOAP protocols and deployed using a suitable application server. There are numerous application servers available which provide web service support including those from IBM and Microsoft.

In the test case described here, the Jakarta Tomcat server [Tomcat] with Axis [Axis] from the Apache Foundation was used. Tomcat is a Java Servlet Container, allowing Java classes to be made available to remote machines. The connection to the Tomcat server is through a defined port and utilises the HTTP protocol. Axis provides an implementation of the SOAP standard and is built on top of the Tomcat server to allow SOAP messages to be created and read within the services.

In this setup a Java program is created and compiled using the Axis libraries. To be run as a service the class is deployed to the server. This is achieved by creating a WSDD (Web Service Deployment Descriptors) file which specifies the class and the methods of the class to be made available via the service. Tomcat allows the processing of this file and adds a service relating to the Java class. It is possible from the server to find all the available services and to generate a WSDL file to describe any service that is available.

7.2.2 *Web Service Client*

To call the service a client of some description is required. Clients may be created on any platform and in any language that supports the SOAP protocol. There are SOAP libraries for most major languages.

For this example a SOAP library which can handle "soap with attachments" is required for sending files via web services. This is required as the XML is essentially a text based format. The Axis libraries are used to implement the client side functionality.

For testing purpose a small Java client was created which allows a file to be selected to be uploaded to the web service where it is dimensionally reduced. When the file is returned the client allows it to be saved as a new file. This client remained the same as the software on the server was changed.

7.2.3 *Full Demonstration*

The web service was publicly demonstrated at the VIVACE project's Forum 1 event [VIVACE]. In this demo a client was developed within the FIPER [FIPER] system.

FIPER allows processes to be designed using a drag and drop interface, and in this case a dimensional reduction process was created. FIPER also has the ability to call web services, and so it was possible to make the design analysis process call the dimensional reduction web service instead of a local program. With the process created it was possible to drop the dimensional reduction process into a workflow. In the demo a Unigraphics part file was modified live, and the edited file had its profile extracted as a STEP file and automatically sent to the dimensional reduction service which was running on a server in the Queen's University of Belfast. The returned STEP file was then used automatically by the next phase in the process.

8 Analysis of a mixed dimensional idealisation

This section details the comparison of three different analysis models for a simple aero engine component. This investigation was used to quantify the advantages of mixed dimensional modelling for Finite Element Analysis.

The first model is a 3D solid model of the component, which was used as the reference model. The component is created as a single CAD feature produced by revolving a sketch around an axis, and is therefore suitable for dimensional reduction using the techniques described in previous sections. To ensure accurate analysis results the 3D model was meshed very densely. The second is a mixed dimensional model produced using the web page described in section 7, and the third is an idealised model produced by a skilled analyst using the current industrial practice.

8.1 Model Creation

All models had the same material properties and were analysed using MSC NASTRAN. The interfaces for the mixed dimensional model were coupled using the RSSCON MPCs native to NASTRAN.

The sketch used to create the model is shown in Figure 20 and its mixed dimensional idealisation is shown in Figure 21. Table 1 shows a comparison of the model data for each model. The shell model is 2 orders of magnitude smaller than the 3D model, whilst the mixed dimensional model is less than 5 times bigger than the shell model.

Figure 20. Sketch revolved to create component.

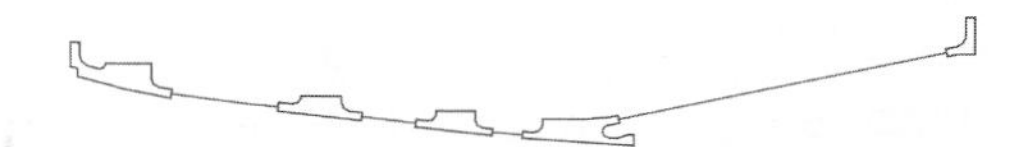

Figure 21. Dimensionally reduced sketch.

Table 1. Comparison of model data.

	3D Model	Shell model	Mixed Model
No. of nodes	198000	1807	8520
Elements	137880	2100	4860
Element Details	137880 HEX	420 BAR 1680 QUAD4	540 QUAD 300 WEDGE 4020 HEX 480 MPCs

8.2 Modal Analysis

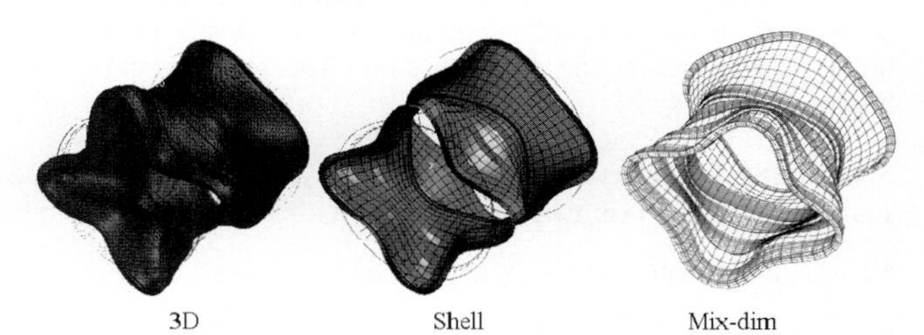

Figure 22. Mode shapes (modes 19 - 20).

Table 2 details the results for the modal analyses of the models. During the analysis the first 30 modes were retained. The 3D model was used as the reference model to calculate the error in the idealisations.

Table 2. Modes and errors for models.

	3D	Shell		Mix-dim	
Modes	Freq. Hz	Freq. Hz	Error %	Freq. Hz	Error %
7-8	83.95	77.99	-7.11	84.51	0.66
9-10	96.43	90.39	-6.26	97.27	0.87
11-12	224.00	216.49	-3.35	226.07	0.93
13-14	310.29	283.28	-8.70	311.97	0.54
15-16	371.05	336.63	-9.28	376.00	1.33
17-18	418.48	369.24	-11.77	420.39	0.46
19-20	503.57	455.61	-9.52	509.30	1.14
21-22	514.42	520.08	1.10	522.91	1.65
23-24	622.08	520.21	-16.38	626.14	0.65
25-26	650.80	545.57	-16.17	657.82	1.08
27-28	664.42	642.26	-3.34	663.00	-0.21
29-30	747.96	660.28	-11.72	762.00	1.88

All of the mode shapes matched for each model. Figure 22 shows the mode shapes for the three models for modes 19-20, whilst

Table 3 shows the time taken to complete the modal analysis on each model.

Table 3. Analysis times.

	3D	Shell	Mix-dim
Analysis Time (s)	1212.578	6.375	20.796

8.3 Stiffness Analysis

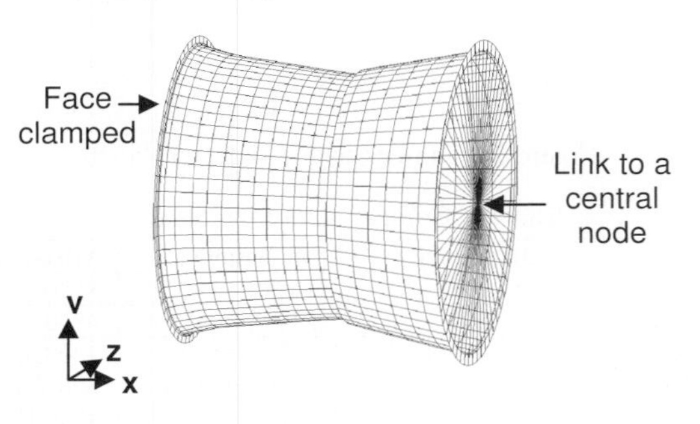

Figure 23. Stiffness analysis setup.

The second type of analysis carried out on the models was a stiffness analysis. This involved clamping the face of the component at one end and applying different loads to a node linked to the face at the opposite end. The setup is shown in Figure 23.

Six different loads were applied to the node. These were:

- 1N force in X, Y and Z directions
- 1Nm moment about X, Y and Z axis

For each load case the displacement and rotations of the centre node are measured, and taken to correspond to the stiffness of the model.

Table 4 lists the errors in the idealisations relative to the 3D model for each load applied. The analysis time for each model is shown in Table 5.

Table 4. Error in idealisation relative to 3D model.

	Shell error (%)	Mix-dim error (%)
Force X (1N)	-44.43	1.12
Force Y (1N)	-18.14	1.62
Force Z (1N)	-18.14	1.62
Moment X (1Nm)	-20.61	1.02
Moment Y (1Nm)	-49.55	0.91
Moment Z (1Nm)	-49.55	0.91

Table 5. Stiffness analysis time.

	3D	Shell	Mix-dim
Analysis Time (s)	496.203	1.75	7.812

9 Discussion

There are two main topics for discussion arising from the work detailed, namely the advantages and disadvantages offered by the procedures so far, and how to advance the automatic idealisation of general 3D CAD models.

9.1 Current procedures

The procedures detailed in this paper have been shown to produce 3D-2D idealised models that can be analysed in run times that are more than 80% smaller than those of the corresponding 3D model for a modal analysis. The error in the results for the idealisation is shown to be less than 2% relative to the 3D model. Clearly the reduction in analysis time is significant. Table 1 shows that the size of the model created by the procedures is smaller than the 3D model; for example it only contains 8520 nodes compared to 198000 for the 3D model. However, the idealisation created using the current best practice is smaller still with only 1807 nodes.

The analysis time for the mixed dimensional idealisation is 4.5 times longer for a modal analysis than an idealised model created using industrial best practice, and over three times longer for a stiffness analysis. In both the stiffness and the modal analysis the model created using the current best practice had a maximum error of 17% and 50% respectively relative to the 3D model. The maximum error for the mixed dimensional model created using the procedures in this paper is < 2%. This is because elements of full dimension are used where required i.e. in chunky regions where the solution is varying in a complex way. Reduced dimensional elements are used where they are appropriate i.e. in thin sheets of material away from the stress concentrations caused by chunky regions.

The creation of a mixed dimensional model using the procedures detailed in this paper is much simpler and less error prone than the current practice.

It is worth noting the mixed dimensional model for which the analysis results are shown in section 8 was created using a critical aspect ratio of 2. Increasing this value may improve the accuracy, but will also increase the analysis time. As the accuracy is already acceptable the disadvantages of increasing the critical aspect ratio outweigh the advantages.

The web implementation of the procedures allows the dimension reduction of planar profiles in the STEP format. This allows remote access to the procedures detailed, without having to install software locally. It also allows the procedures to be integrated into any workflow managed by a web service enabled system.

9.2 Future work

To reap the full rewards of mixed dimensional modelling a robust technique for automatically idealising a general 3D component is necessary. The two areas where future work is needed are the improvement of the procedures in this paper, and the use of 3D geometric data to allow the idealisation of non feature based models.

9.2.1 Improvements to current procedures

As has been previously mentioned the procedures outlined in this paper could not be fully implemented due to problems with how Boolean operations and topological information are exposed through the scripting interface of CATIA V5. Solutions to these problems would allow more advanced idealisation techniques to be developed.

To increase the range of models that can be idealised a more sophisticated method of interrogating and interpreting the CAD model feature tree must be established.

Feature specific techniques could be extended to consider other features known to produce thin sheets e.g. the shelling of solids. Some CAD modellers currently create mid surface models, but they do not treat chunky regions properly, as 3D solids embedded in global shell models.

In section 5.2 procedures are described that use Boolean operations to identify the differences between a dimensionally reduced feature and the component model. These techniques are also applicable for more general situations, for example in keeping track of the differences between a geometric model and an analysis model which may have undergone extensive de-featuring and modification.

One disadvantage of the feature based approach outlined in this paper is that it requires a model to be in its native CAD format, and contain a feature tree. It cannot be used therefore in packages that do not support a feature based design approach, or models exported from their native CAD package in a neutral format (e.g. STEP or IGES). A robust technique for interrogating the "dumb" geometry is required.

9.2.2 Idealising dumb 3D geometries

The 3D MAT could be used to identify slender regions in a 3D model in much the same way as a 2D MAT was used for a planar surface. In this implementation the medial surfaces in the model could be used to determine where thin sheets of material occur, based on an aspect ratio taking account of lateral dimensions relative to the diameter of the inscribed sphere. A similar approach to that used for 2D could be used to partition the body.

If the 3D MAT were used to idealise a model using an extension of the 2D procedures an appropriate idealisation could still be created even if the implementation is not robust for all complex geometries. If some thin sheets are not identified by whatever method is used for geometric reasoning, they will be analysed in 3D meaning that the analysis will be slower but still accurate.

The need to identify thin sheets of material in a solid exists even if a full 3D analysis is used. [Yin et al. 2005] describe a procedure for finding thin sheets for use with 3D p-adaptive finite element analysis. Their approach identifies the surface triangles that exist on the opposite faces of thin regions. It achieves this using information about the MAT, namely the pairs of triangles associated with medial surface points, distance, adjacency, and classification information. This approach is used to create corresponding surface meshes on the opposite faces of thin solids and special p-element meshes in the thin solids which are much more efficient.

[Ahmad et al 1968] discuss the use of thick shell elements, the implementation of which is becoming more common in commercial finite element packages. The advantage of using thick shell elements are that no coupling is required, as the nodes of the thick shells correspond with those of the adjacent 3D bodies, and no dimensional reduction of the geometry is required. Once the component has been partitioned into slender and chunky regions, Figure 6, a process similar to that used by [Yin et al. 2005] can be used to generate the thick shell meshes.

[Lee et al. 2005] detail an approach where coarse global models are analysed. The global models are created by suppressing small details which have negligible effect on the global model and allow a coarse mesh to be used for the body. The small features are analysed separately using local sub models for which a denser mesh is applied. If a more sophisticated approach were found for interrogating the feature tree of a CAD model it might be possible to combine their approach with the procedures detailed in this paper.

10 Conclusions

Procedures to automate the dimensional reduction of a CAD component comprised of sketch based sweep and/or revolve features have been described.

The mixed dimensional models produced using these procedures provide accurate results in a fraction of the run time and model size of 3D models. The run time is longer than for models created using current industry best practice, but the accuracy is greatly improved. Suitable models can be created using a critical aspect ratio of 2.

The techniques are also less error prone since the appropriate idealisations are chosen based on quantitative assessment of aspect ratio as opposed to human judgement.

The web service implementation of the dimensional reduction of planar profiles in STEP format was straightforward. It greatly facilitated testing and dissemination of the research ideas.

Future effort will be directed at improving the ability to use CAD model feature tree information, and a method for interrogating dumb 3D geometries.

11 Acknowledgements

Thanks to the VIVACE project AIP3-CT-2003-502917 of the European FP6_Aerospace programme for supporting this work and allowing publication of the paper.

12 References

AHMAD, S., IRONS B.M. IRONS AND ZIENKIEWICZ, O.C. 1968. Curved thick shell and membrane elements with particular reference to axi-symmetric problems. In L. Berke, R.M. Bader, W.J. Mykytow, J.S. Przemienicki and M.H. Shirk (eds), *Proc. 2nd Conf. Matrix Methods in Structural Mechanics,* Volume AFFDL-TR-68-150, pp. 539-72, Air Force Flight Dynamics Laboratory, Wright Patterson Air Force Base, OH, October 1968.

ALTAIR, http://www.altair.com/software/hw_hm.htm, 10 March 2006

ANG, P. Y. AND ARMSTRONG, C. G. 2002. Adaptive shape-sensitive meshing of the medial axis, *Engineering with Computers*, 18, 3, 253-264.

ARMSTRONG, C. G., BRIDGETT, S. J., DONAGHY, R. J., MCCUNE, R. W., MCKEAG, R. M. AND ROBINSON, D. J. 1998. Techniques for Interactive and Automatic Idealisation of CAD Models,

Numerical Grid Generation in Computational Field Simulations, University of Greenwich, 643-662.

AXIS. http://ws.apache.org/axis/. 10 November 2005

BLUM, H. 1973. Biological shape and visual science, *Journal of Theoretical Biology*, 38,205-287.

BLUM, H. 1967. A transformation for extracting new description of shape, *Models for the Perception of Speech and Visual Form*, W. Wathen-Dunn.(editor), MIT Press pp362-380.,

DONAGHY, R. J., ARMSTRONG, C. G. AND PRICE, M.A. 2000. Dimensional reduction of surface models for analysis, *Engineering with Computers*, 16, (1), 24-35.

FIPER. http://www.engineous.com/product_FIPER.htm

GOODIER, J. N. 1933. On the problems of the beam and the plate in the theory of elasticity, Trans. Roy. Soc. Canada Sect. III, 65-88.

GURSOY, H.N. AND PATRIKALAKIS, N. M. 1991. Automated interrogation and adaptive subdivision of shape using medial axis transform. *Adv Eng Software*, 13(5/6):287-302.

HANNIEL, I., MUTHUGANAPATHY, R., ELBER, G. AND KIM, M. 2005. Precise Voronoi Cell Extraction of Freeform Rational Planar Closed Curves, *Proceedings of the ACM Symposium on Solid and Physical Modeling 2005*: 51-59

IDEAS, http://www.ugs.com/products/nx/docs/fs_ideas_surfacing_set.pdf, 10 March 2006

LEE, K. Y., ARMSTRONG, C. G., PRICE, M. A. AND LAMONT, J. H. 2005. A small feature suppression/unsuppression system for preparing B-rep models for analysis, *Proceedings of the 2005 ACM symposium on Solid and Physical modeling*, Cambridge, Massachusetts, 113-124.

LEE, T., KASHYAP, R. L. AND CHU, C. 1994. Building skeleton models via 3-D medial surface/axis thinning algorithms, *CVGIP: Graphical Models and Image Processing,* 56, (6), 462-478.

LEE, Y. AND LEE, K. 1997. Computing the medial surface of a 3-D boundary representation model, *Adv Eng Software*, 28, (9), 593-605.

LIN, C. 2005. *Optimal Design of Aluminium Extrusion Dies using a Novel Geometry based Approach*, Ph.D. thesis, University of Wales, Swansea.

MCCUNE, R. W., ARMSTRONG, C. G. AND ROBINSON, D. J. 2000. Mixed-dimensional coupling in finite element models, *Int J Numer Methods Eng,* 49, 6, 725-750.

MONAGHAN, D. J., LEE, K. Y., ARMSTRONG, C. G. AND OU, H. 2000. Mixed dimensional finite element analysis of frame models, *Proceeedings of the 10th International Offshore and Polar Engineering Conference*, May 28-Jun 2 2000, Seattle, WA, USA, Vol. 4, 263-269.

MSC, http://www.mscsoftware.com/assets/2810_data_sheet_Patran_2004.pdf, 10 March 2006

PTC, http://www.ptc.com/appserver/wcms/relnotes/note.jsp?&im_dbkey=26696&icg_dbkey=826, 10 March 2006

QUB. http://www.fem.qub.ac.uk/research/newAreas/cadscript/dimred.php, 10 March 2006

RAMANATHAN, M. AND GURUMOORTHY, B. 2003: Constructing medial axis transform of planar domains with curved boundaries. *Computer-Aided Design* 35(7): 619-632 (2003)

SHEEHY, D. J., ARMSTRONG, C. G. AND ROBINSON, D. J. 1996. Shape description by medial surface construction, *IEEE Trans Visual Comput Graphics*, 2, 1, 62-72.

SHERBROOKE, E. C., PATRIKALAKIS, N. M. AND BRISSON, E. 1996. Algorithm for the medial axis transform of 3D polyhedral solids*, IEEE Trans Visual Comput Graphics*, 2, 1, 44-61.

SHIM, K. W., MONAGHAN, D. J. AND ARMSTRONG, C. G. 2002. Mixed Dimensional Coupling in Finite Element Stress Analysis, *Engineering with Computers*, 18, 3, 241-252.

SURESH, K. 2003. Automating the CAD/CAE dimensional reduction process*, Eighth ACM Symposium on Solid Modeling and Applications*, Jun 16-20 2003, Seattle, WA, United States, 76-85.

TAM, T. K. H. AND ARMSTRONG, C. G. 1991. 2D finite element mesh generation by medial axis subdivision, *Adv Eng Software*, 13, 5-6, 313-324.

TOMCAT. http://tomcat.apache.org/, 10 Novermber 2005

TRANSCENDATA. http://www.transcendata.com/cadfix.htm, 10 November 2005.

UNIGRAPHICS http://www.ugs.com/products/nx/simulation/prod_apps/scenario.shtml, 10 March 2006.

VIVACE. http://www.vivaceproject.com, 10 Novermber 2005

W3 (1). http://www.w3.org/XML/, 10 Novermber 2005

W3 (2). http://www.w3.org/TR/soap/, 10 Novermber 2005

W3 (3). http://www.w3.org/TR/wsdl/, 10 Novermber 2005

YIN, L., LUO, X. AND SHEPHARD, M. S. 2005. Identifying and Meshing Thin Sections of 3-d Curved Domains, *Proceedings of the 14th International Meshing Roundtable*, San Diego, USA. 33-54

Holoimages

Xianfeng Gu[*]
Computer Science
Stony Brook University

Song Zhang[†]
Mathematics Department
Harvard University

Liangjun Zhang[‡]
Computer Science
UNC-Chapel Hill

Ralph Martin[§]
School of Computer Science
Cardiff University

Peisen Huang[¶]
Mechanical Engineering
Stony Brook University

Shing-Tung Yau[‖]
Mathematics Department
Harvard University

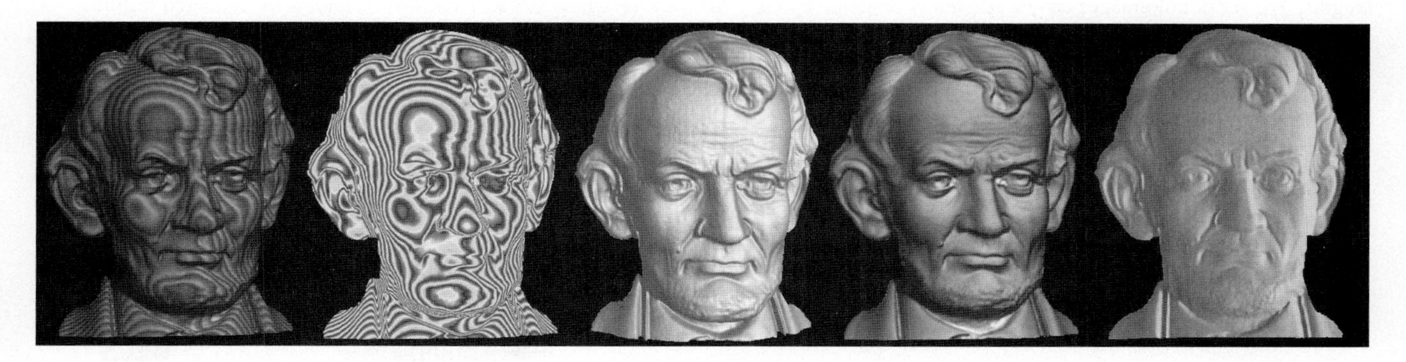

Figure 1: **Holoimage representing both geometry and shading.** The first image is a 24-bit holoimage with 512 × 512 resolution, with spatial frequency 64Hz, and 80° projection angle. All the other geometry and images are deduced from it, and are, in order, the phase map, the geometric surface, the shaded image, and the normal map.

Abstract

We introduce a novel geometric representation called the *holoimage*, which encodes both shading and geometry information within the same image, based on the principles of wave optics. 'Image' referes to the representation and records the amplitude of the lighting; 'holo' means that it encodes phase, and hence, three-dimensional information. Compared to conventional geometry images or depth images, the holoimage has much higher geometric accuracy. Thus, 3D information can readily be stored and transmitted using the common 24-bit image format.

Holoimages can be efficiently rendered by modern graphics hardware; rendering speed is independent of the geometric complexity and only determined by the image resolution. Rendering holoimages requires no meshes, only textures.

Holoimages allow various geometric processing tasks to be performed simply using straightforward image processing methods, including such tasks such as embossing and engraving, geometric texture extraction, and surface deformation measurement.

Conventional geometric representations, such as meshes, point clouds, implicit surfaces and CSG models, can be easily converted to holoimages using conventional rendering techniques in real time. The opposite process, converting holoimages to geometry in the form of a depth map is accomplished efficiently accomplished by graphics hardware.

Furthermore, holoimages can be easily captured from the real world with a projector and a camera at video frame rate.

CR Categories: I.3.5 [Computer Graphics]: Computational Geometry and Object Modeling —Curve, surface, solid, and object representations I.3.3 [Computer Graphics]: Picture/Image Generation—Viewing algorithms ; I.3.3 [Computer Graphics]: Picture/Image Generation—Digitizing and scanning ;

Keywords: Geometry image, Holoimage, Wave optics, Fringe projection, Phase shifting, Geometric data acquisition.

1 Introduction

Light is an electromagnetic wave with both amplitude and phase. However, human eyes and most cameras can only perceive amplitude, and thus miss the phase information. Traditional computer graphics is based on geometric optics and normally considers amplitude only. The current work differs from traditional image representations, and image formation methods in that it is based on wave optics. It hence emphasizes the phase which conveys geometric information about the scene.

This paper proposes a new geometric representation, the *holoimage*, which is a combination of a traditional image, including shading, texture and silhouettes, with a fringe texture projected by structured light, which encodes phase information. Through the use of a single image, both shading and 3D geometry can be recovered (see Figure 1).

Holoimages have the following advantages, which are valuable for real applications:

- *High Geometric Accuracy.* Compared to conventional geometry images and depth images, holoimages have much higher 3D accuracy. In his work on geometry images [Gu et al. 2002], the first author found it unworkable to use common

[*]e-mail: gu@cs.sunysb.edu
[†]e-mail:szhang@fas.harvard.edu
[‡]e-mail:zlj@cs.unc.edu
[§]e-mail:ralph@cs.cf.ac.uk
[¶]e-mail:peisen.huang@sunysb.edu
[‖]e-mail:yau@math.harvard.edu

SPM 2006, Cardiff, Wales, United Kingdom, 06–08 June 2006.
© 2006 ACM 1-59593-358-1/06/0006 $5.00

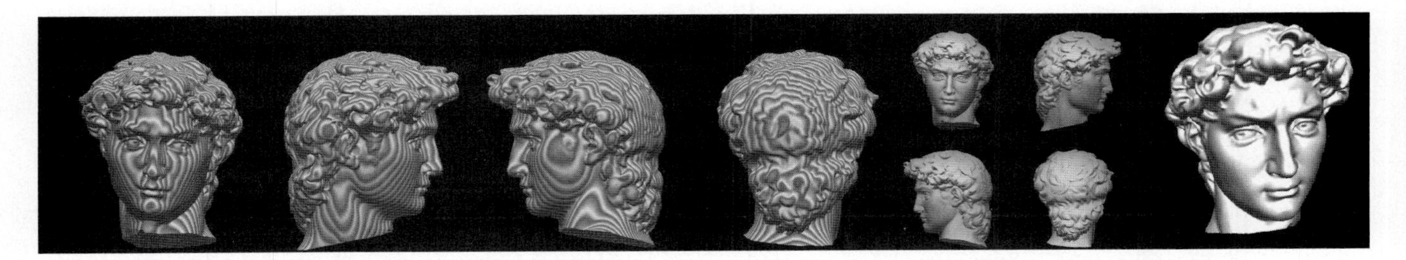

Figure 2: **Holoimage-based rendering.** The David head model represented as holoimages viewed from different directions at two wavelengths. The set of holoimages can be efficiently rendered by GPU to cover the whole surface. The last column is the rendered result.

24-bit images to represent a geometry image, as each channel requires at least 12 bits. Holoimages overcome this disadvantage of geometry images. They can be stored and transmitted using 24-bit images. The reason for the higher 3D accuracy is further explained in Section 2.4.

- *Efficient Mesh-Free Rendering.* Holoimages represent both geometry and shading within the same image, allowing them to be rendered efficiently by modern graphics hardware without meshes, using only textures (see Figure 2). This has the potential to simplify the architecture of graphics hardware.
- *High Acquisition Speed.* Holoimages can be captured from the real world with a simple setup involving just a projector and a camera. The reconstruction algorithms are simple enough to be implementable on modern graphics hardware. A sequence of holoimages can be captured at video frame rates for dynamic geometric data acquisition. We have built a system for the capture of dynamic human facial expressions where acquisition and reconstruction are performed in real time at up to 30 frames per second. A description of the system and reconstructed surface data can be found at [Gu et al. 2004].
- *Direct Manipulation.* Holoimages allow image processing techniques to be applied directly to carry out geometric processing, producing such effects as embossing and engraving, geometric texture extraction and deformation measurement. It can be very conveniently converted to and from other geometric representations by hardware automatically (see Figure 10).

Holoimages are based on the principles of wave optics. They inherit some drawbacks from this optical basis:

- *Occlusion.* A holoimage can only represent the part of a surface visible from a single viewpoint. In order to represent the whole surface, several holoimages are required. (Geometry images, however, for example, can represent the whole surface).
- *Special Materials.* Holoimages of dark or glossy materials can not be captured accurately because of their reflectance properties.

The rest of paper is organized as follows: Section 2 introduces the theoretical background, Section 3 reviews related work, Section 4 explains the generation of holoimages in details, Section 5 demonstrates several important real applications, and finally, Section 6 concludes the paper.

2 Background

Light is an electromagnetic wave represented by its electric field. In the special case where a plane wave propagates in the direction of a unit vector $\mathbf{n}$, the expression describing the field at an arbitrary time t and point with position vector $\mathbf{r} = (x,y,z)$ is given by

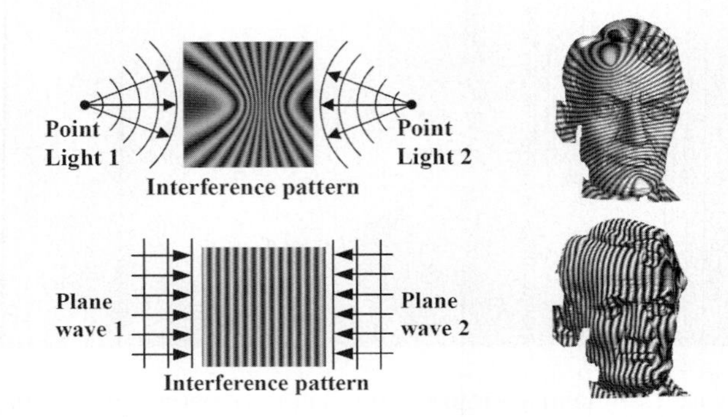

Figure 3: **Principle of holoimages.** Phase information is encoded in the interference fringe pattern. Geometric information is encoded in the phase information.

$$\psi(\mathbf{r},t) = Ue^{i\phi(\mathbf{r})}e^{-i2\pi\nu t}, \qquad \phi(\mathbf{r}) = \mathbf{n}\cdot\mathbf{r}/\lambda + \delta,$$

where λ is the wavelength, ν the frequency, δ the initial phase, and U the amplitude of the field. The wavefront is parallel to the phase planes satisfying $\mathbf{n}\cdot\mathbf{r} = const$. Since we are interested in the spatial distribution of the field in this research, only the spatial complex amplitude $u = Ue^{i\phi}$ is considered.

There are two general approaches to capturing phase information in image format: interference based methods and fringe projection methods. The latter can be considered to be a special case of the former.

2.1 Interference

Interference occurs when two (or more) waves overlap each other in space. Assume two waves described by $u_1 = U_1e^{i\phi_1}$, $u_2 = U_2e^{i\phi_2}$ overlap. Electromagnetic wave theory indicates that the resulting field is $u = u_1 + u_2$. The intensity is given by

$$I = |u|^2 = |u_1 + u_2|^2 = I_1 + I_2 + 2\sqrt{I_1 I_2}\cos\Delta\phi,$$

where $\Delta\phi = \phi_1 - \phi_2$.

Figure 3 illustrates interference patterns formed by two coherent light sources on a plane, and on a complicated surface. In the first row, the light sources are point lights; in the second row, the light sources are planar lights.

By taking one or more images bearing interference patterns (assuming the surface is not dark or glossy, which would prevent the patterns from being clearly visible) the phase difference $\Delta\phi$ can be estimated with an ambiguity of $2j\pi$ for some integer j. By assuming continuity of the surface, this ambiguity can be eliminated, and the depth map of the objects in the scene can be recovered [Ghiglia

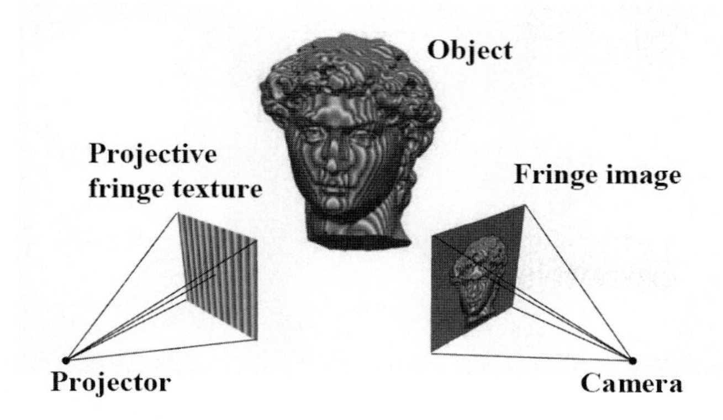

Figure 4: **System setup.**

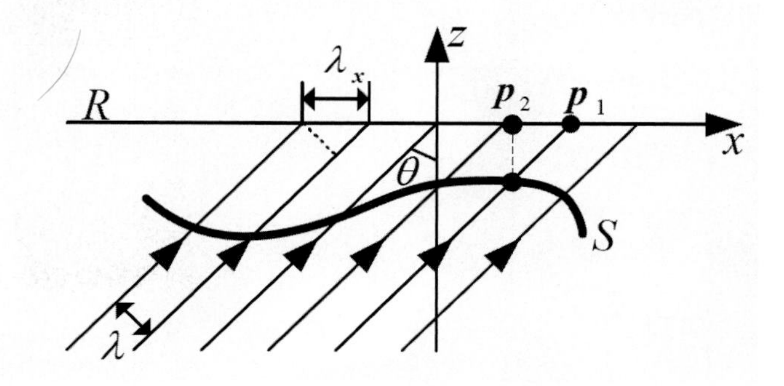

Figure 5: **Canonical Configuration.**

and Pritt 1998]. The depth map can be converted to geometry once the system is calibrated [Zhang and Huang 2006; **?**]. Interference methods have been widely applied in interferometry and metrology for geometric profile measurement [Gåsvik 2002].

2.2 Fringe Projection

Consider the case shown in the second row of Figure 3. The interference pattern caused by parallel planar light sources can be obtained in an alternative manner using just a single light source of varying intensity, by orthographically projecting a fringe pattern of sinusoidally varying intensity onto the surface in a direction parallel to the wave front as we now show.

If the two planar light sources are located at $x = \pm c$, and that the wave propagation directions are $(\mp 1, 0, 0)$ respectively, then the waves are given by $u_1 = U\exp[2\pi i(x-c)/\lambda]$, $u_2 = U\exp[-2\pi i(x-c)/\lambda]$. The interference wave is thus $u = u_1 + u_2 = 2U\cos[2\pi(x-c)/\lambda]$, and has intensity of the interfering light sd

$$I = 2U^2\left(\cos[4\pi(x-c)/\lambda]+1\right).$$

Thus, instead of making coherent lights interfere on the test surface, the interference fringe pattern can be projected directly to the surface along the z-axis using a planar light with intensity given by the above formula. The fringes can then be viewed from a different angle by a camera. This method is called *fringe projection* [Malacara 1992].

Figure 4 illustrates the basic setup. A sinusoidal fringe pattern is projected onto a geometric object by a projector and an image of the surface with fringes distorted by the geometry is captured by a camera.

Projective Texture We now consider the fringe patterns used to generate a holoimage. A standard computer display projection system may be used to project the fringe pattern onto the object being captured. A texture image is input to the projector. To be able to separate and recover the effects of shape, shading, and ambient light in the fringe patterns, we need to make three independent intensity measurements at each position (x, y), as explained shortly in Section 2.3. We use the red, green and blue channels to store three separate textures to be projected simultaneously; these are 3 sinusoidal fringe patterns with phase differences horizontally in space of $-\frac{2\pi}{3}, 0, \frac{2\pi}{3}$. Each of these ideal sinusoidal fringe textures, having a given spacing, can be described as

$$T_k(u,v) = \frac{1}{2}\left(\cos\left(2\pi u/\lambda + \frac{2(k-1)\pi}{3}\right)+1\right), \quad k=0,1,2, \quad (1)$$

where (u, v) are the texture coordinates, λ is the fringe period or inter-fringe distance, k is the channel number. The resulting three-channel holoimage is now captured using a camera, as described next.

Depth recovery from phase We now consider recovery of depth information from the captured fringe patterns, in the form of a depth map $z(x, y)$. We assume a particular geometric setup as follows: the image plane is the xy-plane, a reference plane is at $z = 0$, the optical axis of the camera is the z-axis; the optical axis of the projector is on the xz-plane at an angle θ to the z-axis. The u axis of the projective texture is in the xz-plane; the v axis is parallel to the y axis, so the projected fringes are parallel to the y-axis.

The fringe period in the reference plane along x-axis is thus given by $\lambda_x = \lambda/\cos\theta$. The surface being imaged can be represented as a depth map $z(x, y)$. The intensity at each pixel (x, y) on the fringe image is given by

$$I(x,y) = a(x,y) + r(x,y)\cos(\psi(x,y)). \quad (2)$$

Here, $a(x, y)$ is the *ambient light* intensity, and $r(x, y)$ the *reflectivity* of the surface, which is very similar to the bidirectional reflection distribution function (BRDF) and depends on the direction of lighting, the normal to the surface, and its color, material, and texture. ψ is the *phase* and proportional to the depth $z(x, y)$. A fringe originally positioned at p_1 on a reference plane in front of the object will be displaced to a position p_2 due to the difference in depth between the reference plane and the actual surface (illustrated in Figure 5). The relative depth can recovered using the relationship of

$$z(x,y) \approx \frac{(\psi_2(x,y) - \psi_1(x,y))\lambda}{2\pi\sin\theta}. \quad (3)$$

2.3 Phase Shifting

We measure intensity, and wish to use Equation 2 to recover three unknowns, $a(x, y)$, $r(x, y)$ and $\psi(x, y)$ (and hence depth). Therefore, we need three independent intensity measurements at a given point to do so. We project three separate fringe textures, $k = 0, 1, 2$, with the same wavelength with phase shifts, $-\frac{2\pi}{3}, 0, \frac{2\pi}{3}$, leading to corresponding intensities of

$$I_k(x,y) = a(x,y) + \frac{1}{2}r(x,y)(1+\cos(\psi(x,y) + \frac{2(k-1)\pi}{3})),$$

The phase ψ, reflectively r, and the ambient light a can, in principle, be computed from these intensities using

$$\psi = \tan^{-1}(\sqrt{3}\frac{I_0 - I_2}{2I_1 - I_0 - I_2}), \quad (4)$$

$$r = 2\sqrt{3(I_0 - I_2)^2 + (2I_1 - I_0 - I_2)^2}, \quad (5)$$

$$a = (I_0 + I_1 + I_2)/3 - r/2, \quad (6)$$

and $a + \frac{r}{2}$ is the reconstructed shading.

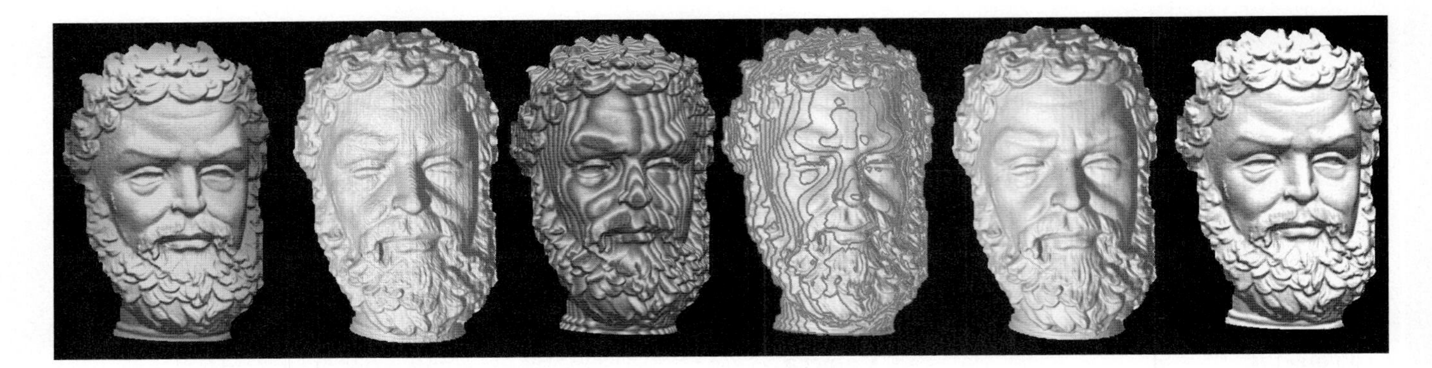

Figure 6: **Two-wavelength phase unwrapping.** The first image is a holoimage with spatial frequency 1Hz; the coarse reconstructed geometry is shown in the second image does not need phase unwrapping. This is used as a reference to phase unwrap holoimages with denser fringes. The 3rd image is a holoimage with spatial frequency 64Hz. The 4th image is the reconstructed geometry without phase unwrapping; the red contours show phase jumps. The 5th image is the reconstructed geometric surface after phase unwrapping to depth consistency with the 2nd image. The last image is a shaded image.

Phase Ambiguity The above process is called *phase wrapping*. The recovered phase using Equation 4 has a 2π phase ambiguity.This phase ambiguity introduces depth jumps on the reconstructed geometry as shown in the 4th image of Figure 6. In order to reconstruct the correct smooth geometry, a *phase-unwrapping* algorithm has be to used. Phase unwrapping intends to remove the artifacts caused by2π discontinuities by adding or subtracting appropriate multiples of 2π.

If the surface is smooth, then the places where phase jump discontinuities occur (the red contours in the 4th image in Figure 6) can easily be located and removed by comparing the phase values with those at neighboring pixels. Although this process is simple, it is unsuited to GPU architecture. Therefore, in our approach, we usually use a two-wavelength method [Creath 1987]. We select 2 spatial periods λ_1, λ_2 for projected textures, such that λ_1 is big enough to cover the whole range of the scene in one period. In this case, there is no phase ambiguity, but the reconstructed geometric accuracy is low. λ_2 is much smaller, and there is phase ambiguity, but the geometric accuracy is much higher. By requiring depth consistency between the two holoimages, the phase of the second one can be *unwrapped*.

Figure 6 illustrates the computation process for the Zeus sculpture with dense geometric features using the two-wavelength method. We also show the Stanford bunny surface with an imposed texture, converted to a holoimage, in Figure 8. Note that both the geometry and the shading with surface texture are reconstructed. Because the original triangulation of the surface is not dense, the flat triangulation structure can clearly be seen on the reconstructed surface. This indicates that the method works robustly for surfaces with complicated textures, and that the geometric resolution of a holoimage can be higher than traditional triangle meshes.

2.4 Error Analysis

The accuracy of the reconstructed geometry is determined by many factors. The direct factors are: the image resolution $m \times m$, the pixel bit-depth n, the fringe spatial frequency λ, and the projection angle θ. Assume there is a one-bit error at a specific pixel in the holoimage. The reconstructed geometric error is

$$\delta z \propto \frac{1}{\lambda 2^n m \sin\theta}. \qquad (7)$$

This formula makes it evident that accuracy can be increased by increasing the projection fringe spatial frequency. In reality, the spatial frequency has an upper limit determined by the resolution of the projector and the resolution of the camera. Similarly, increasing the projection angle θ improves the accuracy of the reconstructed geometry but increases the likelihood of occluded regions. The effects of different parameters are illustrated in Figure 7.

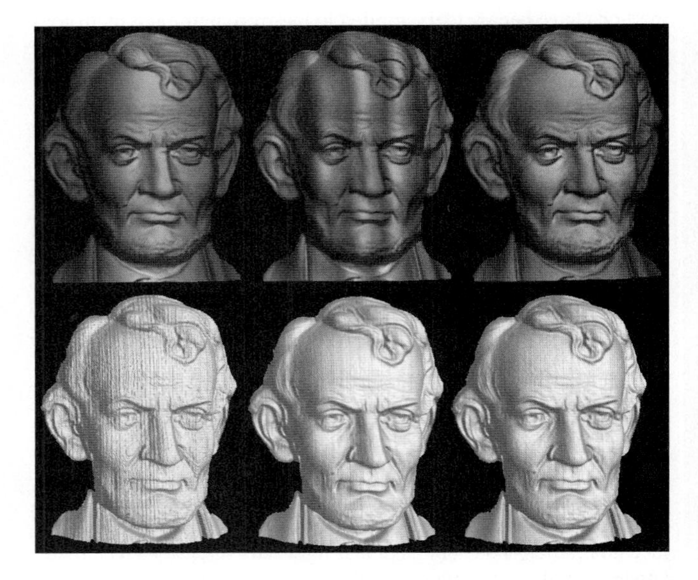

Figure 7: **Holoimage quality is proportional to projection angle and fringe spatial frequency.** The first two holoimages have the same projection angle, 15°, but different spatial frequencies, 2Hz and 8Hz, respectively. The first and the third column holoimages have the same spatial frequency, 2Hz, but different projection angles, 15° and 80°, respectively.

Comparison with Geometry Images Conventional geometry images [Gu et al. 2002] and depth images [Shade et al. 1998] encode the depth information using color information. The geometric error can be formulated as $\delta z \propto 1/m2^n$. Compared with Formula 7, it is obvious that if $\lambda \sin\theta \gg 1$, the geometric accuracy of holoimages is much higher. In practice, we choose λ to be 128, and θ is around 30 degrees. Therefore, holoimage has much higher geometric accuracy than geometry images.

We did some experiements to compare geometry images and holoimages. The Lincoln statute model in figure 1 are converted to both an geometry image and a holoimage. The holoimage is represented as a 24 bits image, and is enough to reconstruct high quality

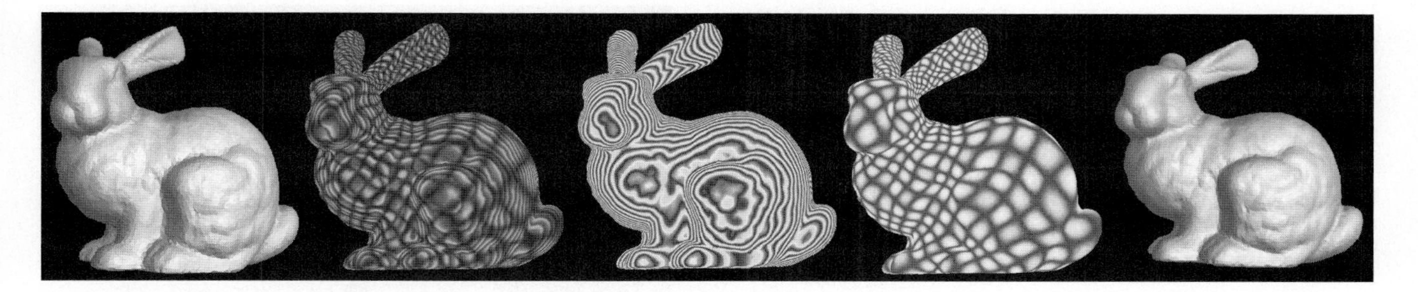

Figure 8: **Holoimage of texture mapped surfaces.** The original geometric surface is flat shaded in the first image, the flat triangles are visible. The second image is the 24-bit holoimage for the textured bunny surface. The third image shows the phase map. The fourth image shows the texture reconstructed from the holoimage. The reconstructed geometry is shown in the last image, the flat triangles are recovered.

surface. But for the geometry image, it requires 36 bits image.

For the stanford bunny surface shown in figure 8,one holoimage can only represent one side of the surface, but a single geometry image can represent the whole surface. Therefore, we use two holoimages to cover the whole surface, the size is comparable to a single geometry image.

3 Related Work

A holoimage is a representation of both geometry and shading in a single image format, which is closely related to the idea of a geometry image. Holoimages can be used to acquire geometric data using simple setups. Thus we should consider the geometric data acquisition literature, and related phase-based methods which have been applied broadly in metrology and interferometry [Gåsvik 2002; Malacara 1992].

3.1 Geometric Representation

Layered depth images are introduced in [Shade et al. 1998], which associate each pixel with a depth value and can be used for efficient rendering.

Geometry images were first introduced in the seminal work of Gu et al. [Gu et al. 2002], which aims to represent geometry using regularly sampled grids. Such grids allow efficient traversal, random access, convolution, composition, down-sampling, compression, and synthesis. Praun and Hoppe generalized geometry images using spherical parameterization [Praun and Hoppe 2003], and applied them to shape compression [Hoppe and Praun 2005]. Smooth geometry images were explored by Losasso et al. in [Losasso et al. 2003], which models general genus zero surfaces by a single spline surface patch. Multi-chart geometry images were introduced by Sander et al. in [Sander et al. 2003] to reduce parametric distortion and improve geometric fidelity. Geometry clipmaps have been utilized for terrain rendering by Losasso and Hoppe in [Losasso and Hoppe 2004].

Geometry images suffer from various disadvantages compared to holoimages. In order to convert conventional irregular meshes to geometry images, sophisticated topological operations must be used, and time-consuming surface parameterization is required. Shading information cannot be encoded in geometry images directly. Furthermore, it is not possible to directly capture geometry images from real objects.

Holoimages are an extension of the idea behind geometry images: representing geometry in an image format. Holoimages can very easily be synthesized in real time using current graphics rendering techniques. The parameter domain is just the image plane, so it is intuitive and thus straightforward to use for editing operations. Shading and texture information can be simultaneously encoded in a holoimage as well as geometry. Furthermore, holoimages can be directly acquired from real objects using a digital camera and a projector.

3.2 Geometric Data Acquisition

Traditional geometric data acquisition methods are mainly based on geometric optics, where depth information is usually recovered by intersecting two rays. A number of methods have been proposed including stereo, laser stripe scanning, and time or color-coded structured light [Salvi et al. 2004]. Among all existing ranging techniques, stereovision is probably the most studied method. However, the shortcoming of a stereo-based method is that the matching of stereo images is usually time-consuming. It is therefore difficult to realize real-time 3D shape reconstruction from stereo images.

There are essentially two approaches toward real-time 3D shape measurement. One approach is to use a single pattern, typically a color pattern [Harding 1991; Huang et al. 1999]. Because they use color to code the patterns, the shape acquisition result is affected, to varying degrees, by the variations of the object surface color. Rusinkiewicz et al. developed a real-time 3D model acquisition system that utilizes four patterns coded with stripe boundary codes [Rusinkiewicz et al. 2002]. The data acquisition speed achieved was 15 fps. However, like any binary coding method, the spatial resolution is relatively low in comparison with phase-shifting based methods. Huang et al. proposed a high-speed 3D shape measurement based on a rapid phase-shifting technique [Huang et al. 2003]. They use three phase-shifted, sinusoidal grayscale fringe patterns to achieve pixel-level resolution.

Zhang and Huang developed a 3D shape acquisition system with a speed of up to 40 fps [Zhang and Huang 2004]. The geometric shape acquisition, reconstruction, and display was simultaneously realized at 40 fps by using a fast three step phase-shifting algorithm [Huang and Zhang 2006; Zhang 2005]. Moreover, if the system is calibrated by a novel structured light system calibration method proposed by [Zhang and Huang 2006], the 3D geometry measurement error is RMS 0.10-0.22 mm over a volume of of 342(H) $\times$ 376(V) $\times$ 700(D) mm. [Zhang and Huang 2004] research works on 3D shape measurement laid the foundations for quickly capturing holoimages from real objects, since recovering the geometry is simple enough to be accomplished in hardware.

3.3 Phase-Dependent Method

Phase-based methods have been broadly applied in engineering for purposes such as optical testing, real time wavefront sensing for active optics, distance measuring using interferometry, surface contouring, and microscopy [Gåsvik 2002; Malacara 1992]. The underlying techniques are based on wave optics. Moiré fringes may be used for the analysis of deformations of materials [Post et al. 1994] as well as for the acquisition of 3D object shapes [Takasaki

1970]. Holoimages can also be generated using traditional Moiré-based methods [Malacara 1992].

4 Generating Holoimages

Holoimages can be easily captured from real objects using structured light, but they can also readily be generated by straightforward rendering methods.

4.1 Generated Holoimages

It is very easy to synthesize a holoimage using a modern graphics pipeline. As exemplified by Figure 4, three sinusoidal fringe patterns can be precomputed and stored as a 3-channel 24-bit color texture image. In order to simplify the analysis of the holoimage, a canonical configuration is preferred, where both the projective texture and the camera use orthogonal projection, and the geometric object is normalized to be inside a unit cube.

In the computation, if we only care about geometric information and do not wish to represent any shading or texture for the surface, we can set the OpenGL texture environment mode to *replace*. If we want to encode both geometry and shading information, we should set the texture environment mode to *modulate*. If a texture is also to be rendered on the surface, we need to use a multitexturing technique to generate the holoimage. Our current methods can only encode monochromatic shading information into a holoimage. If colour shading information is required, we can keep a separately rendered image without any fringe texture for use as a texture image for the reconstructed geometry.

Figure 8 illustrates a holoimage which represents geometry shaded with a complicated texture. The holoimage was synthesized using the following procedure: first, we parameterized the Stanford bunny surface using conformal parameterizations (any parameterization method could have been used), then we texture mapped an arbitrary monochromatic bitmap onto it, and finally we shaded color sinusoidal fringes on the bunny using projective texture mapping. These two textures were combined using multitexturing. Specifically, we used the *ARB_multitexture* extension of OpenGL on a Nvidia GeForce 6800 graphics card, setting the texture environment mode to modulate, to blend the surface texture color and the fringe pattern.

4.2 Captured Holoimages

To capture holoimages from real objects, we use a Digital-Light-Processing (DLP) projector (Plus U2-1200) to generate the fringe patterns, and a high-speed digital CCD camera (Dalsa A-D6-0512) synchronized with the projector so that we can capture each projected color channel separately. A DLP projector is a digital projector that has a unique projection mechanism make it brighter than a LCD projector. Since its digital nature, it makes the contrast of the projected image larger than a LCD projector. Moreover, a single chip DLP projector projects three color channels sequentially, we take advantages of this projection mechanism and realized real-time 3-D measurement. By sequentially capturing RGB channels of the projected image, we realize the functionality of a color fringe pattern while eliminating color-matching problems.

Data can be acquired at a speed limited by those of the projector (120Hz) and camera used; in practice we can achieve 30Hz. This is adequate to capture general moving surface deformations, such as human expressions. Geometric 3D data illustrated in Figures 1, 6, 14, 15, and 16 were captured from real objects using our system. Photographs of our system, further captured holoimages, and reconstructed surfaces are available at [Gu et al. 2004].

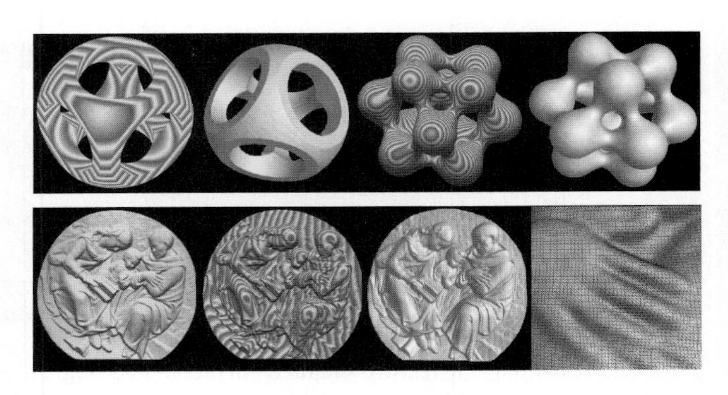

Figure 10: **Conversion.** Other geometric representations can be easily converted to holoimages using projective texture. The top row shows conversion from a CSG model and an implicit surface. The bottom row shows conversion from a point cloud.

4.3 Differences

There are fundamental differences between a synthesized holoimage and one captured from real life: the projective texture mapping of a synthetic holoimage does not include shadows or self-occlusion. Therefore,the projection angle can be as large as a right angle to improve the accuracy of the reconstructed geometry. For captured holoimages, in order to avoid shadows caused by projector, the projection angle must usually be quite small.

Another difference is related to color. Three monochromatic fringe projective textures can be combined into one color projective texture and a color 24-bit holoimage can be synthesized. However, using color holoimages for geometry capture is undesirable since the measurement accuracy is affected by the color of the object. Therefore, three monochromatic fringe images are usually needed to reconstruct the geometry.

5 Applications

Holoimages can be easily produced from other geometric representations as shown in Figure 10. Holoimages are valuable for many important applications.

Most surface models in this paper are captured by our holoimage system, such as 1,6,7,15,16,14 and the second row of 10.

5.1 Mesh-free Rendering

A holoimage represents both shading and geometry within the same image. Recent development of graphics hardware makes it feasible to efficiently render holoimages directly using its programmable pipeline. For rendering purposes, we believe holoimages have the potential to replace traditional meshes, and therefore, simplify the architecture of GPU.

5.1.1 Surface Reconstruction using Holoimages

The architecture of our surface reconstruction algorithm can be readily mapped on to graphics processors, promising a significant acceleration over CPU based methods. The surface reconstruction algorithm requires computationally intensive and accurate numerical computations, on a complete grid pattern, and fully satisfies the requirements for an SIMD computation [Bolz et al. 2003; Purcell et al. 2002]. Therefore, our reconstruction algorithm can be efficiently accelerated on high performance graphics processors (GPUs), such as nVIDIA GeForce 6800 and ATI Radeon 9800, which expose a flexible SIMD programming interface with powerful floating computational capacity.

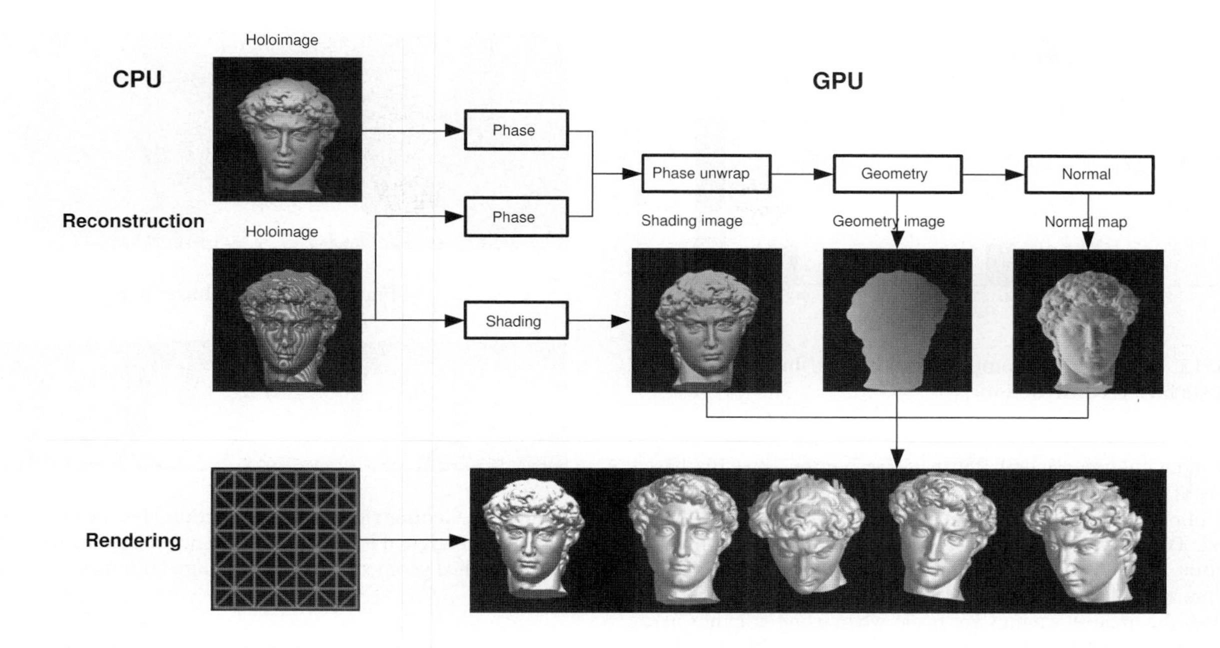

Figure 9: **Pipeline of surface reconstruction and rendering for holoimages on GPU.**

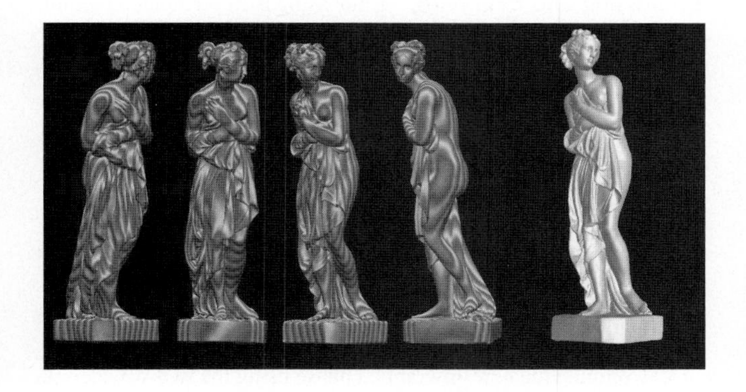

Figure 11: **Holoimage-based Rendering.**

We map our surface reconstruction to graphics hardware as in Figure 9. The original holoimages are first uploaded to the GPU as textures. Then, a series of GPU computation-intensive kernels are used, based on *pixel shaders*, including a *Phase Kernel* which computes Equation 4, a *Shading Kernel* for Equations 6 and 5, a *Phase Unwrap Kernel*, a *Geometry Kernel* for Equation 3, and a *Normal Kernel*, are sequentially applied to the original and intermediate generated textures. Each computational kernel processes every texel in parallel and produces exactly one element of the intermediate texture for the next processing stage. The results of our reconstruction algorithms, including the geometry image, the shaded image, and the normal map, remain in the GPU memory for further rendering if desired.

Besides performing computation on a regular grid, the algorithm also has the attractive property of local memory access. None of the computational kernels here need to access any adjacent texels, apart from the normal kernel, which only needs to access the 8-connected neighbors. Because streaming hardware can effectively hide the latency of memory access, our GPU-based algorithm shows distinct advantages over the CPU-based version [Purcell et al. 2002].

5.1.2 Holoimages Rendering

Holoimages which have just been captured can be rendered efficiently because they do not need to transferred from GPU to CPU and back for rendering, thus overcoming a common performance bottleneck within GPU accelerated systems [Carr and Hart 2004]. To render the recovered geometry, the system constructs a dummy grid mesh in the GPU, with the same connectivity as the object surface but fake geometric and shading information. The texture coordinates assigned to each input vertex encode its grid indices. The *vertex shader*, which on current GPUs can directly access the texture memory, then fetches the true geometry position, shading color, and normal vector for each vertex using the pre-assigned texture coordinates, and computes the transformed position and lighting color. In practice, the rendering can be further optimized by drawing indexed primitives, which can reduce the number of executions of vertex shaders.

Adaptive real-time rendering can easily be performed using holoimages, allowing a smooth transition between texture and geometry. When the rendered object is far away from the synthetic camera, the system can simply map the shading texture to a rectangle. When the object gets closer, the system adds the normal maps to the textured rectangle. Finally, once the object is close enough, the geometry is rendered using the texture and normal maps.

5.1.3 Performance

We have implemented our GPU-based reconstruction and rendering algorithms; Figure 11 shows a further example of rendering, while Figure 12 shows times taken by various steps of our algorithm for CPU and GPU based implementations, at different resolutions. The overall time includes the time required for each kernel and system overheads, which include the time for texture loading and system setup. The GPU-accelerated system was implemented using OpenGL and the Cg Toolkit to do the high-level GPU programming. The computation kernel for reconstruction was implemented as a pixel shader. To fully utilize the channels for each texel, the phase kernel and shading kernel were merged, and only one intermediate texture was generated. All tests were performed on a 3.00GHz Intel Processor, with an nVIDIA GeForce 6800 graphics card.

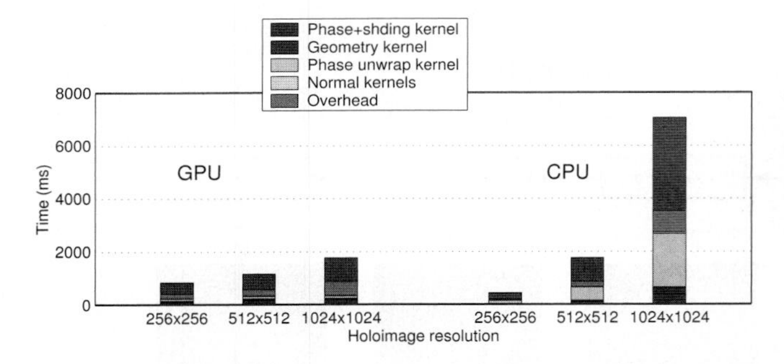

Figure 12: **Performance comparison for GPU-based and CPU-based surface reconstruction.**

For input holoimages with 256×256 resolution, the CPU-based version is faster. However, for larger holoimages, the GPU-based implementation significantly outperforms the CPU-based method. Typically, to reconstruct a surface (200K faces) from high-resolution (1024×1024) holoimages, the GPU-based system is much more efficient than the CPU implementation. Figure 12 indicates that the greatest savings are made when using graphics hardware during normal kernel compuations, due to the GPU-based normal kernel using built-in hardware vector operations, such as cross product and normalization.

Figures 2 and 6, showing David and Zeus, demonstrate that our reconstruction algorithms are insensitive to complexity of the input geometry.

Moreover, our experiments showed that within each kernel, only a small percentage of time is occupied by arithmetic instructions, while most of the time is used for texture lookups. This suggests our GPU-based reconstruction algorithm may be further accelerated.

Our efficient GPU-based methods for surface reconstruction and rendering based on holoimages leads us to believe that by representing geometry, shading and normal information of surfaces with images, holoimages may be used to replace triangle meshes for rendering purposes. Also, holoimages have the potential to simplify the architecture of modern graphics hardware.

5.2 Geometric Processing

By using holoimages to represent geometry on a regular grid, many geometric processing tasks can be accomplished by image processing techniques, which can also be accomplished efficiently by graphics hardware.

Since geometry is encoded in the phase information, geometric processing can be performed using phase information directly. Suppose the phase of a holoimage is $\psi(x,y)$. A *phase map* is a map from the image plane to the unit circle, $\sigma : R^2 \to S^1, \sigma(x,y) = e^{i\psi(x,y)}$. Two phase maps can be composed together by pixel-wise multiplication or division. Then the set of phase maps forms a group

$$\Sigma = \{e^{i\psi(x,y)}\}, \sigma_1 \circ \sigma_2(x,y) = \sigma_1(x,y)\sigma_2(x,y).$$

By composing the phase maps of two surfaces, it is very easy to generate embossing and engraving effects using GPU processing of holoimages.

In our implementation, we find it is sufficient to use only 16 bits per image to represent a phase map. The red and blue channels are used to encode the real and the imaginary parts of the phase map,

$$r(x,y) = 128\cos\psi(x,y) + 128, \quad b(x,y) = 128\sin\psi(x,y) + 128$$

Then the wrapped phase $\psi(x,y)$ can be simply computed as

$$\psi(x,y) = \tan^{-1}\frac{b(x,y) - 128}{r(x,y) - 128}.$$

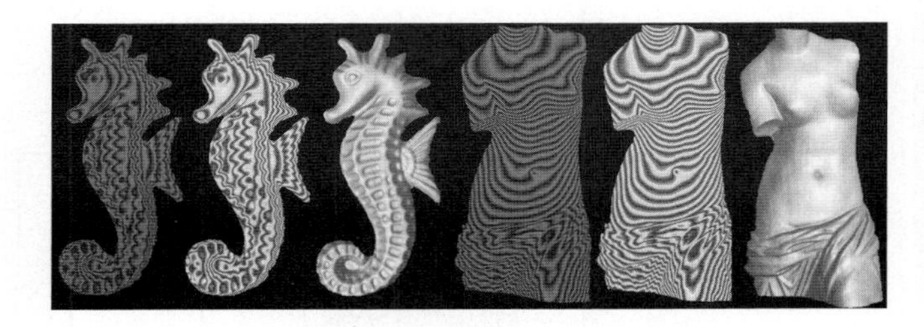

Figure 14: **16-bit phase maps.**

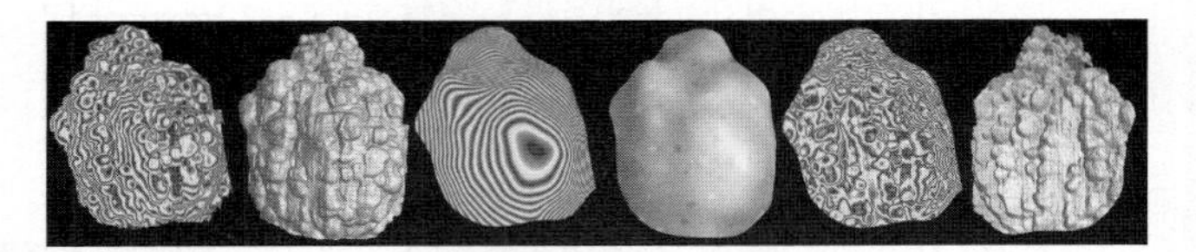

Figure 15: **Geometric texture extraction.** The images from left to right are an original holoimage and associated geometry, holoimage and smoothed geometry, extracted texture holoimage and geometry.

Figure 14 illustrates 16-bit holoimages and the corresponding geometries for the Seahorse and Venus surfaces.

Embossing and Engraving Embossing is implemented by multiplying two phase maps. The following algorithm uses the following steps for the embossing process:

1. Choose the camera position and zoom factor for the two surfaces.
2. Render the holoimages of the two surfaces with two wavelengths.
3. Compute the wrapped phase of each holoimage and convert them to phase maps.
4. Multiply the corresponding phase maps of different surfaces.
5. Compute the new unwrapped phase using the two-wavelength algorithm.
6. Convert the phase function to a geometric surface.

All the operations in the algorithm can be performed pixel-wise without requiring information from neighboring pixels, and can thus be readily implemented on a GPU. Figures 13 illustrates examples of our embossing and engraving algorithm.

Geometric Texture Extraction Textures are important for geometric modeling. Although it is easy to capture image textures, it is generally difficult to capture geometric textures. It is desirable to extract geometric texture from real surfaces with complicated profiles and transfer the texture to other geometric models.

By using holoimages, we can extract geometric textures from real objects easily. First a complex surface is scanned and the geometry reconstructed. By using conventional geometric smoothing techniques, we remove the geometric texture from the surface. Then, by subtracting the smooth surface from the original surface, the geometric texture can be extracted. The process is illustrated in Figure 15.

Deformation Measurement It is useful to detect and measure deformation of surfaces for many applications, for example to model human facial expressions and muscle movements. Figure 16 shows an interesting application of holoimages, 'a smile without a face'. We captured a human face with different expressions. Two

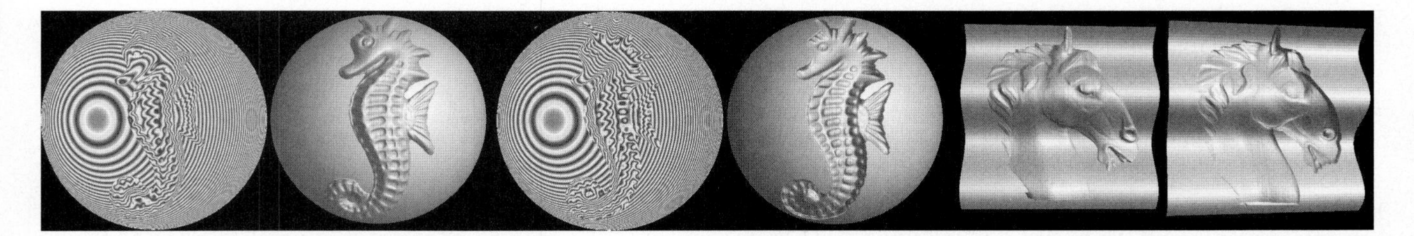

Figure 13: **Geometry editing using holoimage.** The seahorse surface is used to make an embossed or engraved relief on a sphere, showing phase maps and geometric results. Further images show embossing and engraving of a horse's head on a wave surface.

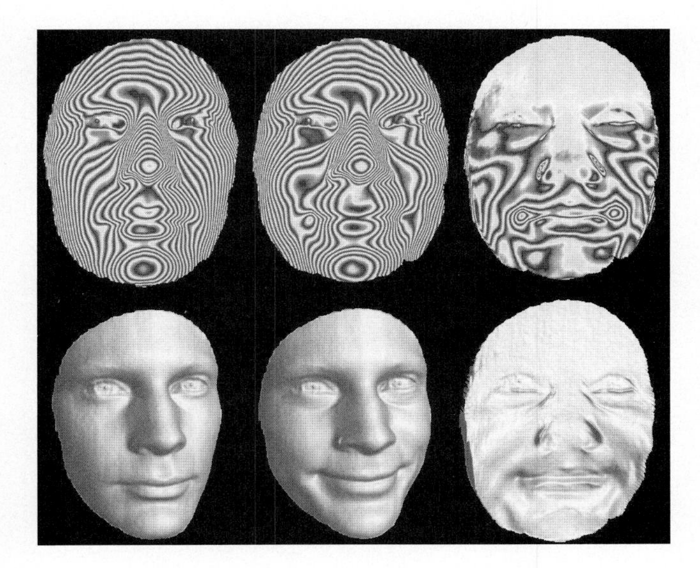

Figure 16: **Distorsion measurement by holoimage.** The phase maps and the geometric surfaces of the calm face, the smiling face and the smile itself are shown. On the smile phase map, the distortion is proportional to the density of the fringes. All the subtleties of the smile are accurately measured.

frames were selected, one face being neutral, the other with a smile. By converting both of them to phase maps, computing the difference phase map, and reconstructing the geometry, 'a smile without a face' can be obtained. From the deformation phase map, it is clear that the lips of the mouth have changed the most, the muscles on the cheeks have changed moderately, and the forehead and the nose bridge have remained unchanged.

6 Conclusions

This paper has introduced a novel geometric representation called the *holoimage*, which encodes both amplitude and phase information in a single image. Compared to conventional geometry images and depth images, holoimages have much higher geometric accuracy, and can be stored and transmitted using common image formats.

Holoimages can be synthesized using a conventional graphics pipeline or captured from real life with a simple hardware setup, in both cases using fast algorithms. Furthermore, the algorithms to reconstruct a depth map from a holoimage are simple and can be accomplished by graphics hardware.

Holoimages can be efficiently rendered using GPU hardware. Holoimage based rendering is mesh-free and has the potential to lead to simpler GPU architecture.

Holoimages can be applied to a range of geometric processing tasks using image processing methods, again utilizing graphics hardware. These include embossing, engraving, geometric texture extraction and distortion measurement.

In the future, extending holoimage based methods to better handle occlusions will be explored. Acquiring holoimages from glossy or dark surfaces will also be investigated.

Acknowledgement

We thank Stanford university and RWTH Aachen University for the surface models.

This work was partially supported by the NSF CAREER Award CCF-0448339 and NSF DMS-0528363 to X. Gu.

References

BOLZ, J., FARMER, I., GRINSPUN, E., AND SCHRÖDER, P. 2003. Sparse matrix solvers on the gpu: conjugate gradients and multigrid. In *Proceedings of SIGGRAPH 2003*, ACM Press / ACM SIGGRAPH, 917–924.

CARR, N. A., AND HART, J. C. 2004. Painting detail. In *Proceedings of SIGGRAPH 2004*, ACM Press / ACM SIGGRAPH, 845–852.

CREATH, K. 1987. Step height measurement using two-wavelength phase-shifting interferometry. *Appl. Opt. 26*, 2810–2816.

GÅSVIK, K. J. 2002. *Optical Metrology*, 3rd ed. John Wiley and Sons, Inc.

GHIGLIA, D. C., AND PRITT, M. D. 1998. *Two-Dimensional Phase Unwrapping: Theory, Algorithms, and Software*. John Wiley and Sons, Inc.

GU, X., GORTLER, S., AND HOPPE, H. 2002. Geometry images. In *Proceedings of SIGGRAPH 2002*, ACM Press / ACM SIGGRAPH, Computer Graphics Proceedings, Annual Conference Series, ACM, 355–361.

GU, X., ZHANG, S., AND HUANG, P. 2004. University of new york at stony brook geometric surface archive. http://metrology.eng.sunysb.edu/holoimage/index.htm.

HARDING, K. G. 1991. Phase grating use for slop discrimination in moiré contouring. In *Proc. SPIE*, vol. 1614 of *Optics, illumination, and image sensing for machine vision VI*, 265–270.

HOPPE, H., AND PRAUN, E. 2005. Shape compression using spherical geometry images. In *Advances in Multiresolution for Geometric Modelling*, N. Dodgson, M. Floater, and M. Sabin, Eds., 27–46.

HUANG, P. S., AND ZHANG, S. 2006. A fast three-step phase shifting algorithm. *Appl. Opt. (under press)*.

HUANG, P. S., HU, Q., JIN, F., AND CHIANG, F. P. 1999. Color-encoded digital fringe projection technique for high-speed three-dimensional surface contouring. *Opt. Eng. 38*, 1065–1071.

HUANG, P. S., ZHANG, C., AND CHIANG, F. P. 2003. High-speed 3-d shape measurement based on digital fringe projection. *Opt. Eng. 42*, 1, 163–168.

LOSASSO, F., AND HOPPE, H. 2004. Geometry clipmaps: terrain rendering using nested regular grids. *ACM Trans. Graph. 23*, 3, 769–776.

LOSASSO, F., HOPPE, H., SCHAEFER, S., AND WARREN, J. 2003. Smooth geometry images. In *Eurographics Symposium on Geometry Processing*, 138–145.

MALACARA, D., Ed. 1992. *Optical Shop Testing*. John Wiley and Songs, NY.

POST, D., HAN, B., AND IFJU, P. 1994. *High-Sensitivity Moiré: Experimental Analysis for Mechanics and Materials*. Springer-Verlag, Berlin.

PRAUN, E., AND HOPPE, H. 2003. Spherical parametrization and remeshing. *ACM Trans. Graph. 22*, 3, 340–349.

PURCELL, T. J., BUCK, I., MARK, W. R., AND HANRAHAN, P. 2002. Ray tracing on programmable graphics hardware. In *Proceedings of SIGGRAPH 2002*, ACM Press / ACM SIGGRAPH, 703–712.

RUSINKIEWICZ, S., HALL-HOLT, O., AND LEVOY, M. 2002. Real-time 3d model acquisition. *ACM Trans. Graph. 21*, 3, 438–446.

SALVI, J., PAGES, J., AND BATLLE, J. 2004. Pattern codification strategies in structured light systems. *Pattern Recogn. 37*, 4, 827–849.

SANDER, P. V., WOOD, Z. J., GORTLER, S. J., SNYDER, J., AND HOPPE, H. 2003. Multi-chart geometry images. In *Eurographics Symposium on Geometry Processing*, 146–155.

SHADE, J. W., GORTLER, S. J., WEI HE, L., AND SZELISKI, R. 1998. Layered depth images. In *Proceedings of SIGGRAPH 98*, 231–242.

TAKASAKI, H. 1970. Moiré topography. *Appl. Opt. 9*, 6, 1467–1472.

ZHANG, S., AND HUANG, P. 2004. High-resolution, real-time 3-d shape acquisition. In *IEEE Computer Vision and Pattern Recognition Workshop (CVPRW'04)*, vol. 3, 28–37.

ZHANG, S., AND HUANG, P. S. 2006. A novel structured light system calibration. *Opt. Eng. (under press)*.

ZHANG, S. 2005. *High-Resolution, Real-Time 3-D Shape Measurement*. PhD thesis, Stony Brook University, State University of New York.

Interactive Spline-Driven Deformation for Free-Form Surface Styling

Li Han[*, Δ, #]   Raffaele De Amicis[†]   Giuseppe Conti[‡]

[*] [†‡] Graphitech, Salita Dei Molini, 2, 38050, Villazzano (TN), Italy
[Δ] Information and Communication Technology Faculty, Trento University, Italy
[#] College of Computer and Information Technology of Liaoning Normal University, Dalian, China

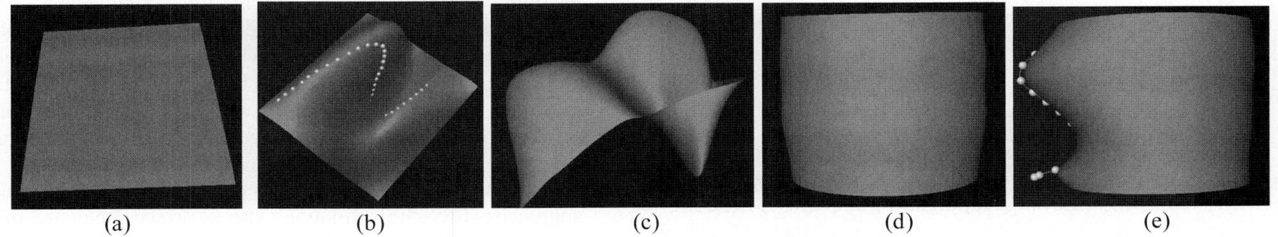

(a) (b) (c) (d) (e)

Figure 1: Surface styling by spline-driven method
(a) (b) (c) the plane is deformed by sequent splines sculpting (d) (e) a cylinder is restyled by spline-driven local deformation method

Abstract

This paper tries to answer to the increasing demand for more intuitive methods for creating and modifying free-form curves and surfaces which is emerging in the field of conceptual design. We present a novel approach which adopts a simple 3D sketching technique together with a finite element deformation method to create free-form models. The method proposed lets the user interactively sculpt splines to modify a surface in a predicable way. Our algorithm automatically extracts the key points from the sketched "target curve" and it adaptively distributes the external-force constraints which impose the forces on the corresponding vertices along their normals. We have limited the influence of these constraints to a localized area by attaching an influence factor to each vertex of the parent surface. The smoothing function introduced later allows good control over the transition intervals and the symmetry features. The proposed method is finally implemented within a 3D environment and the results of its use show how the designers can intuitively and exactly control the shape of the sketched surface.

CR Categories: I.3.5 [Computer Graphics]: Computational geometry and Object Modeling — Curve, surface, solid, and object representations; I.3.6 [Computer Graphics]: Methodology and Techniques —Interaction techniques

[*] email: han@dit.unitn.it
[†] email: raffaele.de.amicis@graphitech.it
[‡] email: Giuseppe.conti@graphitech.it

SPM 2006, Cardiff, Wales, United Kingdom, 06–08 June 2006.
© 2006 ACM 1-59593-358-1/06/0006 $5.00

Keywords: free-form surface, spline-driven method, adaptive constraint, local deformation, dynamic control

1 Instruction

Efficient and intuitive shape manipulation techniques are vital to the success of geometric modeling, computer animation, physical simulation and other computing areas. Recently considerable achievements have been reached through the adoption of Free-Form Deformation (FFD) and Extended Free-form Deformation (EFFD). Such methods embed the whole object into a tensor product volume. This can be indirectly deformed by acting accordingly upon splines' control points. Unfortunately though, manipulation of splines is not intuitive. Some authors propose alternative physically-based manipulation techniques to improve the natural operation. An example of these is the new Medial Axial Deformation method (AxDf) proposed by [Lazarus et al. 1994] aims at achieving better control over deformations. Nevertheless the degree of freedom available to control the shape is still limited.

The technique presented in this paper supports fully interactive and intuitive shape control, ranging from free-form surface creation to predictable shape deformation. Our system provides adequate interpretation of the user's freehand sketches. The proposed algorithm automatically extracts a series of key points on a "target" spline and it imposes adaptive "forces" to relocate corresponding vertices on a "parent" surface. Furthermore, a series of linear influence functions are introduced in order to improve the continuity and the symmetry.

The paper is organized as follows: in section 2 we present the related works. Section 3 we detail the relevant mathematical concepts necessary to the description of the process of surface creation. In section 4 we describe the implementation of our algorithm showing how to control the deformation process and how to consequently vary the shape. Section 5 describes some experimental results. Finally in section 6 we conclude with a

2 Previous works

The first modeling deformation technique introduced into the CAD/CAM field was the method relying on global and local deformations. This method and its further improvements [Güdükbay 1990] can provide support for regular deformations such as twisting, tapping, bending, rotating and scaling. However the method does not easily yield arbitrary shapes, further, it involves tedious and unintuitive geometrical operations. So-called Free-Form Deformation (FFD) [Sederberg 1986] methods tackle the issue of generating more complex objects through control points. That is, geometrical elements such as points, lines or a plane are associated to the geometry through specific weight factors. Control points are moved, while the coordinates of the object's local points remain unchanged. As a result the topological structure of the deformed object remains unchanged [Bechmann 1994]. This way it is possible to induce deformations within complex shapes.

In fact, FFD is one of the most versatile and powerful deformation technique. However it is not easy for the user to exactly predict the resulting deformation and thus to precisely reach the desired effect. Further accurate placement of control points is hard to achieve. Several authors [Kalra 1992; Lamousin 1994; Griessmair 1989; Coquillart 1991; Chadwick 1989] have proposed several FFD techniques improved in terms of shape control functions. The results of these research works have been exploited across several domains including human body animations and dynamic flexible deformations. However, while these approaches increase a degree of flexibility in terms of control, on the other hand they are based on the solution of complex nonlinear equations through numerical methods.

An alternative approach which promotes an easier and more intuitive interface is the so-called Extended Free-Form Deformation (EFFD) technique proposed by [Coquillart 1990]. The EFFD replaces the parallelepiped-shaped set of control points used in FFD with a lattice of control points which takes into account the approximate shape of the deformed shape. Consequently the method requires the user to know the general shape before starting to model.

Alternatively other users have used deformable surfaces for geometric modeling. [Georger 1991] applied the finite element method to minimize surface energy while meeting certain constraints. These ideas were later extended by Terzopoulos and Qin [Demetri 1994] to make use of NURBS, and in [Qin et al. 1998] they make use of Catmull-Clark subdivision surfaces. Nevertheless the resulting interface is still a direct representation of the underlying mathematics.

[Borrel et al. 1994] present an effective technique for local deformation. This was further improved by [Xiaogang, 2000] to support, not only the deformation of point constraints, but also lines and surfaces constraints. Léon et al. [Léon, 1997; Léon, 1995] have linked the control polyhedron of a surface to the mechanical equilibrium of a bar network by using the concept of force density. Although its effect on the deformation is very satisfactory in terms of aesthetic feeling, the technique often needs the solution of high-order systems of linear equations with the resulting computational load.

The deformation technique described in this paper belongs to this group. We have incorporated the concept of curve and point-surface constraint with the control polyhedron of a NURBS surface to obtain a sequence of desired models. Moreover, we further introduce a method for optimal distribution of external forces which reduces the complexity of the required computations.

3 Surface creation and relevant concepts

The definition of a surface is usually based on the use of incident curves. The constraints defining the curve-surface incidence are expressed as linear systems of equations. These have to consider the degrees of freedom (DOF) of the surface as well as the design parameters - the control points of 3D curve [Michalik 2004]. In this case the parameterization of a NURBS surface F and the curve G in the domain of F are given and fixed. Further we can assume that a 3D sketched curve H' (which serves as a shape parameter) is mapped onto a surface curve $H = F(G)$. Then the control points of this surface, which satisfies $H' = F(G)$, can be determined as a solution of a system of linear equations.

However, if G is a general domain curve, it is not obvious how to formulate the equations which constrain the incidence of an arbitrary curve. On the other hand, in contrast to traditional methods, usually the constraints do not uniquely determine the surface. In particular this involves huge computations to determine the rank of the system matrix.

In this work we instead propose an adaptive discretisation approach where the continuous curve-surface incidence problem is discretized by considering many point-surface constraints ordered along a given 3D curve thus improving the accuracy and validity of surface generation. During the implementation we further present a fairness measure in order to obtain the effective key points from the sketched target spline. Finally we demonstrate the distribution of adaptive forces among a "parent surface".

In the following sections we will show how we have interpreted 3D free-hand sketches to obtain the desired "parent surface". Then we will illustrate how our algorithm incorporates both external forces and the distributions of internal deformations to implement the process of interactive surface restyling.

3.1 The creation of the parent surface

The so-called "parent surface" is the surface which will be affected by the spline-driven deformation process. As mentioned earlier, the system supports free sketching of splines. These are initially drawn onto a default plane (see Figure 2).

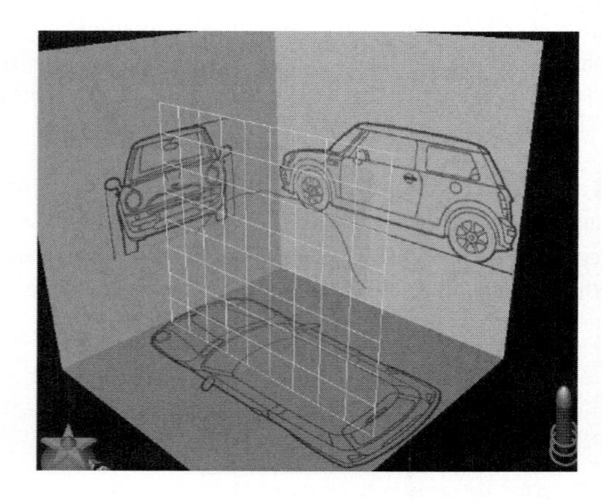

Figure 2: Sketching (red curve) on a default plane (the grid)

Each pen-stroke results in a set of ordered points on the current plane, which are automatically translated into 3D vector descriptions. When the user is drawing, the free-form spline is adaptively approximated to Cubic B-spline curve by sampling the points which are generated by considering both curvature and speed features [Li 2005].

The system allows users to sketch subsequent pen-strokes, and each Nurbs curve is attached to one plane. Proceeding this way one or several 3D curves can be created. The conversion of these curves' into NURBS surfaces depends on the selected design mode. This can be selected by the user among the following ones:

1. *Geom-filling mode*: the designer is allowed to sketch two or more 3D curves which serve as the constrained boundary of the surface.
2. *Skinning mode*: the designer sketches a surface by using the well-known concept of extrusion. He/she first draws a free-form 3D curve. The curve is then attached to the pointer and when this is moving, the process of surface generation starts and the shape is immediately shown.
3. *Revolving mode*: a surface is generated by revolving a sketched curve around a pre-defined axis.
4. *Sweeping mode*: one input spline is interpreted as a profile while a second one, so-called "path", is sculpted by further pen-strokes inside of an orthogonal plane which is perpendicular to the profile curve. The surface can be obtained by sweeping the profile along the path.
5. *Sculpting mode:* The user draws one spline (target curve) to affect the curves incidence on a selected surface.

As a result, the NURBS parent surface is constructed and represented by a "multi-patch". This, as shown in formula 1, is composed of a compatible network of iso-parametric curves, and these iso-parametric curves are represented by a series of vertices sampled by given u or v. where *numRow* and *numCol* represent the number of iso-parametric curves *in U* and *V* directions.

$$S\left(\sum_{i=1}^{numRow} C_i(u) ; \sum_{j=1}^{numCol} C_j(v) \right); \qquad (1)$$

In the following section, we will further describe how the constraint-based resultant surface is reconstructed by combining multi-patch use with the so-called "physical force" distribution technique.

3.2 Relevant concepts

Before illustrating further details of the approach we first introduce the mathematical representation behind the process presented in the following sections.

Bounding curve and target curve

Let C: □ *(u, v) = 0* be a sketched closed-curve in 3D space. This will be used for deciding the region to be deformed. The influence factor $E(Q_{i,j})$ $(0 \le i, j \le Nt)$ is attached to each vertex $Q_{i,j}$ within the parent surface S, where Nt is the number of the vertex on the parent surface. If the vertex $Q_{i,j}$ lies inside the bounding curve, E is equal to 1 and it can be influenced by force constraints. Otherwise E is set to 0 and it will thus keep a "static" status (see Formula 2). We initialize the default value $E(Q_{i,j}) = 1$, which implies that the system will automatically proceed to global deformation operations when the bounding curve is not sketched.

$$E(Q_{i,j}) = \begin{cases} 1 & \varphi(u_i, v_j) \le 0 \\ 0 & \varphi(u_i, v_j) > 0 \end{cases} \qquad (2)$$

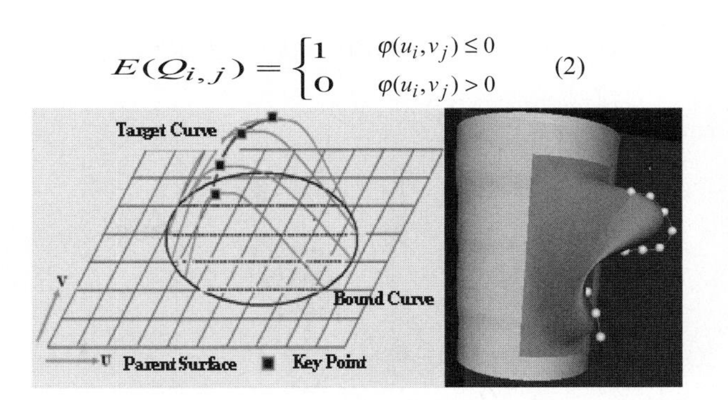

Figure 3 : The target curve influences only the area which is inside of the bounding curve

Likewise the sketched target curve is used to define the resultant shape by □ *(u, v) = 0.* A series of forces are adaptively produced through the key points on this curve (see Figure 3).

The next section will illustrate how we effectively obtain these forces and how they influence the whole parent surface.

Linear force constraint $f(K_t, P)$.

We define K_t as the key points on the target curve and $D(K_t)$ as the projection distance from K_t to the parent surface S along the normal N_t (see Figure 4 - left). $Q_{i,j}$ is the closest vertex to the projected point P which is used for determining the corresponding curve on the parent surface. In this way, each force $f(K_t, P)$ will be distributed among the vertices on the corresponding curve. Therefore, the parent surface will be gradually approximated to the leading target curve (see Figure 4 - right).

$$f(K_t, P) = E(Q_{i,j}) D(K_t) = \begin{cases} D(K_t) \\ 0 \end{cases} \qquad (3)$$

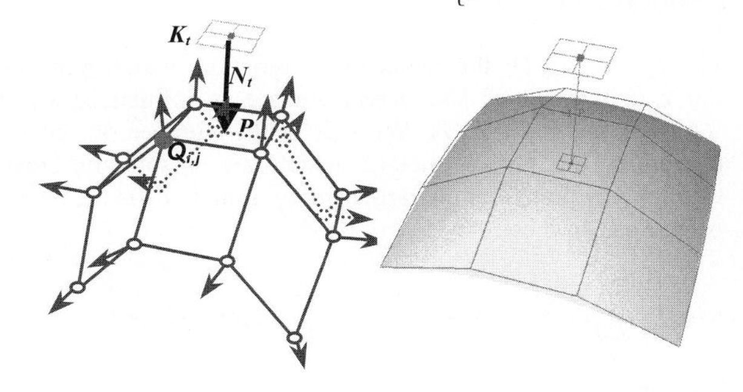

Figure 4: (Left) the multi-patch structure of a surface. K_t is the tth key point which imposes the force f to the patch, and $Q_{i,j}$ is the closest vertex to the projected point P. (Right) the resultant surface under the influence of the force f.

The force intensity "α"

Within our model, α represents the contribution of the force to the parent surface S. If the projected point P lies in one patch, the force will be distributed among the neighboring four vertices (see Figure 5);

$$\alpha(Q_{i,j}) = \alpha_0 \, x_b \, y_b \qquad 0 \le x_a, x_b, y_a, y_b \le 1 \qquad (4)$$

$$F(Q_{i,j}) = \alpha(Q_{i,j}) f(K_t, P) = \alpha(Q_{i,j}) E(Q_{i,j}) D(K_t) \quad (5)$$

$$\delta F(C(u)) = \sum_{i=1}^{Np} F(Q_{i,j}) \quad (6)$$

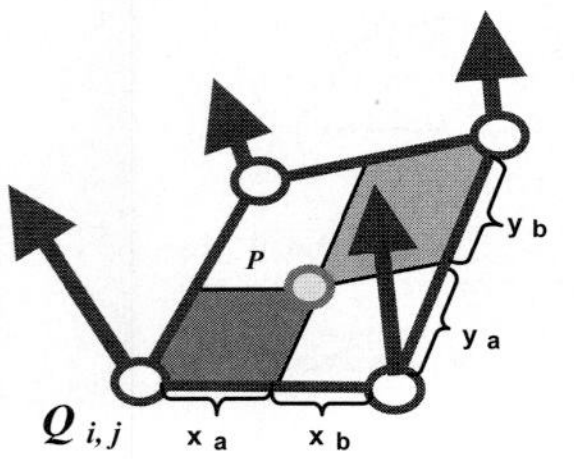

Figure 5 : Force Intensity distribution in one patch. *P* is the projected point from key point K_t.

Here x_a, x_b, y_a and y_b are defined as the distance from *P* to the four neighbor vertices, and the unit of the intensity is set to $\alpha_0 = 1$. Then the force exercises its effect inversely to the extent of the area. Therefore the force's influence on the vertex $Q_{i,j}$ can be described as $F(Q_{i,j})$ (see formula 4, 5). Likewise, when $Q_{i,j}$ is the exact vertex where the key point K_t is projected on, the force intensity $\alpha(Q_{i,j})$ yields $\alpha_0 = 1$, i.e. the force is completely imposed on this vertex.

The vertex $Q_{i,j}$ determines two curves respectively in *U* and *V* directions. Our approach adopts the prediction of the motion tendency of the target curve to decide which curve is going to response the force effect. This process will be detailed in the section 4. In formula 6, we here exemplify $\delta F(C(u))$ as the force's influence along the *U*-direction curve, where the *Np* is the number of vertex on this curve. In the following formulas we also assume that the forces are only imposed on the U direction curves.

Resultant surface

Finally we call D_L the replacement function, which represents the extent to which the parent surface is influenced by the forces *f* (see formula 7). We call *m* the number of sensitive curves in the parent surface which are adaptively created through key points on the target spline. Finally N_t is the number of vertices on these curves.

$$D_L(u,v) = \sum_{t=1}^{m} A(\Delta\omega, \Delta\beta) \bullet \delta F(C_t(u)) = \sum_{t=1}^{m} A(\Delta\omega, \Delta\beta) \bullet f(K_t, P)$$

$$= \sum_{t=1}^{m}\sum_{i=1}^{Nt} (A(\Delta\omega, \Delta\beta) \bullet \alpha(Q_{i,j}) \bullet D(K_t)) \bullet E(Q_{i,j}) \quad (7)$$

We introduce a dynamic factor *A* to monitor the state of current target spline which is formulated through the orientation constraint $\Delta\beta$ and the translation constraint $\Delta\omega$. Whenever the target spline is adjusted in 3D space, it will be used to dynamically update the forces *f* and to then redistribute the effect of these forces on the resulting surface. This way a number of shape variations are available to users as it will be detailed in the following section 4.5.

4 Spline-driven deformation process

4.1 The deformation algorithm

The details of the algorithm are presented through the help of a pseudo-code description.

Initialization: $E(Q_{i,j}) = 1$; $A(\Delta\omega = 0, \Delta\beta = 0)$

Step1: The user creates a free form surface *S* in the preferred mode (e.g. by geom-filling, skinning or revolving and so on);

IF (Local deformation)

{

The user draws a bounding curve □ *(u, v)* and consequently the system calculates the influence factor $E(Q_{i,j})$ for each vertex.

}

Step2: The user sketches the target curve □ *(u, v)*;

Step3:

IF (Over-constrained) *then Goto* Step *5.*

IF (Under-constrained) *or* (Well-constrained)

1. The system predicts the motion tendency of the target curve "*_DR*" and it determines the number of force constrains "*m*"
2. Switch (*_DR*)

   Case (U direction):

   The curves in *V* direction evolve repositioning their vertices according to the symmetrical distribution of the forces.

   Case (V direction):

   Likewise, the curves in *U direction* evolve by repositioning their vertices.
3. The system further resolves the transition intervals and improves the boundary features.

Step4: The system finally renders the resultant surface

IF (Sensor -checking is true)

{Update (Dynamic factor $A(\Delta\omega, \Delta\beta)$;

Update (External force $f(K_t, P)$);

Goto *Step3*;

}

Else Goto *Step5.*

Step5: End

Departing from what described in [Karan 98], our algorithm supports more interactive shape control through user-applied sketching operations which range from shape creation to shape edition. Especially we have optimized the spline-driven deformation process by using adaptive force distribution. Instead of evaluating every vertex in the parent surface we impose the force only to the corresponding curve in *U* or *V* direction by predicting the motion tendency of the target spline. Then the energy will be symmetrically distributed along this curve. Meanwhile the influence factor $E(Q_{i,j})$ attached to each vertex will effectively localize the influence of the forces on the surface, thus greatly reducing the computational requirements.

In the following sections we will detail how we obtain the key points on the sketched target curve and how we classify three constraint configurations (Over-constrained, under-constrained and well-constrained). Finally we will also describe how to improve the boundary features of the resultant surface.

4.2 The determination of the number of force constraints *“m”*

Since the designer’s sketching activity produces only an approximation of the desired shape, it is important that the resultant surface captures the “shape” features of the target curve. However, in the free-form domain, the number of constraints is usually unknown. Most current approaches provide only a solution that is the result of a pre-determined criterion. We instead propose a method which adaptively provides such criteria through the prediction of the motion of the target curve (see Figure 6).

For this we adopt the partial derivatives θ_1 and θ_2 (see equation 8). As shown in Figure 6 we can easily get the points P_s (u_s, v_s) and P_e (u_e, v_e) by projecting K_s and K_e onto the parent surface S. Then we extract the span of the patches where Cs and Ce define respectively the curve position in the V direction while Rs and Re describe the curve position in the U direction.

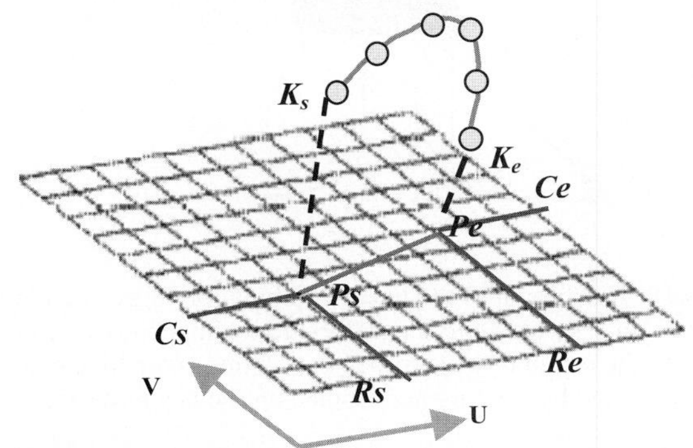

Figure 6: The target curve (in orange) and its projected line $H=\overrightarrow{P_sP_e}$ onto the parent surface. The yellow circles represent the key points which are adaptively produced by considering the orientation of H.

$$\theta_1 = \frac{\partial H}{\partial u};\ \theta_2 = \frac{\partial H}{\partial v};$$
$$m = \begin{cases} Ce - Cs & ;\ \theta_1 \geq \theta_2 \\ Re - Rs & ;\ \theta_1 \prec \theta_2 \end{cases}$$
$$Re, Rs, Cs, Ce \in Integer \qquad (8)$$

When $\theta_1 \geq \theta_2$, the target curve is leading towards the V direction. Therefore the number of key points (constraints number) on the target curve m is determined by the difference between Ce and Cs. Vice versa, when $\theta_1 < \theta_2$, m is calculated by the difference between Re and Rs. In this way the key points on the target curve will be proportionally produced and they will impose the force’s spring to the surface.

4.3 Improvement of the boundary feature of the resultant surface

During the process of local deformation, we have excluded the option of having all the vertices outside the bounding curve fixed and having to operate only on those inside. However this choice could still result in an inaccurate and insufficient deformed shape around the bounding curve. Furthermore, the leading target curve may result over-constrained or just show unacceptable undulations. To avoid these issues, we propose two ways of improving the quality of the deformation. First, we classify the constraints into three cases:

- Over-constrained: if the target curve completely lies outside the bounding curve.
- Under-constrained: if the target curve partly lies inside the bounding curve.
- Well-constrained: if the target curve lies well inside the bounding curve.

When the configuration is over-constrained the parent surface is not be affected. Conversely when the configuration is well-constrained, we use the aforementioned Formula 8 to get the adaptive constraints. And in the case of under-constrained, we adopt and four extremes (see Figure 7), it is easier to obtain the intersection part between the target curve and bounding curve. In this way the effective span of the target curve can be calculated;

Secondly, we introduce two factors to resolve the undulations near the bounding region as follows.

1) Approximation Scale

As detailed in Formula 9, we provide the scale factor λ [Li, 2005], through which the users can interactively adjust the degree of approximation to the target curve.

$$\lambda\ f(K_t,P) = \lambda\ E(Q_{i,j})D(K_t) = \begin{cases} \lambda\ D(K_t) \\ 0 \end{cases} \qquad 0 \leq \lambda \leq 1 \quad (9)$$

2) Relaxation Interval

The so-called Relaxation Interval is used to provide the transition parts from the two ending points on the target curve to the parent surface. We define the transition parts through computing the minimum bounding box of bounding curve. Then we calculate the following four extremes: *MinRow*, *MaxRow*, *MinCol* and *MaxCol*. As shown in Figure 7-top, the relaxation intervals are calculated according to the patches’ span $|Re - MaxRow|$ and $|Rs - MinRow|$. The force $f(K_s,P)$ and $f(K_e,P)$ will gradually decrease to reach zero within these two parts as it is shown in Formula 10 and Formula 11.

$$\{F(C_r(u))\}_{r=Rs}^{MinRow} = \sum_{r=Rs}^{MinRow} \sum_{i=1}^{Np} \alpha_i \frac{f(K_s,P)(r - MinRow)}{|Rs - MinRow|} \qquad (10)$$

$$\{F(C_r(u))\}_{r=Re}^{MaxRow} = \sum_{r=Re}^{MaxRow} \sum_{i=1}^{Np} \alpha_i \frac{f(K_e,P)(MaxRow - r)}{|Re - MaxRow|} \qquad (11)$$

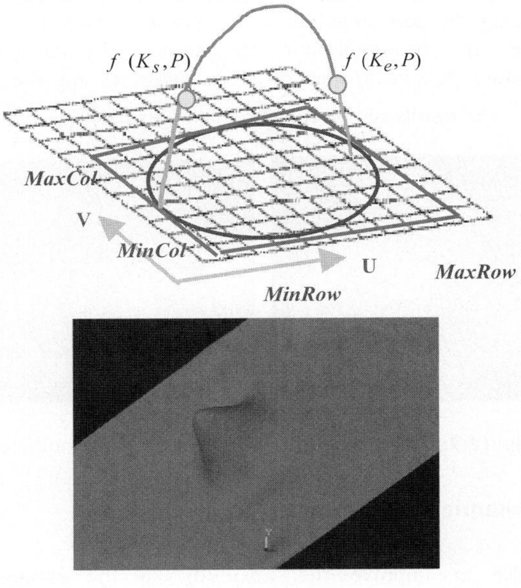

Figure 7: (Top) the Relaxation Interval in green line. (Bottom) an example of dealing with relaxation interval.

4.4 The smoothing function used to improve symmetry and continuity

Since the target curve is used to drive the surface's deformation process this might be characterized by a sharp line behavior as shown on Figure 9-left. For this we propose a smoothing function which improves the symmetry of the deformed surface. This provides strong visual impact in terms of quality of the surface. Without the need for any new patches insertion, we maintain the same topology by symmetrically distributing the external force influence to the corresponding curve (see Figure 8).

$$Tolerance = \frac{Np}{2\|C(u)\|} \quad (12)$$

$$\delta F_i(C(u)) = \sum_{t=1}^{Np} F(Q_t) = \sum_{t=1}^{r} \frac{\lambda E(Q_t)D(K_i)\ \alpha_t\ t}{Tolerance} + \sum_{t=r}^{Np} \frac{\lambda E(Q_t)D(K_i)\ \alpha_t\ t}{Tolerance}$$

$$Q_t \in C(u) \quad 1 \leq t,\ r \leq Np \quad (13)$$

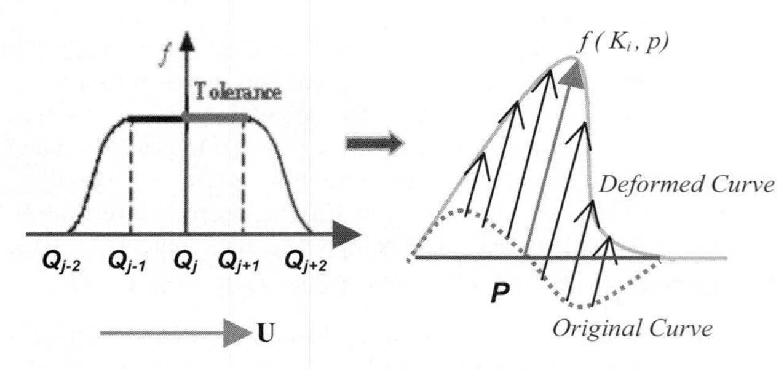

Figure 8: The force f is symmetrically distributed along the curve. (Left) The "*Tolerance*" serves as a step. (Right) the deformed curve is produced by distribution of symmetrical force.

The details are shown in formula 12 where the value $\|C(u)\|$ is the length of curve C, while Np is the number of vertices on each curve. The *Tolerance* factor is used to determine the distribution step along the corresponding curve. From formula 13, we can achieve symmetric deformation by symmetrically and gradually distributing $f(K_i,P)$ to the different vertex Q_t on the parent surface. The results can be compared in Figure 9.

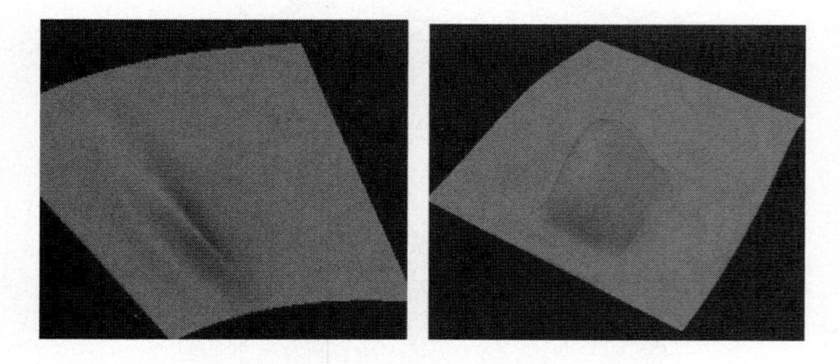

Figure 9: Before smoothing (left) and after smoothing (right)

4.5 Dynamic 3D sensor

In order to improve the flexibility of the spline-driven deformation, we adopt a 3D sensor and a 3D dragger which provides dynamic orientation and translation constraint.

As described before, in our system the curve is sketched in a separated plane and then we attach a sensor to each plane. Whenever the curve is changed the incident surface will be trimmed.

All the planes are freely controlled by a 3D dragger, that implements grab, drag and rotate operations. This way the attached curve is adjusted according to the position and the orientation (see Figure 10). By moving the 3D dragger the user controls the plane's position and orientation. Whenever the dragger is moved the 3D sensor attached to this plane will start to record the dragger's position and it will thus relocate the key points on the target spline. Subsequently the displacements of the forces will be used to update the resulting surface.

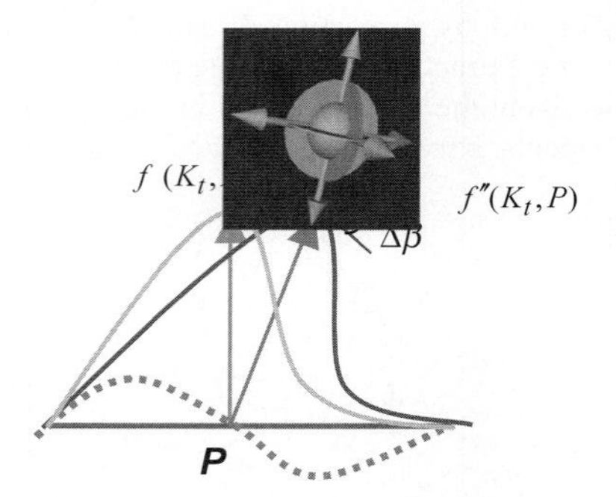

Figure10: Illustration of the linear combination of translation and orientation. The blue curve is deformed by imposing the force f (K_t, P) to the original curve (green dotted line). The blue spline is obtained by rotating and displacing the key point K_t.

At any moment during the manipulation phase, our system computes a space warp that takes as input the original positions of the point constraints C and the current position of 3D dragger C' The former is defined by a local coordinate system U, V, W while the latter is defined by the system U', V', W' and we assume that $W = U \times V$ and $W' = U' \times V'$. We wanted a smooth space warp that could move the starting position to the ending position. For this to be satisfied three translation constraints and three rotation constraints (see Figure 11) are required. Since the dragger controls the whole plane, we can use unified scale to update all the key points. This way the constraint $C\ (U, V, W)$ represents the plane's initial state. We assume then that $[K_1, K_2, \ldots\ K_n]$ are the set of original constraining points in target spline while A $(\Delta\omega(\kappa), \Delta\beta(\kappa))$ with $\kappa \in \neg^3$, are the warped positions. Specifically $\Delta\omega(\kappa)$ and $\Delta\beta(\kappa)$ represent respectively the variation in the translation and the orientation caused by the user. When $\Delta\beta(\kappa)$ is equal to NULL the deformation is a pure translation and therefore the distance D can be calculated by the formula 14. Then all the forces $f''(K',P)$ produced by constraining points $[K'_1, K'_2, \ldots\ K'_n]$ can be described by the formula 16.

$$D = UU' \times VV' + VV' \times WW' + WW' \times UU'; \quad (14)$$

$$\Delta\omega(\kappa) = D / |D|; \quad (15)$$

$$f''(K',P) = f(K,P) + \Delta\omega(\kappa) \quad (16)$$

It is obvious that $\Delta\omega(\kappa)$ determines the displacement of the key points and it thus affects the intensity of the force applied to the surface. On the other hand, the orientation $\Delta\beta(K)$ will adjust the responding curves on the original surface as described in following formula 17, 18. Here K represents the original key points on the target curve, while K' is their final position. This way the sensitive curves on the surface determined by these key points will be updated. And then the force distribution in the parent surface will be adjusted as well.

$$\Delta\beta = 2\sin^{-1}\left(\frac{|UU'|}{2|D \times U|}\right) \quad (17)$$

$$K'(U', V', W') = K(U, V, W) \times \tan(\Delta\beta) \quad (18)$$

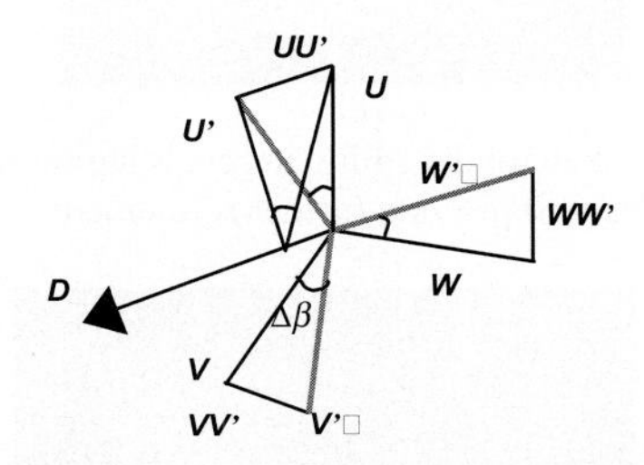

Figure 11: Computation of orientation and translation

Therefore, once the sensor detects the dragger's movement, the system automatically updates the dynamic factor *A*. Consequently *G(S)* is recalculated, in this way we can get the sequence of the deformations (see Figure 12).

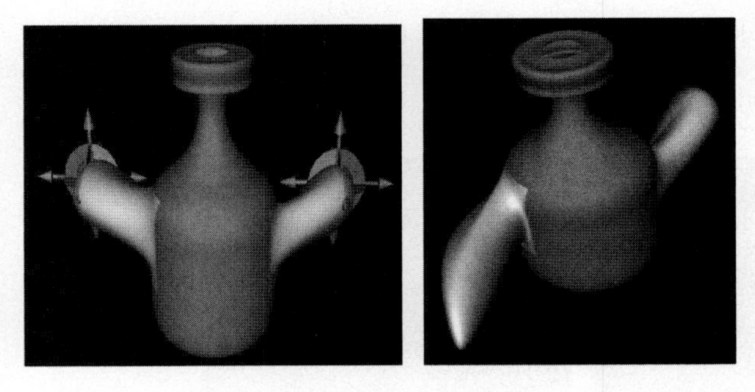

Figure 12: (Left) Two target splines controlled by 3D draggers produce series of force springs on the surface and they only impose strain to predefined local regions. (Right) When the 3D draggers are moving, the result is changed based on the updated force intensity and redistribution.

5 Experiments and results

We have implemented our method in C++ with OpenGL and OpenInventor 4.0 on a Pentium 4 1.6GMhz with 512MB of RAM. This implementation provides real-time feedback (approx 20 frames per second for average 30,000 vertices) with a sequence of deformations. In order to improve the interaction for the required shape, we have developed a 3D dragger which can be freely controlled in 3D space. This is used for handling the plane where the object lies. The user is thus capable of dynamically controlling the target curve and parent models to reproduce a series of results. In our application the 3D dragger allows change of orientation and translation by simply grabbing and dragging operation. In order to preserve the smoothness of the deformed surface we allow the setting of a sensor to each node. However, as described before, whenever the sensor is activated the space-warp operator is starting to recalculate the resulting surface. Since this behavior would require expensive numerical computation we have attached a time sensor which triggers the resulting calculations only at specific intervals. The interval was experimentally set to 0.02 sec. to achieve good visual feedback with appropriate numerical computation.

Furthermore, we apply two methods to test the influence of the force to the surface. First the local area is directly obtained by projecting the target curve onto the parent model. This way a series of springs are produced in the parent model. This are going to respond to the energy strains from the target curve. In the second method, we directly define a local region by sketching a bounding curve as aforementioned. The comparison is shown in the following figures.

Our experiments indicate that our method is intuitive and effective for creating and editing a large variety of free form shapes (see Figure 13, Figure 14, Figure 15 and Figure 16).

6 Conclusions and future work

In this paper, we present a spline-driven modelling and deformation method. When working with our method, the designer does not need to manipulate some non-intuitive mathematical shape parameters, such as control points and control vectors. Instead, he/she can work with the point constraints and spline-based constraints, therefore designers can easily and intuitively control the resulting shape.

In our system, the function is centred around the interactive surface sculpting and the intuitive spline-driven deformation. Compared with other methods, this approach has the following advantages: intuition, locality and simplicity of use since it combines shape creation and deformation. Finally, it is possible to use it for modelling of various free-form shapes and for their manipulation.

In future works, we will further investigate intelligent operations for modelling multi-surfaces based on 3D sketching operation such as surface splitting and stitching. We also plan to improve the connectivity and continuity between different surfaces based on declarative constraints.

Acknowledgements: The research presented in this paper is supported by the EU project "IMPROVE" and of the Part of the PAT project InSIDe.

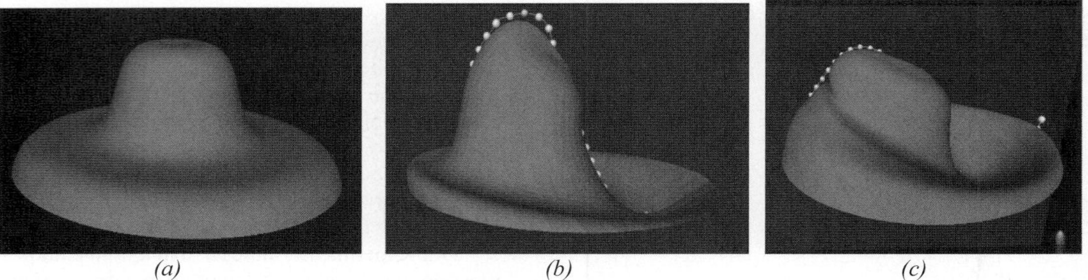

(a) (b) (c)

Figure 13: Design of a hat through spline-driven global deformation (a) the hat is created by sketching a spline in revolving mode. (b) The target spline imposes the force effects on the surface (c) The adjusted target spline will provide further refinement.

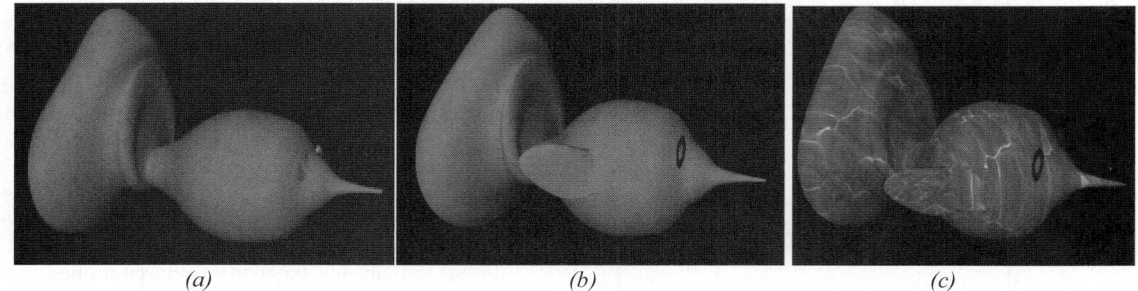

(a) (b) (c)

Figure 14: Local deformation by spline-driven method. (a) A Fish body is generated by drawing a 4-strokes spline. (b) The fins are created by two target curves with a predefined local region. (c) The texture is applied and its mapping is dynamically changed through the user's interaction.

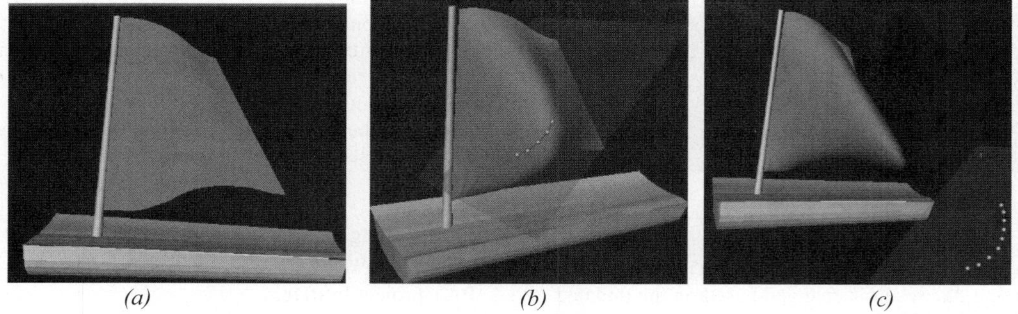

(a) (b) (c)

Figure 15: Dynamical shape control through the spline-driven deformation method. (a) This boat is created by using pre-defined drawing mode (7 strokes); and the red sail is selected as the sensitive surface which is going to be deformed. (b) The local area is automatically obtained by projecting a target curve (blue spline) onto the surface. (c) When the target curve is moving away from the parent model the surface will be dynamically adjusted in response to the energy distribution.

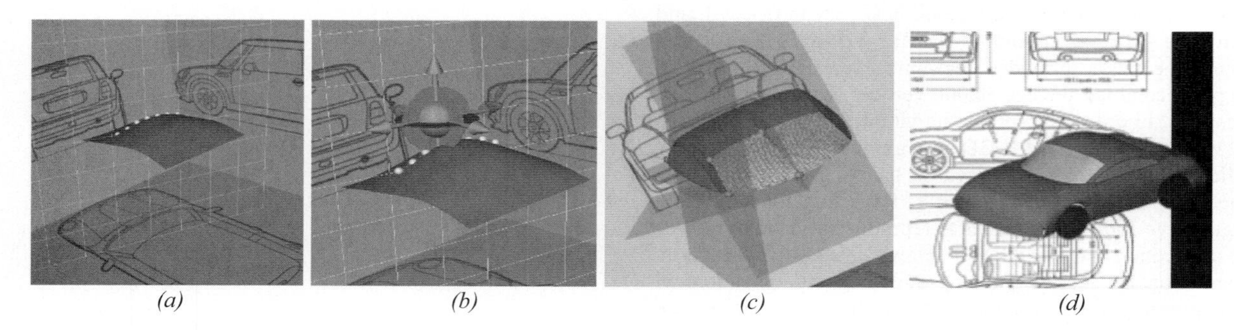

(a) (b) (c) (d)

Figure 16: Car styling through the spline-based technique. (a) The sketched splines are interpreted into the corresponding surface. (b) The surface is restyled by a target spline which is controlled by a 3D dragger (c) More surfaces are incrementally generated according to the user's drawing. (d) The final car model.

7 References

Barr, A. H. 1984. Global & Local Deformations of Solid Primitives. *Computer Graphics*, 18(3), 21-30.

Bechmann, D. 1994. Space Deformation Models Survey. *Computer & Graphics*, 18(4), 571-586.

Borrel, P. and Rappoport, A. 1994. Simple Constrained Deformations for Geometric Modeling and Interactive Design. *ACM Transactions on Graphics*, 13(2), 137-155.

Chadwick, J. E., Haumann, D. and Parent, R. E. 1989. Layered Construction for Deformable Animated Characters. *Computer Graphics*, 23(3), 243-252.

Chesutet, V., Catalano, C.E. and Pernot, J. P. 2004. 3D Sketching with Fully Free Form Deformation Features for Aesthetic Design. In *EUROGRAPHICS Workshop, Sketch-based Interfaces and Modeling*, 9-18.

Coquillart, S. 1990. Extended Free-form Deformation: A Sculpting Tool for 3D Geometric Modeling. *Computer Graphics*, 24(4), 187-196.

Coquillart, S. and Jancene, P. 1991. Animated Free-form Deformation: An Interactive Animation Technique. *Computer Graphics*, 25(4), 23-26.

De Araujo, B., Jorge, J. A. 2003. Blobmaker: Free Form Modelling with Variational Implicit Surfaces. In *Proc. of the 12th Portuguese Computer Graphics Meeting,* 17–26.

Demetri, T., Kurt, F. 1988. Modeling Inelastic Deformation: Viscolelasticity, Plasticity, Fracture. *ACM SIGGRAPH Computer Graphics*, v.22 n.4, 269-278, Aug.

Demetri, T. and Hong, Q. 1994. Dynamic NURBS with Geometric Constraints for Interactive Sculpting. *ACM Transactions on Graphics*, April, 13(2):103–136,

Dietz, U. 1998. Creation of Fair B-Spline Surface Fillets. *In Creating Fair and Shape Preserving Curves and Surfaces. B.G. Teubner, Stuttgart*, 2, 3, 8

Fontana, M., Giannini, F. and Meirana, F. 2000. Free Form Features for Aesthetic Design. *Int. Jou. Shape Modelling, vol. 6, n°2,* 273-302.

Georger, C. and Dave, G. 1991. Deformable Surface Finite Elements for Free-form Shape Design. *Computer Graphics*, July 1991, 25(4):257–266,

Gibson, S. F. F., Mirtich, B. 1997. A Survey of Deformable Modeling. *In Computer Graphics.* Tech. Rep. TR-97-19, Mitsubish Electric Research Laboratoy.

Griessmair, J. and Purgathofer, W. 1989. Deformation of Solids with Trivariate B-spline. *In Proc. EUROGRAPHICS'89*, 137-148.

Güdükbay, U. and Üzgüç, B. 1990. Free-form Solid Modeling Using Deformations. *Computer Graphics*, 14(3/4), 491-500.

Hong, Q., Chhandomay, M. and Baba, C. V. 1998. Dynamic Catmull-clark Subdivision Surfaces. *IEEE Transactions on Visualization and Computer Graphics*, 4(3):215–229.

Igarashi, T., Masuoka, S., and Tanaka, H. 1999. Teddy: A Sketching Interface for 3d Freeform Design. In *Proc. of SIGGRAPH '99*, 409–416.

Kalra, P., Mangili, A. and Thalmann, N. 1992. Simulation of Facial Muscle Actions based on Rational Free-form Deformation. *Computer Graphics Forum*, **2**(3), 59-69.

Karan, S., Eugene, F., 1998. Wires: A Geometric Deformation Technique. *In Proceedings of the 25th annual conference on computer graphics and interactive techniques*, 405-414.

Karpenko, O., Hughes, J.F. and Raskar, R. 2002. Free-Form Sketching with Variational Implicit Surfaces. *Computer Graphics Forum*, Volume 21, Issue 3.

Lamousin, H. J. and Waggenspack, W. N. 1994. NURBS based Freeform Deformation. *IEEE Computer Graphics & Applications*, 14(6), 59-65.

Lazarus, F., Coquillart, S., and Jancene, P. 1994. Axial Deformations: an Intuitive Deformation Technique. *Computer-Aided Design,* 26(8): 607-613, August.

Léon, J. C. and Trompette, P. 1995. A New Approach Towards Freeform Surfaces Control. *C.A.G.D*, 12(4), 395-416.

Léon, J. C. and Veron, P. 1997. Semiglobal Deformation and Correction of Free-form Surface Using a Mechanical Alternative. *The Visual Computer*, 13(3), 109-126.

Li, H., Giuseppe C., and Raffaele. D. A. 2005. Freehand 3D Curve Recognition and Oversketching. *Eurographics UK Chapter,* 187-193.

Michalik, P., and Brüderlin,B.D. 2004. Constraint-based Design of B-spline Surface from Curves. *ACM Symposium on Solid Modeling and Applications*, 213-220.

Naya, F., Jorge, J. A., Conesa, J. 2002. Direct modeling: from Setches to 3d Models. In *Proc. of the 1st Ibero-American Symposium in Computer Graphics*, 109–117.

Sederberg, T. W. and Parry, R. 1986. Free-form Deformations of Solid Geometric Models. *ComputerGraphic,* 20(4), 151-160.

Shi-Min, H., Hui, Zh. , Chiew-Lan, T. 2000. Direct Manipulation of FFD: Efficient Explicit Solutions and Decomposable Multiple Point Constraints. *The Visual Computers*, vol. 17, No. 6, 370-379.

Xiaogang, J., Youfu, L. and Qunsheng, P. 2000. General Constrained Deformation based on Generalized Metaballs. *Computer & Graphics,* 24(2), 219-231.

Yan, Zh. and Canny, J. 1999. Real-time Simulation of Physically Realistic Global Deformation. *IEEE Vis'99. San Francisco, California*. October 24-29.

Yan, Zh. and Canny, J. 2000. Haptic Interaction with Global Deformations. In *Proceedings of the 2000 IEEE International Conference on Robotics & Automation.* 2428-2433.

Pen-based Styling Design of 3D Geometry Using Concept Sketches and Template Models

Levent Burak Kara*, Chris M. D'Eramo†, Kenji Shimada‡
Mechanical Engineering Department
Carnegie Mellon University
Pittsburgh, Pennsylvania 15213

Abstract

This paper describes a new approach to industrial styling design that combines the advantages of pen-based sketching with concepts from variational design to facilitate rapid and fluid development of 3D geometry. The approach is particularly useful for designing products that are primarily stylistic variations of existing ones. The input to the system is a 2D concept sketch of the object, and a generic 3D wireframe template. In the first step, the underlying template is aligned with the input sketch using a camera calibration algorithm. Next, the user traces the feature edges of the sketch on the computer screen; user's 2D strokes are processed and interpreted in 3D to modify the edges of the template. The resulting wireframe is then surfaced, followed by a user-controlled refinement of the initial surfaces using physically-based deformation techniques. Finally, new design edges can be added and manipulated through direct sketching over existing surfaces. Our preliminary evaluation involving several industrial products have demonstrated that with the proposed system, design times can be significantly reduced compared to those obtained through conventional software.

CR Categories: H.5.2 [User Interfaces]: Graphical User Interfaces (GUI)—Pen-based interaction; I.3.5 [Computational Geometry and Object Modeling]: Curve, surface, solid, and object representations—Physically based modeling

Keywords: Pen computing, style design, 3D sketching, camera calibration, physically-based deformation, surfacing.

1 Introduction

Advances in 3D shape modeling have resulted in a myriad of sophisticated software available for a range of different applications. Most commercial systems, while versatile, are tedious to use due to their intricate interface, and rely heavily on users' knowledge of the underlying mathematics for representing, creating and manipulating 3D shapes. Moreover, these systems are typically tailored toward the later stages of the design where ideas have sufficiently matured and expected alterations to the design concept are not too severe. Hence, these tools are typically used by computer modelers downstream in the design cycle, who have little or no control on the concept development. For years, this shortcoming has forced a gap in the design practice where concepts and computer models are developed in different media by different personnel. Typically, designers spend a considerable amount of time generating concept sketches on paper, which are then handed over to computer modelers who use these sketches as a visual reference during computer modeling. This separation often results in conflicts between the intended form and the digital model, requiring multiple iterations between the two parties until the desired results are achieved. Nevertheless, a key advantage of the conventional software is that it is highly suitable for 'variational design,' where the design process mostly involves a modification of an earlier design. That is, rather than forcing the designer to start from scratch, these systems help designers use existing computer models as a starting point in their current tasks.

To alleviate the shortcomings of conventional software, some recent research has focused on "user-centered" techniques that aim to provide more intuitive interfaces and interaction tools. These systems allow users to quickly create and manipulate 3D forms, often through the use of a digital pen and a tablet, while freeing the user from most mathematical details. Researchers have successfully demonstrated the utility of these systems in various domains [Zeleznik et al. 1996; Igarashi et al. 1999; Karpenko et al. 2002; Tsang et al. 2004; Bourguignon et al. 2004; Das et al. 2005; Masry et al. 2005]. While these systems greatly facilitate 3D shape development through an intelligent use of computer vision, human perception and new interaction techniques, they are often limited to simple shapes, or they impose constraints not suitable for industrial styling design.

This work describes a new approach to industrial styling design that combines the advantages of pen-based computer interaction with the efficacy of variational design. The proposed method attempts to improve the current practice by allowing designers to utilize their paper sketches in conjunction with existing computer models to facilitate rapid and fluid development of 3D geometry. The method is designed to enable those with sufficient drawing skills to easily operate on 3D form without having to know much about the underlying details. The input to the system is a scanned or digitally-created concept sketch, and a generic 3D wireframe model, called a template, that has the basic form of the object. In the first step, the template is aligned with the digital sketch, bringing the projection of the template as close as possible to the object in the sketch. Next, using a digital pen, the user traces, in a moderately casual fashion, the feature edges of the sketch on the computer screen. User's 2D strokes are processed and interpreted in 3D to give the desired form to the template located underneath the image. In a similar way, other sketches exposing different vantage points can be utilized to modify different parts of the template. Alternatively, the user can abandon the use of the input sketches and continue sketching directly on the template. Once the desired form is achieved, the resulting template is surfaced to produce a solid model, followed by a user-controlled refinement of the initial surfaces. While our current surface modification tools support a limited class of deformations, they are designed to be simple enough to allow users to explore alternative surface shapes in a controllable and predictable way. Fi-

*e-mail: lkara@andrew.cmu.edu
†e-mail: cderamo@andrew.cmu.edu
‡e-mail: shimada@cmu.edu

SPM 2006, Cardiff, Wales, United Kingdom, 06–08 June 2006.
© 2006 ACM 1-59593-358-1/06/0006 $5.00

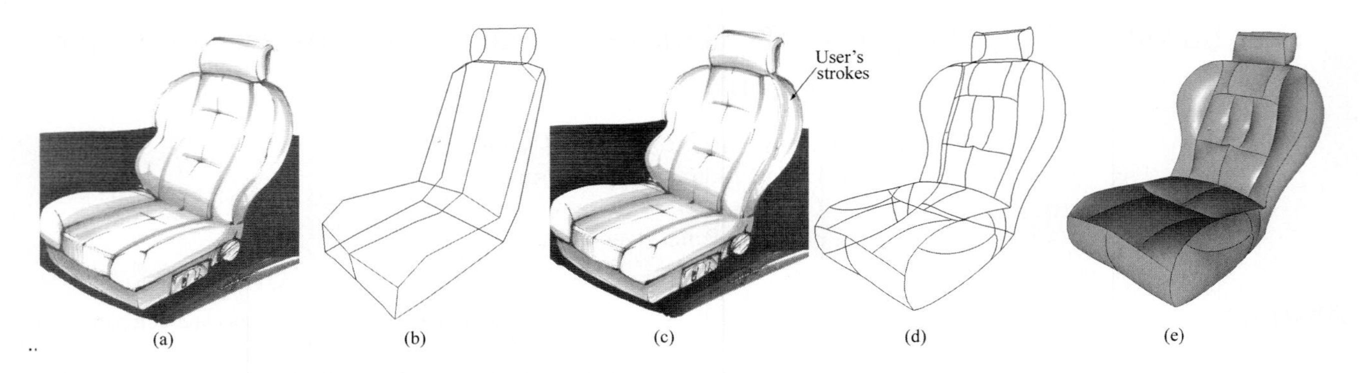

Figure 1: Overview of the modeling process. The input to the system consists of: (a) User's 2D concept sketch, and (b) A 3D template model. (c) After aligning the template with the sketch, the user draws directly on the sketch. The strokes modify the underlying template. (d) The final shape obtained after modifying the template and adding new feature edges. (e) The final solid model after surface creation and modification. The final surfaces are designed using a pressure-based deformation tool.

nally, the model can be further enhanced by sketching new feature edges on existing surfaces and manipulating them as necessary.

The proposed approach is particularly useful in styling design where the new product is primarily a stylistic variation of an existing one. For instance, the approach is well-suited to car body design wherein a company can use previous years' sedan models as templates to facilitate the design of a new sedan. In many cases, the repository of past designs serves as a natural pool of templates, eliminating the need for extra investment in template generation. Likewise, this approach is useful in cases where the basic form is readily dictated by universal artifacts arising from social or ergonomic reasons. In car industry for example, seats customarily consist of a headrest, a backrest, and a base. Hence, a generic template model embodying these main parts can be used in the design of a wide variety of seat models. However, to accommodate designs that are markedly different than their predecessors, it is conducive to generate and maintain a variety of different templates in the database, and choose the most suitable one as needed.

The remainder of the paper is organized as follows. Section 2 reviews existing approaches to 3D shape development. Section 3 describes the user interaction and an overview of the proposed approach. Section 4 details the alignment of the template model with the input sketch. Section 5 explains the 3D modification of the template based on users' strokes. Section 6 describes the generation of surfaces from the designed template and the associated modification tools. Section 7 provides examples and discussions. Section 8 presents conclusions.

2 Related Work

Over the past years, 3D shape interpretation and modeling techniques have evolved in parallel with enabling interaction technologies. In 3D interpretation from 2D input, the well-known issue of one-to-many mapping (thus the lack of a unique solution) has resulted in the development of various constraint and optimization based methods. To date, much work has focused on interpreting line drawings of polyhedral objects [Grimstead and Martin 1995; Turner et al. 1999; Varley 2003; Varley 2004; Masry et al. 2005]. These methods typically use some form of a line-labeling algorithm, followed by an optimization step, to produce the most plausible interpretation. Results are shown to be improved by the use of various image regularities such as symmetry, edge junctions, parallelism and intersection angles. The difficulty of the problem setting (usually a single drawing constructed from an unknown arbitrary viewpoint) makes these methods most suitable to objects with flat faces and simple edge geometries. In our case, 3D interpretation is facilitated through the use of templates thus alleviating some of the difficulties faced by those systems. Similar to our approach, some recent systems exploit topological templates to extend existing interpretation principles to curved objects. [Mitani et al. 2000] use a six-faced topological template for interpretation. The nature of the template, however, limits the scope of the method to objects topologically equivalent to a cube. [P.A.C.Varley et al. 2004] present an approach most similar to ours. In their approach, a template is first created by interpreting a drawing of a polyhedral object using line labeling techniques. Next, curves are added by bending the edges of the template through sketching. Since there are infinitely many 3D configurations of a curve corresponding to the 2D input, the best configuration is determined based on the assumption that the modification produces a symmetrical distortion across the object's major plane of symmetry. The idea of templates has also been explored by [Yang et al. 2005] who use 2D templates to recognize and convert users' sketches into 3D shapes. The recognition and 3D geometry construction algorithms make this approach suitable to a limited number of objects with relatively simple geometry.

Some researchers have explored alternative methods to space curve construction. [Cohen et al. 1999] exploit shadows to facilitate 3D interpretation. In their system, a space curve drawn in a 2D interface is complemented with a sketch of the same curve's shadow on a plane. However, this approach relies on user's ability to accurately visualize and depict a curve's shadow. [Das et al. 2005] describes an approach for free-form surface creation from a network of curves. Their solution to 3D interpretation from 2D input seeks to produce 3D curves with minimum curvature. This choice is justified on the grounds that the resulting 3D curve will be least surprising when viewed from a different viewpoint. As described in Section 5, our formulation of the best 3D interpretation in based on a similar rationale, except we minimize the *spatial deviation* from the original template as opposed to curvature. [Tsang et al. 2004] present an image-guided sketching system that uses existing images for shape construction. Users create 3D wireframe models by tracing 2D profiles on images that reside on orthonormal construction planes. While their work is similar to ours in the way existing images are used, their approach relies primarily on the use of top, side and front construction planes as opposed to an arbitrary viewpoint. Additionally, profile creation uses a click-and-drag interaction instead of a sketch-based interaction. Nevertheless, as described in Section 5, our approach recognizes the utility of active contours [Kaas et al. 1988] for curve manipulation similar to theirs.

From an interaction point of view, various gesture-based interfaces have been developed for 3D shape creation [Zeleznik et al. 1996; Eggli et al. 1995] and modification [Hua and Qin 2003; Draper and Egbert 2003]. In these systems, designers' strokes are used primarily for geometric operations such as extrusion or bending, rather than for depicting the shape. Silhouette-based approaches [Igarashi et al. 1999; Karpenko et al. 2002; Bourguignon et al. 2004; Schmidt et al. 2005; Cherlin et al. 2005] enable free-form surface generation. In these methods, users' strokes are used to form a 2D silhouette representing an outline or a cross-section, which is then extruded, inflated or swept to give 3D form. These systems are best suited to creating cartoon-like characters or similar geometries. Systems such as [Cheutet et al. 2004; Nealen et al. 2005] allow users to directly operate on existing surfaces to deform or add features lines using a digital pen. The key difference of these systems compared to gesture-based interfaces is that users' strokes are directly replicated in the resulting shape. However, these systems are most useful during detail design where the main geometry is already available.

In addition to pen and tablet based systems, a number of virtual reality based systems have also been developed. Systems proposed by [Bimber et al. 2000; Wesche and Seidel 2001; Diehl et al. 2004; Fleisch et al. 2004] allow users to construct 3D wireframes in a virtual environment using specialized input devices and a head mounted display. Once the wireframe is created, surfaces covering the wireframe are added to produce a solid model. Inspired by tape drawing commonly used in automotive industry, [Grossman et al. 2002] describe a system for constructing 3D wireframe models using a digital version of the tape drawing technique. [Llamas et al. 2005] describe a method for deforming 3D shapes based on a virtual ribbon. The ribbon serves as a flexible spline controlled by the user that, when attached to the solid object and deformed, allows the object to be deformed in parallel. A common difficulty in such systems is that users' unfamiliarity with the input devices and interaction techniques makes these methods less attractive to those accustomed to traditional tools.

In addition to the above 3D shape modeling techniques, a number of techniques have been devised to *understand* users' hand-drawn sketches [Kara and Stahovich 2004; Gennari et al. 2004; Alvarado and Randall 2004; Hammond and Davis 2004; Shilman and Viola 2004; LaViola and Zeleznik 2004]. Given the context in which the sketches are created, the primary goal in these methods is to have the computer reliably parse and identify the objects suggested by the pen strokes. While the work in sketch understanding has made the pen an attractive alternative for inputting graphical information, to date, most studies have focused on 2D scenes, with little or no concern for the stylistic or aesthetic aspect of users' strokes.

3 User Interaction and Overview

The main input device is a pressure sensitive digitizing tablet with a cordless pen. The drawing surface of the tablet is an LCD display, which allows users to see digital ink directly under the pen. Users' strokes are collected as time sequenced (x, y) coordinates sampled along the stylus' trajectory. The user interface consists of a main drawing region, and a side toolbar for accessing commonly used commands.

Figure 1 summarizes the main steps of the approach. In a typical session, the user begins by loading a scanned or digitally-created sketch of the design object, and an appropriate template model. The input sketch and template are independent in that the creation of one does not require the knowledge of the other, and vice versa. However, given the design sketch, the user must choose the appropriate template from the database. This means, if the design object in question is a car seat, the user must choose a car seat template, if it is a sedan car, a sedan template must be chosen etc. If there are multiple candidate templates that could be used with the sketch, it is preferable to choose the template that most closely resembles the sketch.

Since input sketches may have been drawn from arbitrary vantage points, the first step involves geometrically aligning the template with the sketch until the projection of the template to the image plane matches the sketch. This requires a reliable identification of the camera properties suggested in the sketch. These properties correspond to the position, orientation, and lens parameters of a virtual camera that, if directed at the design object, would produce the image in the sketch. To uncover these parameters, the user is asked to sketch a virtual bounding box enclosing the object in the sketch and mark its eight corners. 2D coordinates of these eight corners trigger a calibration algorithm that determines the unknown camera parameters. The computed parameters are then applied to the template model, thus bringing the template in close correspondence with the sketch.

The above alignment sets the stage for an image-guided sketching process in which the user replicates the input sketch (or parts of it) by retracing its characteristic edges with the digital pen. Each edge can be retraced using any number of strokes, drawn in any direction and order, thus accommodating casual sketching styles. After retracing an edge, the user invokes a command that processes the accumulated strokes to tacitly modify the corresponding edge of the template in 3D (details are presented in Section 5). At any point, the user may reveal the underlying template by hiding the sketch, and continue sketching directly on the template. This feature is useful when bulk of the template has been modified by retracing the reference sketch, but further touches are necessary to obtain the final shape. In such cases, with the template visible, the user can change the vantage point and modify the edge from the new view. When necessary, the reference sketch and the associated camera properties can be recalled to realign the template with the sketch. Alternatively, other sketches drawn from different vantage points may also be used. When using a new sketch, however, the user must perform camera calibration to orient the template according to the new vantage point. If desired, the program will preserve symmetry across a user-specified plane. This allows users to work solely on a single side of a symmetrical object; the geometry is automatically replicated on the other side. Note that although this work advocates the use of existing concept sketches to facilitate modeling, their use is optional. Without loss of benefits, designers may elect to work directly on the template by navigating around the model and modifying its edges as necessary.

Once the desired form is achieved, the newly designed wireframe is automatically surfaced, followed by a user-controlled modification and refinement of the initial surfaces. The surface modification tool, inspired by the physical deformation of a thin membrane under a pressure force, allows the user to inflate or flatten a surface by a controllable amount. The intuitive nature of this deformation tool enables different surface shapes to be quickly and straightforwardly explored. Surfaces may be further refined using a method inspired by mechanical springs. This method works to minimize the variation of mean curvature, producing surfaces that are fair and aesthetically pleasing. Finally, with the new surfaces in place, the user can enhance the model by sketching new design edges directly on the surfaces and manipulating them as necessary.

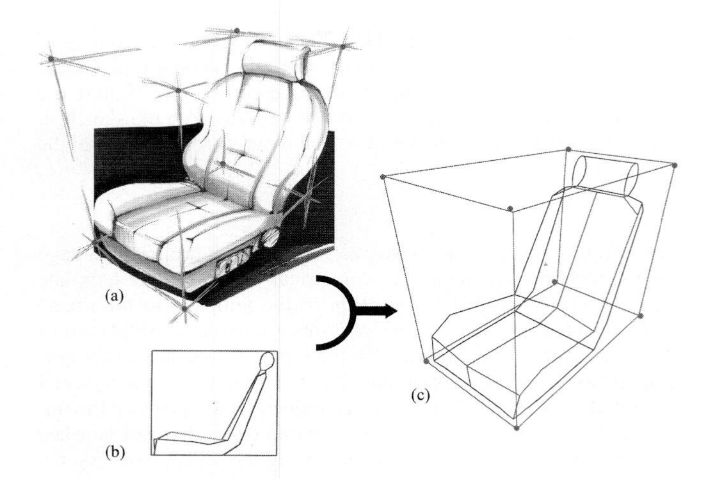

Figure 2: Template alignment. (a) User draws an approximate bounding box on the sketch and marks its eight corners. (b) The default configuration of the template is an orthographic side view. (c) The camera calibration algorithm closely aligns the template with the sketch. In an ideal alignment, the corners of the template bounding box would exactly match the red dots marked by the user.

4 Template Alignment

To align the template with the sketch, the user is asked to draw a bounding box enclosing the object in the sketch, and mark its eight corners. It is assumed that the sketch is depicted from a general viewpoint with all eight corners of the bounding box distinctly revealed. The 2D coordinates of the eight corner points set up a camera calibration algorithm that aligns the user-drawn bounding box with that of the template[1]. Adopting the convention used in [Forsyth and Ponce 2003], the camera model relating a homogeneous world coordinate $\mathbf{P} = [x\ y\ z\ 1]^T$ to a homogeneous image coordinate $\mathbf{p} = [u\ v\ 1]^T$ can be described as follows:

$$\mathbf{p} = \frac{1}{s}\mathbf{K}\left[\begin{array}{cc}\mathbf{R} & \mathbf{t}\end{array}\right]\mathbf{P}$$

where s is an unknown scale factor. $\mathbf{R}$ and $\mathbf{t}$ are the extrinsic camera properties corresponding to the rotation and translation matrices. $\mathbf{K}$ is the camera intrinsic matrix and is given by:

$$\mathbf{K} = \begin{bmatrix} \alpha & -\alpha\cot(\theta) & u_0 \\ 0 & \dfrac{\beta}{\sin(\theta)} & v_0 \\ 0 & 0 & 1 \end{bmatrix}$$

with (u_0, v_0) the coordinates of the camera center, α and β the scale factors in image u and v axes, and θ the skew angle in radians between the two image axes. Given the eight corner points indicated by the user and the corresponding eight world coordinates of the template corners, the goal is to reliably identify matrices $\mathbf{K}$, $\mathbf{R}$, $\mathbf{t}$, and the unknown scalar s. Once these parameters are determined, they can be applied to the virtual camera directed at the template to align the projection of the template with the sketch.

The calibration process can be decomposed into two parts [Forsyth and Ponce 2003]: (1) Computing a 3x4 projection matrix $\mathbf{M}$; the product $(1/s)\cdot\mathbf{K}\cdot[\mathbf{R}\ \mathbf{t}]$, (2) Estimating intrinsic and extrinsic parameters from $\mathbf{M}$.

[1]Drawing the bounding box provides a visual guide to the user and is optional. Only the eight corner points are required by the calibration algorithm.

For the solution of the first part, we use the Linear Direct Transform method [Abdel-Aziz and Karara 1971]. In this method, n image coordinates and their corresponding n world coordinates yield a system of $2n$ homogenous linear equations in the twelve unknown coefficients of $\mathbf{M}$. When $n \geq 6$ (in our case n is 8), the twelve coefficients can be obtained in the least-squares sense as the solution of an eigenvalue problem.

The second step extracts s, $\mathbf{K}$, $\mathbf{R}$ and $\mathbf{t}$ from $\mathbf{M}$. This is facilitated by the fact that the rows of the rotation matrix have unit length and are orthonormal. Without presenting the details, the unknown camera parameters can be found as follows (see [Forsyth and Ponce 2003] for details):

Rewrite $\mathbf{M} = [\mathbf{A}_{3x3}\mathbf{b}_{3x1}]$. Note that $\mathbf{A}$ and $\mathbf{b}$ are trivially determined from $\mathbf{M}$. Let $\mathbf{a}_1$, $\mathbf{a}_2$ and $\mathbf{a}_3$ be the three column vectors of $\mathbf{A}$. Let $\mathbf{r}_1$, $\mathbf{r}_2$ and $\mathbf{r}_3$ be the three unknown column vectors of $\mathbf{R}$. Then:

$$\begin{aligned}
s &= \pm 1/||\mathbf{a}_3||, \\
\mathbf{r}_3 &= s\,\mathbf{a}_3, \\
u_0 &= s^2(\mathbf{a}_1\cdot\mathbf{a}_3), \\
v_0 &= s^2(\mathbf{a}_2\cdot\mathbf{a}_3), \\
\cos(\theta) &= -\frac{(\mathbf{a}_1\times\mathbf{a}_3)\cdot(\mathbf{a}_2\times\mathbf{a}_3)}{||\mathbf{a}_1\times\mathbf{a}_3||\cdot||\mathbf{a}_2\times\mathbf{a}_3||}, \\
\alpha &= s^2||\mathbf{a}_1\times\mathbf{a}_3||\cdot\sin(\theta), \\
\beta &= s^2||\mathbf{a}_2\times\mathbf{a}_3||\cdot\sin(\theta), \\
\mathbf{r}_1 &= \frac{\mathbf{a}_2\times\mathbf{a}_3}{||\mathbf{a}_2\times\mathbf{a}_3||}, \\
\mathbf{r}_2 &= \mathbf{r}_3\times\mathbf{r}_1, \\
\mathbf{t} &= s\cdot\mathbf{K}^{-1}\mathbf{b}
\end{aligned}$$

Two $\mathbf{R}$ matrices can be computed depending on the sign of s. Typically, the sign of t_z is known in advance; it is positive if the origin of the world coordinate system is in front of the camera. An inspection of the computed $\mathbf{t}$ vector thus allows a unique selection of $\mathbf{R}$. Figure 2 shows the calibration result for a car seat.

Compared to alternative methods, such as a manual alignment of the template, this calibration method has the advantage that extrinsic and intrinsic camera parameters can be simultaneously computed. While manually adjusting the position and orientation of the template might be feasible, a manual calibration of the intrinsic parameters is not trivial. One key issue in the above approach, however, is whether designers can accurately portray the bounding box in the sketch, and if not, how sensitive the approach is to such inaccuracies. Our informal observations involving several users have indicated that most can draw bounding boxes accurately enough, especially for sketches exhibiting conventional vantage points. Nevertheless, even if the user's depiction of the bounding box is quite inaccurate, the least-squares nature of the calibration yields satisfactory results in most practical settings.

Once the template is aligned with the sketch using this method, the resulting intrinsic and extrinsic camera properties can be saved in a text file, and can later be recalled with a single button click. As the results are available for later use, the user has to perform calibration only once through out the design cycle. This is especially useful where multiple sketches depicting different vantage points are used for design. In such cases, the user may calibrate the template separately once for each sketch. During the design process, individual calibration data can be quickly retrieved, thus allowing a fluid switching between different sketches.

While the bounding box provides a suitable set of eight points that facilitates calibration, the approach can be extended to a more general setting using a more obvious set of calibration points. For instance, instead of using the bounding box corners of a car seat, one may elect to use the bottom four corners of the seat base, the intersection points of the base and the backrest, the intersection points of the backrest and the headrest etc. Likewise, in car body design, the centers of the four wheels, the corners of the roof, the intersection points between the hood and the windshield etc., can be used as suitable calibration points. As long as six or more such points can be identified (preferably dispersed around the entirety of the object to achieve the best overall fit), the above algorithm can be readily applied without any modification. For a given design object, however, the user must know which points are used by the algorithm so as to be able to mark the corresponding points on the sketch. Currently, all of our design objects use the bounding box corners as the calibration points.

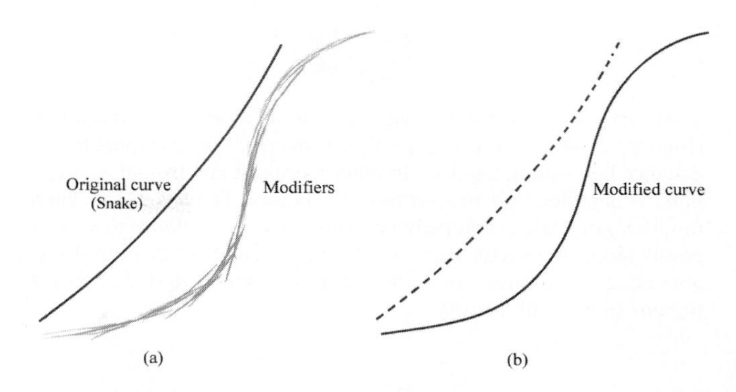

Figure 3: Modification of a 2D curve. (a) Original curve and the modifier strokes. (b) New shape of the target curve. The original curve is shown dashed for reference.

5 Pen-based 3D Shape Modification

After the template is aligned with the sketch, the user begins tracing the edges in the sketch as if the sketch was recreated on the computer. While sketching is a purely 2D operation, the key here is that input strokes are used to modify the template in 3D. The challenge is to compute a modification that results in the best match with the sketch, while generating the most appropriate 3D form. To facilitate analysis, the approach is designed such that edges are modified one at a time, with freedom to return to an earlier edge. At any point, the edge that the user is modifying is determined automatically as explained below, thus allowing the user modify edges in arbitrary order. After each set of strokes, the user presses a button that processes accumulated strokes, and modifies the appropriate template edge. For the purposes of discussion, we shall call users' input strokes as *modifiers*, and the template edge modified by those modifiers as the *target edge* or *target curve*.

Modification of the template is performed in three steps. In the first step, edges of the template are projected to the image plane resulting in a set of 2D curves. It is assumed that the user intends to modify the edge whose projection lies closest to the input strokes. Hence, in this step, the projected curve that lies nearest to the modifiers is taken to be the target curve. The spatial proximity between a projected curve and the modifiers is computed by sampling a set of points from the curve and the modifiers, and calculating the aggregate minimum distance between the two point sets. In the second step, the target curve is deformed in the image plane using an energy minimization algorithm until it matches well with the modifiers. In the third step, the modified target curve is projected back into 3D space. The following sections detail curve modification in the image plane and projection back into 3D space.

5.1 Curve modification in 2D image plane

Given the set of modifiers and the target curve, this step deforms the target curve in the image plane until it closely approximates the modifiers. The solution is facilitated by the use of energy minimizing splines based on active contour models [Kaas et al. 1988]. Active contours (also known as *snakes*) have long been used in image processing applications such as segmentation, tracking, and registration. The principal idea is that a snake moves and conforms to certain features in an image, such as intensity gradient, while minimizing its internal energy due to bending and stretching. This approach allows an object to be extracted or tracked in the form of a continuous spline.

Our method adopts the above idea for curve manipulation. Here, the target curve is modeled as a snake, whose nodes are sampled directly from the target curve. The nodes of the snakes are connected to one another with line segments making the snake geometrically equivalent to a polyline. The set of modifier strokes, on the other hand, is modeled as an unordered set of points (point cloud) extracted from the input strokes. This allows for an arbitrary number of modifiers, drawn in arbitrary directions and order, thus accommodating casual drawing styles. With this formulation, the snake converges to the shape of the modifiers, but locally resists excessive bending and stretching to maintain smoothness (Figure 3). Mathematically, this can be expressed as an energy functional to be minimized:

$$E_{snake} = \sum_i E_{int}(\mathbf{v}_i) + E_{ext}(\mathbf{v}_i)$$

where $\mathbf{v}_i = (x_i, y_i)$ is the i'th node coordinate of the snake. E_{int} is the internal energy arising from the stretching and bending of the snake. It involves first and second order derivatives of $\mathbf{v}_i$ with respect to arc length. Since minimizing E_{int} in its original form is computationally intensive, we resort to a simpler approximate solution of applying a restitutive force $\mathbf{F}_{rest}$ which simply moves each snake node toward the barycenter of its neighboring two nodes (Figure 4). However, to prevent the snake from shrinking indefinitely, its two ends are pinned to the two extremal points of the modifiers prior to modification.

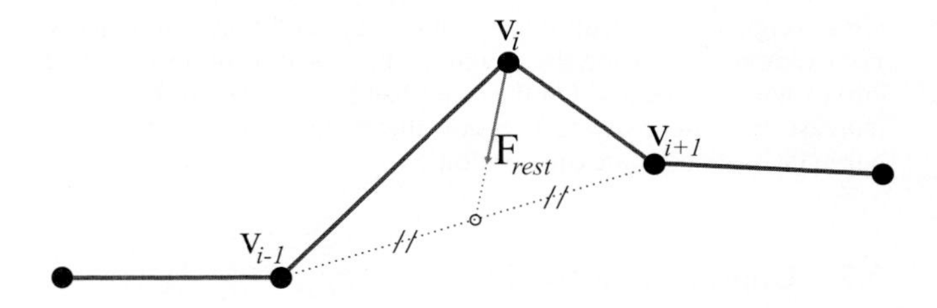

Figure 4: Internal energy due to stretching and bending is minimized approximately by moving each snake node to the barycenter of its neighbors similar to Laplacian smoothing.

External energy E_{ext} describes the potential energy of the snake due to external attractors, which arise in the presence of modifiers. The modifiers' influence on the snake consists of two components: (1) location forces, (2) pressure forces. The first component moves the snake toward the data points sampled from the modifiers. For each snake node $\mathbf{v}_i$, a force $\mathbf{F}_{loc}(\mathbf{v}_i)$ is computed corresponding to the influence of the location forces on $\mathbf{v}_i$:

$$\mathbf{F}_{loc}(\mathbf{v}_i) = \sum_{n \in k\_neigh} \frac{\mathbf{m}_n - \mathbf{v}_i}{||\mathbf{m}_n - \mathbf{v}_i||} \cdot w(n)$$

where $\mathbf{m}_n$ is one of the k closest neighbors of $\mathbf{v}_i$ in the modifiers (Figure 5). $w(n)$ is a weighting factor inversely proportional to the distance between $\mathbf{m}_n$ and $\mathbf{v}_i$. In other words, at any instant, a snake node $\mathbf{v}_i$ is pulled by k nearest modifier points. The force from each modifier point $\mathbf{m}_n$ is inversely proportional to its distance to $\mathbf{v}_i$, and points along the vector $\mathbf{m}_n - \mathbf{v}_i$. Using k neighbors has the desirable effect of suppressing outliers, thus directing the snake toward regions of high ink density.

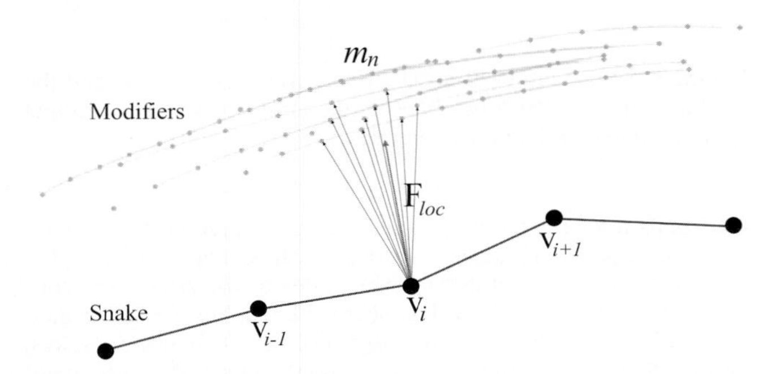

Figure 5: Location force on a node.

The second component of E_{ext} is related to pressure with which strokes are drawn. The force created due to this energy pulls the snake toward sections of high pressure. The rationale behind considering the pressure effect is based on the observation that users typically press the pen harder to emphasize critical sections while sketching. The pressure term exploits this phenomenon by forcing the snake to favor sections drawn more emphatically. For each snake node $\mathbf{v}_i$, a force $\mathbf{F}_{pres}(\mathbf{v}_i)$ is computed as:

$$\mathbf{F}_{pres}(\mathbf{v}_i) = \sum_{n \in k\_neigh} \frac{\mathbf{m}_n - \mathbf{v}_i}{||\mathbf{m}_n - \mathbf{v}_i||} \cdot p(n)$$

where $p(n)$ is a weight factor proportional to the pen pressure recorded at point $\mathbf{m}_n$.

During modification, the snake moves under the influence of the two external forces while minimizing its internal energy through the restitutive force. In each iteration, the new position of $\mathbf{v}_i$ is determined by the vector sum of $\mathbf{F}_{rest}$, $\mathbf{F}_{loc}$ and $\mathbf{F}_{pres}$, whose relative weights can be adjusted to emphasize different components. For example, increasing the weight of $\mathbf{F}_{rest}$ will result in smoother curves with less bends. On the other hand, emphasizing $\mathbf{F}_{pres}$ will increase the sensitivity to pressure differences with the resulting curve favoring high pressure regions.

5.2 Unprojection to 3D

In this step, the newly designed 2D curve is projected back into 3D space. Theoretically, there is no unique solution because there are infinitely many 3D curves whose projections match the 2D curve. Therefore the best 3D configuration must be chosen based on certain constraints. The nature of the problem, however, provides some insights into these constraints. Trivially, the 3D curve should appear right under the user's strokes. Additionally, if those strokes occur precisely over the original target curve (i.e, the strokes do not alter the curve's 2D projection), the target curve should preserve its original 3D shape. Finally, if the curve is to change shape, it must maintain a reasonable 3D form. By "reasonable," we mean a solution that the designer would accept in many cases, while anticipating it in the worst case.

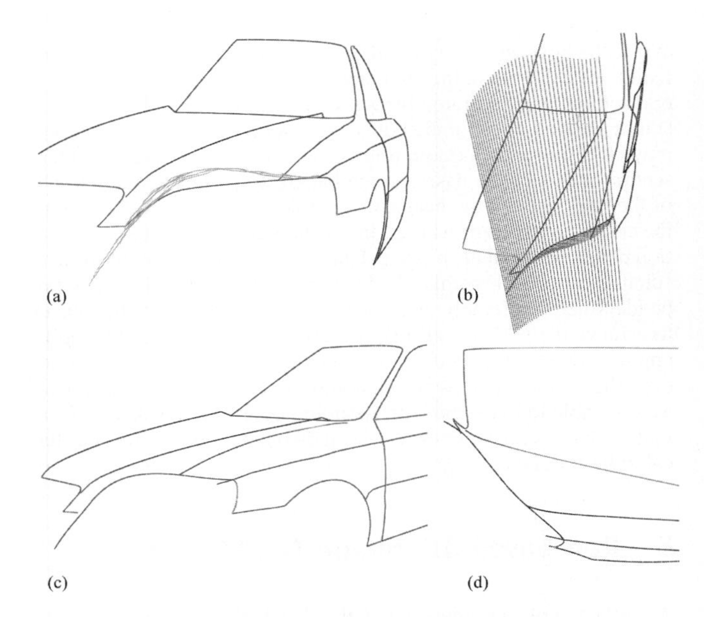

Figure 6: Modification of a car hood near the headlight. (a) User's strokes. (b) Surface S created by the rays emanating from the user's eyes and passing through the strokes, and the minimum-distance lines from the original curve. (c) Resulting shape. (d) Resulting shape from the top view.

Based on these premises, we choose the optimal configuration as the one that minimizes the spatial deviation from the original target curve. That is, among the 3D curves whose projections match the newly designed 2D curve, we choose the one that lies nearest to the original target curve. This can be formulated as follows:

Let $\mathbf{C}$ be a curve in $\mathbb{R}^3$ constrained on a surface S.[2] Let $\mathbf{C}^{orig}$ be the original target curve in $\mathbb{R}^3$ that the user is modifying. The new 3D configuration $\mathbf{C}^*$ of the modified curve is computed as:

$$\mathbf{C}^* = \underset{\mathbf{C}}{argmin} \sum_i ||\mathbf{C}_i - \mathbf{C}_i^{orig}||$$

where $\mathbf{C}_i$ denotes the i'th vertex of $\mathbf{C}$. With this criterion, $\mathbf{C}^*$ is found by computing the minimum-distance projection points of $\mathbf{C}_i^{orig}$ onto S (Figure 6b).

The rationale behind this choice is that, by remaining proximate to the original curve, the new curve can be thought to be "least surprising" when viewed from a different viewpoint. Put differently, under normal usage, it is difficult to generate a curve unexpectedly different from the original curve. The only exception is when the user attempts to modify a curve from an odd viewpoint (e.g., trying to draw the waistline of a car from the front view). However, such situations are not frequently encountered as users naturally tend to choose suitable viewpoints. Nevertheless, if the new curve is not satisfactory, users can redraw the curve from other viewpoints. Because the curve deviates minimally between modifications, each step will introduce its own changes while preserving the changes made in the previous steps. This allows the desired shape to be obtained in a relatively few steps.

[2] S is the surface subtended by the rays emanating from the user's viewpoint and passing through the newly designed 2D curve. This surface extends into 3D space and is not visible from the original viewpoint.

The above processes may cause originally connected curves of the template to be disconnected. In these cases, the user may invoke a "trim" command that merges curve ends that lie sufficiently close to one another. However, instead of simply extending or shortening the curves at their ends, each curve is translated, rotated, and scaled in its entirety until its ends meet with other curves. This eliminates kinks that could otherwise occur near the ends. Additionally, by manipulating the curve as a whole, it preserves the shape established by the user without introducing unwarranted artifacts. Note that since edge ends to be trimmed are usually sufficiently close to one another, scaling effects are hardly noticeable. At the end, a well-connected wireframe is obtained that can be subsequently surfaced. Details of the surfacing process are described in the next section.

After surfacing, new feature curves can be added to the model by directly sketching on existing surfaces. In these cases, the 3D configurations of the input strokes are trivially computed using the depth buffer of the graphics engine. Once created, the new curves can be manipulated the same way original template curves are manipulated. Additionally, they can be used in the construction of new surfaces.

6 Surface Creation and Modification

In the last step, the newly designed wireframe model is surfaced to obtain a solid model. Once the initial surfaces are obtained, the user can modify them using simple deformation tools. The following sections detail these processes.

6.1 Initial surface creation

Given the wireframe model, this step creates a surface geometry for each of the face loops in the wireframe. In this work, it is assumed that the wireframe topology is already available with the template model and therefore all face loops are known apriori[3]. Each face loop may consist of an arbitrary number of edge curves. For each face loop, the surface geometry is constructed using the method proposed in [Inoue 2003]. In this method, each curve of the wireframe is represented as a polyline, and the resulting surfaces are polygonal surfaces consisting of purely triangular elements.

Figure 7 illustrates the creation of a surface geometry on a boundary loop. In the first step, a vertex is created at the centroid of the boundary vertices. Initial triangles are then created that use the new vertex as the apex, and have their bases at the boundary. Next, for each pair of adjacent triangular elements, edge swapping is performed. For two adjacent triangles, this operation seeks to improve the mesh quality by swapping their common edge (Figure 8a). The mesh quality is based on the constituent triangles' quality. For a triangle, it is defined as the radius ratio, which is the radius of the inscribed circle divided by the radius of the circumscribed circle. Next, adjacent triangles are subdivided iteratively, until the longest edge length in the mesh is less than a threshold (Figure 8b). Between each iteration, edge swapping and Laplacian smoothing is performed to maintain a regular vertex distribution with high quality elements. At the end, the resulting surface is refined using a physically-based mesh deformation method, called the V-spring operator. This method, which will be presented in detail in Section 6.2.2, iteratively adjusts the initial mesh so that the total variation of curvature is minimized. Once the initial surfaces are created in this way, new feature curves can be added to the model by direct sketching, as described in the previous section. Figure 9 shows the final surface generated for the car seat. Note that new feature curves are added, which were not part of the original template model .

[3]If the topology is unknown, it has to be computed automatically, or it must be manually specified by the user. Currently, we are working toward automatically computing the wireframe topology.

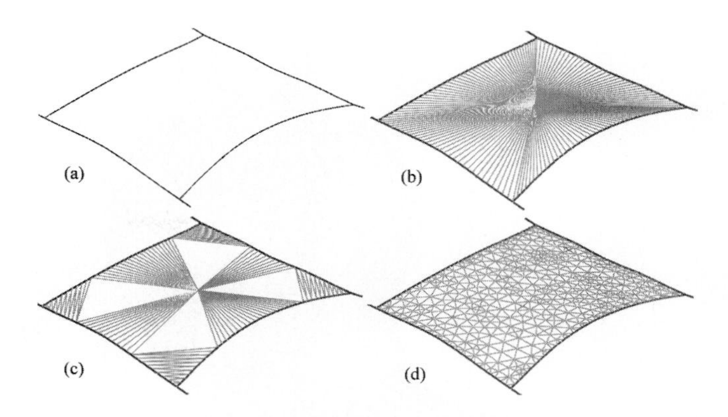

Figure 7: Surface creation. (a) Initial boundary loop consisting of four curves. (b) Preliminary triangulation using a vertex created at the centroid. (c) Edge swapping (d) Final result after face splitting and mesh smoothing using V-spring method.

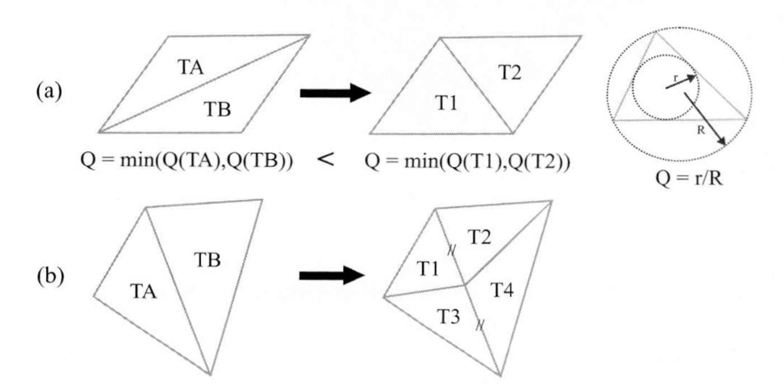

Figure 8: (a) Edge swapping. Diagonals of adjacent triangles are swapped if minimum element quality increases. (b) Triangle subdivision.

6.2 Surface modification

Often times, the designer will need to modify the initial surfaces to give the model a more aesthetic look. In this work, we adopt a simple and intuitive modification scheme that allows users to explore different surface alternatives in a controllable and predictable way. Unlike most existing techniques, our approach operates directly on the polygonal surface without requiring the user to define a control grid or a lattice structure.

Our approach consists of two deformation methods. The first method uses pressure to deform a surface. With this tool, resulting surfaces look rounder and inflated, with more volume. The second method is based on the V-spring approach described by [Yamada et al. 1999]. In this method, a network of mechanical springs work together to minimize the variation of surface curvature. A discussion of the practical utility of this type of surface can be found in [Hou 2002]. In both methods, deformation is applied to the interior of the surface while keeping the boundaries fixed. This way, the underlying wireframe geometry is preserved, with no alterations to the designed curves.

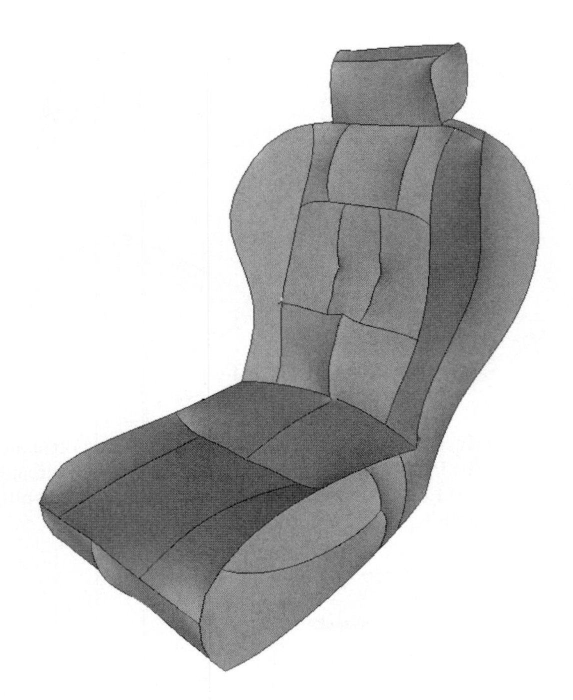

Figure 9: The initial surfaced model.

6.2.1 Surface modification using pressure force

This deformation tool simulates the effect of a pressure force on a thin membrane. The tool allows surfaces to be inflated or flattened in a predictable way. The extent of the deformation depends on the magnitude of the pressure, which is controlled by the user through a slider bar. Different pressure values can be specified for individual surfaces, thus giving the user a better control on the final shape of the solid model.

The equilibrium position of a pressurized surface is found iteratively. In each step, each vertex of the surface is moved by a small amount proportional to the pressure force applied to that vertex. The neighboring vertices, however, resist this displacement by pulling the vertex toward their barycenter akin to Laplacian smoothing. The equilibrium position is reached when the positive displacement for each node is balanced by the restitutive displacement caused by the neighboring vertices. Figure 10 illustrates the idea.

The algorithm can be outlined as follows. Let p be the pressure applied to the surface. Until convergence do:

for each vertex $\mathbf{v}_i$

- Compute the unit normal $\mathbf{n}_i$ at vertex $\mathbf{v}_i$
- Compute the pressure force on $\mathbf{v}_i$

  $$F_i = p \cdot A_i^{voronoi}$$

- $\Delta \mathbf{v}_i^{pres} = F_i \cdot \mathbf{n}_i$
- $\Delta \mathbf{v}_i^{laplc} = (\frac{1}{K} \sum_{j=1}^{K} \mathbf{v}_{ij}) - \mathbf{v}_i$, where $\mathbf{v}_{ij}$ is one of the K adjacent

vertices of $\mathbf{v}_i$

- $\mathbf{v}_i \leftarrow \mathbf{v}_i + ((1-\xi)\Delta \mathbf{v}_i^{pres} + \gamma \Delta \mathbf{v}_i^{laplc})$

end for

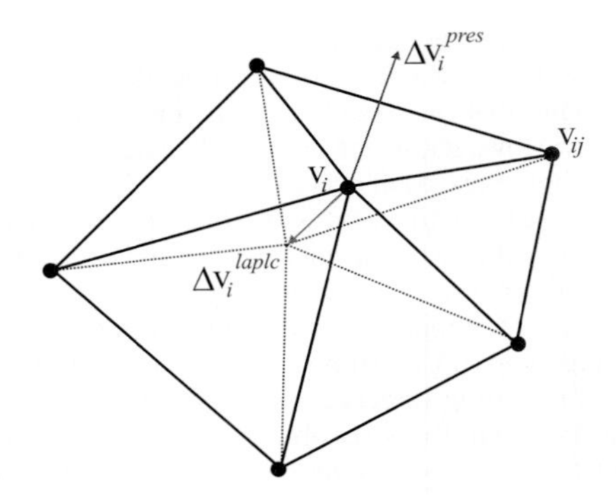

Figure 10: A pressure force applied to a vertex moves the vertex along its normal direction. The neighboring vertices, however, pull the vertex back toward their barycenter. Equilibrium is reached when displacements due to the pressure force and the neighbors are balanced.

The vertex normal $\mathbf{n}_i$ is updated in each iteration and is computed as the average of the normals of the faces incident on $\mathbf{v}_i$, weighted by the area of each face. $A_i^{voronoi}$ is the Voronoi area surrounding $\mathbf{v}_i$. It is obtained by connecting the circumcenters of the faces incident on $\mathbf{v}_i$ with the midpoints of the edges incident on $\mathbf{v}_i$ (Figure 11). ξ and γ are damping coefficients that control the convergence rate. Too low values of ξ or γ may cause instability in convergence.

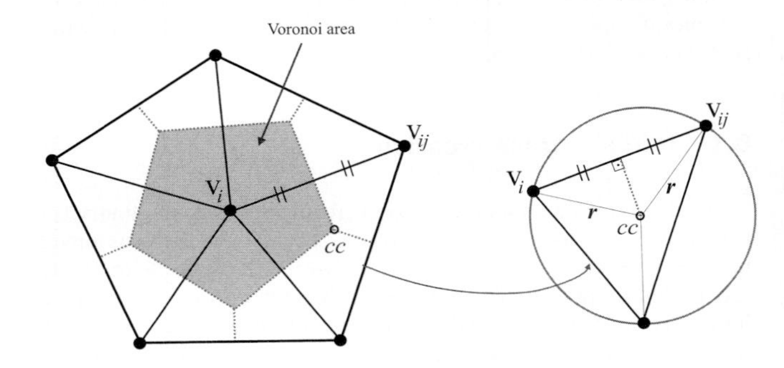

Figure 11: Voronoi area surrounding vertex $\mathbf{v}_i$.

The above algorithm is applied to all surface vertices while keeping the boundary vertices fixed. Figure 12 shows an example on the seat model. If necessary, negative pressure can be applied to form concavities.

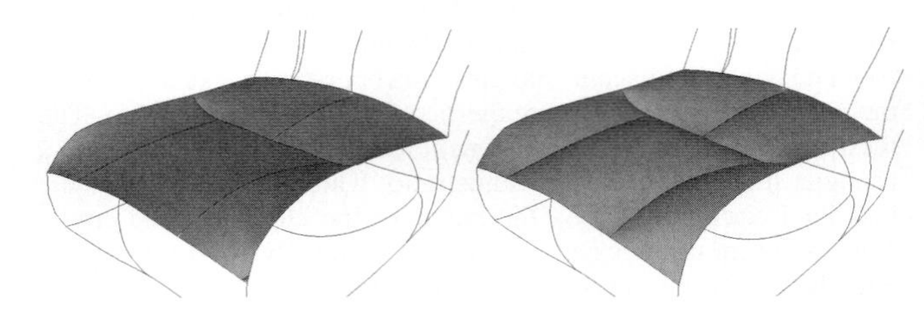

Figure 12: Modification of a seat base using pressure force.

6.2.2 Surface modification using V-spring method

This method creates surfaces of minimized curvature variation based on a discrete spring model. This scheme produces fair surfaces that vary smoothly, which is known to be an important criterion for aesthetic design purposes [Nickolas S. Sapidis 1994]. Additionally, when applied to a group of adjacent surfaces, it reduces sharp edges by smoothing the transition across the boundary curves.

In this method, a spring is attached to each surface vertex. Neighboring springs usually form a "V" shape, thus giving the name to the method. The spring length approximately represents the local curvature. During modification, the springs work together to keep their lengths equal, which is equal to minimizing the variation of curvature (Figure 13). Each vertex thus moves under the influence of its neighbors until the vertices locally lie on a sphere.

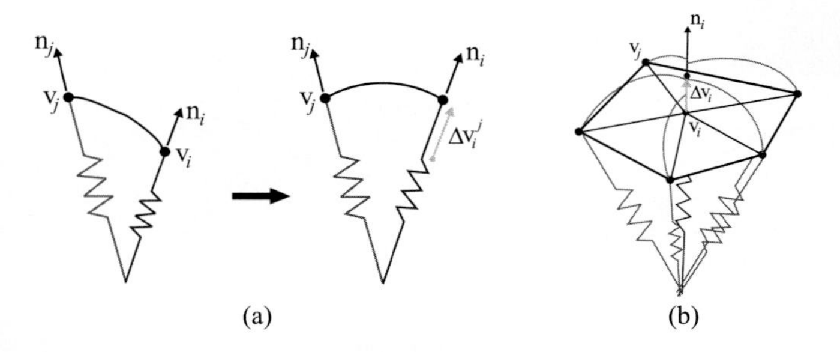

Figure 13: V-spring. (a) Displacement of $\mathbf{v}_i$ due to $\mathbf{v}_j$. (b) Net displacement due to all neighbors.

Based on this model, the displacement of $\mathbf{v}_i$ due to a neighboring vertex $\mathbf{v}_j$ is given as follows (see [Yamada et al. 1999] for details):

$$\Delta\mathbf{v}_i^j = \frac{1}{||\mathbf{v}_j - \mathbf{v}_i||}\left[\frac{(\mathbf{v}_j - \mathbf{v}_i)\cdot(\mathbf{n}_i + \mathbf{n}_j)}{1 + (\mathbf{n}_i \cdot \mathbf{n}_j)}\right]\mathbf{n}_i$$

where $\mathbf{n}_i$ and $\mathbf{n}_j$ are unit normal vectors at $\mathbf{v}_i$ and $\mathbf{v}_j$. The total displacement of $\mathbf{v}_i$ is computed as the average of displacements due to neighboring vertices. However, to maintain a regular vertex distribution throughout iterations, each vertex is also moved horizontally along its current tangent plane toward the barycenter of its neighbors. In each iteration, the positions and normals of the vertices are updated. The iterations are continued until the net displacement of each vertex is less than a threshold. Figure 9 shows the output of this method on the seat model. The initial surfaces are by default created using this scheme. Since surfaces are treated independently, however, transitions across boundary edges are not smoothed.

7 Example and Discussions

Figure 14 illustrates the proposed approach applied to car body design. Given the input sketch and the template, it took about 70 minutes to obtain the surfaced model at the bottom. About 50% of the time was spent during template modification, 20% for surface generation and refinement, and 30% for adding and modifying new design edges.

Our informal tests have shown that it takes a relatively proficient user about three to four hours in Discreet®3ds max®to create a comparable model. While these findings are not conclusive, we believe they are useful to the demonstrate the utility of our approach.

Currently, the snake-based curve modification algorithm assumes that a target curve is modified in its entirety. That is, local modifications to a curve are not permitted. Likewise, the current approach does not allow two or more curves connected in series to be modified by a single set of modifiers. Hence, for a successful modification, the user needs to know the start and end points of each template edge. Currently, this information is conveyed visually by rendering the ends of the template edges in a slightly different color. This helps the user easily distinguish different edges of the template. However, when the user is working through the input sketch, the template becomes invisible. In such cases, the user may temporarily need to hide the sketch to identify the edge ends. While this causes some inconvenience, we have not found it to be too constraining. Nevertheless, we plan to alleviate this difficulty by extending our snake-based modification algorithm to enable local modifications to a single curve, and global modifications to two or more curves.

Another observed difficulty involved the selection of a target curve. When the template contains a large number of edges, the target curve selection scheme becomes fragile. This is because during the curve modification in the image plane, too many candidates occur near the set of modifiers, which makes it difficult to identify the intended curve based on spatial distance. To alleviate this difficulty, we added an option of "focused design" to the interface, which allows the user to mark a specific curve and work exclusively on that curve as long as desired. When the program is in this mode, all modifiers affect this selected curve, regardless of how far the modifiers occur from the curve in the image plane. We have found this feature to greatly facilitate the design process.

8 Conclusions

In this work, we presented a new technique for 3D styling design that uses a pen-based computer interface as the main interaction medium. The main novelty of the proposed method is that it allows designers to utilize their concept sketches in conjunction with existing computer models to facilitate rapid and fluid development of 3D geometry. The approach is particularly useful for styling design, where the new product is a stylistic variation of an existing one, or some other canonical shape.

At the heart of our approach is a shape modification algorithm that uses 2D input strokes to modify a 3D template. In a typical operation, the template and the digital sketch are first aligned using a camera calibration algorithm. Next, to create the 3D form, the user simply traces the feature edges in sketch. The input strokes are interpreted to appropriately modify the corresponding edge of the template in 3D. This work has shown that within the scope of the problem, the best 3D interpretation of a 2D curve is the one that minimizes the spatial deviation from the curve's original 3D configuration. After the template edges are modified using this principle, the newly designed template is surfaced to produce a solid model. Finally, the user can refine the initial surfaces using two physically-based deformation methods, and add new feature edges as desired.

Our experience so far, and the feedback from external users have indicated that the proposed system is a viable alternative to existing style design tools. In the near future, we plan to conduct field studies to further assess its performance.

9 Acknowledgements

We would like to thank our colleagues Tomotake Fruhata, Dr. Soji Yamakawa, and Miguel Vieira for useful discussions.

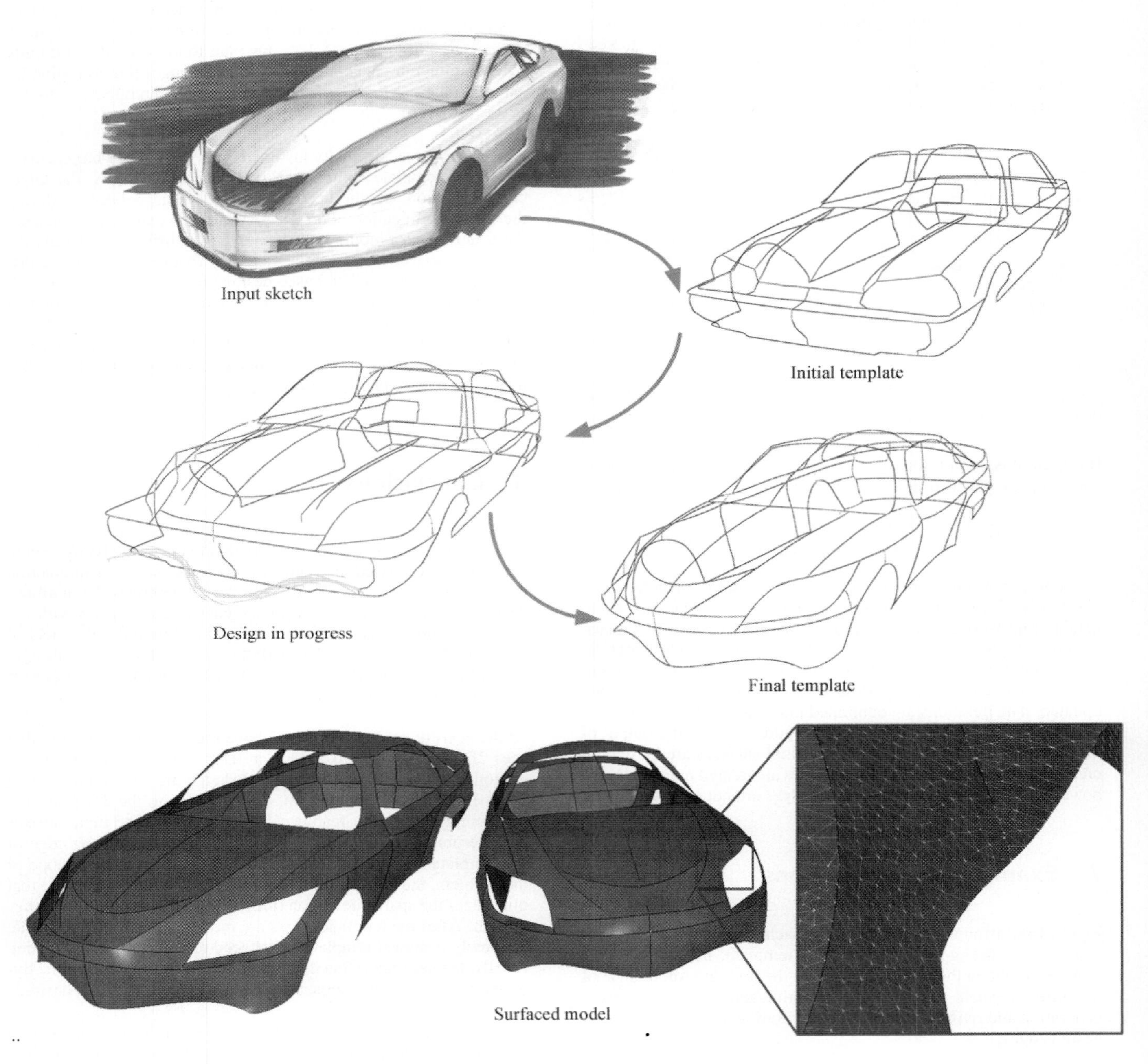

Figure 14: Design of a car using our system. Top left: Input sketch. Top right: Initial template model. Middle left: Design in progress. The sketch is hidden to reveal the template. Middle right: Resulting wireframe model. Bottom: Surfaced model. Surfaces are refined using pressure force and V-spring. Also shown is a close up of the triangular mesh near the headlight. Notice the strong feature edge near the front grill, which is also apparent in the original sketch.

References

ABDEL-AZIZ, Y., AND KARARA, H. 1971. Direct linear transformation from comparator coordinates into object space coordinates in close-range photogrammetry. In *Symposium on Close-Range Photogrammetry, American Society of Photogrammetry*, 1–18.

ALVARADO, C., AND RANDALL, D. 2004. Sketchread: a multi-domain sketch recognition engine. In *User Interface Software Technology (UIST)*.

BIMBER, O., ENCARNAO, L. M., AND STORK, A. 2000. A multi-layered architecture for sketch-based interaction within virtual environments. *Computer Graphics 24*, 6.

BOURGUIGNON, D., CHAINE, R., CANI, M.-P., AND DRETTAKIS, G. 2004. Relief: A modeling by drawing tool. In *EUROGRAPHICS Workshop on Sketch-Based Interfaces and Modeling*.

CHERLIN, J. J., SAMAVATI, F., SOUSA, M. C., AND JORGE, J. A. 2005. Sketch-based modeling with few strokes. In *SCCG '05: Proceedings of the 21st spring conference on Computer graphics*, ACM Press, 137–145.

CHEUTET, V., CATALANO, C., PERNOT, J., FALCIDIENO, B., AND GIANNINI, F. 2004. 3d sketching with fully free form deformation features (d-f4) for aesthetic design. In *EUROGRAPHICS Workshop on Sketch-Based Interfaces and Modeling*.

COHEN, J. M., MARKOSIAN, L., ZELEZNIK, R. C., HUGHES, J. F., AND BARZEL, R. 1999. An interface for sketching 3d curves. In *SI3D '99: Proceedings of the 1999 symposium on Interactive 3D graphics*, ACM Press, 17–21.

DAS, K., DIAZ-GUTIERREZ, P., AND GOPI, M. 2005. Sketching free-form surfaces using network of curves. In *EUROGRAPHICS Workshop on Sketch-Based Interfaces and Modeling*.

DIEHL, H., MLLER, F., AND LINDEMANN, U. 2004. From raw 3d-sketches to exact cad product models concept for an assistant-system. In *EUROGRAPHICS Workshop on Sketch-Based Interfaces and Modeling*.

DRAPER, G., AND EGBERT, P. 2003. A gestural interface to free-form deformation. In *Graphics Interface 2003*, 113–120.

EGGLI, L., BRUDERLIN, B. D., AND ELBER, G. 1995. Sketching as a solid modeling tool. In *SMA '95: Proceedings of the third ACM symposium on Solid modeling and applications*, ACM Press, 313–322.

FLEISCH, T., BRUNETTI, G., SANTOS, P., AND STORK, A. 2004. Stroke-input methods for immersive styling environments. In *2004 International Conference on Shape Modeling and Applications*, 275–283.

FORSYTH, D. A., AND PONCE, J. 2003. *Computer Vision: a Modern Approach*. Prentice Hall.

GENNARI, L., KARA, L. B., AND STAHOVICH, T. F. 2004. Combining geometry and domain knowledge to interpret hand-drawn diagrams. In *AAAI Fall Symposium Series 2004: Making Pen-Based Interaction Intelligent and Natural*.

GRIMSTEAD, I. J., AND MARTIN, R. R. 1995. Creating solid models from single 2d sketches. In *SMA '95: Proceedings of the third ACM symposium on Solid modeling and applications*, ACM Press, 323–337.

GROSSMAN, T., BALAKRISHNAN, R., KURTENBACH, G., FITZMAURICE, G., KHAN, A., AND BUXTON, B. 2002. Creating principal 3d curves with digital tape drawing. In *CHI '02: Proceedings of the SIGCHI conference on Human factors in computing systems*, 121–128.

HAMMOND, T., AND DAVIS, R. 2004. Automatically transforming symbolic shape descriptions for use in sketch recognition. In *19th National Conference on Artificial Intelligence (AAAI-2004)*.

HOU, K.-H. 2002. *A Computational Method for Mesh-based Free-form functional Surface Design*. Ph.d., Carnegie Mellon University.

HUA, J., AND QIN, H. 2003. Free-form deformations via sketching and manipulating scalar fields. In *SM '03: Proceedings of the eighth ACM symposium on Solid modeling and applications*, ACM Press, 328–333.

IGARASHI, T., MATSUOKA, S., AND TANAKA, H. 1999. Teddy: a sketching interface for 3d freeform design. In *SIGGRAPH '99: Proceedings of the 26th annual conference on Computer graphics and interactive techniques*, 409–416.

INOUE, K. 2003. *Reconstruction of Two-Manifold Geometry from Wireframe CAD Models*. Ph.d., Univerity of Tokyo.

KAAS, M., WITKINS, A., AND TERZOPOLUS, D. 1988. Snakes: active contour models. *International Journal of Computer Vision 1*, 4, 312–330.

KARA, L. B., AND STAHOVICH, T. F. 2004. Hierarchical parsing and recognition of hand-sketched diagrams. In *User Interface Software Technology (UIST)*.

KARPENKO, O., HUGHES, J. F., AND RASKAR, R. 2002. Free-form sketching with variational implicit surfaces. In *Eurographics*.

LAVIOLA, J., AND ZELEZNIK, R. 2004. Mathpad2: A system for the creation and exploration of mathematical sketches. In *ACM Transactions on Graphics (Proceedings of SIGGRAPH 2004)*, vol. 23, 432–440.

LLAMAS, I., POWELL, A., ROSSIGNAC, J., AND SHAW, C. 2005. Bender: A virtual ribbon for deforming 3d shapes in biomedical and styling applications. In *ACM Symposium on Solid and Physical Modeling 2005*.

MASRY, M., KANG, D. J., AND LIPSON, H. 2005. A freehand sketching interface for progressive construction of 3d objects. *Computers and Graphics 29*, 4, 563–575.

MITANI, J., SUZUKI, H., AND KIMURA, F. 2000. 3d sketch: Sketch-based model reconstruction and rendering. In *Workshop on Geometric Modeling 2000*, 85–98.

NEALEN, A., SORKINE, O., ALEXA, M., AND COHEN-OR, D. 2005. A sketch-based interface for detail-preserving mesh editing. *ACM Transactions on Graphics 24*, 3, 1142–1147.

NICKOLAS S. SAPIDIS, E. 1994. *Designing Fair Curves and Surfaces: Shape Quality in Geometric Modeling and Computer-Aided Design*.

P.A.C.VARLEY, Y.TAKAHASHI, J.MITANI, AND H.SUZUKI. 2004. A two-stage approach for interpreting line drawings of curved objects. In *EUROGRAPHICS Workshop on Sketch-Based Interfaces and Modeling*.

SCHMIDT, R., WYVILL, B., SOUSA, M. C., AND JORGE, J. A. 2005. Shapeshop: Sketch-based solid modeling with blobtrees.

In *EUROGRAPHICS Workshop on Sketch-Based Interfaces and Modeling*.

SHILMAN, M., AND VIOLA, P. 2004. Spatial recognition and grouping of text and graphics. In *EUROGRAPHICS Workshop on Sketch-Based Interfaces and Modeling*.

TSANG, S., BALAKRISHNAN, R., SINGH, K., AND RANJAN, A. 2004. A suggestive interface for image guided 3d sketching. In *CHI '04: Proceedings of the SIGCHI conference on Human factors in computing systems*, 591–598.

TURNER, A., CHAPMAN, D., AND PENN, A. 1999. Sketching a virtual environment: modeling using line-drawing interpretation. In *VRST '99: Proceedings of the ACM symposium on Virtual reality software and technology*, ACM Press, 155–161.

VARLEY, P. A. C. 2003. *Automatic Creation of Boundary-Representation Models from Single Line Drawings*. Ph.d. thesis, Cardiff University.

VARLEY, P. A. C. 2004. Using depth reasoning to label line drawings of engineering objects. In *9th ACM Symposium on Solid Modeling and Applications SM'04*, 191–202.

WESCHE, G., AND SEIDEL, H.-P. 2001. Freedrawer: a free-form sketching system on the responsive workbench. In *VRST '01: Proceedings of the ACM symposium on Virtual reality software and technology*, 167–174.

YAMADA, A., FURUHATA, T., SHIMADA, K., AND HOU, K. 1999. A discrete spring model for generating fair curves and surfaces. In *PG '99: Proceedings of the 7th Pacific Conference on Computer Graphics and Applications*.

YANG, C., SHARON, D., AND PANNE, M. V. D. 2005. Sketch-based modeling of parameterized objects. In *EUROGRAPHICS Workshop on Sketch-Based Interfaces and Modeling*.

ZELEZNIK, R. C., HERNDON, K. P., AND HUGHES, J. F. 1996. Sketch: an interface for sketching 3d scenes. In *SIGGRAPH '96: Proceedings of the 23rd annual conference on Computer graphics and interactive techniques*, 163–170.

Constructing Smooth Branching Surfaces from Cross Sections

N.C. Gabrielides *
NTUA

A.I. Ginnis†
SDL-NTUA

P.D. Kaklis‡
SDL-NTUA

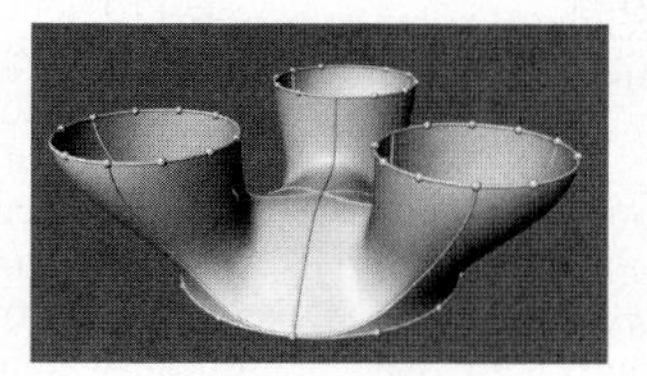

Abstract

This paper proposes a framework for constructing G^1 surfaces that interpolate data points on parallel cross sections, consisting of simple disjoined and non-nested contours, the number of which may vary from plane to plane. Using appropriately estimated cross tangent vectors at the given points, we split the problem into a sequence of local Hermite problems, each of which can be one of the following three types: *"one-to-one"*, *"one-to-many"* or *"many-to-many"*.

The solution of the *"one-to-many"* branching problem, where one contour on the i-plane is to be connected to $\mathscr{M}$-contours on the (i+1)-plane, is based on combining skinning with trimming and hole filling. More specifically, we firstly construct a G^1 *surrounding curve* of all $\mathscr{M}$-contours on the (i+1)-plane, consisting of contour portions connected with linear segments, the so-called *bridges*. Next, we build a surface that skins the i-plane contour with the (i+1)-plane surrounding curve and trim suitably along the bridges. The resulting multi-sided hole is covered with quadrilateral Gordon-Coons patches that possess G^1 continuity. For this purpose, we develop a hole-filling technique that employs shape-preserving *guide curves* and is able to preserve data symmetries. The *"many-to-many"* problem is handled by combining the *"one-to-many"* methodology with a zone-separation technique, that achieves to split the configuration into two *"one-to-many"* problems. The methodology, implemented as a C++ Rhino v3.0 plug-in, is illustrated via a synthetic example.

CR Categories: I.3.5 [Computational Geometry and Object Modeling]: Curve, surface, solid and object representations J.6.0 [Computer-aided design (CAD)] G.1.1 [Interpolation]: Spline and piecewise polynomial interpolation

Keywords: design, reconstruction, cross sections, branching surfaces, G^1 surfaces, skinning, trimming, hole filling, shape preserving interpolation

*NTUA: Nat. Techn. Univ. Athens e-mail: ngabriel@deslab.ntua.gr
†SDL: Ship Design Laboratory e-mail:ginnis@naval.ntua.gr
‡e-mail:kaklis@deslab.ntua.gr

SPM 2006, Cardiff, Wales, United Kingdom, 06–08 June 2006.
© 2006 ACM 1-59593-358-1/06/0006 $5.00

1 Introduction

The request for designing or reconstructing objects from planar cross sections arises in various applications, ranging from CAD [Bentamy et al. '05], to GIS [Fujimura & Kuo '99], [Sirakov & Muge '01], and medical imaging [Schmitt et al. '04], [Moschino et al. '06].

In the *reconstruction* case the density of input data is likely to be very high, in terms of the densities of captured cross sections and points per contour, thus orienting research toward developing fast algorithms for constructing C^0 planar triangular interpolants of the cross-sectional data. On the contrary, when we *design* a surface from parallel cross sections, the available data information is limited - only a small number of contours is usually available - which imposes the need for data interpolants that possess adequate smoothness (at least G^1) and whose analytic type enables their easy and robust transfer to the surface modeler used by the designer.

The method presented herein fits to the *design* context, though it is also readily applicable for *smooth reconstruction* purposes. Furthermore, we focus on handling the general case of planar cross sections that consist of disjoint smooth Jordan curves, henceforth referred to as contours, whose number can vary from plane to plane. We are thus able to handle, besides the standard *"one-to-one"* case, the *"one-to-many"* and *"many-to-many"* configurations, guaranteeing a G^1 interpolation surface, composed from polynomial patches of principally quadrilateral topology, that branches suitably at/through an intermediate, designer-specified, level/zone, respectively.

There is a plethora of papers dealing with the *"one-to-one"* problem in the design context, employing mainly B-Spline surfaces and NURBS, e.g., [Park & Kim '96], [Jaillet '97], [Johnstone & Sloan '98], , [Hohmeyer & Barsky '91], [Piegl & Tiller '96]. Part of this literature is concerned with controlling the shape of the outcome surface by, e.g., minimizing its twist [Goodman & Greig '98], or preserving its sectional curvature between shape similar contours, [Jüttler '97], [Kaklis & Ginnis '96].

On the other hand, lots of references deal with developing algorithms for solving the *local "one-to-one"* or *tilling,* according to [Meyers et al. '92], problem for reconstruction purposes. Such algorithms yield usually C^0 triangular surfaces with the aid of global optimality criteria, such as bounding the maximum volume [Keppel '75], or minimizing the area of the surface [Fuchs et al '77], [Sloan & Painter '88], [Meyers et al. '92]. In these works the sought for triangulation is thought as path in a graph. Computing this path can also be accomplished locally, node-by-node, by imposing local criteria on the path nodes, such as minimizing the edge length of each triangle that is being generated after adding a new node to the path [Christiansen & Sederberg '78], [Batnitzki et al. '81] [Herbert et al.

'95], [Sederberg et al. '97]. Note that in contrary to their complexity, these local approaches introduce a non-local step, namely the definition of the first node. There also exist hybrid schemes, which use local weights for satisfying global criteria [Ekoule et al. '91], [Welzl & Wolfers '93]. Finally, the graph model of [Keppel '75], that works for polygonal contours, is extended in [Shinagawa & Kunii '91] for continuous parametric contour representations, enabling its application to quadrilateral surfaces; see also [Cohen et al. '97],[Goodman & Greig '98].

The literature devoted to the *"one-to-many"* problem can be classified into four main families. The family of *contour-connection* methods attempt to artificially render a *"one-to-many"* problem into *"one-to-one"* by connecting the disjoint contours with line [Christiansen & Sederberg '78] or triangular facet bridges [Meyers et al. '92]. The first choice succeeds only for very simple configurations while the second one constrains unnaturally the saddle points of the branching surface to lay on the plane containing the disjoint contours.

The so-called *intermediate-contour* methods are based on introducing an intermediate contour, thus splitting the original problem into two problems, an *"one-to-one"* and a new *"one-to-many"*. The second problem is further simplified into $\mathscr{M}$ *"one-to-one"*by suitably partitioning the intermediate contour into $\mathscr{M}$ parts. This technique was introduced by [Shinagawa & Kunii '91] and was first accomplished by [Ekoule et al. '91] and [Jeong et al. '99].

The family of *partial contour connection and hole filling* methods, outlined in [Barequet & Sharir '96], [Bajaj et al.'96] and [Barequet et al. '00], is characterized by matching partially the disjoint contours with the single contour of the neighboring plane, thus leaving a number of 3D holes to be filled in the final step of such a scheme. In [Goodman et al. '93], where the case *"one-to-two"*is being treated, the single hole is filled by a appropriate hyperboloid.

Finally, the family of *implicit* schemes relies on the assumption that we possess implicit representations of the contours composing the cross sections. Then an implicit interpolant can be obtained by taking convex combination of the contour representations [Bedi '92], or employing the notion of distance function [Jones & Chen '94], [Floater & Westgaard '95].

The herein presented method for handling the *"one-to-many"* problem belongs to the class of partial contour connections and hole filling schemes. Its apparent novelties pertain to:

1. Ensuring G^1 parametric continuity,
2. Providing as final outcome a spline polynomial surface, composed from patches of mainly quadrilateral topology, thus enabling portability to contemporary CAD systems,
3. Intensive use of shape-preserving interpolation technique for all 2D/3D curves encountered within the surface construction scheme, in order to control the shape quality of the final outcome,
4. Introducing the concepts of *surrounding curve* and surrounding surface (§4.2),
5. Developing a hole-filling technique that, though grounded on [Hahn '89], is able to preserve data symmetries (§4.3), and
6. Constructing the guide curves as shape-preserving polynomial splines, that can intrinsically incorporate the geometry of the hole boundary (§4.3).

Last but not least, and to the best of authors knowledge, pertinent literature is lacking of works handling the *"many-to-many"*problem in G^1 setting, which is achieved by the technique presented in §5 of this paper.

The rest of the document is structured as follows. Starting with a formulation of the (global) problem (see Problem 2.1), §2 ends with its decomposition to a sequence of local Hermite problems; see Problem 2.2. In §3 we briefly present the technique we adopt for constructing the contour curves and the tangent ribbons along them. Section 4 deals with the *"one-to-many"*configuration, being the kernel of the paper and consisting of three subsections. In §4.1 we introduce the notions and describe the construction of *bridges* and *surrounding curve/surface*. §4.2 outlines the trimming process on the surrounding surface, that leaves us with a hole-filling request, which is handled in some detail in §4.3. Section 5 outlines the method for reducing the *"many-to-many"* case to two *"one-to-many"*problems. The tangent-vector estimates, that are necessary for decomposing the global problem into a sequence of local ones, are obtained as described in §6. The paper ends with illustrating the proposed methodology for a synthetic data set, representing a *"one-to-three"* branching problem; see §7.

2 Problem decomposition

As stated in the Introduction, we aim to develop a method for modeling or reconstructing an object from its cross sections with a set of parallel planes, say $z = z_i, i = 1, ..., N$. It is permitted that the boundary of each cross section may consist of one or more disjoint Jordan curves, C_{ij}, $j = 1, \ldots, \mathscr{M}_i$, $i = 1, \ldots, N$, referred to as *contours*, the number $\mathscr{M}_i$ of varying generally from section to section. Further, and in conformity with the case that is likely to arise in practice, it is assumed that we possess only discrete approximations of contours C_{ij} in the form of ordered point sets $\mathscr{P}_{ij} \in C_{ij}$. Then, we proceed to formulate the following

Problem 2.1 *Construct a composite G^1 parametric surface $\mathbf{S}(u,v)$ that interpolates the contour point sets $\mathscr{P}_{ij} = \{\mathbf{P}_{ijk} \in \mathbb{E}^3,\ k = 0, \ldots, m_{ij} - 1\}$, $j = 1, \ldots, \mathscr{M}_i$, $i = 1, \ldots, N$, and whose isoparametrics $v = z_i$ lay on the planes $z = z_i$, $i = 1, \ldots, N$, respectively.*

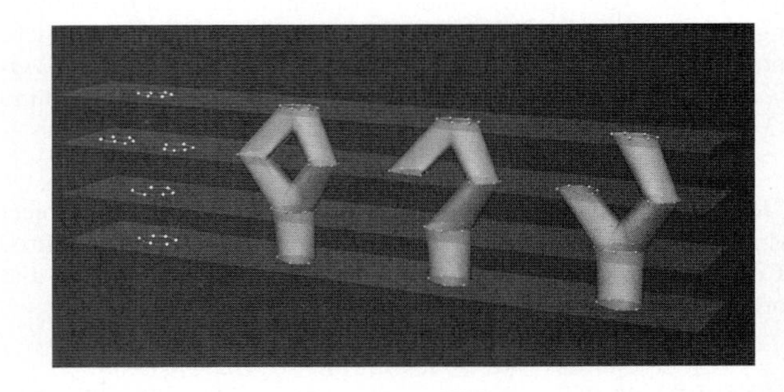

Figure 1: One data set, three different correspondence choices

Regarding the well posedness of the above problem, one can readily observe that, if $\mathscr{M}_i > 1$ for some i−plane, then the *correspondence* question: *which of the contours on the i−plane should be connected to which of the contours on the $(i+1)$−plane,* admits more than one, topologically legitimate, answers; see Fig. 1. The solution of it can be represented by a *graph* structure, $\mathscr{G}$, whose vertices are the contours while edges indicate their correspondences [Meyers et al. '92]. In what follows, we shall assume that $\mathscr{G}$ is user-defined, which is a definitely reasonable assumption in the design context. In general, the correspondence problem remains an important issue; see [Nonato et al. '05] and the references therein. Now, if we consider the correspondence issue locally, then the connections between the contours of the two neighboring planes, that represent topologically different configurations, can be represented by a set of isolated subgraphs $\mathscr{G}_{i\ell}$ of $\mathscr{G}$, $\ell = 1,...,L_i$, whose structure can be of the following three types:

I. A list of two elements, consisting of two vertices, each on the $i-$ and $(i+1)-$ planes, and an edge connecting them.

II. A two-level tree, consisting of one vertex on the $i-$plane, V_{i+1} vertices on $(i+1)-$plane and V_{i+1} edges connecting them.

III. $\mathscr{G}_\ell$ has $V_i > 1$ vertices on the $i-$plane and $V_{i+1} > 1$ vertices on the $(i+1)-$plane.

Each of the above local-subgraph cases leads to a local interpolation subproblem, which will be referred to as: (I) the *"one-to-one"* subproblem, (II) the *"one-to-many"* or *branching* subproblem and (III) the *"many-to-many"* or *multiple branching* subproblem, respectively.

We can then proceed to decompose Problem 2.1 into a sequence of local subproblems of the above three types, provided we can eventually secure the validity of the G^1 global continuity condition. For this purpose, we shall assume that all data points $\mathbf{P}_{ijk}$ are enhanced with tangent vector estimates $\mathbf{T}_{ijk}$ along the v parametric direction. A way to obtain these estimates, that resides a C^0 version of the methodology described in the ensuing three sections, is described in §6. The sought for decomposition of Problem 2.1 can then take the following form:

Problem 2.2 *Let be given the contour point sets $\mathscr{P}_{ij}$, their corresponding tangent-vector estimates $\mathscr{T}_{ij} = \{\mathbf{T}_{ijk} \in I\!R^3, k = 0,\dots,m_{ij}-1\}$, $j = 1,\dots,\mathscr{M}_i$, $i = 1,\dots,N$, and the graph $\mathscr{G}$.*

(i) Construct planar parametric G^1 curves $\mathbf{C}_{ij}(u)$ and tangent-vector ribbons $\mathbf{T}_{ij}(u)$ that interpolate $\mathscr{P}_{ij}$ and $\mathscr{T}_{ij}$, respectively.

(ii) Construct $\mathbf{S}(u,v)$ of Problem 2.1 as the union of all composite G^1 parametric patches $\mathbf{S}_i(u,v)$, $z_i \le v \le z_{i+1}$, that solve the local subproblems defined by the Hermite data prepared in (i) and the local subgraphs $\mathscr{G}_{i\ell}$ of $\mathscr{G}$, $\ell = 1,...,L_i$, $i = 1,\dots,N-1$.

3 Planar contours and their tangent ribbons

In order to construct the contour curves $\mathbf{C}_{ij}$ and the tangent ribbons $\mathbf{T}_{ij}$ referred to in Problem 2.2(i) we first have to supply the data-sets $\mathscr{P}_{ij}$ with lists $\mathscr{U}_{ij}$ of parametric nodes.

We adopt the method proposed in [Marsan & Dutta '99] where for convex point-sets $\mathscr{P}_{ij}$ the parametric node of each point is taken equal to its polar angle with respect to the center of the polygon $\mathscr{P}_{ij}$. In case $\mathscr{P}_{ij}$ defines a non-convex polygon we employ a technique introduced by [Ekoule et al. '91] which can roughly be described through the example shown on Fig. 2. The first step is to decompose the polygon into elementary convex hulls. This leads to a general planar tree structure. Then, starting from the last level of it, we project the internal vertices of the concavity onto the edge of the above level and continue with the next level until all vertices are projected onto the convex hull.

Having defined the knot vectors $\mathscr{U}_{ij}$ and exploiting the shape-preserving curve interpolation scheme, in [Kaklis & Karavelas '97] we compute the contour curves $\mathbf{C}_{ij}$ and the tangent-vector distributions $\mathbf{T}_{ij}$ along them. Then, the solution of the *"one-to-one"* problem follows readily by using the standard skinning surface scheme.

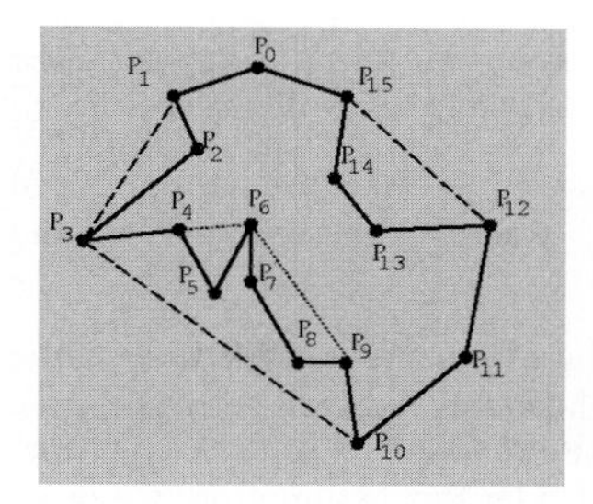

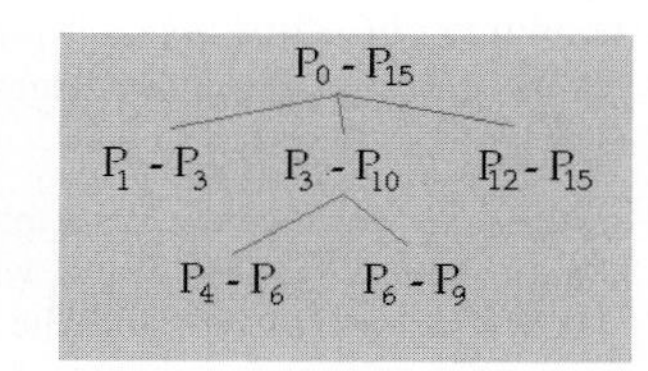

Figure 2: Decomposing a non convex polygon into a tree of its elementary convex hulls.

4 The *"one-to-many"* Hermite problem

4.1 Surrounding curve/surface

Before proceeding with a detailed construction of the surrounding curve let us denote by H, the convex hull of all point sets $\mathscr{P}_{i+1,j}$, $j = 1,\dots,\mathscr{M}$ on the (i+1)-plane, $\mathscr{H}$, the boundary of H and H^C, the complement with respect to H of the closed domains bounded by the simple polygons formed by the ordered point sets $\mathscr{P}_{i+1,j}$. The complement H^C is the union of the open disjoined domains Q_k, $k = 1,\dots,R$, each of which may be multiply connected, if it contains at least one of the point sets $\mathscr{P}_{i+1,j}$, or simply connected otherwise.

Definition 4.1 *Let $\mathscr{Q}_k$, $k = 1,\dots,R$, be the boundary (resp. outer boundary) of a simply (resp. multiply) connected domain Q_k. If Q_k shares common points with only one of the polygons formed by $\mathscr{P}_{i+1,j}$, then Q_k will be called outer residual, otherwise it will be referred to as inner residual.*

The above definition is illustrated in Fig. (3).

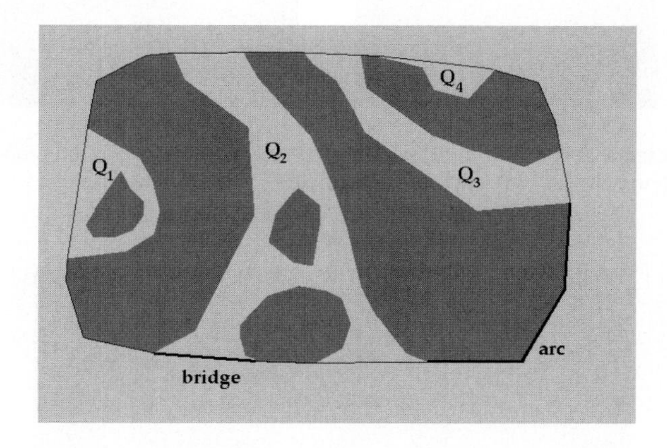

Figure 3: A set of $\mathscr{M} = 6$ coplanar contours and their convex hull. The complement H^C consists of the multiply connected outer residual, Q_1, the multiply connected inner residual, Q_2, the simply connected inner residual, Q_3 and the simply connected outer residual, Q_4.

In the sequel we shall restrict ourselves to H^C consisting only of simply connected inner and outer residuals.

Definition 4.2 *Let H^S be the domain, obtained by subtracting from H all the outer residuals. The boundary of H^S is a closed polygonal line,which will be called the surrounding polygon of $\mathscr{P}_{i+1,j}$, $j = 1,\dots,\mathscr{M}$ and denoted by $\tilde{\mathscr{C}}$. The segments of $\tilde{\mathscr{C}}$ that do not belong* *entirely on any of the polygons formed by $\mathscr{P}_{i+1,j}$ will be called bridges, the rest of them being referred to as arcs.*

Note that arcs are in general polygonal portions of the polygonal formed by $\mathscr{P}_{i+1,j}$, while bridges are always line segments linking different $\mathscr{P}_{i+1,j}$; see Fig 3.

In order to simplify the description of our method for solving the *"one-to-many"* problem and without any loss of generality, we shall henceforth assume that the $\mathscr{M}$ point-sets $\mathscr{P}_{i+1,j}$, $j = 1 \dots \mathscr{M}$, form only one inner residual $\mathscr{Q}$. Under this assumption, we order the contours in such a way that the jth bridge connects the polygons formed by $\mathscr{P}_{i+1,j}$ and $\mathscr{P}_{i+1,j+1}$. This in turn implies that each polygon, defined by $\mathscr{P}_{i+1,j}$, meets two bridges that split it into two parts: one that lies on the boundary of $\mathscr{Q}$ and one belonging to the surrounding polygon $\tilde{\mathscr{C}}$. Analogously constructed contour curves $\mathbf{C}_{i+1,j}$ and tangent ribbons $\mathbf{T}_{i+1,j}$, are split into two parts: the outer one, corresponding to $\tilde{\mathscr{C}}$, and the inner part corresponding to the inner residual.

Next, we provide $\tilde{\mathscr{C}}$ with a parameterization $\tilde{\mathscr{U}}$ by employing the method described in §3. We reparameterize the outer part of $\mathbf{C}_{i+1,j}$ according to $\tilde{\mathscr{U}}$ and finally we construct a C^1 connection between the outer parts of $\mathbf{C}_{i+1,j}$ and $\mathbf{C}_{i+1,j+1}$. This procedure defines the so called *surrounding curve*, $\tilde{\mathbf{C}}$. Working analogously with the tangent ribbons $\mathbf{T}_{i+1,j}$ we define a tangent vector distribution on it, $\tilde{\mathbf{T}}$.

The so-called *surrounding surface*, can now be constructed by employing the standard Hermite skinning technique for interpolating the contour $\mathbf{C}_i$ on the i-plane and the surrounding $\tilde{\mathbf{C}}$ on the (i+1)-plane along with their corresponding tangent ribbons, $\mathbf{T}_i$ and $\tilde{\mathbf{T}}$, and a process that is being illustrated in Figs.4-6.

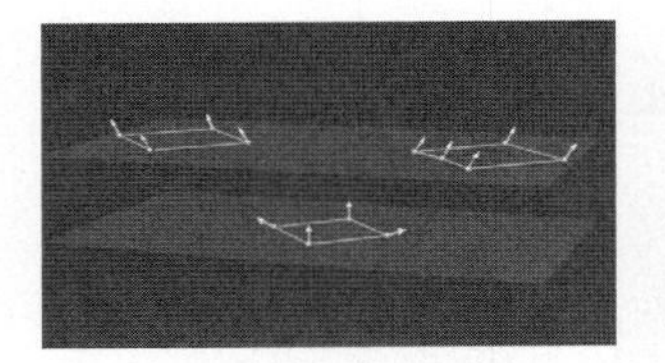

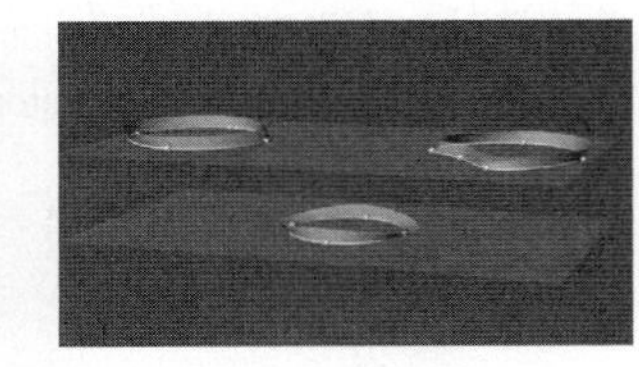

Figure 4: Data from a *"one-to-two"* example and the initial contours and tangent ribbons on them.

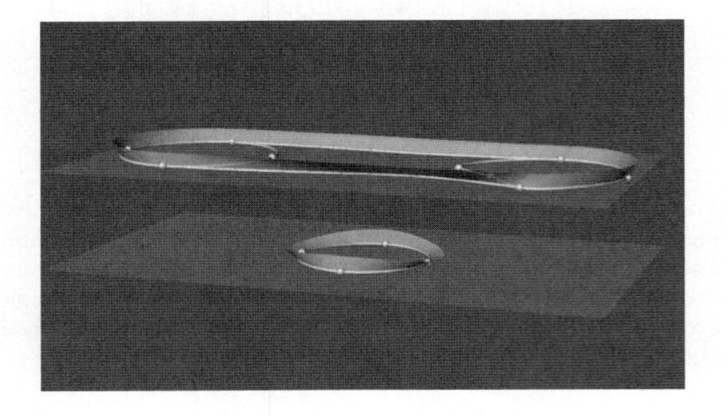

Figure 5: The surrounding curve $\tilde{\mathbf{C}}$ and the contour $\mathbf{C}_i$ along with their tangent ribbons $\tilde{\mathbf{T}}$ and $\mathbf{T}_i$, respectively.

As may one observe in Fig., in order for the final surface to interpolate $\mathscr{P}_{i+1,j}$, the surrounding surface has to be trimmed near the

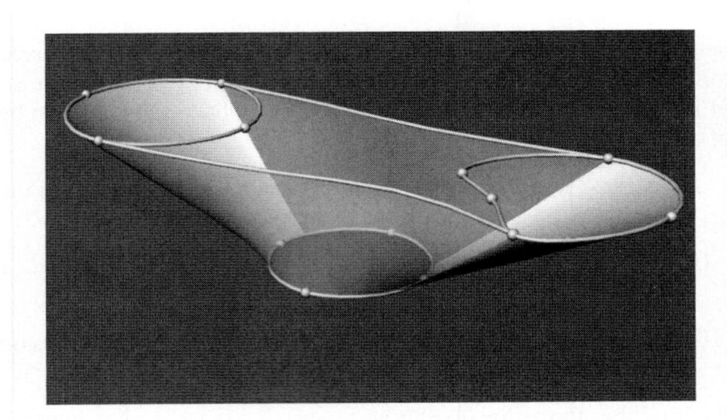

Figure 6: The surrounding surface interpolating $\mathbf{C}_i$ and $\tilde{\mathbf{C}}$. The transparent parts correspond to the two bridges.

bridges. The resulting new bounds of the trimmed surrounding surface along with the inner parts of the contour curves, which have not been interpolated yet, form a closed 3D hole, which remains to be filled in order to complete the interpolation proccess. The following two paragraphs deal with these two problems.

4.2 Trimming

Let $\mathbf{S} : [u_S, u_E] \times [0,1] \longrightarrow I\!R^3$ be a surrounding surface patch corresponding to a bridge of the surrounding polygon. We aim to trim $\mathbf{S}(u,v)$ so that the trimmed patch will still interpolate the bridge end points $\mathbf{B}^S = \mathbf{S}(u_S, 1)$ and $\mathbf{B}^E = \mathbf{S}(u_E, 1)$.

The domain curve $\mathbf{X}(t)$, $t \in [0,T]$, is selected to be a C^1 quadratic spline in order to satisfy Hermite boundary conditions at bridge end-points, reduce the number of hole sides and keep as low as possible the degree of the trimming curve $\mathbf{Y}(t) = \mathbf{S} \circ \mathbf{X}(t)$.

Eventually, $\mathbf{X}(t)$ is set to be symmetric with respect to the middle of the interval $[u_S, u_E]$, consisting of two linear and quadratic segments as shown in Fig. 7. The extent of the linear segments defines a free parameter $r \in (0,1)$, which will be used in the hole filling construction. Note that C^1 continuity of $\mathbf{X}(t)$ dictates dependency between T and r, which in our case takes the form $T = 2(r+1)$.

Along the so-constructed trim curve $\mathbf{Y}(t)$ (see Fig. 8) we need to define a cross-tangent vector distribution that is needed for filling the hole with G^1-continuity along its boundary. This distribution, say $\mathbf{Y}^{(1)}(t)$, can be expressed as a linear combination of the first partial derivatives of $\mathbf{S}(u,v)$: $\mathbf{Y}^{(1)}(t) = a_S(t)\frac{\partial \mathbf{S}(u,v)}{\partial u} + b_S(t)\frac{\partial \mathbf{S}(u,v)}{\partial v}$. For the linear part, $t = [0, 1-r]$, of the domain curve $\mathbf{X}(t)$, we take $\mathbf{Y}^{(1)}(t) = \frac{\partial \mathbf{S}(u,v)}{\partial u}$, whereas for the quadratic part, $t \in [1-r, 1+r]$ we use cubic blending functions $a_S(t)$, $b_S(t)$ fulfilling the Hermite boundary conditions:

$$a_S(1-r) = 1,\ a_S(1+r) = 0, \quad b_S(1-r) = 0,\ b_S(1+r) = 1,$$

$$a'_S(1-r) = a'_S(1+r) = b'_S(1-r) = b'_S(1+r) = 0.$$

where prime denotes differentiation with respect to t. It is worth noticing that it can be easily proved that the so constructed cross-tangent vector distribution is nowhere collinear to the tangent of the trimming curve.

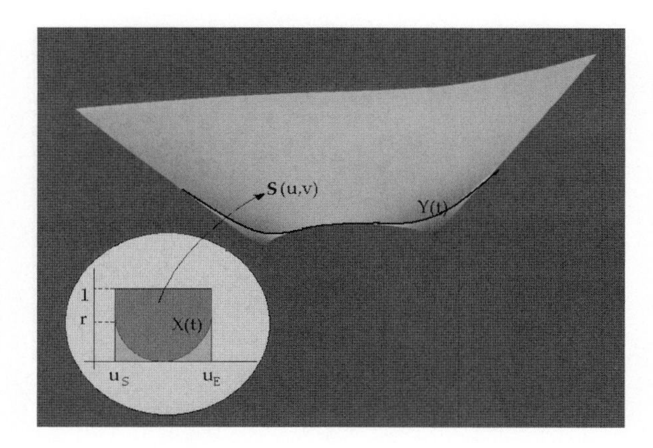

Figure 7: The domain curve $\mathbf{X}(t)$ and its map $\mathbf{Y}(t)$ on $\mathbf{S}(u,v)$.

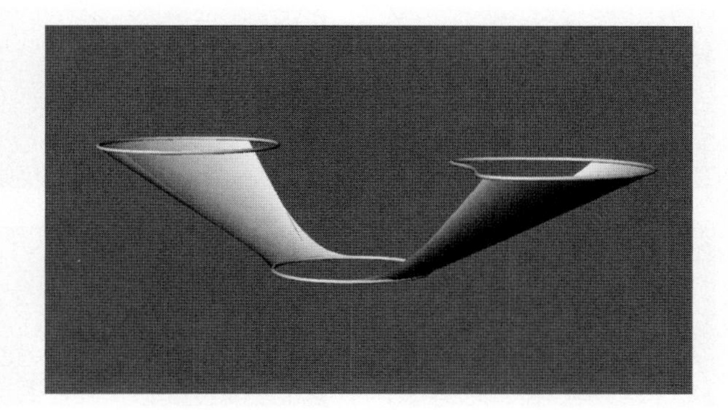

Figure 8: The trimmed surrounding surface.

4.3 Hole filling

Trimming of the surrounding surface provides us with a 3D closed hole with $2\mathcal{M}$ C^1-continuous sides, which remains to be filled in order to give a complete solution to the *"one-to-many"*problem.

Our method adopts the methodology presented in [Hahn '89] according to which, given a user-defined center $\mathbf{K}$ and a set of guide curves $\mathbf{G}_K(t)$, $t \in [0,t_K]$, $k = 1,\ldots,2\mathcal{M}$, that connect $\mathbf{K}$ with the middle parameter point $\mathbf{M}_K$ on each side, and are endowed with appropriately defined derivative distributions on them, the hole can be filled by a Gordon-Coons patchwork; see Fig. 9. In our case we are restricted to G^1 continuity, offering nevertheless the additional advantages of permitting $\mathbf{G}_K$ to be splines and preserve data symmetries in the following sense: *if* $\mathbf{G}_K$ *is planar and the cross tangent vectors at the end points of* $\mathbf{G}_K$ *are symmetric with respect to this plane, then the tangent ribbons in between should be symmetric as well.*

The hole center $\mathbf{K}$ is user-defined, lying between the $i-$ and $(i+1)$-planes with its xy-projection set initially to be equal to the centroid of the polygon formed by those $\mathbf{M}_K$ lying on the $(i+1)$-plane, in case it is convex, or just averaging them, otherwise. The user-defined z-coordinate $\mathbf{K}_z$ of $\mathbf{K}$, $z_i < \mathbf{K}_z < z_{i+1}$, is linked with the parameter r, used in §4.2 for controlling the extent of the linear parts of the domain curves $\mathbf{X}(t)$, as follows:

$$r = \frac{\mathbf{K}_z - z_i}{z_{i+1} - z_i}. \tag{4.1}$$

A natural choice for the tangent plane through $\mathbf{K}$ is the plane $z = \mathbf{K}_z$, on which we define the so-called star vectors $\mathbf{v}_K$ [Hahn '89], that will be used as boundary tangent vectors of the guide curves $\mathbf{G}_K$ emanating from $\mathbf{K}$. The direction of $\mathbf{v}_K$ is determined by projecting the vector connecting $\mathbf{K}$ with $\mathbf{M}_K$ onto $z = \mathbf{K}_z$ while its length is taken as the average of the lengths of the tangent vectors of the neighboring sides at $\mathbf{M}_{K-1}$ and $\mathbf{M}_{K+1}$. Finally, we force the surface to exhibit plane behaviour in the vicinity of $\mathbf{K}$ by imposing zero second-order derivatives at $\mathbf{K}$.

As a preprocessing step, implied by the Gordon-Coons scheme, we have to reparameterize pairs of contour segments (segments $\mathbf{M}_{K-1}\mathbf{L}_K$, $\mathbf{M}_{K+1}\mathbf{R}_K$ in Fig.10) which bound neighboring patches that are *topologically opposite* to their common edge (segment $\mathbf{KM}_K$ in Fig.10), so that eventually they are defined over the same parametric interval. Employing linear reparameterization and requesting minimization of the length variation of their end tangent vectors, we are lead to a well posed least-squares problem.

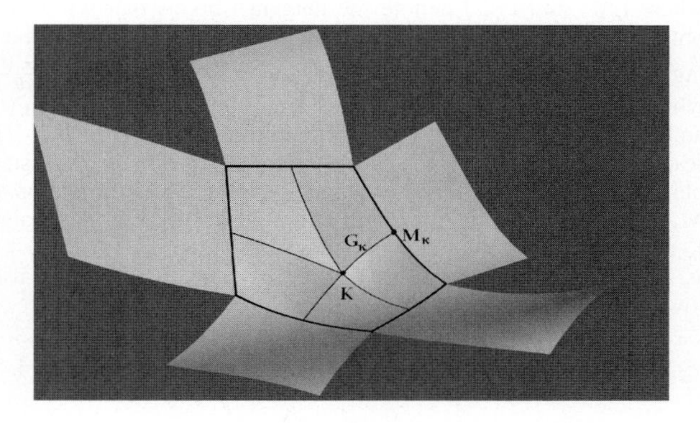

Figure 9: The Gordon-Coons patches covering a five-sided hole.

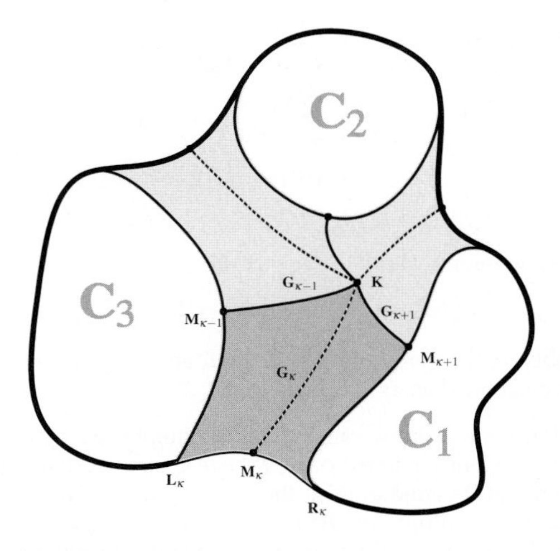

Figure 10: Top view of three contours, $\mathbf{C}_1$, $\mathbf{C}_2$ and $\mathbf{C}_3$, on the (i+1)-plane forming one inner residual and the Gordon-Coons patchwork covering the six-sided hole.

Then, we proceed to construct the family of guide curves $\mathbf{G}_K$, which is divided into two sets, those $\mathbf{G}_K$ that connect $\mathbf{K}$ with contour segments (solid curves in Fig.10) and those connecting $\mathbf{K}$ with trimming curves (dashed curves in Fig.10). Guide curves, belonging to the first set are constructed as shape preserving 3D polynomial curves [Gabrielides & Kaklis '00] that satisfy the boundary conditions implied by the corresponding star vectors and the requested planar behaviour at $\mathbf{K}$, as well as positional and tangential information readily available at $\mathbf{M}_K$. For the second set of guide curves, we use shape-preserving interpolation splines with interpolation points whose xy-projection is obtained from the middle isoparametric of the linear skinning surface interpolating the topologically opposite contour segments. This choice serves our aim to embed the shape of the contributing contours to the guide curve. The z-coordinate of the interpolation points is calculated with the aid of the cubic Hermite polynomial that connects $\mathbf{K}$ with $\mathbf{M}_K$.

It remains to build the tangent ribbons $\mathbf{L}_K$ and $\mathbf{R}_K$ that are to be imposed on every pair of neighboring patches, sharing the same guide curve $\mathbf{G}_K$ as common boundary. These are expressed as

$$\mathbf{L}_K(t) = \lambda_K(t)\left(\mathbf{l}_K(t) - \mathbf{r}_K(t)\right) + \mu_K(t)\frac{d\mathbf{G}_K(t)}{dt} \tag{4.2}$$

and

$$\mathbf{R}_K(t) = \nu_K(t)\left(\mathbf{r}_K(t) - \mathbf{l}_K(t)\right) + \xi_K(t)\frac{d\mathbf{G}_K(t)}{dt} \tag{4.3}$$

where $\mathbf{l}_K(t)$ and $\mathbf{r}_K(t)$ denote the tangent ribbons that we would obtain if step 6 in [Hahn '89] had been applied to the two neighboring patches. Obviously, $\mathbf{R}_K(t)$ and $\mathbf{L}_K(t)$ are coplanar with $\frac{d\mathbf{G}_K(t)}{dt}$ while $\lambda_K(t)$, $\mu_K(t)$, $\nu_K(t)$ and $\xi_K(t)$ are functions such that $\mathbf{L}_K(t)$ and $\mathbf{R}_K(t)$ share the same boundary conditions with $\mathbf{l}_K(t)$ and $\mathbf{r}_K(t)$, respectively. The computed star vectors $\mathbf{v}_K$ along with the requested planar behaviour at $\mathbf{K}$ and the available geometrical information at $\mathbf{M}_K$ provide a set of boundary conditions that uniquely determine $\mathbf{l}_K$ and $\mathbf{r}_K$ as cubic polynomials. As for the functions $\lambda_K(t)$, $\mu_K(t)$, $\nu_K(t)$ and $\xi_K(t)$, it can be proved that they should satisfy the following constraints:

$$\lambda_K(0) = \frac{\mathbf{v}_{K-1} \times \mathbf{v}_K}{(\mathbf{v}_{K-1} - \mathbf{v}_{K+1}) \times \mathbf{v}_K}, \quad \lambda_K(t_K) = \frac{1}{2}, \tag{4.4}$$

$$\mu_K(0) = \frac{\mathbf{v}_{K+1} \times \mathbf{v}_{K-1}}{\mathbf{v}_K \times (\mathbf{v}_{K-1} - \mathbf{v}_{K+1})}, \quad \mu_K(t_K) = 0, \tag{4.5}$$

$$\nu_K(0) = \frac{\mathbf{v}_{K+1} \times \mathbf{v}_K}{(\mathbf{v}_{K+1} - \mathbf{v}_{K-1}) \times \mathbf{v}_K}, \quad \nu_K(t_K) = \frac{1}{2}, \tag{4.6}$$

$$\xi_K(0) = \frac{\mathbf{v}_{K-1} \times \mathbf{v}_{K+1}}{\mathbf{v}_K \times (\mathbf{v}_{K+1} - \mathbf{v}_{K-1})}, \quad \xi_K(t_K) = 0, \tag{4.7}$$

$$\lambda_K'(0) = -\nu_K'(0), \quad \mu_K'(0) = \xi_K'(0), \tag{4.8}$$

$$\lambda_K'(t_K) = \mu_K'(t_K) = \nu_K'(t_K) = \xi_K'(t_K) = 0, \tag{4.9}$$

Choosing cubic polynomials and setting

$$\lambda_K'(0) = \mu_K'(0) = \nu_K'(0) = \xi_K'(0) = 0, \tag{4.10}$$

these polynomials are uniquely defined and thus $\mathbf{L}_K$ and $\mathbf{R}_K$ can now be calculated using (4.2) and (4.3).

The filling hole process can be now accomplished by employing the Gordon-Coons scheme once we can guarantee the validity of the compatibility conditions at the corner points of the hole. This is achieved by multiplying the tangent ribbon along each trimming curve with a suitably defined cubic with vanishing end derivatives.

5 The *"many-to-many"* Hermite problem

The *"many-to-many"*problem is handled via a direct extension of the *"one-to-many"*methodology described in the previous section. In fact the two methodologies are identical up to the construction step of the surrounding surface; see Fig. 12.

Next, by taking two v-isoparametric curves on the surrounding surface we create a separating zone (see Fig. 13) between the i- and the (i+1)-planes, that reduces the problem into two independent to each other *"one-to-many"*subproblems, for which the necessary surrounding surfaces are already there; see Fig. 14.

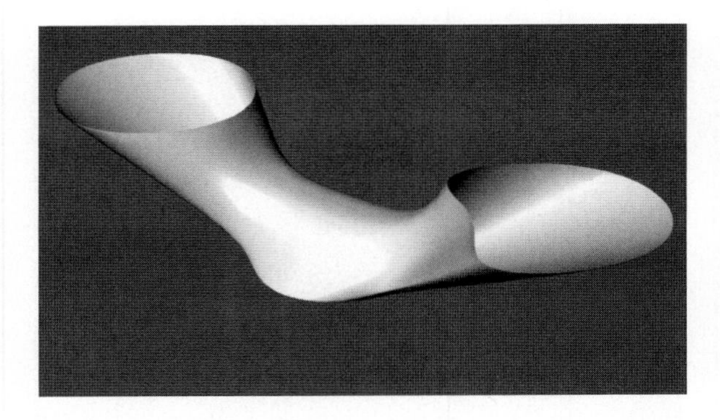

Figure 11: The final *"one-to-two"* surface.

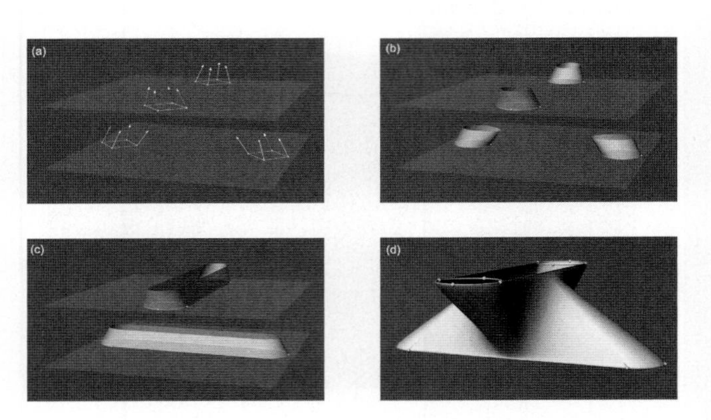

Figure 12: Handling the *"many-to-many"* problem up to the surrounding-surface step: data sets (a), contours along with tangent ribbons (b), surrounding contours with tangent ribbons (c), surrounding skinning surface (d)

6 Tangent-vector estimation

Tangent vectors, especially those on the intermediate contours, are rarely given in real world applications, thus one has further to provide a way for estimating them. For this purpose we use a C^0 version of the so far developed methodology in the following sense: skinning and hole filling does not take into account any tangential information.

Suppose we want to obtain an estimate of $\mathbf{T}_{ijk}$ at $\mathbf{P}_{ijk}$. We propose to take as $\mathbf{T}_{ijk}$ the tangent of a z-parameterized parabola, that passes through $\mathbf{P}_{ijk}$ and two points, $\mathbf{P}_u$ and $\mathbf{P}_l$, lying on the two u-isoparametrics of the neighboring patches of the C^0 surface, meeting at $\mathbf{P}_{ijk}$.

It is reasonable to expect that the estimation varies continuously with respect to u. To meet this constraint we take $\mathbf{P}_u$ and $\mathbf{P}_l$ on the intersection of the u-isoparametric with the furthest, with respect with the i-plane, v-isoparametrics for which continuity with regard to u is preserved.

So in the simplest case when the neighboring surfaces are both skinning (see Fig. 16), $\mathbf{P}_u$ and $\mathbf{P}_l$ are taken on the upper and lower contours, respectively. In the more elaborate case of Fig. 17, it can be proved that $\mathbf{P}_l$ lies on the v-isoparametric with $v = \mathbf{K}_z$, namely the z-coordinate of the hole center $\mathbf{K}$.

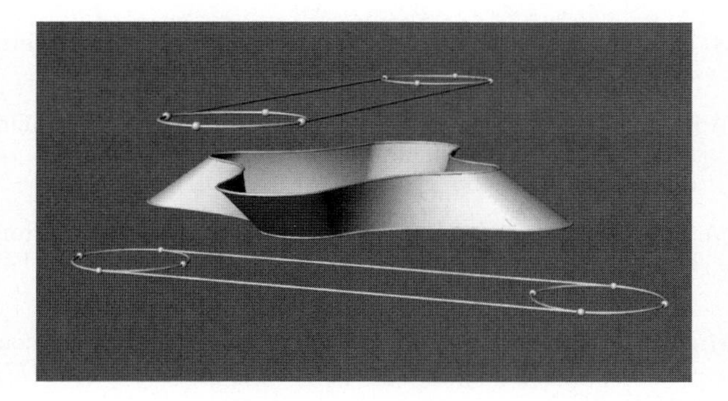

Figure 13: Formatting of the separating strip.

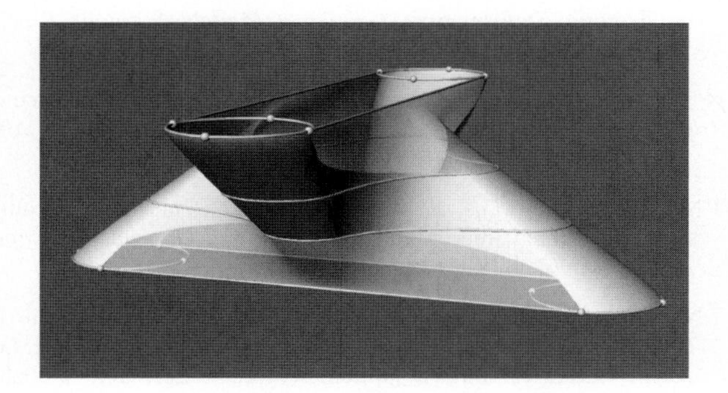

Figure 14: The surrounding surface after trimming.

7 Numerical test

The kernel of the presented methodology, which is the solution of the *"one-to-many"* branching problem, is illustrated here through a synthetic example representing an air duct. The contour point sets lie on two planes, $z = 0$ and $z = 10$ and form a *"one-to-three"* problem with one inner residual, as shown in Fig.18.

The data point sets are taken from two circles of radius 5 and 8.5 units respectively and are arranged in such a way that their formation exhibits symmetry with respect to three planes meeting along the z-axis (see Fig.19). Furthermore, the imposed tangent vectors are such that they do not violate this symmetry (i.e. are parallel to z-axis and share the same length).

The steps of the method, as described in section 4, namely skinning, trimming and hole-filling, are illustrated through Figures 20, 21 and 22, respectively. Finally, Fig.23 reveals the capability of the filling hole procedure to preserve data symmetry, by depicting the Gaussian curvature of the final surface.

We are currently working towards customizing the constructed generic Rhino plug-in, so that we can handle branching problems from the areas of mechanical engineering and medical imaging.

8 Discussion

The algorithm solving the problem *"one-to-many"* is the core of the paper. It seems to have many advantages over other algorithms in literature, due to the analytic structure of the final surface patchwork, the G^1 smoothness and its ability to preserve the shape and symmetry of the input data. The method was designed to be easily implemented in an ordinary CAD system, thus it uses structures

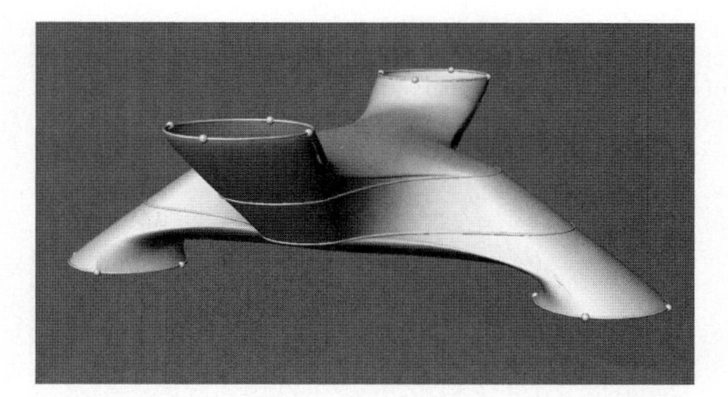

Figure 15: The surface after hole filling.

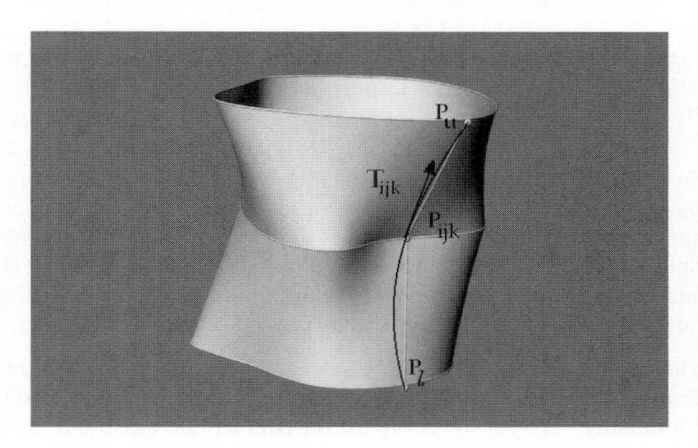

Figure 16: Tangent vector estimation on an intermediate contour between skinning surfaces.

(such as polynomials, rectangular surface patches) which constitute the basis of most of the these systems. Since the proposed method is aiming to provide a framework for designing branching surfaces, it leaves to the involved designer a number of free parameters up to his alley. The definition of the trimming curve, or the construction of the guide curves or the heuristic rules for the information on the hole center, can be freely modified in order to apply the method to a particular application more efficiently. In the following paragraphs we provide some specific comments and a brief discussion on open questions regarding the algorithm.

To start with, the methodology adopted for computing the parametrization of the point-sets is fast and stable. Moreover, it provides an acceptable automatic solution to the parameter-matching problem for constructing the necessary skinning surfaces. On the other hand, the algorithm used for constructing the contour curves, [Kaklis & Karavelas '97], guarantees the construction of convexity-preserving curves independently of the chosen parametrization. Note also that the same algorithm can guarantee that no intersections between different contour curves may occur, provided the polygons formed by $\mathscr{P}_{ij}$ do not intersect one another.

According to the Definition 4.2, the surrounding curve can be constructed rapidly and robustly from the point-sets $\mathscr{P}_{i+1,j}$ and the tangential information supplied by the already constructed contour curves. On the other hand, we should note an apparent topological instability of the proposed algorithm, with respect to the relative position of the above point-sets to their convex hull. Let us assume, e.g., that the polygons formed by those point-sets all contribute to the convex hull, i.e., there is one (or more) simply connected residual(s). Then, a small perturbation of one of the polygons may result in the formation of a multiply connected inner/outer residual, a situation which is not treated by the algorithm. One can overcome

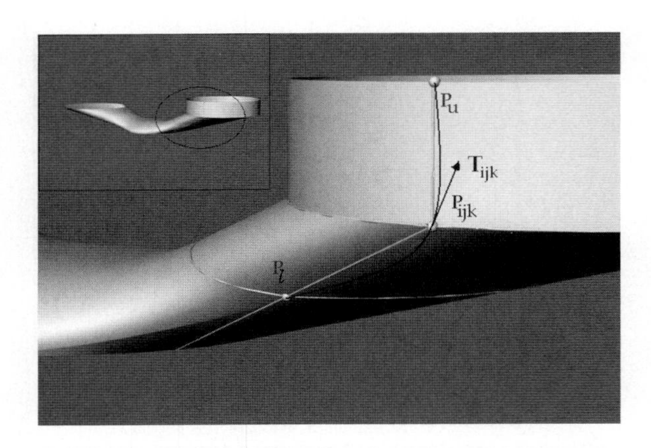

Figure 17: Tangent vector estimation of an intermediate contour which bounds at least one Gordon-Coons patch.

this instability by giving an alternative definition of the surrounding curve, which would not make use of the convex hull notion and connect all contours on the (i+1)-plane. In this case no multiply connected residuals would be produced, thus permitting the next steps of the *"one-to-many"* algorithm to handle a much wider range of topological configurations of cross sectional data.

The rest of the method involves skinning surfaces and Gordon-Coons interpolation, which are both widely used by the CAGD community. They also enable us to easily extend the algorithm in order to incorporate NURBS curves. Finally, based on these underlying structures, we can increase the order of continuity of the resulting surface, at the cost of enriching the data sets with higher order information.

The solution to the *"one-to-many"* problem led us give an analogously reliable solution to the problem *"many-to-many"*, a case been handled already by very few algorithms.

References

C.L. Bajaj, E.J. Coyle and K.-N. Lin, Arbitrary topology shape reconstruction from planar cross sections, *Graphical Models & Image Proc.*, Vol.58 (1996), pp.524-543.

G. Barequet, D. Shapiro and A. Tal, Multilevel sensitive reconstruction of polyhedral surfaces from parellel slices", *Visual Computer*, Vol.16 (2000), pp.116-133.

G. Barequet & M. Sharir, Piecewise-linear interpolation between polygonal slices, *Comp. Vision & Image Underst.*, Vol.63 (1996), pp.251-272.

S. Batnitzki, H.I. Price, P.N. Cook, L.T. Cook and S.J. Dwyer III, Three-dimensional computer reconstruction from surface contours for head CT examinations, *J. of Computer Assist. Tomogr.*, Vol.5 (1981), pp.60-67.

A. Bentamy, F. Guibault and J.-Y. Trépanier, Cross-sectional design with curvature constraints, *CAD*, Vol.37, (2005), pp.1499-1508.

S. Bedi, Surface design using functional blending, *CAD*, Vol.24, (1992), pp.505-511.

H.N. Christiansen & T.W. Sederberg, Conversion of complex contour lines into polygonal element mosaics, In *Computer Graphics (SIGGRAPH '78 Proceedings)*, R.L. Phillips (ed), Vol.12, 1978, pp.187-192.

S. Cohen, G. Elber and R. Bar-Yehuda, Matching of free-form curves, *CAD*, Vol.29, (1997), pp.369-378.

A.B. Ekoule, F.C. Peyrin and C.L. Odet, A triangulation algorithm from arbitrary shaped multiple planar contours, *ACM Trans. on Graph.*, Vol.10 (1991), pp.182-199.

M.S. Floater & G. Westgaard, Smooth surface reconstruction from cross-sections using implicit surfaces, Sintef, Report No STF42 A96023,1996.

H. Fuchs, Z.M. Kedem and S.P. Uselton, Optimal surface reconstruction from planar contours, *Commun. ACM*, Vol.20 (1977), pp.693-702.

K.Fujimura and E. Kuo, Shape Reconstruction from Contours Using Isotopic Deformation, *Graphical Models & Image Proc.*, Vol.61 (1999), pp.127--147.

N.C. Gabrielides & P.D. Kaklis, C^4 interpolatory shape-preserving polynomial splines of variable degree, *Computing Suppl.*, Vol. 14 (2001), pp. 119-134.

T.N.T. Goodman & G.T.D. Greig, Matching and choice of parameter in sectional interpolation, *International Journal of Shape Modeling*, Vol.4 (1998), pp.197-208.

T.N.T. Goodman, B.H. Ong and K. Unsworth, Reconstruction of C^1 closed surfaces with branching, In *Geometric Modelling*, G. Farin, H. Hagen and H. Noltemeier (eds), Springer, Vienna, 1993, pp.101-115.

J. Hahn, Filling polygonal holes with rectangular patches, In *Theory and practice of geometric modeling*, W. Straer and P.-H. Siedel (eds), Springer, Berlin, 1989, pp.81-91.

M.J. Herbert, C.B. Jones and D.S. Tudhope, Three-dimesional reconstruction of geoscientific objects from serial sections,*Visual Computer*, Vol.11 (1995), pp.343-359.

M. Hohmeyer & B.A. Barsky, Skinning rational B-spline curves to construct an interpolatory surface, *Graphical Models & Image Proc.*, Vol.53 (1991), pp.511-521.

F. Jaillet, B. Shariat and D. Vandorpe, Periodic B-spline surface skinning of anatomic shapes, In 9th *Canadian Conference in Computational Geometry*, Kingston, Canada, 1997, pp.199-210.

J. Jeong, K. Kim, H. Park, H. Cho and M. Jung, B-Spline surface approximation to cross sections using distance maps, *Adv. Manufact. Techn.*, Vol.15 (1999), pp.876-885.

J.K. Johnstone & K.R. Sloan, A philishophy for smooth contour reconstruction, *Comp. Suppl.*, Vol.13 (1998), pp.153-163.

M.W. Jones & M. Chen, A new approach to the construction of surfaces from contour data. *Computer Graphics*, Vol.13 (1994), pp.75-84.

B. Jüttler, Sectional curvature-preserving interpolation of contour lines, In *"Curves and Surfaces with Applications in CAGD"*, A. Le Méhauté, C. Rabut and L.L. Schumaker (eds), Vanderbilt University Press, Nashville TN, 1997, pp.203-210.

P.D. Kaklis & A.I. Ginnis, Sectional-curvature preserving skinning surfaces, *CAGD*, Vol.13 (1996), pp.601-619.

P.D. Kaklis & M.I. Karavelas, Shape-preserving interpolation in $\mathbb{R}^3$, *IMA J. Numer. Anal.*, Vol.17 (1997), pp.373-419.

E. Keppel, Approximating complex surfaces by triangulation of contour lines, *IBM J. Res. Devel.*, Vol.19 (1975), pp.2-11.

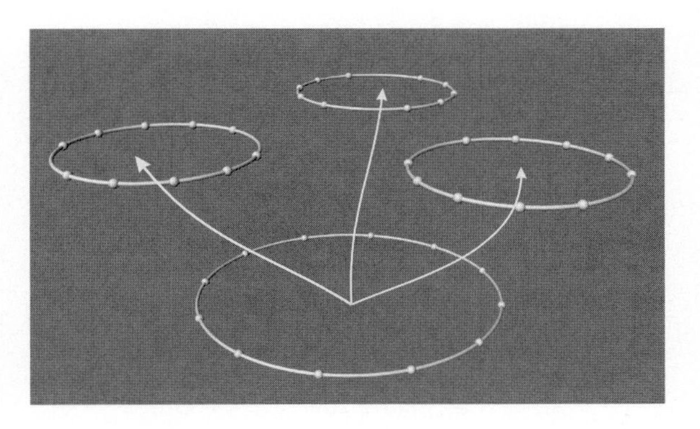

Figure 18: The data points along with the graph representing the *"one-to-three"* correspondence.

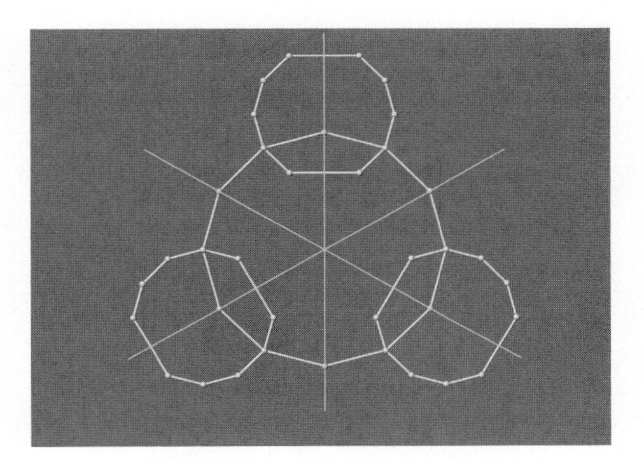

Figure 19: Top view of the data points and the trace of the planes of symmetry on the *xy* plane.

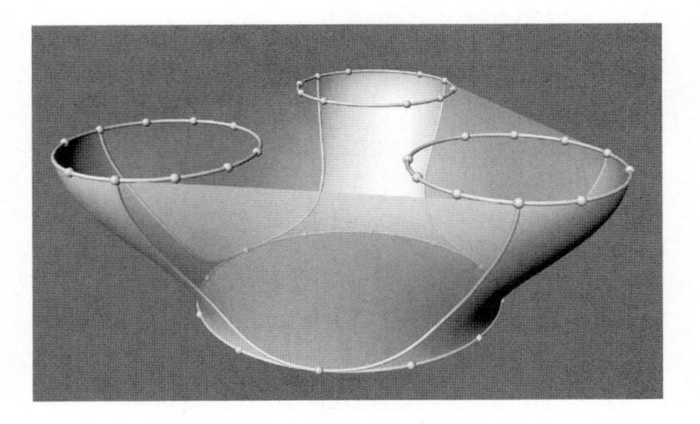

Figure 20: The surrounding surface, with the trimmed area shown transparently.

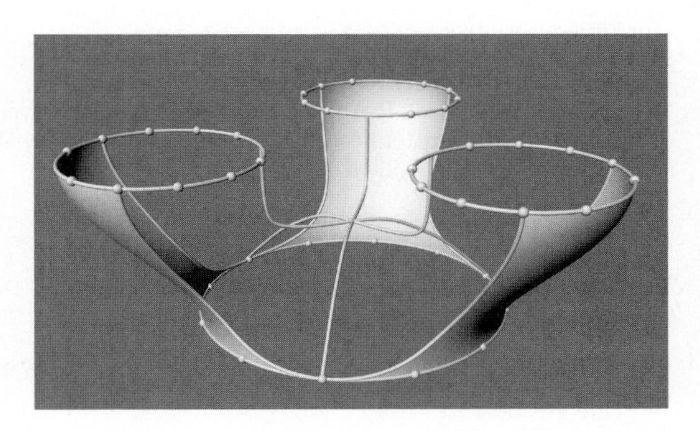

Figure 21: The boundary of the six-sided hole, along with the guide curves which form six four-sided patches.

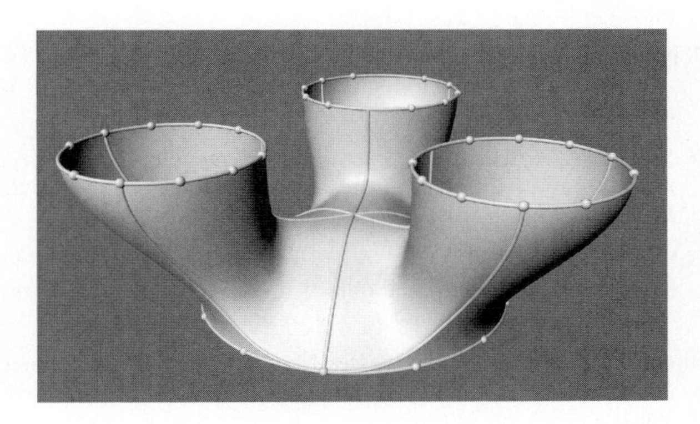

Figure 22: The surface after the hole-filling.

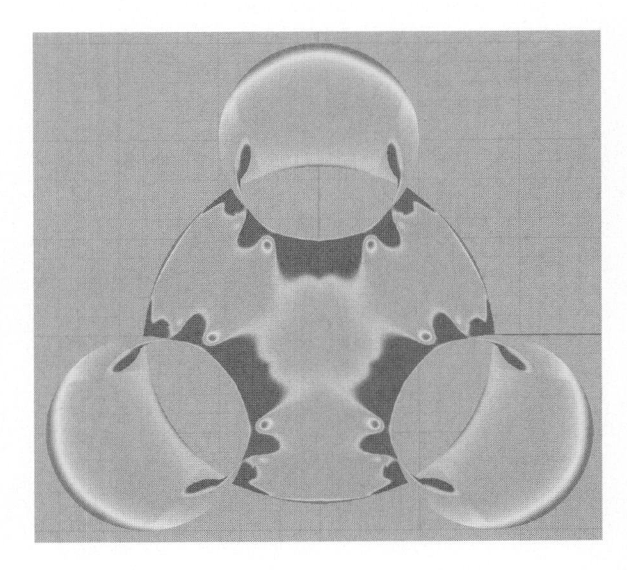

Figure 23: Top view of the surface with its Gaussian curvature mapped on it revealing the capability of the method to preserve data symmetry.

A.L. Marsan & D. Dutta, Computational techniques for automatically tiling and skinning branched objects,*Computers & Graphics*, Vol.23 (1999), pp.111-126.

E. Moschino, Y. Maurin and P. Andrey, Joint registration and averaging of multiple 3D anatomical surface models,*Comp. Vision & Image Underst.*, Vol.101 (2006), pp.16-30.

D. Meyers, S. Skinner and K. Sloan, Surfaces from contours, *ACM Trans. on Graph.*, Vol.11 (1992), pp.228-258.

L.G. Nonato, A.J. Guardos-Vargas, R. Minghim and M.C.F. De Oloveira, *ACM Trans. on Graph.*, Vol.24 (2005), pp.1239-1258.

H. Park & K. Kim, Smooth surface approximation to serial cross-sections, *CAD*, Vol.28, (1996), pp.995-1005.

L. Piegl & W. Tiller, Algorithm for approximate NURBS skinning, *CAD*, Vol.28, (1996), pp.699-706.

S. Schmitt, J.F. Evers, G. Duch, M. Scholz and K. Obermeyer, New methods for the computer-assisted 3D reconstruction of neurons form confocal image stacks, *NeuroImage*, Vol.23 (2004), pp.1283-1298.

T.W. Sederberg, K.S. Klimaszewski, M. Hong and K. Kaneda, Triangulation of branching contours using area minimization, *Intern. J. of Comp. Geom. & Appl.*, Vol.8 (1998), pp.389-406.

Y. Shinagawa & T.L. Kunii, The homotopy model: A generalized model for smooth surface generation from cross sectional data, *Visual Computer*, Vol.7 (1991), pp.72-86.

N.M. Sirakov and F.H. Muge, A system for reconstructing and visualising three-dimensional objects, *Computers & Geosciences*, Vol.27 (2001), pp.59-69.

Sloan & Painter, Pessimal Guesses May Be Optimal: A Counter-intuitive Search Result, *IEEE Trans. Pattern Anal. Mach. Intell.*, Vol.10 (1988), pp.949-955.

E. Welzl & B. Wolfers, Surface reconstruction between simple polygons, In *Proceedings, 1st Annual European Symposium on Algorithms (ESA '93), Lecture Notes in Computer Science*, Vol. 726, Springer-Verlag, Berlin/New York, 1993, pp.397-408.

Parameterization of Mesh-Models: Theory, Implementation and Applications

Bruno Levy

Mesh parameterization has been a very active research topic for the last several years, since it appeared as a fundamental tool for a wide class of problems in Digital Geometry. Its applications include texture mapping, geometry processing, remeshing, pasting, morphing, compression and reconstruction to name but a few. Computing a parameterization of a mesh model means finding a set of 2d coordinates associated to each vertex of the mesh. These coordinates are defined by the solution of a linear system, or correspond to the minimizer of a (possibly non-linear) energy functional. We will first explain the basic concepts and theorems that transform the parameterization problem into a numerical problem, and review the most popular methods. We will then explain how to solve the numerical optimization problem in practice. The presentation will be illustrated by demonstrations and examples of applications in the industry. The source-code corresponding to the demos is available online, in the OpenNL library (http://www.loria.fr/~levy/software/). To conclude, we will review some on-going research about the parameterization of meshes with arbitrary genus, currently tackled by the Digital Geometry research community.

SPM 2006, Cardiff, Wales, United Kingdom, 06–08 June 2006.
© 2006 ACM 1-59593-358-1/06/0006 $5.00

Bruno Levy is a Researcher with INRIA. He is the head of the ALICE research group. He develops the "Numerical Geometry" approach, i.e. new formalisms to define geometrical operators acting on discretized objects. His main contributions concern texture mapping and parameterization methods for triangulated surfaces.

He obtained his Ph.D in 1999, from the INPL (Institut National Polytechnique de Lorraine). His work, entitled "Computational Topology: Combinatorics and Embedding", was awarded the SPECIF price in 2000 (best French Ph.D. thesis in Computer Sciences). After his Ph.D., he did a Post-Doc in Stanford, in the SCCM Dept. headed by G. Golub (applied math. and numerical analysis), and in the Earh Sciences Dept. (with K. Aziz and A. Journel).

His main results have been transferred to the industry: His parameterization method is the kernel of the gridding tools available in the Gocad modeler, commercialized by the Earth Decision Sciences company, and used by major oil companies. His texture mapping algorithms are implemented in several 3D modelers, including Maya, Silo and Blender.

Generalized Penetration Depth Computation

Liangjun Zhang [1] Young J. Kim [2] Gokul Varadhan [1] Dinesh Manocha [1]

[1] *Dept. of Computer Science, University of North Carolina at Chapel Hill, USA, {zlj,varadhan,dm}@cs.unc.edu*

[2] *Dept. of Computer Science and Engineering, Ewha Womans University, Korea, kimy@ewha.ac.kr*

http://gamma.cs.unc.edu/PDG

Abstract

Penetration depth (PD) is a distance metric that is used to describe the extent of overlap between two intersecting objects. Most of the prior work in PD computation has been restricted to *translational PD*, which is defined as the minimal translational motion that one of the overlapping objects must undergo in order to make the two objects disjoint. In this paper, we extend the notion of PD to take into account both translational and rotational motion to separate the intersecting objects, namely *generalized PD*. When an object undergoes rigid transformation, some point on the object traces the longest trajectory. The generalized PD between two overlapping objects is defined as the minimum of the longest trajectories of one object under all possible rigid transformations to separate the overlapping objects.

We present three new results to compute generalized PD between polyhedral models. First, we show that for two overlapping convex polytopes, the generalized PD is same as the translational PD. Second, when the complement of one of the objects is convex, we pose the generalized PD computation as a variant of the convex containment problem and compute an upper bound using optimization techniques. Finally, when both the objects are non-convex, we treat them as a combination of the above two cases, and present an algorithm that computes a lower and an upper bound on generalized PD. We highlight the performance of our algorithms on different models that undergo rigid motion in the 6-dimensional configuration space. Moreover, we utilize our algorithm for complete motion planning of polygonal robots undergoing translational and rotational motion in a plane. In particular, we use generalized PD computation for checking path non-existence.

CR Categories: I.3.5 [Computer Graphics]: Computational Geometry and Object Modeling.

Keywords: Penetration depth

1 Introduction

Calculating a distance measure is a fundamental problem that arises in many applications such as physically-based modeling, robot motion planning, virtual reality, haptic rendering and computer games. Typical distance measures employed in these applications include separation distance, Hausdorff distance, spanning distance and penetration depth [Lin and Manocha 2003]. Among these measures, penetration depth (PD) is used to quantify the extent of inter-penetration between two overlapping, closed, geometric objects.

SPM 2006, Cardiff, Wales, United Kingdom, 06–08 June 2006.
© 2006 ACM 1-59593-358-1/06/0006 $5.00

PD computation is important in a number of applications. In rigid body dynamics, inter-penetration between simulated objects is often unavoidable due to the nature of discrete, numerical simulation. As a result, several response algorithms like penalty-based simulation methods need the PD information to compute the non-penetration constraint force [Mirtich 2000; Stewart and Trinkle 1996]. The PD is also used to estimate the time of contact to apply impulsive forces in impulse-based methods [Kim et al. 2002a]. Sampling-based motion planning techniques perform PD computation between the robot and the obstacles to generate samples in narrow passages in the configuration space [Hsu et al. 1998]. Many 6-DOF haptic rendering algorithms use penalty-based methods to compute a collision response and need to compute the PD at haptic update rates [Kim et al. 2003]. Other applications include tolerance verification, where PD could be used to estimate the extent of interference between the parts of a machine structure [Requicha 1993].

Most of the prior work on PD computation has been restricted to *translational PD*. The translational PD between two overlapping objects is often defined as the minimum translational distance needed to separate the two objects. Many good algorithms to estimate the translational PD between convex and non-convex polyhedra are known [van den Bergen 2001; Kim et al. 2002b; Kim et al. 2002a]. However, translational PD computation is not sufficient for many applications as it does not take into account the rotational motion. For example, in rigid body dynamics simulations, objects undergo both translational and rotational motion due to external forces and torques. In order to compute an accurate collision response, we also need to take into account rotational motion during PD computation. Similarly in 6-DOF haptic rendering, the rotational component in penalty forces, such as torque, should be considered in order to compute the response force. Also, since the configuration space of a rigid polyhedral model is a 6-dimensional space, the rotational PD metric is important for motion planning.

In this paper, we take into account the translational and rotational motion to describe the extent of two intersecting objects and refer to that extent of inter-penetration as the *generalized penetration depth*. When an object undergoes rigid transformation, some point on the object traces the longest trajectory. The generalized PD between two overlapping objects is defined as the minimum of the longest trajectories of one object under all possible rigid transformations to separate the overlapping objects. To the best of our knowledge, there is no prior work on generalized PD computation between polyhedral models.

In general, computing the generalized PD between two non-convex polyhedra is more difficult than computing translational PD due to the non-linear rotational term embedded in the definition. In case of translational PD, the problem reduces to computing the closest point from the origin to the boundary of the Minkowski sum of the primitives. The combinatorial complexity of Minkowski sum can be as high as $O(m^3n^3)$ for non-convex polyhedra, where m and n are the number of features in the two polyhedra. However, no similar formulation is known to compute the generalized PD. Computing generalized PD can be viewed as minimizing a distance metric in the configuration space. The configuration space for the case of generalized PD computation in 3D is 6-dimensional and the problem reduces to computing an arrangement of $O(n^2)$ five-dimensional contact hyper-surfaces. The combinatorial complexity of the arrangement is $O(n^{12})$ [Halperin 2005].

1.1 Main Results

We present a formulation of generalized PD and present novel results related to computing PD between polyhedral models. These include:

- We propose a novel definition for generalized penetration depth to quantify both the translational and rotational amount of inter-penetration between two overlapping polyhedra.
- We prove that for convex models, their generalized PD is the same as translational PD.
- We present an approximate algorithm to compute the generalized PD for non-convex models. We compute a lower bound on the generalized PD by using convex-covering techniques and computing the translational PD for each pair of convex polytopes.
- We reduce the problem of computing an upper bound for the generalized PD to a variant of 3D convex containment problem using linear programming.

We have implemented our algorithm and applied to many non-convex 3D models undergoing rigid motion in 6-dimensional configuration space. The running time varies based on model complexity and the relative configuration of two objects. In practice, our algorithm takes about 2 ms to 6 ms on 2.8 GHz PC to compute the lower bound on generalized PD, and 21 ms to 1.02 sec for the upper bound. We also use our algorithms to perform *C-obstacle query* for complete and collision-free motion planning of planar robots. We use this query as part of a sampling-based complete motion planning algorithm and use the generalized PD computation to accelerate the check for path non-existence.

1.2 Organization

The rest of the paper is organized in the following manner. Section 2 briefly surveys the previous work on PD computation and distance metrics in configuration space. Section 3 presents our formulation of generalized PD and highlights many of its properties. In Section 4, we show that the generalized PD computation between convex polytopes is same as translational PD computation. Section 5 highlights the relationship between generalized PD computation and the containment problem, which enables us to estimate the generalized PD between non-convex models in Section 6. Section 7 and 8 present our experimental results and an application to complete motion planning.

2 Previous Work

In this section, we give a brief overview of prior work on PD computation and distance metrics in configuration space (C-space).

2.1 PD and Arrangement Computation

Given a finite set of hypersurfaces $\mathscr{S}$ in R^d, their arrangement $\mathscr{A}(\mathscr{S})$ is the decomposition of R^d into cells $\mathscr{C}$ of dimensions $0,1,\ldots,d$. Here, a k-dimensional cell $\mathscr{C}^k$ in $\mathscr{A}(\mathscr{S})$ is a maximal connected set contained in the intersection of a subset of the hypersurfaces in $\mathscr{S}$ that is not intersected by any other hypersurfaces in $\mathscr{S}$ [Halperin 1997]. It is well known that the worst case combinatorial complexity of an arrangement of n hypersurfaces in R^d is $O(n^d)$.

Since both the translational and generalized PD can be formulated in C-space, the complexity of computing both the PD's is governed by that of C-space boundary. In case of polyhedral objects in 3D, their C-space can be computed by enumerating their contact surfaces and computing their arrangement [Latombe 1991]. As a result, one can calculate the PDs by computing the arrangement of contact surfaces. However, the combinatorial complexity of the arrangement is $O(n^{12})$ [Halperin 2005]. Moreover, in practice, robust computation of arrangements is known to be a hard problem [Raab 1999].

2.2 Translational Penetration Depth

The translational PD, PD^t, is defined as a minimum translational distance to make two objects disjoint. This definition can be formulated in terms of the *Minkowski sum* of two objects [Dobkin et al. 1993]. Several algorithms have been proposed for exact or approximate computation of PD^t. Bergen proposes a quick lower bound estimation to PD^t between two convex polytopes by iteratively expanding a polyhedral approximation of the *Minkowski sum* [van den Bergen 2001]. Kim *et al.* [2002b] presents an incremental algorithm to estimate a tight upper bound on PD^t between convex polytopes by walking to a "locally optimal solution". They have also presented an algorithm to compute an approximation of global PD^t between two general polyhedral models by using hierarchical refinement [2002a]. The hierarchical refinement approach decomposes the non-convex objects into convex polytopes and uses a bounding volume hierarchy to recursively refine the estimation of PD^t. Redon *et al.* [2005] describe a fast method to compute an approximation of the local penetration depth between two general polyhedral models using graphics hardware. The best known theoretical algorithm to compute PD^t between convex polytopes is given in [Agarwal et al. 2000] and its running time is $O(m^{\frac{3}{4}+\varepsilon}n^{\frac{3}{4}+\varepsilon}+m^{1+\varepsilon}+n^{1+\varepsilon})$ for any positive constant ε, where m and n denote the number of features in the two polytopes. However, we are not aware of any implementation of this algorithm. In case of general polyhedral models, it is known that the computational complexity of PD^t computation can be as high as $O(m^3n^3)$ [Kim et al. 2002a].

2.3 Generalized Penetration Depth

To the best of our knowledge, there is no prior published work on generalized PD computation for either convex or non-convex polyhedral objects. If we view the problem of separating the object A from B as placing A into $\bar{B}$ - the complement space of B, the most closely related work to generalized PD is the 2D polygon containment algorithms [Chazelle 1983; Milenkovic 1999; Grinde and Cavalier 1996; Avnaim and Boissonnat 1989; Agarwal et al. 1998] and rotational overlapping minimization [Milenkovic 1998; Milenkovic and Schmidl 2001].

The standard 2D polygon containment problem is to check whether a polygon Q with n vertices can contain another polygon P with m vertices. For general non-convex polygons, the time complexity of this problem is $O(m^3n^3log(mn))$ [Avnaim and Boissonnat 1989]. When restricted to convex objects, the time complexity of the 2D containment problem can be significantly improved. [Chazelle 1983] proposed an enumerative algorithm with an $O(mn^2)$ time complexity. [Milenkovic 1999; Grinde and Cavalier 1996] used mathematical programming techniques to compute an optimal solution.

Given an overlapping layout of polygons inside a container polygon, the rotational overlapping minimization problem is to compute the translation and rotation motion to minimize their overlap. [Milenkovic 1998] pose this problem as constraint-solving and employ mathematical programming methods to solve it. By using the non-overlapping property as a hard constraint, [Milenkovic and Schmidl 2001] minimize a quadratic function of the position and orientation of objects to compute a non-overlapping layout based on quadratic programming.

2.4 Distance Metric in Configuration Space

The configuration space of an object is the space for all possible placements of this object in environment [Latombe 1991; LaValle 2006]. For example, if a rigid object in 3D can translate and rotate, its C-space is 6-dimensional. A configuration is called *free* if the placement of the object at that configuration does not result in a collision with other obstacles in the environment. Otherwise, it is a *colliding* configuration. Essentially, the PD is a distance metric in C-space to represent a shortest distance from a given, colliding configuration to all the free configurations.

When only translation is allowed in the distance metric, the corresponding configuration space can be formulated using the Minkowski sum, which has $O(n^2)$ combinatorial complexity for two convex polytopes (with n features), and $O(n^6)$ for non-convex polyhedra [Halperin 2002]. Since only translation is allowed, we can use the Euclidean distance between two configurations as a distance metric. Therefore, the translational penetration depth computation, which finds a nearest point on the surface of Minkowski sum to the origin, has the same combinatorial complexity as the Minkowski sum formulation.

When both translation and rotation are allowed (i.e., 6-DOF C-space), the corresponding C-space becomes more complex and its combinatorial complexity is $O(n^{12})$ for 3D non-convex polyhedra [Halperin 2005]. Moreover, in 6DOF C-space, it is difficult to define a meaningful distance metric that can encode both translational and rotational movement than the one in 3-DOF, Euclidean space [Kuffner 2004; Amato et al. 2000].

The L_p $(p \geq 1)$ metric is one of the important family of metrics in 6DOF C-space [LaValle 2006]. Another important distance metric is the displacement metric; this is the minimum Euclidean displacement distance between all the points on the model when it is at two different configurations.[LaValle 2006].

3 Generalized Penetration Depth

The translational PD, PD^t is defined as a minimum translation distance to separate two overlapping objects A and B:

$$\text{PD}^t(A,B) = \min(\{\| \mathbf{d} \| \,|interior(A+\mathbf{d}) \cap B = \emptyset\}), \quad \mathbf{d} \in \mathscr{R}^3. \quad (1)$$

In our work, we extend the notion of PD^t by taking into account translational as well as rotational motion to separate the overlapping objects. Before proceeding to the definition of generalized PD, we first introduce our notation that is used throughout this paper.

3.1 Notation

We use a bold face letter, such as the origin $\mathbf{o}$, to distinguish a vector quantity from a scalar quantity. We use a sextuple $(x, y, z, \phi, \theta, \psi)$ to encode the 6-dimensional *configuration* of a 3D object, where x, y and z represent the translational components, and ϕ, θ and ψ are an Euler angle representation for the rotational components. The rotation component can be also represented as a rotation vector $r = (r_1, r_2, r_3)^T = \alpha\hat{\mathbf{a}}$, where α is the rotation angle and $\hat{\mathbf{a}}$ is the rotation vector. $A(\mathbf{q})$ is a placement of an object A at configuration $\mathbf{q}$, and $\mathbf{p}(\mathbf{q})$ is the corresponding position of a point $\mathbf{p}$ on A.

3.2 Distance Metric D_g In C-Space

In order to define generalized PD, PD^g, we first introduce a distance metric D_g defined in configuration space (or C-space). We use this metric to measure the distance of an object A at two different configurations.

Let l_i be a curve in C-space, which connects two configurations $\mathbf{q}_0$ and $\mathbf{q}_1$ (Fig. 1-(a)) and is parameterized in t. When the configuration of A changes along the curve l, any point $\mathbf{p}$ on A will trace out a trajectory in 3D Euclidean space shown in Fig. 1-(b). This trajectory can be represented as $r = \mathbf{p}(l(t))$, and its arc-length $\mu(\mathbf{p}, l)$, which is denoted as *trajectory length*, can be calculated as:

$$\mu(\mathbf{p}, l) = \int ||\dot{\mathbf{p}}(l(t))||\, d(l(t)).$$

As Fig. 1-(a) shows, there can be multiple curves connecting two configurations $\mathbf{q_0}$ and $\mathbf{q_1}$. When A moves along any such curve, some point on A corresponds to the longest *trajectory length* as compared all other points on A. For each C-space curve connecting $\mathbf{q_0}$ and $\mathbf{q_1}$, we consider the corresponding longest *trajectory length*. We define the distance metric $D_g(\mathbf{q_0}, \mathbf{q_1})$ as the minimum over all longest *trajectory lengths* (Fig. 1-(c)):

$$D_g(\mathbf{q_0}, \mathbf{q_1}) = min(\{max(\{\mu(\mathbf{p}, l)|\mathbf{p} \in A\})|l \in L\}), \quad (2)$$

where L is a set of all the curves connecting $\mathbf{q_0}$ and $\mathbf{q_1}$.

Properties of D_g metric. The distance metric defined above has the following properties [LaValle 2006]:

- **Non-negativity:** $D_g(\mathbf{q_0}, \mathbf{q_1}) \geq 0$,
- **Reflexivity:** $D_g(\mathbf{q_0}, \mathbf{q_1}) = 0 \iff \mathbf{q_0} = \mathbf{q_1}$,
- **Symmetry:**$D_g(\mathbf{q_0}, \mathbf{q_1}) = D_g(\mathbf{q_1}, \mathbf{q_0})$,
- **Triangle inequality:** $D_g(\mathbf{q_0}, \mathbf{q_1}) + D_g(\mathbf{q_1}, \mathbf{q_2}) \geq D_g(\mathbf{q_0}.\mathbf{q_2})$.

Lower bound on $D_g(\mathbf{q_0}, \mathbf{q_1})$. Let $\text{DISP}(\mathbf{q_0}, \mathbf{q_1})$ be the displacement metric, which is defined as the maximum Euclidean displacement of points on a model at these two configurations. It follows that $\text{DISP}(\mathbf{q_0}, \mathbf{q_1})$ is a lower bound of $D_g(\mathbf{q_0}, \mathbf{q_1})$:

$$D_g(\mathbf{q_0}, \mathbf{q_1}) \geq \text{DISP}(\mathbf{q_0}, \mathbf{q_1}).$$

Upper bound on $D_g(\mathbf{q_0}, \mathbf{q_1})$. In order to compute an upper bound for a 3D rigid object with translational and rotational DOFs, we first consider computing the D_g of by only varying a single DOF. Then, when we vary all the DOFs simultaneously, the final D_g would be less than or equal to the sum of the D_g's computed with respect to each DOF [Schwarzer et al. 2005; LaValle 2006].

When Euler angles are used to represent rotation, the upper bound on D_g can be calculated as:

$$D_g(\mathbf{q_0}, \mathbf{q_1}) \leq \Delta(\mathbf{q}_x) + \Delta(\mathbf{q}_y) + \Delta(\mathbf{q}_z) + R_\phi \Delta(\mathbf{q}_\phi) + R_\theta \Delta(\mathbf{q}_\theta) + R_\psi \Delta(\mathbf{q}_\psi), \quad (3)$$

where the Lipshitz constants R_ϕ, R_θ, R_ψ are the maximum Euclidean distances from any point on A to X, Y, Z axes in the local coordinate system, respectively. Δ denotes the difference of each DOF between these two configurations.

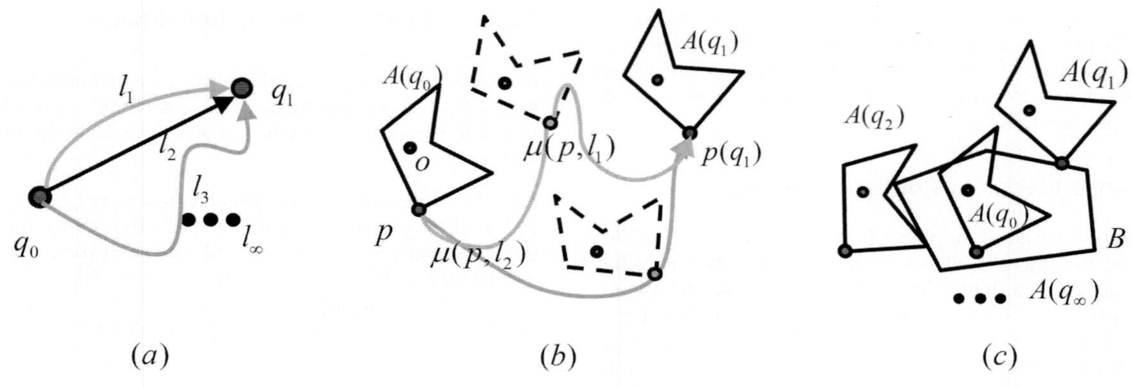

Figure 1: *Generalized penetration depth* PD^g *definition : (a) In the C-space, there are an infinite number of curves, such as* l_1, l_2, *that connect two configurations* $\mathbf{q}_0$ *and* $\mathbf{q}_1$. *(b) When the configuration of the object A changes along any curve l, any given point on A will trace out a distinctive trajectory in the 3D Euclidean space. This sub-figure shows the trajectories traced by* $\mathbf{p}$ *when A travels along* l_1 *and along* l_2, *while* $\mu(\mathbf{p},l_1)$ *and* $\mu(\mathbf{p},l_2)$ *are the arc-lengths of these trajectories, respectively. For each curve l, some point on A corresponds to the longest trajectory length as compared all other points on A. The distance metric* $D_g(\mathbf{q}_0,\mathbf{q}_1)$ *is defined as the minimum over longest trajectory lengths over all curves connecting* $\mathbf{q}_0$ *and* $\mathbf{q}_1$. *(c)* PD^g *is defined as the minimum of* $D_g(\mathbf{q}_0,\mathbf{q})$ *over all free configurations, which do not intersect with B (such as* $\mathbf{q}_1$, $\mathbf{q}_2$*).*

If the rotation vector is used to represent rotation, the upper bound can be calculated by:

$$D_g(\mathbf{q_0},\mathbf{q_1}) \leq \Delta(\mathbf{q}_x)+\Delta(\mathbf{q}_y)+\Delta(\mathbf{q}_z)+R\sum_{k=1}^{3}\Delta r_k, \tag{4}$$

where the constant R is the maximum Euclidean distance from the origin of A to every point on A. In section 6, we use these two upper bound formulae to compute an upper bound on PD^g.

3.3 PD^g Definition

Using D_g metric, we define our generalized PD, PD^g as:

$$\mathrm{PD}^g(A,B) = min(\{D_g(\mathbf{q}_0,\mathbf{q})|interior(A(\mathbf{q}))\cap B=\emptyset\}) \tag{5}$$

where $\mathbf{q}_0$ is the initial configuration of A, and $\mathbf{q}$ is in C-space (Fig. 1).

The translational PD^t defined by Eq. (1) is essentially a special case of PD^g. When an object A can only translate, all the points on A traverse the same distance. As a result, the distance metric $D(\mathbf{q}_0,\mathbf{q}_1)$ is equal to the Euclidean distance $\| \mathbf{q}_0-\mathbf{q}_1 \|$. In this case, Eq. (5) can be simplified to Eq. (1). Our generalized PD formulation has a geometric interpretation in C-space. PD^g is realized by some configuration $\mathbf{q}$ on the boundary of free space whose distance (D_g) to the given configuration $\mathbf{q}_0$ is the minimum.

In terms of handling general non-convex polyhedra, it is difficult to compute PD^g. This is due to the high combinational complexity of C-Space arrangement computation, which can be as high as $O(n^{12})$. However, reducing the problem to only dealing with convex primitives can significantly simplify the problem. In the following sections, we show that if both the input polyhedra are convex, their PD^g is equal to PD^t. Furthermore, if the complement of one of the polyhedra is convex, we reduce PD^g to a variant of a convex containment problem. In case of general non-convex polyhedra, we treat them as a combination of above two cases to compute a lower bound and an upper bound on PD^g.

4 PD^g Computation between Convex Objects

In this section, we consider the problem of computing generalized PD between two convex objects. In this case, we prove that PD^g is equal to PD^t. As a result, the well known algorithms to compute PD^t between convex polytopes [van den Bergen 2001; Kim et al. 2002b] are directly applicable to PD^g.

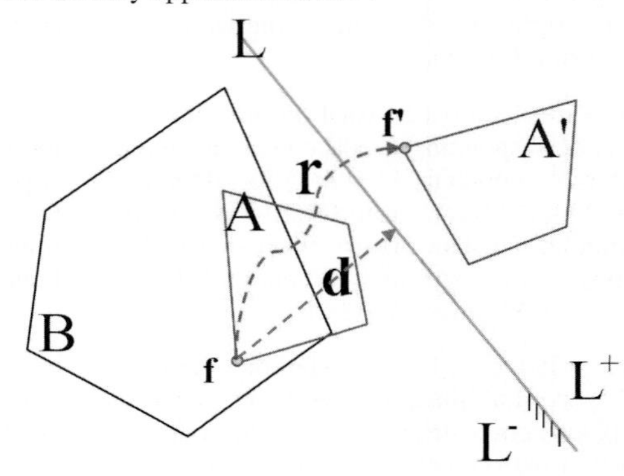

Figure 2: *Proof for* $\mathrm{PD}^t(A,B) = \mathrm{PD}^g(A,B)$ *for convex objects A and B. Let A′ a placement of A which realizes* PD^g. *L is an arbitrary separating plane between A′ and B, which divides the space into two half-spaces* L^- *and* L^+. *For any L, there always exists a point* $\mathbf{f}$ *on A on* L^- *side with* $||d|| \geq \mathrm{PD}^t(A,B)$. *As a result, we cannot move A towards L+ side with a traveling distance that is less than* $\mathrm{PD}^t(A,B)$ *even when rotational DOFs are allowed. Therefore, generalized PD is equal to translational PD for convex objects.*

Theorem 1 *Given two convex objects A and B, we have*

$$\mathrm{PD}^g(A,B) = \mathrm{PD}^t(A,B)$$

Proof Let us assume that A and B intersect, otherwise it is trivial to show that $\mathrm{PD}^g = \mathrm{PD}^t = 0$.

First of all, we can say that $\mathrm{PD}^g \leq \mathrm{PD}^t$, as PD^g is realized under more DOFs than PD^t . Next we show that $\mathrm{PD}^g < \mathrm{PD}^t$ is not possible and therefore, we can conclude $\mathrm{PD}^g = \mathrm{PD}^t$. We use a proof by contradiction.

Suppose $\mathrm{PD}^g < \mathrm{PD}^t$. Let us call A' as the placement of A that realizes PD^g, implying that A' is disjoint from B (Fig. 1). Since A' and B are convex, there exists a separating plane L that separates A' and B. Moreover, let L divide the entire space into two half-spaces: L^-, which contains B and L^+, which contains A'. Let $\mathbf{f}$ be the farthest point on A on L^- side from the separating plane L and $\mathbf{d}$ be

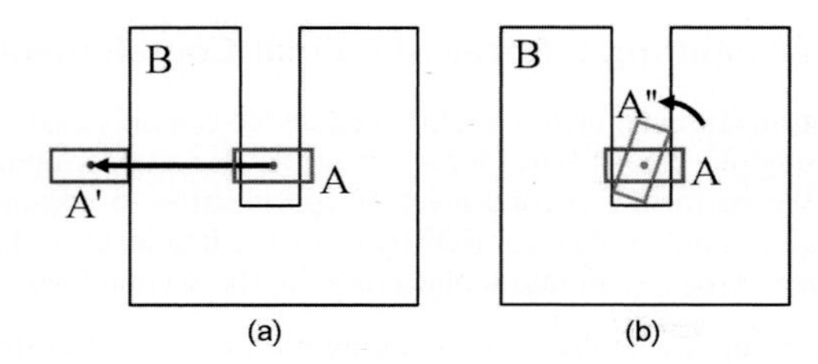

Figure 3: *An example of* $PD^g < PD^t$ *between convex A and non-convex B. The trajectory length that A travels is much shorter when both translation and rotation transformation are allowed (b) than the length when only translation is allowed (a).*

the vector from $\mathbf{f}$ to its nearest point on L. As a result, $||\mathbf{d}|| \geq PD^t$. Otherwise, we could separate A and B by translating A by $\mathbf{d}$, which would result in a smaller PD^t (i.e. $\mathbf{d}$) and this contradicts the definition of PD^t in Eq. 1.

Since $\mathbf{f}$, which is on L^- side, is at least PD^t far away from L, $\mathbf{f}$ must travel at least by PD^t to reach the new position $\mathbf{f}'$, which can be lying on L or contained in L^+. However, according to the definition of PD^g in Eq. (5) and the assumption of $PD^g < PD^t$, there must exist a trajectory l connecting $\mathbf{f}$ and $\mathbf{f}'$, whose arc-length is less than PD^t. This means that $\mathbf{f}$ could be moved to L or within L^+ by less than the amount of PD^t, which is contradictory to the earlier observation that $\mathbf{f}$ must travel at least by PD^t. Therefore, we conclude that L can not be a separating plane between A' and B.

The above deduction shows under the assumption that $PD^g < PD^t$, no separating plane can exist. This contradicts the fact that there must exist a separating plane when convex objects are disjoint. Therefore, $PD^g < PD^t$ is not possible and hence $PD^g = PD^t$. Q.E.D.

Corollary 1 *For two convex objects A and B, their generalized PD is commutative; i.e.,*

$$PD^g(A,B) = PD^g(B,A).$$

Proof For convex objects, $PD^g(A,B) = PD^t(A,B)$ and $PD^g(B,A) = PD^t(B,A)$). Since PD^t is commutative such that $PD^t(A,B) = PD^t(B,A)$, it follows that $PD^g(A,B) = PD^g(B,A)$. Q.E.D.

Non-Convex objects. Note that, for non-convex objects, $PD^g(A,B)$ is not necessarily equal to $PD^t(A,B)$. Figs. 3 and 4 show such examples. In Fig. 3, $PD^g(A,B) < PD^t(A,B)$, because the trajectory length that any point on A travels is shorter when both translation and rotation transformation are allowed (b) than its corresponding length when only translation is allowed (a). In Fig. 4, an object B, which could be infinitely large with a hole inside, can contain A only when A adjusts its initial orientation. Hence, the $PD^t(A,B) = \infty$ (i.e. the height of B), but $PD^g(A,B)$ is not ∞ (i.e. is much smaller than the height). So, $PD^g(A,B) < PD^t(A,B)$. We can also see that $PD^g(A,B)$ is not necessarily equal to $PD^g(B,A)$ in this example. If B is movable, the D_g metric for B at any two distinctive orientations is always ∞, because B is unbounded. Therefore, $PD^g(B,A)$ is ∞ in this case, while $PD^g(A,B)$ is not ∞.

5 PD^g Computation between a Convex Object and a Convex Complement

In this section, we show how to pose the generalized PD computation as a containment problem. Using this formulation, we investigate a special case of generalized PD where a movable object A and the complement of a fixed object B (i.e. $\bar{B}$) are both convex (as shown in Fig. 4). Instead of computing an exact solution, we compute an upper bound of PD^g by using a two-level optimization algorithm based on linear programming.

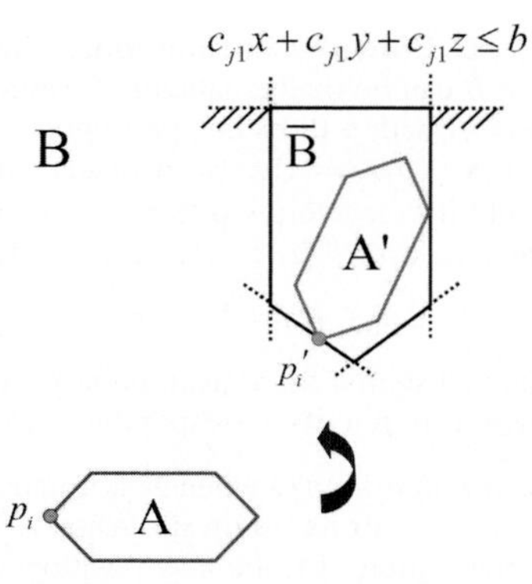

Figure 4: PD^g *between the convex object A and the object B whose complement -* $\bar{B}$ *is convex. In this case, the* $PD^g(A,B) \neq PD^t(A,B) = \infty$ *and* $PD^g(A,B) \neq PD^g(B,A) = \infty$*. We compute an upper bound on* PD^g *by reducing the problem to a variant of the convex containment problem by using linear programming.*

5.1 Relationship between PD^g and Object Containment

The general *object containment problem* can be stated as follows: given two objects P and Q, determine whether Q can contain P by performing translation and rotation transformation on P. The PD^g definition in Eq. (5) is closely related to the object containment problem. That is, testing $interior(A(\mathbf{q})) \cap B = \emptyset$ in Eq.(5) can be reduced to a containment problem: whether $\bar{B}$ can contain A, as shown in Fig. 4. However, there are a few differences between these two problems. The object containment problem finds one instance of a placement of A that can fit inside of $\bar{B}$, whereas PD^g computation needs to search through all valid containment configurations to find a configuration that minimizes the objective function D_g in Eq. (5).

The standard object containment problem is known to be difficult even for 2D polygonal models. However, if the primitives are convex, computational complexity of containment reduces from $O(m^3n^3log(mn))$ to $O(mn^2)$ for polygons with m and n vertices [Chazelle 1983; Avnaim and Boissonnat 1989]. As a result, we consider the case when a movable polyhedron A is convex and the complement of a fixed polyhedra B is convex as well. To compute an upper bound of PD^g for this case, our algorithm performs two levels of optimizations:

1. We compute a configuration $\mathbf{q_1}$ for A such that the convex container $\bar{B}$ contains $A(\mathbf{q_1})$. This is performed by minimizing their overlap. The valid containment yields an upper bound of PD^g, which may not be tight.
2. We iteratively compute a configuration, $\mathbf{q_2}$, to yield a tighter upper bound of PD^g by setting the upper bound of D_g metric in Eq. (4) as the objective function for optimization.

5.2 Computing a Containment

In this section, we introduce the formulation of the convex containment problem, and extend the 2D optimization-based algorithm described in [Milenkovic 1999; Grinde and Cavalier 1996] to 3D objects, both of which serve as a foundation of finding a locally-optimal containment.

Formulation of 3D Convex Containment. To check whether A fully lies inside $\bar{B}$ can be mathematically formulated as follows. The convex object, $\bar{B}$ with n faces is represented as an intersection of n half-spaces $\mathbf{c}_j\mathbf{x} \leq b_j, j = 1,...,n$. A placement of A lies fully inside $\bar{B}$ if and only if every vertex $\mathbf{p}_i(i = 1,...,m)$ on A lies inside all the half-spaces, i.e. $\mathbf{c}_j\mathbf{p}_i \leq b_j, i = 1,...,m, j = 1,...,n$, or:

$$C\mathbf{p}_i \leq \mathbf{b}, i = 1,...,m,. \tag{6}$$

Here $\mathbf{c}_j$ is normalized so that for a given point $\mathbf{p}$, $|\mathbf{c}_j \cdot \mathbf{p} - b_j|$ is the Euclidean distance from $\mathbf{p}$ to its corresponding face j.

Denote R as the *rotation matrix* when A is rotated around an arbitrary axis with respect to its origin $\mathbf{o}$. When A is rotated by R, followed by the translation of $\mathbf{t}$, the new position of $\mathbf{p}$ in A can be calculated as:

$$\mathbf{p}' = R(\mathbf{p} - \mathbf{o}) + \mathbf{o} + \mathbf{t}. \tag{7}$$

Using the above notation, the 3D containment problem now can be stated as finding a solution to the following system:

$$\mathbf{C}\mathbf{p}'_\mathbf{i} \leq \mathbf{b}, i = 1,...,m. \tag{8}$$

Linearizing 3D Convex Containment Problem. The 3D containment computation is a non-linear problem, as the rotation matrix R is embedded with non-linear terms. These non-linear terms could be linearized by using a *small-angle approximation* [Milenkovic and Schmidl 2001]. When A is rotated by α around an arbitrary axis $\hat{\mathbf{a}}$, its rotation vector $\mathbf{r}$ is equal to $\alpha\hat{\mathbf{a}}$. If the variation of a rotation angle α is small enough, we can get a linearized approximation for Eq. (7) can be obtained:

$$\tilde{\mathbf{p}} \approx \mathbf{p} + \mathbf{r} \times (\mathbf{p} - \mathbf{o}) + \mathbf{t}. \tag{9}$$

By replacing $\mathbf{p}'$ by its approximation $\tilde{\mathbf{p}}$, the non-linear system in Eq. 8 is simplified to a linear one:

$$g_{ij} = \mathbf{c}_j \cdot \mathbf{t} - (\mathbf{c}_j \times (\mathbf{p}_i - \mathbf{o})) \cdot \mathbf{r} + (\mathbf{c}_j \cdot \mathbf{p}_i - b_j) \leq 0, \ \forall i, j. \tag{10}$$

Here $\mathbf{t}$ and $\mathbf{r}$ are the unknown vectors. g_{ij}, which is called as *containment function*, is defined for each pair of the vertex on A and the face of $\bar{B}$:

In order to solve the linear system defined in Eq. (10), a slack variable d_{ij} is introduced to represent the distance from $\tilde{\mathbf{p}}_i$, the approximate position of $\mathbf{p}_i$ on A after it is transformed, to the j'th face on $\bar{B}$. In this case, the 3D convex containment constraint for A and $\bar{B}$ can be approximated as a linear programming problem (LP1):

$$\begin{aligned} &\min \ Z = \sum_{i=1}^{m} \sum_{j=1}^{n} d_{ij}, \\ &\text{subject to } \ g_{ij}(\mathbf{t}, \mathbf{r}) - d_{ij} \leq 0 \ \forall i, j. \end{aligned} \tag{11}$$

If $Z = 0$ for this optimization problem, we end up computing a solution to Eq. (10).

Containment Computation. Given A and $\bar{B}$, we construct a linear programming problem defined as in Eq. (11) and apply the standard linear programming technique to optimize its objective function Z. We compute the solution, say $(\mathbf{t}, \mathbf{r})$, and place A at A'. A new linear programming formulation (like LP1) is constructed for A' and solved iteratively until a local minimum for Z is computed. As the algorithm iterates, the small-angle approximation for the rotation matrix R becomes more accurate. When the objective Z approaches zero, a valid containment of A at configuration $\mathbf{q}_1$ has been found.

5.3 Computing a Locally-Optimal Containment

The optimization algorithm highlighted above can only find a valid containing placement $A(\mathbf{q}_1)$ for A, which yields an upper bound for PD^g. We perform a second level of optimization to compute an even tighter upper bound for PD^g by using the first level containing placement $A(\mathbf{q}_1)$ as an initial placement for the second level.

Let $\mathbf{q}_0 = (\mathbf{t}_0, \mathbf{r}_0)$ be the initial configuration of A used in the first level optimization. Let $\mathbf{q}_1 = (\mathbf{t}_1, \mathbf{r}_1)$ be the configuration for the containing placement A as a result of the first level optimization. Our goal is to compute $\Delta\mathbf{q} = (\Delta\mathbf{t}, \Delta\mathbf{r})$, such that $\mathbf{q}_2 = (\mathbf{t}_1 + \Delta\mathbf{t}, \mathbf{r}_1 + \Delta\mathbf{r})$ yields another containing placement of A while $D_g(\mathbf{q}_0, \mathbf{q}_2) < D_g(\mathbf{q}_0, \mathbf{q}_1)$.

We perform the second level optimization by setting the upper bound on the D_g metric in Eq. (4) as an optimization objective function. Here we do not choose Eq. (3), because the 3D containment computation uses the notations of rotation vector. By imposing that A needs to be contained by $\bar{B}$ as a hard constraint, we get the system:

$$\begin{aligned} &\min \ Z = \sum_{k=1}^{3} |\Delta t_k + t_{1,k} - t_{0,k}| + R\sum_{k=1}^{3} |\Delta r_k + r_{1,k} - r_{0,k}|, \\ &\text{subject to } \ g_{ij}(\Delta\mathbf{t}, \Delta\mathbf{r}) \leq 0 \ \forall i, j, \end{aligned} \tag{12}$$

where $t_{0,k}$ and $r_{0,k}$ are, respectively, the kth translational and rotational DOF for an initial configuration of A, and similarly $t_{1,k}$ and $r_{1,k}$ are the kth DOF for a configuration as a result of the first level optimization, and Δt_k, Δr_k are the variables. Note, now the *containment function* g_{ij} is computed from every vertex of $A(\mathbf{q}_1)$ (instead of $A(\mathbf{q}_0)$) and each face of $\bar{B}$. Let us further set $u_k = \Delta t_k + t_{1,k} - t_{0,k}$ and $v_k = \Delta r_k + r_{1,k} - r_{0,k}$. In this case, we can rewrite the second level optimization problem in Eq. (12) as:

$$\begin{aligned} &\min \ Z = \sum_{k=1}^{3} |u_k| + R\sum_{k=1}^{3} |v_k|, \\ &\text{subject to } \ g^1_{ij}(\mathbf{u}, \mathbf{v}) \leq 0 \ \forall i, j, \end{aligned} \tag{13}$$

where g^1_{ij} is obtained from g_{ij} in Eq. 12 by the change of variables: $\mathbf{u} = \Delta\mathbf{t} + \mathbf{t}_1 - \mathbf{t}_0$ and $\mathbf{v} = \Delta\mathbf{r} + \mathbf{r}_1 - \mathbf{r}_0$.

The objective function in the optimization system (Eq. (13)) contains absolute arithmetic operations. We replace $|u_k|$ with $u^+_k + u^-_k$ in the objective function, and u_k with $u^+_k - u^-_k$ in the containment function g^1_{ij}, where $u^+_k, u^-_k \geq 0$ for $k = 1,2,3$. A similar replacement is performed for v_k. After the replacement, finally we formulate this optimization problem as a linear programming problem:

$$\begin{aligned} &\min \ Z = \sum_k (u^+_k + u^-_k) + R(v^+_k + v^-_k), \\ &\text{subject to } \ g^2_{ij}(u^+, u^-, v^+, v^-) \leq 0 \ \forall i, j \\ &u^+_k, u^-_k, v^+_k, v^-_k \geq 0, k = 1,2,3. \end{aligned} \tag{14}$$

where g^2_{ij} is obtained from g^1_{ij} by the change of variables.

By solving Eq. (14), we get $u^+_k, u^-_k, v^+_k, v^-_k$. Using the solution, we can compute $\Delta t^+_{1,k}, \Delta t^-_{1,k}, \Delta r^+_{1,k}$ and $\Delta r^-_{1,k}$, which yields a new configuration $\mathbf{q}_2$ to replace $\mathbf{q}_1$. This process is iterated until the objective Z in Eq. (14) converges to a local minimum. At this stage, since $u^+_k, u^-_k, v^+_k, v^-_k$ are zeroes, our small-angle approximation becomes accurate and A is forced to be disjoint from B. After computing an optimal containing placement $\mathbf{q_2}$ of A, we compute an upper bound on PD^g using Eq. (3).

6 PD^g Estimation for Non-Convex Objects

In this section, we present our algorithm to efficiently compute a lower bound and an upper bound on PD^g between non-convex objects. Our algorithm is built on the properties of PD^g, presented in Section 4 and Section 5.

6.1 Lower Bound on PD^g

Our algorithm to compute a lower bound on PD^g is based on the fact that PD^g is equal to PD^t for convex polyhedra. As a result, we compute a lower bound of PD^g by first computing the inner-convex covers for each input models The inner-convex cover refers to a set of convex pieces whose union is a subset of the original model [Milenkovic 1998; Cohen-Or et al. 2002]. Next, we take the maximum value of PD^t_i's between all pairwise combinations of convex pieces. The overall algorithm proceeds as:

1. As a preprocessing, compute inner-convex covers for A and B i.e., $\cup A_i \subseteq A$ and $\cup B_i \subseteq B$ where A_i, B_i are convex sets, but are not necessarily disjoint from each other.
2. During the run-time query, place A_i at the configuration $\mathbf{q}$, i.e. compute $A_i(\mathbf{q})$.
3. For each pair of $(A_i(\mathbf{q}), B_j)$ where $i = 1, \ldots, M$ and $j = 1, \ldots, N$,
   (a) Perform collision detection to check for overlaps.
   (b) If the pair overlaps, let $PD^g_k = PD^t((A_i(\mathbf{q}), B_j)$; otherwise $PD^g_k = 0$, where $k = 1, \ldots, MN$.
4. Finally, $PD^g = \max(PD^g_k)$ for all k.

6.1.1 Translational Penetration Depth Computation

In our method, the lower bound on generalized PD^g computation is decomposed into a set of PD^t queries among convex primitives. The PD^t between two convex polyhedra can be computed using the algorithms presented in [Cameron 1997; van den Bergen 2001; Kim et al. 2002b]. These methods compute PD^t by calculating the minimum distance from the origin to the surface of the Minkowski sum of the two convex polyhedra.

Since we are computing a lower bound to PD^g, this imposes that the PD^t computation algorithm used by our method should compute an exact value or a lower bound to the PD^t. In particular, the algorithm proposed by Cameron [1997] satisfies this requirement and Gino's algorithm [2001] also provides a tight lower bound.

6.1.2 Acceleration using Bounding Volume Hierarchy

Our lower bound to PD^g computation can be accelerated by employing a standard bounding volume hierarchy. For two disjoint convex pieces, their PD^t corresponds to zero. Typically there are many disjoint pairwise combinations of convex pieces (A_i, B_j). We detect such disjoint pairs using an oriented bounding box (OBB) [Gottschalk et al. 1996] hierarchy and prune them away.

6.1.3 Analysis

The computational complexity of the lower bound PD^g is determined by the number of convex pieces decomposed from the robot A and the obstacle B, and the geometric complexity of these convex pieces, which is determined by the total number of features of the resulting convex pieces. Let m, n be the number of the convex pieces of A and B, respectively. Let the geometric complexity of the convex pieces of A and B be a and b, respectively. Then, the average numbers of features in each piece of A and B are $\frac{a}{m}$ and $\frac{b}{n}$, respectively. Using computational complexity of translational PD, we can derive that the computational complexity of PD^g for 2D rigid objects is $O(an + bm)$, and for 3D rigid objects is $O(ab)$.

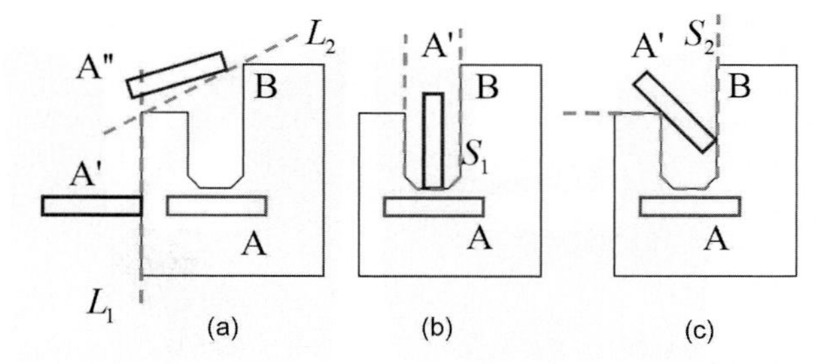

Figure 5: *Separating plane, convex separator and non-convex separator: (a). L_1 and L_2 are separating planes, which separate A' and B, and A'' and B respectively. (b). S_1 is a separator, which is composed by a set of piece-wise linear plane. S_1 separates A' from B. A separator is called convex (i.e. S_1), if it lies on the boundary of its convex hull. (c). A non-convex separator S_2 separates A from B'.*

6.2 Upper Bound on PD^g

One simple way to compute an upper bound to PD^g for general non-convex objects is to compute the PD^t between their convex hulls. This corresponds to an upper bound because $PD^g(A,B) \leq PD^g(\mathrm{CH}(A), \mathrm{CH}(B))$, and the latter is equal to $PD^t(\mathrm{CH}(A), \mathrm{CH}(B))$, thanks to Theorem 1. In practice, this upper bound is relative simple to compute. However, this algorithm could be overly conservative for non-convex models, as shown in Figs. 3 and 4.

$PD^t(A,B)$ is also an upper bound on $PD^g(A,B)$. However, this can result in a conservative upper bound in practice. Since the computational complexity of exact computation of $PD^t(A,B)$ for non-convex models can be high, current approaches typically compute an upper bound of $PD^t(A,B)$ [Kim et al. 2002a].

We present an algorithm to compute an upper bound on PD^g for non-convex polyhedra by reducing this problem to a set of containment optimization sub-problems (as defined in Section 5).

6.2.1 Algorithm Overview

Given two disjoint non-convex objects A and B, there is either a single separating plane between the objects (as shown in Fig. 5-(a)) or there is a set of piecewise linear surfaces, which is called a *separator* [Mount 1992]. (Figs. 5 -(b) and -(c)). More precisely, the separator is defined as a simple piece-wise linear surface that divides the space into two half-spaces. The separator can be an open surface or a closed surface. A separator S is convex if and only $S \subset \partial(CH(S))$, as shown in Fig. 5-(b). Otherwise, the separator is non-convex, as shown in Fig. 5-(c). A single separating plane can be regarded as a special case of a separator. However, we specifically use the term separator to refer to the non-plane separator.

Our upper bound $PD^g(A,B)$ computation algorithm proceeds as follows: during the preprocessing phase, we enumerate all possible separating planes and convex separators by analyzing the convexity of the boundary of B. During the query phase, for each separating plane L (or each convex separator S), we compute an upper bound on D_g distance when A is separated from B with with respect to the separating plane L (or separator S) using the technique described in Sec. 5. The minimum over all these upper bounds yields a global upper bound on PD^g. Now we explain how to efficiently enumerate L and S as part of the preprocessing step.

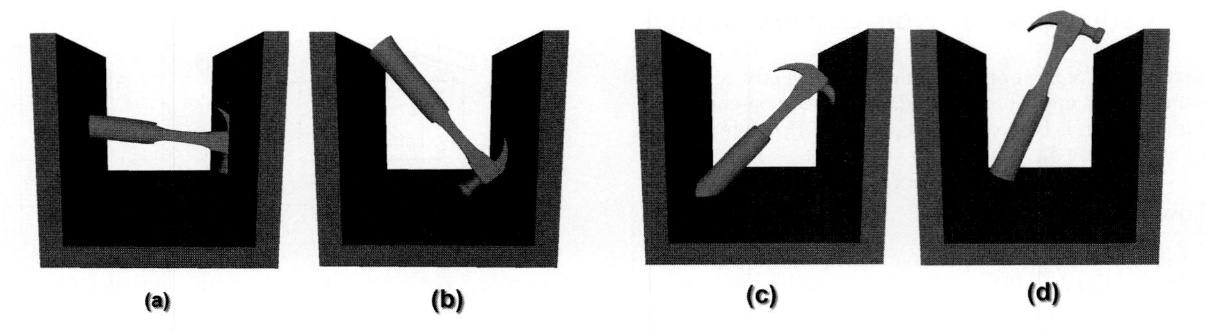

Figure 6: *The 'hammer' example: (a) When the 'hammer' is at time t=0, it collides with the 'notch'. (b) The collision-free placement of the 'hammer' for scenario (a). We use our containment optimization algorithm to get this free configuration, which realizes the* $UB_1(\mathrm{PD}^g)$*. (c) The 'hammer' at time t=0.5. (d) The collision-free placement is computed for scenario to get the* $UB_1(\mathrm{PD}^g)$

6.2.2 Separating Planes

The set of all possible separating planes is included in the complement of the convex hull of B. According to Theorem 1, $\mathrm{PD}^g = \mathrm{PD}^t$ for convex objects, and the minimum D_g distance with respect to all these separating planes is $\mathrm{PD}^t(\mathrm{CH}(A), \mathrm{CH}(B))$. This means that the computation of $\mathrm{PD}^t(\mathrm{CH}(A), \mathrm{CH}(B))$ implicitly takes into account all possible separating planes. Therefore, we need not enumerate any separating planes explicitly during the preprocessing phase.

6.2.3 Convex Separators

Any separator S divides the whole space into two half-spaces. One half-space would include the object B. We can regard the other half-space as a container. Placing A inside the container is equivalent to making A and B disjoint with respect to each separator S. Therefore, the computation of the minimum D_g distance for S can be regarded as a 3D convex containment optimization problem. By applying two levels of linear programming optimization algorithm, discussed in Sec. 5, we compute an upper bound of PD^g for each convex separator S. The minimum of all PD^g over all enumerated convex separators yields an upper bound on PD^g.

6.2.4 Convex Separators Enumeration

Enumerating convex separators of B can be performed as a preprocessing. This step can be regarded as computing a convex covering of the complement space of B. Given the fact that we are computing an upper bound of PD^g, the conservativeness of the separator enumeration does not affect the correctness of our algorithm.

We use the surface convex decomposition for the complement space of B [Ehmann and Lin 2001]. We discard the surface with one face from the surface decomposition, since these planes have been processed as separating planes.

Moreover, if the geometry of input A and B is very complex, i.e. high polygon count or a number of features, we compute a simplification of each primitive to compute a coarser model A', B'. If $A \subseteq A'$ and $\bar{B} \subseteq \bar{B}'$, it is easy to prove that $\mathrm{PD}^g(A,B) \le \mathrm{PD}^g(A',B')$. Therefore, we can compute the upper bound by applying our algorithm on these simplified models.

6.2.5 Separator Culling

We can cull some of the separators by making use of the currently known upper bound on PD^g during any stage of the algorithm. If the separator is farther away from the object A than the current upper bound, we can discard this separator. We use the PD^t between the two convex hulls of input models as an initial upper bound of PD^g.

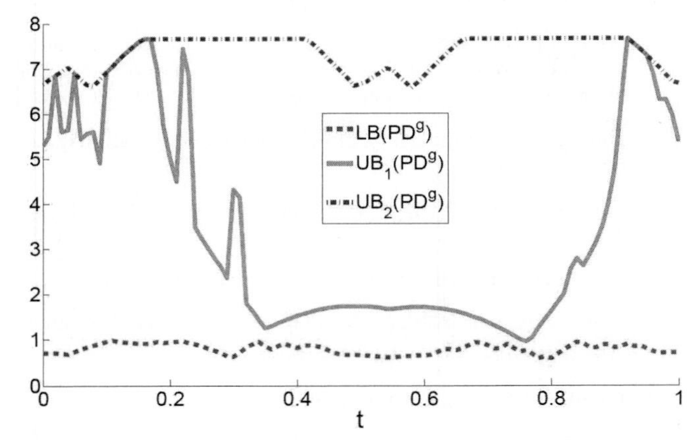

Figure 7: *Comparison of lower and upper bounds on* PD^g *for 'hammer' example. The lower and upper bounds on* PD^g *between the 'hammer' and the 'notch' models are computed over all interpolated configurations. The dash-dot blue curve* $LB(\mathrm{PD}^g)$ *stands for the lower bound of* PD^g *by computing pairwise translational PD. The dashed red curve* $UB_2(\mathrm{PD}^g)$ *stands for the upper bound of* PD^g *computed by the translational PD of their convex hull. The solid green curve* $UB_1(\mathrm{PD}^g)$ *highlights the upper bound of* PD^g *by using our containment optimization, which always lies between* $LB(\mathrm{PD}^g)$ *and* $UB_2(\mathrm{PD}^g)$*. In this example,* $UB_1(\mathrm{PD}^g)$ *is less than* $UB_2(\mathrm{PD}^g)$ *for most of time* t.

7 Implementation and Performance

We have implemented our lower and upper bound computation algorithms for generalized PD computation between non-convex polyhedra. We have tested our algorithms for PD^g on a set of benchmarks, including 'hammer' (Fig. 6), 'hammer in narrow notch' (Fig. 9), 'spoon in cup' (Fig. 8) and 'pawn' (Fig. 10) examples. All the timings reported in this section were taken on a 2.8GHz Pentium IV PC with 2 GB of memory.

7.1 Implementation

Lower bound on PD^g**.** In our implementation, the convex covering is performed as a preprocessing step. Currently, we use the surface decomposition algorithm proposed by [Ehmann and Lin 2001], which can be regarded as a special case of convex covering problem. In order to compute the PD^t between two convex polytopes, we use the implementation available as part of SOLID [van den Bergen 2001]. In order to accelerate this algorithm, we precompute

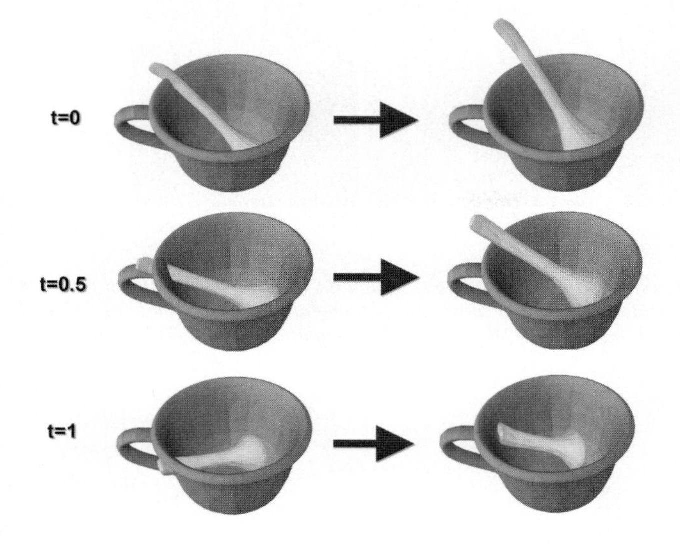

Figure 8: *The 'cup' example. The left column shows the placements of the 'spoon' in the 'cup', when t=0.0, t=0.5, and t=1.0, respectively. At all of these placements, the 'spoon' collides with the 'cup'. The right column shows the collision-free configurations which are realized for $UB_1(\mathrm{PD}^g)$ at each t.*

an OBB hierarchy [Gottschalk et al. 1996] and use the bounding volumes to conservatively cull convex pairs that do not intersect with each other.

Upper bound on PD^g. The preprocessing step of convex separator enumeration can be regarded as convex decomposition of the complement of the input model. In our implementation, we used the surface decomposition algorithm to generate a set of convex surfaces [Ehmann and Lin 2001] and discard the surfaces that have only one face. For each convex separator, we use the containment optimization technique developed in Sec. 5 to compute an upper bound on PD^g. Moreover, we use the *QSopt* [1] package to solve the linear programming problems. In order to accelerate the upper bound computation, we conservatively cull the convex separators that are farther away than the current upper bound on PD^g.

7.2 Performance

We use different benchmarks to test the performance. Our experimental setup is as follows. Each benchmark includes two polyhedral models A and B, where A is movable and B is fixed. The model A is assigned a staring configuration $\mathbf{q}_0$ and an end configuration $\mathbf{q}_1$. We linearly interpolate between these two configurations with n intermediate configurations (i.e. n samples). For each interpolated configuration $\mathbf{q} = (1-t)\mathbf{q}_0 + t\mathbf{q}_1, t \in [0,1]$, we compute various bounds for PD^g between $A(\mathbf{q})$ and B, including:

1. $LB(\mathrm{PD}^g)$: The lower bound on PD^g based on pairwise translational PD^t computation.
2. $UB_1(\mathrm{PD}^g)$: The upper bound on PD^g computed by containment optimization.
3. $UB_2(\mathrm{PD}^g)$. The upper bound on PD^g based on the translational PD^t computation between their convex hull.

[1] http://www2.isye.gatech.edu/~wcook/qsopt/

In order to get accurate timing profiling, we run our PD algorithms for each configuration with a batch number b. The average time for each bound computation is the total running time on all samples over the product of the number of samples and the batch number b.

'Hammer' example. Fig. 6, and Tab. 1 and Fig. 7 show the results and timings for the 'hammer' example. In this case, the 'hammer' model has 1,692 triangles, which is decomposed into 214 convex pieces. The 'notch' model has 28 triangles, which is decomposed into 3 convex pieces and there is a notch (i.e. convex separator) in the center of the 'notch' model. Initially (at t=0), the 'hammer' intersects with the 'notch' as shown in Fig. 6(a). Fig. 6(b) shows a collision-free placement of the 'hammer', which corresponds to the position after moving by $UB_1(PD^g)$. According to Fig. 7, the value is $UB_1(\mathrm{PD}^g) = 4.577083$, which is greater than $LB(\mathrm{PD}^g)$ (0.744020) and less than $UB_2(\mathrm{PD}^g)$ (6.601070).

For this example, we generate 101 samples for the 'hammer' when it is rotated around the Z axis. The rotation motion is linearly interpolated from the configuration $(0,0,0)^T$ to $(0,0,\pi)^T$. Fig. 6(c) shows the placement of the 'hammer' at $t = 0.5$. Fig. 6(d) is the corresponding collision-free placement, which realizes the $UB_1(\mathrm{PD}^g)$.

We also compare the lower and upper bounds on PD^g over all the configurations. In Fig. 7, the solid green curve highlights the value of $UB_1(\mathrm{PD}^g)$ between the 'hammer' and the 'notch' over all interpolated configurations. The dashed red curve, which corresponds to $UB_1(\mathrm{PD}^g)$, always lies between $LB(\mathrm{PD}^g)$ and $UB_2(\mathrm{PD}^g)$. In this example, $UB_1(\mathrm{PD}^g)$ is less than $UB_2(\mathrm{PD}^g)$.

The timing for this example is shown in Tab. 1. We run the PD^g algorithm 5 times (b=5) for all the configurations (n=101). The average timing for $LB(\mathrm{PD}^g)$, $UB_1(\mathrm{PD}^g)$, and $UB_2(\mathrm{PD}^g)$ is 1.901ms, 21.664ms and 0.039ms respectively.

'Hammer in narrow notch' example. We perform a similar experiment on 'Hammer in narrow notch' example (Fig. 9) to test the robustness of our algorithm. This example is modified from the 'hammer' example, where the size of the notch is decreased such that there is only narrow space for the 'hammer' to fit inside. Our algorithm can robustly compute the lower and upper bounds on PD for this example. Fig. 11 compare the lower and upper bounds on PD^g over all sampled configurations (n=101). The third row of Tab. 1 shows the performance of our algorithm for this example.

'Spoon in cup' example. We apply our algorithm on more a complex scenario such as shown in Fig. (8). In this example, the 'spoon' model has 336 triangle and is decomposed into 28 convex pieces. The 'cup' model has 8,452 triangles. We get 94 convex pieces and 53 convex separators after simplifying the original model to 1,000 triangles.

In Fig. 8, the left column shows the placements of the 'spoon' in the 'cup', corresponding to $t = 0.0$, $t = 0.5$, and $t = 1.0$, respectively. At all these placements, the 'spoon' collides with the 'cup'. The right column of this figure shows the collision-free configurations that are computed based on $UB_1(\mathrm{PD}^g)$ in each case. We also compare our computed lower bound and upper bounds over all the samples (n=101), which is shown in Fig. 8. The timing performance for this example is also listed on Tab. 1.

'Pawn' example. The last benchmark used to demonstrate the performance of our algorithm is the 'pawn' example. As Fig. 10 shows, the large 'pawn' is fixed, while the small one is moving. The

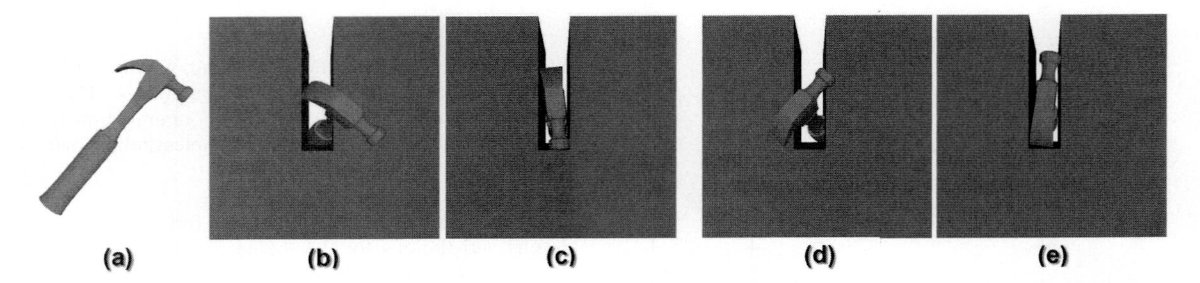

Figure 9: *The 'hammer in narrow notch' example. This example is modified from the 'hammer' example, where the size of the notch is decreased such that there is only narrow space for the 'hammer' to fit inside. (b) and (d) shows the placement of the 'hammer' at t=0 and t=0.5. (c) and (e) are their corresponding configurations respectively, which realize the $UB_1(\mathrm{PD}^g)$. The computed $UB_1(\mathrm{PD}^g)$ is tighter than the $UB_2(\mathrm{PD}^g)$ for most of time t.*

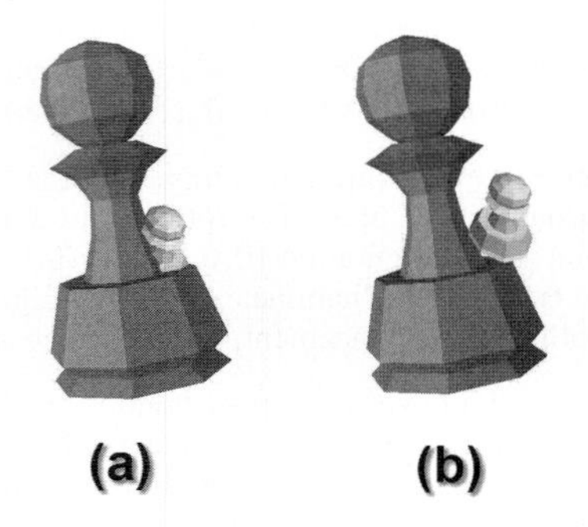

Figure 10: *The 'pawn' example. The large 'pawn' is fixed and the small one is movable. (a) shows the colliding placement of the 'pawn' at t = 0. (b) shows its corresponding collision-free placement, which is computed based on $UB_1(\mathrm{PD}^g)$.*

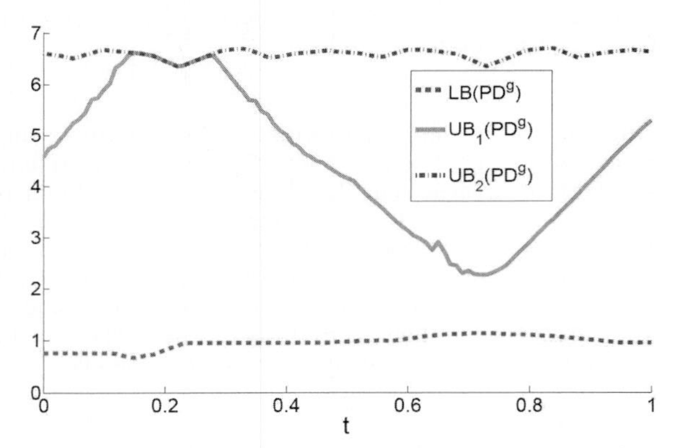

Figure 11: *Comparison of lower and upper bounds on PD^g for the 'hammer in narrow notch' example.*

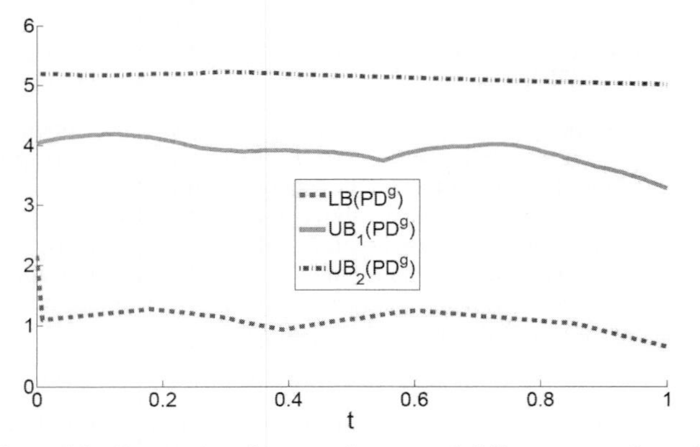

Figure 12: *Comparison between lower and different upper bounds on PD^g for 'cup' example.*

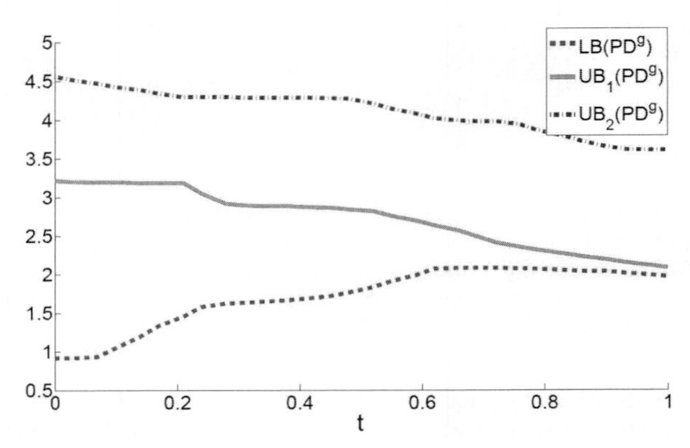

Figure 13: *Comparison of lower and upper bounds on PD^g for the 'pawn' example.*

'pawn' model has 304 triangles and is decomposed into 44 convex pieces. The large 'pawn' has 43 convex separators. Fig. 10(a) shows the colliding placement of the 'pawn' at t = 0. Fig. 10(b) shows its corresponding collision-free placement, which is computed based on $UB_1(\mathrm{PD}^g)$. 13 compares the lower bound and upper bounds over the sampled configuration (n=101). Tab. 1 shows the average time to compute the lower and upper bounds over all configurations.

8 Application to Motion Planning

In this section, we apply our lower bound on PD^g computation algorithm for complete motion planning of planar robots with 3-DOF. The complete motion planning checks for the existence of a collision-free path or reports that no such path exists. It is different from motion planning algorithms based on random sampling, which can not check for path non-existence.

8.1 C-obstacle Query

We mainly use our lower bound on PD^g computation algorithm to perform the *C-obstacle query*. This query for a given C-space is formally defined as checking whether the following predicate P is always true [Zhang et al. 2006b]:

$$P(A,B,Q):\ \forall \mathbf{q} \in Q,\ A(\mathbf{q}) \cap B \neq \emptyset \tag{15}$$

Here, A is a robot, B represents obstacles and Q is a C-space primitive or a cell; $A(\mathbf{q})$ represents the placement of A at the configura-

	Hammer	H2 ⋆	Spoon	Pawn
A	Hammer	Hammer	Spoon	Small
tris #	1,692	1,692	336	304
convex pieces #	215	215	28	44
B	Notch	Notch	Cup	Large
tris #	28	28	8,452	304
convex pieces #	3	3	94	44
separator #	1	1	53	43
sample # (n)	101	101	101	101
batch # (b)	5	5	5	5
t for LB_1 (ms)	1.901	4.300	6.127	4.112
t for UB_1 (ms)	21.664	108.024	1027.014	482.511
t for UB_2 (ms)	0.039	0.053	0.154	0.055

Table 1: *This table highlights the benchmarks used to test the performance of our algorithms. The top rows in the table list the model complexity and the bottom rows report the time taken to compute the lower and upper bounds to* PD^g *on a 2.8GHz Pentium IV PC. 'H2⋆' is the example 'hammer in narrow notch'.*

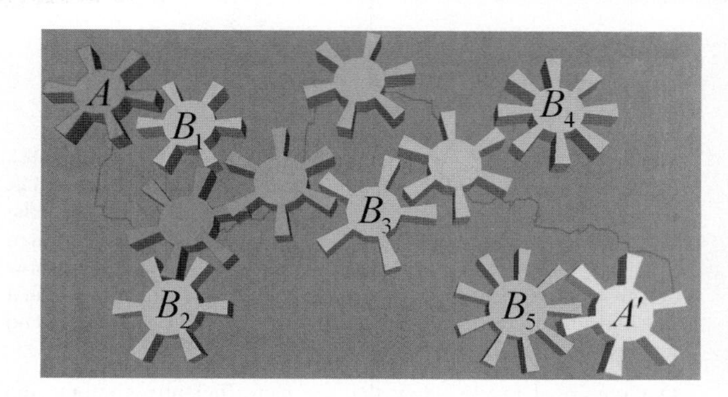

Figure 14: *This figure illustrates an application of our C-obstacle query algorithm to speedup a complete motion planner - the star-shaped roadmap algorithm. In this example, the object Gear needs to move from initial configuration A to goal configuration A′ by translating and rotating within the shaded rectangular 2D region. We show the robot's intermediate configurations for the found path. Using our C-obstacle query, we can achieve about 2.4 times speed up for the star-shaped roadmap algorithm for this example.*

tion **q**. Q may be a line segment, a cell or a contact surface that is generated from the boundary features of the robot and the obstacles.

The C-obstacle query is useful for cell decomposition based algorithms for motion planning [Latombe 1991]. These algorithms subdivide the configuration space into cells and need to check whether a cell is fully contained either in the *free space* or in *C-obstacle space*. The free space is the set of all collision-free configurations of the robot. The C-obstacle space is the complement of the free-space. The *C-obstacle query* checks whether a subset of the C-space (i.e. Q) fully lies in the C-obstacle space.

The *C-obstacle query* also arises in sampling based approaches for motion planning, especially complete motion planning. These include the star-shaped roadmap algorithm [Varadhan and Manocha 2005], which is a deterministic sampling algorithm and subdivides the configuration space into a collection of cells in a hierarchical fashion. Given that the time and space complexity of these methods grows quickly with the level of subdivision, it is important to identify cells that lie in C-obstacle space and no further subdivision is executed.

Another benefit of the *C-obstacle query* is to determine non-existence of any collision-free path. The methods in [Zhang et al. 2006a; Varadhan and Manocha 2005] conclude that no path exists between the initial and goal configurations if they are separated by C-obstacle space. These methods can be performed using the *C-obstacle query* to identify these regions which lie in C-obstacle space.

In order to efficiently perform *C-obstacle query* for any cell in C-space, we compute the PD^g by setting its configuration as the center of the cell. Then we compare it with the maximal motion that the robot can undergo when its configuration is confined within a cell [Schwarzer et al. 2005]. If the lower bound of PD^g is larger than the upper bound of the maximal motion, we conclude that the cell (i.e. Q) fully lies in C-obstacle space [Zhang et al. 2006b].

8.2 Experimental Results

We apply our *C-obstacle query* algorithm to improve the performance of a deterministic sampling motion planning algorithm - the star-shaped roadmap method by [Varadhan and Manocha 2005]. To demonstrate the effectiveness of our C-obstacle cell query, we define the *cell culling ratio* as the number of cells in C-obstacle space identified by our query algorithm over the total number of cells in C-obstacle space.

	Gear
Cell Culling Ratio	75.21%
Time Per Cell Culling(ms)	0.12
Time of Original Method(s)	261.4
Time of Accelerated Method(s)	110.4
Speedup	2.4
Time for C-obstacle Cell Query(s)	13.3

Table 2: *Performance for C-obstacle Cell Query: For the Gear example, our query can identify about 75.21% C-obstacle cells. The average query time is about 0.12ms. Based on* PD^g *computation and C-obstacle query, we improve the performance of the star-shaped motion planning algorithm by 2.4 times in this case.*

Tab. 2 illustrates that our *C-obstacle query* algorithm can achieve 75.21% cell culling ratio in our *Gear* benchmark. Tab. 2 also shows that the average time for each *C-obstacle query* in the *Gear* example is about $0.12ms$. In this complex 2D scenario, the *C-obstacle query* algorithm improves the performance of the motion planning algorithm by 2.4 times.

9 Limitations

Our PD^g computation algorithm has a few limitations. Given the complexity of exact PD^g computation for non-convex polyhedra, we only compute lower and upper bounds and not the exact answer. Moreover, the convex containment optimization algorithm that linearizes the rotational component can not guarantee a global minimum. The bounds computed by our algorithm also depend on convex covering and separator enumeration of the non-convex polyhedra, performed as part of preprocessing step. As a result, we are unable to provide any tight bounds on the approximation to PD^g computed by our algorithm. However, in most practical cases the extent of penetration is small and we expect that our algorithm would compute a good approximation.

10 Conclusions and Future Work

We have addressed the problem of generalized PD computation between non-convex models, which takes into account translational as well as rotational motion. To the best of our knowledge, this is the first algorithm for general 3D polyhedra models. We present three main results related to PD^g computation. Specifically, we show that for convex models, generalized PD is the same as translational PD. We also present practical algorithms to compute the upper and lower bounds on PD^g for non-convex models.

Our empirical results show that we can efficiently compute the lower and upper bounds of generalized PD for non-convex objects. We also use our algorithm for complete motion planning of polygonal robots with 3-DOF C-space.

Future Work. There are many avenues for future work. On a theoretical side, there are two open questions with respect to generalized penetration depth: how to formulate the distance metric D_g and compute the PD^g for non-convex models in a computational tractable way. It would be useful to derive tight bounds on the approximations (i.e. the lower and upper bounds). Furthermore, we would like to use our algorithm for other applications, including motion planning in 6-DOF C-space, dynamic simulation and tolerance verification.

Acknowledgment

This project was supported in part by ARO Contracts DAAD19-02-1-0390 and W911NF-04-1-0088, NSF awards 0400134 and 0118743, ONR Contract N00014-01-1-0496, DARPA/RDECOM Contract N61339-04-C-0043 and Intel. Young J. Kim was supported in part by the grant 2004-205-D00168 of KRF, the STAR program of MOST, the Ewha SMBA consortium and the ITRC program. We would also like to thank the anonymous reviewers for their helpful comments.

References

AGARWAL, P., AMENTA, N., AND SHARIR, M. 1998. Largest placement of one convex polygon inside another. In *Discrete Comput. Geom*, vol. 19, 95–104.

AGARWAL, P., GUIBAS, L., HAR-PELED, S., RABINOVITCH, A., AND SHARIR, M. 2000. Penetration depth of two convex polytopes in 3d. *Nordic J. Computing 7*, 227–240.

AMATO, N., BAYAZIT, O., DALE, L., JONES, C., AND VALLEJO, D. 2000. Choosing good distance metrics and local planners for probabilistic roadmap methods. In *IEEE Transactions on Robotics and Automation*, vol. 16, 442–447.

AVNAIM, F., AND BOISSONNAT, J. 1989. Polygon placement under translation and rotation. In *ITA*, vol. 23, 5–28.

CAMERON, S. 1997. Enhancing GJK: Computing minimum and penetration distance between convex polyhedra. *IEEE International Conference on Robotics and Automation*, 3112–3117.

CHAZELLE, B. 1983. The polygon containment problem. *Advances in Computing Research 1*, 1–33.

COHEN-OR, D., LEV-YEHUDI, S., KAROL, A., AND TAL, A. 2002. Inner-cover of non-convex shapes. In *The 4th Israel-Korea Bi-National Conference on Geometric Modeling*.

DOBKIN, D., HERSHBERGER, J., KIRKPATRICK, D., AND SURI, S. 1993. Computing the intersection-depth of polyhedra. *Algorithmica 9*, 518–533.

EHMANN, S., AND LIN, M. 2001. Accurate and fast proximity queries between polyhedra using surface decomposition. In *Proc. of Eurographics*.

GOTTSCHALK, S., LIN, M., AND MANOCHA, D. 1996. OBB-Tree: A hierarchical structure for rapid interference detection. *Proc. of ACM Siggraph'96*, 171–180.

GRINDE, R., AND CAVALIER, T. 1996. Containment of a single polygon using mathematical programming. In *European Journal of Operational Research*, Elsevier Science, vol. 92, 368–386.

HALPERIN, D. 1997. Arrangements. In *Handbook of Discrete and Computational Geometry*, J. E. Goodman and J. O'Rourke, Eds. CRC Press LLC, Boca Raton, FL, ch. 21, 389–412.

HALPERIN, D. 2002. Robust geometric computing in motion. *International Journal of Robotics Research, 21(3)*.

HALPERIN, D. 2005. Private communication.

HSU, D., KAVRAKI, L., LATOMBE, J., MOTWANI, R., AND SORKIN, S. 1998. On finding narrow passages with probabilistic roadmap planners. *Proc. of 3rd Workshop on Algorithmic Foundations of Robotics*, 25–32.

KIM, Y. J., LIN, M. C., AND MANOCHA, D. 2002. Fast penetration depth computation using rasterization hardware and hierarchical refinement. *Proc. of Workshop on Algorithmic Foundations of Robotics*.

KIM, Y., LIN, M., AND MANOCHA, D. 2002. Deep: Dual-space expansion for estimating penetration depth between convex polytopes. In *Proc. IEEE International Conference on Robotics and Automation*.

KIM, Y. J., OTADUY, M. A., LIN, M. C., AND MANOCHA, D. 2003. Six-degree-of-freedom haptic rendering using incremental and localized computations. *Presence 12*, 3, 277–295.

KUFFNER, J. 2004. Effective sampling and distance metrics for 3d rigid body path planning. In *IEEE Int'l Conf. on Robotics and Automation*.

LATOMBE, J. 1991. *Robot Motion Planning*. Kluwer Academic Publishers.

LAVALLE, S. M. 2006. *Planning Algorithms*. Cambridge University Press (also available at http://msl.cs.uiuc.edu/planning/). to appear.

LIN, M., AND MANOCHA, D. 2003. Collision and proximity queries. In *Handbook of Discrete and Computational Geometry*.

MILENKOVIC, V., AND SCHMIDL, H. 2001. Optimization based animation. In *ACM SIGGRAPH 2001*.

MILENKOVIC, V. 1998. Rotational polygon overlap minimization and compaction. In *Computational Geometry*, vol. 10, 305–318.

MILENKOVIC, V. 1999. Rotational polygon containment and minimum enclosure using only robust 2d constructions. In *Computational Geometry*, vol. 13, 3–19.

MIRTICH, B. 2000. Timewarp rigid body simulation. *Proc. of ACM SIGGRAPH*.

MOUNT, D. 1992. Intersection detection and separators for simple polygons. In *Proc. 8th Annual ACM Sympos. Comput. Geom*, 303–311.

RAAB, S. 1999. Controlled perturbation for arrangements of polyhedral surfaces with application to swept volumes. In *Proc. 15th ACM Symposium on Computational Geometry*, 163–172.

REDON, S., AND LIN, M. 2005. A fast method for local penetration depth computation. *Journal of Graphical Tools*.

REQUICHA, A. 1993. Mathematical definition of tolerance specifications. *ASME Manufacturing Review 6*, 4, 269–274.

SCHWARZER, F., SAHA, M., AND LATOMBE, J. 2005. Adaptive dynamic collision checking for single and multiple articulated robots in complex environments. *IEEE Tr. on Robotics 21*, 3 (June), 338–353.

STEWART, D. E., AND TRINKLE, J. C. 1996. An implicit time-stepping scheme for rigid body dynamics with inelastic collisions and coulomb friction. *International Journal of Numerical Methods in Engineering 39*, 2673–2691.

VAN DEN BERGEN, G. 2001. Proximity queries and penetration depth computation on 3d game objects. *Game Developers Conference*.

VARADHAN, G., AND MANOCHA, D. 2005. Star-shaped roadmaps - a determinstic sampling approach for complete motion planning. In *Proc. of Robotics: Science and Systems*.

ZHANG, L., KIM, Y., AND MANOCHA, D. 2006. A simple path non-existence algorithm for low dof robots. Tech. Rep. 06-006, Department of Computer Science, University of North Carolina at Chapel Hill.

ZHANG, L., KIM, Y., VARADHAN, G., AND D.MANOCHA. 2006. Fast c-obstacle query computation for motion planning. In *IEEE International Conference on Robotics and Automation (ICRA 2006)*.

GEOMETRIC CONSTRAINTS SOLVING: SOME TRACKS

Dominique Michelucci, Sebti Foufou, Loic Lamarque *
LE2I, UMR CNRS 5158, Univ. de Bourgogne, BP. 47870, 21078 Dijon, France

Pascal Schreck
LSITT, Université Louis Pasteur, Strasbourg, France†

Abstract

This paper presents some important issues and potential research tracks for Geometric Constraint Solving: the use of the simplicial Bernstein base to reduce the wrapping effect in interval methods, the computation of the dimension of the solution set with methods used to measure the dimension of fractals, the pitfalls of graph based decomposition methods, the alternative provided by linear algebra, the witness configuration method, the use of randomized provers to detect dependences between constraints, the study of incidence constraints, the search for intrinsic (coordinate-free) formulations and the need for formal specifications.

CR Categories: J.6 [Computer Applications]: Computer Aided Engineering—CAD-CAM; I.3.5 [Computer Graphics]: Computational Geometry and Object Modeling—Geometric algorithms, languages, and systems

Keywords: Geometric Constraints Solving, decomposition, witness configuration, Bernstein base, incidence constraint, randomized prover, rigidity theory, projective geometry

1 Geometric constraints solving

This article intents to present some essential issues for Geometric Constraints Solving (GCS) and potential tracks for future research. For the sake of conciseness and homogeneity, it focuses on problems related to the resolution, the decomposition, and the formulation of geometric constraints.

Today, all geometric modellers in CAD-CAM (Computer Aided Design, Computer Aided Manufacturing) provide some Geometric Constraints Solver. The latter enables designers and engineers to describe geometric entities (points, lines, planes, curves, surfaces) by specification of constraints: distances, angles, incidences, tangences between geometric entities. Constraints reduce to a system of (typically algebraic) equations. Typically, an interactive 2D or 3D editor permits the user to enter a so called approximate "sketch", and to specify geometric constraints (sometimes some constraints are automatically guessed by the software). The solver must correct the sketch, to make it satisfy the constraints.

*e-mail: {dmichel,sfoufou,loic.lamarque}@u-bourgogne.fr
†e-mail:schreck@dpt-info.u-strasbg.fr

SPM 2006, Cardiff, Wales, United Kingdom, 06–08 June 2006.
© 2006 ACM 1-59593-358-1/06/0006 $5.00

Usually, the solver first performs a qualitative study of the constraints system to detect under-, well- and over-constrained parts; when the system is correct, *i.e.* well-constrained, it is further decomposed into irreducible well-constrained subparts easier to solve and assemble. This qualitative study is mainly a Degree of Freedom (DoF) analysis. It is typically performed on some kind of graphs [Owen 1991; Hoffmann et al. 2001; Gao and Zhang 2003; Hendrickson 1992; Ait-Aoudia et al. 1993; Lamure and Michelucci 1998]. This article presents the pitfalls of graph based approaches, and suggests an alternative method. After this qualitative study, if it is successful, irreducible subsystems are solved, either with some formula in the simplest case (*e.g.* to compute the intersection between two coplanar lines, or the third side length of a triangle knowing two other side lengths and an angle, etc), or with some numerical method, *e.g.* a Newton-Raphson or an homotopy, which typically uses the sketch as a starting point for iterations, or with interval methods which can find all real solutions and enclose them in guaranteed boxes (a box is a vector of intervals). Computer algebra is not practicable because of the size of some irreducible systems, and it is not used by nowadays' CAD-CAM modelers. In this article, we will show that in some cases using computer algebra is possible and relevant.

Depending on the context, either users expect only one solution, the "closest" one to an interactively provided sketch; or they expect the solver to give all real roots, and interval methods are especially interesting in this case. For instance, in Robotics, problematic configurations of flexible mechanisms are solutions of a set of geometric constraints: engineers want to know all problematic situations or a guarantee that there is none.

The paper is organized as follow: Section 2 discusses GCS using interval arithmetic and Bernstein bases. Problems related to the decomposition of geometric constraints systems (degree of freedom, scaling, homography, pitfalls of graph based methods, etc.) are discussed in Section 3. This section also provides some ideas on how probabilistic tests such as NPM (Numerical Probabilistic Method) can be used as an efficient alternative for GCS and decomposition when they are used with a good initial configuration (which we refer to as the witness configuration in Section 3.9). Section 4 considers GCS when there is a continuum of solutions and the use of curve tracing algorithms. Section 5 presents the expression of geometric constraints in a coordinate free way and shows how kernel functions can be used to provide intrinsic formulation of constraints. Section 6 discusses the need for formal specifications of constraints and for specification languages. The conclusion is given in Section 7.

2 Interval arithmetic and Bernstein bases

A recurrent problem of interval methods is the wrapping effect [Neumaier Cambridge, 2001]: interval arithmetic loses the dependance between variables, so that the width of intervals increases with the computation depth. Maybe Bernstein bases can help. They are well known in the CAD/CAM world, since Bézier and de Casteljau, but people in the interval analysis seem unaware of

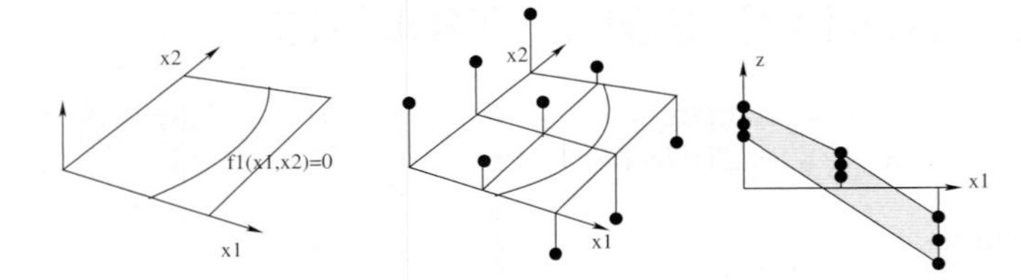

Figure 2: Equation $f_1(x_1,x_2) = 0$ has degree 2 in x and y, and a grid of 3×3 control points. The surface lie inside the convex hull of its control points. Computing the smallest rectangle enclosing the intersection between the plane $z = 0$ and this convex hull is a linear programming problem.

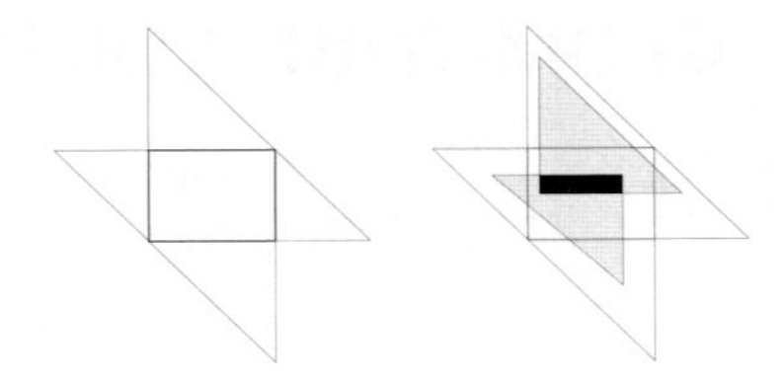

Figure 3: Reduction of a 2D box: it is the intersection of two triangles; reduce in the two triangles and take the bounding box.

them. Fig. 1 permits to compare the naive interval arithmetic with the tensorial Bernstein based one: the same algebraic curve $f(x,y) = 0$ is displayed, with the same classical recursive method, using (above) the naive interval arithmetic and (below) the Bernstein interval arithmetic: clearly, the former needs much more subdivisions than the latter. Transcendental functions are a difficulty, of course. Either we enclose transcendental functions between some polynomials, using for instance a Bernstein-Taylor form as Nataraj and Kotecha [Nataraj and Kotecha 2004], or maybe the Poisson base is a solution [Morin 2001].

2.1 Tensorial Bernstein base

This section gives a flavor of tensorial Bernstein bases on a simple example of a polynomial equation $f(x,y) = 0,\ 0 \leq x,y \leq 1$. We consider $f(x,y) = 0$ as the intersection curve between the plane $z = 0$ and the surface $z = f(x,y)$. Assume f has degree 3 in x and y. Usually the polynomial f is expressed in the canonical base: $(1,x,x^2,x^3) \times (1,y,y^2,y^3)$, but we prefer the tensorial Bernstein base: $(B_{0,3}(x),B_{1,3}(x),B_{2,3}(x),B_{3,3}(x)) \times (B_{0,3}(y),B_{1,3}(y),B_{2,3}(y),B_{3,3}(y))$. The conversion between the two bases is a linear transform:

$$(B_{0,3}(x),B_{1,3}(x),B_{2,3}(x),B_{3,3}(x)) = (1,x,x^2,x^3)\begin{pmatrix} 1 & 0 & 0 & 0 \\ -3 & 3 & 0 & 0 \\ 3 & -6 & 3 & 0 \\ -1 & 3 & -3 & 1 \end{pmatrix} \quad (1)$$

and idem for y. This kind of formula and matrix extends to any degree: for degree n, the Bernstein base $B(t) = (B_{0,n}(t), B_{1,n}(t), \ldots B_{n,n}(t))$ and the canonical base $T = (1,t,t^2 \ldots t^n)$ are related by:

$$\binom{n}{i} t^i = \sum_{j=i}^{n} \binom{j}{i} B_{j,n}(t)$$

and

$$B_{i,n}(t) = \binom{n}{i} t^i (1-t)^{n-i} = \sum_{j=i}^{n} (-1)^{j-i} \binom{n}{j} \binom{j}{i} t^j$$

The surface $z = f(x,y), 0 \leq x,y \leq 1$ has this representation in the Bernstein base:

$$x = \sum_{i=0}^{i=n} \frac{i}{n} B_{i,n}(x);\ \ y = \sum_{j=0}^{j=m} \frac{j}{m} B_{j,m}(y);\ \ z = \sum_{i=0}^{i=n} \sum_{j=0}^{j=m} z_{i,j} B_{i,n}(x) B_{j,m}(y)$$

Points $P_{i,j} = (\frac{i}{n}, \frac{j}{m}, z_{i,j})$ are called control points of the surface $z = f(x,y)$, which is now a Bézier surface. Control points have an intuitive meaning: typically, geometric modelers of curves and surfaces permit the user to interactively move control points and the Bézier curve or surface follows. A crucial property is that the surface patch (the part for $0 \leq x,y \leq 1$) lie inside the convex hull of its control points. Of course the convex hull of $f(0 \leq x \leq 1, 0 \leq y \leq 1)$ is just the interval $[\min z_{i,j}, \max z_{i,j}]$. It gives an enclosing interval for $f(0 \leq x \leq 1, 0 \leq y \leq 1)$, which is often sharper than the intervals provided by other interval arithmetics.

The method to display algebraic curves follows: if $0 \notin [\min_{i,j} z_{i,j}, \max_{i,j} z_{i,j}]$ then the curve does not cut the unit square, otherwise subdivide the square in four; the recursion is stopped at some recursion threshold; the Casteljau subdivision method permits to quickly compute the Bernstein representation (*i.e.* the control points) of the surface for any x interval $[x_0,x_1]$ and any y interval $[y_0,y_1]$, without translation to the canonical base. All that extends to higher dimension and the solving of systems of polynomial equations [Garloff and Smith 2001b; Garloff and Smith 2001a; Mourrain et al. 2004].

To find the smallest x interval $[x^-,x^+]$ enclosing the curve $f(x,y) = 0, 0 \leq x,y \leq 1$, project all control points on the plane x,z; compute their convex hull (it is an easy 2D problem); compute its intersection with the x axis: it is $[x^-,x^+]$. This is visually obvious on Fig. 2. Proceed similarly to find the smallest y-interval. In any dimension d, reducing the box needs only d computations of 2D convex hulls. A variant replaces the 2D convex hull computation by the computation of the smallest and greatest roots of two univariate polynomials, a lowest one and a largest one.

This box reduction is very advantageous when solving [Mourrain et al. 2004; Hu et al. 1996; Sherbrooke and Patrikalakis 1993] an algebraic system $f(x,y) = g(x,y) = 0, 0 \leq x,y \leq 1$ (or a more complex one), since it reduces the search space without subdivision or branching. Box reduction is even more efficient when combined to preconditionning: the system $f(x,y) = g(x,y) = 0$ has the same roots as a linear combination $af(x,y) + bg(x,y) = cf(x,y) + dg(x,y) = 0$; the idea is to use a linear combination such that $af(x,y) + bg(x,y)$ is very close to x and $cf(x,y) + dg(x,y)$ is very close to y: this combination is given by the jacobian inverse at the center of the considered box. It straightforwardly extends to higher dimension. Near a regular root, the convergence of such a solver is quadratic.

2.2 Simplicial Bernstein base

However there is a difficulty in high dimension: the tensorial Bernstein base has an exponential number of coordinates (as the canonical base) and is dense, *i.e.* a polynomial which is sparse in the canonical base becomes dense in the tensorial Bernstein base. For instance, a linear polynomial in d variables is represented by 2^d control points, a polynomial with total degree 2 is represented by

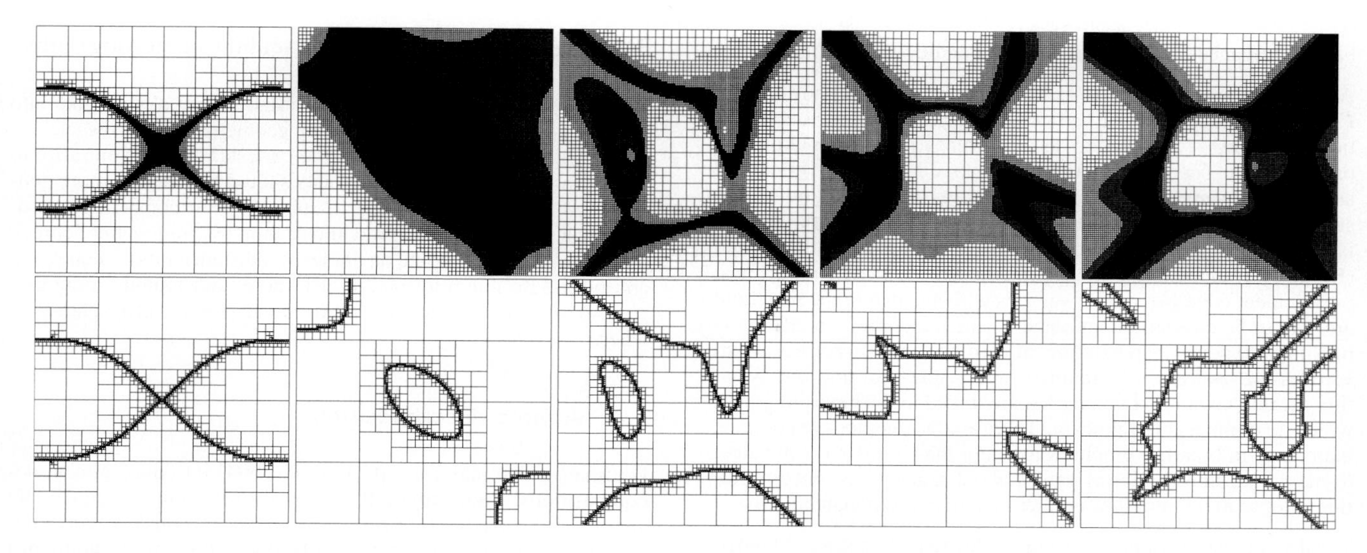

Figure 1: *Above*: naive interval arithmetic. *Below*: Bernstein based arithmetic. *Left to right columns*: Cassini oval: $C_{2,2}(x,y) = 0$ in $[-2,2] \times [-2,2]$, where $C_{a,b}(x,y) = ((x+a)^2+y^2) \times ((x-a)^2+y^2) - b^4$. The curve $f(x,y) = 15/4 + 8x - 16x^2 + 8y - 112xy + 128x^2y - 16y^2 + 128xy^2 - 128x^2y^2 = 0$ on the square $[0,1] \times [0,1]$. Random algebraic curves with total degree 10, 14, 18.

3^d control points, a polynomial with total degree n is represented by $(n+1)^d$ control points. A solution is to use the *simplicial* Bernstein base [Farin 1988] (the previous Bernstein base is the *tensorial* one).

For three variables x, y, z related by $x+y+z = 1$, the simplicial Bernstein base is defined by:

$$(x+y+z)^n = 1^n = \sum_{i+j+k=n} \binom{n}{i,j,k} x^i y^j z^k = \sum_{i+j+k=n} b^{(n)}_{ijk}(x,y,z)$$

and for any number of variables $x_1, \ldots x_d$ related by $x_1 + \ldots x_d = 1$, it is defined by:

$$(x_1+x_2+\ldots x_d)^n = 1^n = \sum_{i_1+\ldots i_d=n} \binom{n}{i_1,\ldots i_d} x_1^{i_1} \ldots x_d^{i_d} = \sum_{i_1+\ldots i_d=n} b^{(n)}_{i_1\ldots i_d}(x_1,\ldots x_d) \quad (2)$$

thus it straightforwardly extends the tensorial base defined by:

$$(x_k+(1-x_k))^n = 1^n = \sum_{i=0}^{i=n} \binom{n}{i} x_k^i (1-x_k)^{n-i} = \sum_{i=0}^{i=n} B_{i,n}(x_k), \quad k = 1,\ldots d$$

In the simplicial Bernstein base, a multivariate polynomial in d variable and with total degree n is represented by $O(d^n)$ control points; thus with total degree 1, 2, 3, etc, there are $O(d)$, $O(d^2)$, $O(d^3)$, etc control points. If the initial system is sparse in the canonical base, adding a logarithmic number of auxiliary unknowns and equations (using iterated squaring), every equation of any total degree $n \geq 2$ is translated into equations with total degree 2: thus with the simplicial Bernstein base the number of control points is polynomial; moreover the good properties of the tensorial Bernstein base still hold with the simplicial one: the convex hull property, the possibility of preconditioning, the possibility of reduction (it only needs several 2D convex hull problems as well), the de Casteljau method. An open question, which seems tractable, is which edge of the simplex to bissect?

2.3 Box reduction

For the moment, nobody uses the simplicial Bernstein base to solve algebraic systems. Perhaps it is due to the fact that domains are no more boxes (vectors of intervals) but simplices, which are less convenient for the programmer. In this respect, the simplicial Bernstein base can be used in a temporarily way, to reduce usual boxes, as follows. See Fig. 3 for a 2D example. A box $B = [x_1^-, x_1^+], \ldots [x_d^-, x_d^+]$ is given, it is cut by an hypersurface $H : h(x_1, \ldots x_d) = 0$; the problem is to reduce the box B as much as possible, so that it still encloses $B \cap H$. In any dimension d, the box is the intersection of two d simplices. Consider the hypercube $[0,1]^d$ for simplicity: a first simplex has $(0,0\ldots 0)$ as vertex, the opposite hyperplane is $x_1 + x_2 + \ldots x_d = d$, its other hyperplanes are $x_i = 0$. The second simplex has vertex $(1,1,\ldots 1)$, the opposite hyperplane is $x_1 + x_2 + \ldots x_d = 0$, its other hyperplanes are $x_i = 1$. To reduce the box, reduce in the two simplices, and compute the bounding box of the intersection between the two reduced simplices.

3 Decomposition related problems

3.1 DoF counting

All graph-based methods to decompose a system of geometric constraints rely on some DoF counting.

It is simpler to explain first the principle of DoF counting for systems of algebraic equations. A system of n algebraic equations is structurally well constrained if it involves n unknowns, also called DoF, and no subset of $n' < n$ equations involves less than n' unknowns, *i.e.* it contains no over-constrained part. For example, the system $f(x,y,z) = g(z) = h(z) = 0$ is not well constrained, because the subsystem $g(z) = h(z) = 0$ over-constrains z; remark that the details of f, g, h do not matter. Second example: the system $f(x,y) = g(x,y) = 0$ is structurally well constrained, *i.e.* it has a finite number of roots for generic f and g; if the genericity assumption is not fulfilled, it can have no solution: $g = f + 1$, or a continuum of solutions: $f = g$. Later we will see that a pitfall of graph based approaches is that the genericity condition is not fulfilled.

A natural bipartite graph is associated to every algebraic system $F(X) = 0$. The first set of vertices represent equations: one equation per vertex. The second set of vertices represent unknowns: one unknown per vertex. An edge links an equation-vertex and an unknown-vertex iff the unknown occurs in the equation. The structural well-constrainedness of a system is equivalent to the existence of a complete matching in the associated bipartite graph (König-Hall theorem): a matching is a set of edges, with at most one incident edge per vertex; vertices with an edge in the matching are said to be covered or saturated by the matching; a matching is maximum when it is maximum in cardinality; it is perfect iff all vertices are saturated. In intuitive words, one can find one equation per unknown (the two vertices are linked in the bipartite graph) which determines this unknown. There are fast methods to compute matchings in bipartite graph. Maximum matchings are equivalent to maximum flows (a lot of papers about graph based decomposition refer to maximum flows rather than maximum matchings).

The decomposition of bipartite graphs due to Dulmage and Mendelsohn also relies on maximum matchings. It partitions the system into a well-constrained part, an over-constrained part, an under-constrained part. The well-constrained part can be further decomposed (in polynomial time also) into irreducible well-constrained parts, which are partially ordered: for example $f(x) = g(x,y) = 0$ is well constrained; it can be decomposed into $f(x) = 0$ which is well constrained, and $g(x,y) = 0$ which is well constrained once x has been replaced by the corresponding root.

Relatively to systems of equations, systems of geometric constraints introduce two complications:

First geometric constraints are (classically...) assumed to be independent of the coordinate system, thus they can, for instance, determine the shape of a triangle in 2D (specifying either two lengths and one angle, or one angle and two lengths, or three lengths) but they can not fix the location and orientation of the triangle relatively to the cartesian frame. This placement is defined by three parameters (an x translation, an y translation, one angle). Thus in 2D, the DoF of a system is the number of unknowns (coordinates, radii, non geometric unknowns) minus 3. The same holds in 3D, where the placement needs six parameters; thus the DoF of a 3D system is the number of unknowns minus 6 — the constant is $d(d+1)/2$ in dimension d. Numerous ways have been proposed for adapting decomposition methods for systems of equations to systems of geometric constraints.

Second, the bipartite graph is visually cumbersome and not intuitive. People prefer the "natural" graph: each vertex represent a geometric unknown (point, line, plane) or a non geometric unknown, and each edge represents a constraint. There is a difficulty for representing constraints involving more than 2 entities; either hyper-arcs are used, or all constraints are binarized. Moreover vertices carry DoF, and edges (constraints) carry DoR: degree of restriction, *i.e.* the number of corresponding equations.

The differences between the bipartite and natural graphs are not essential. In passing, the matroid theory provides yet another formalism to express the same things, but it is not used in the GCS community up to now.

In 2D, a point and a line have 2 DoF; in 3D, points and planes have 3 DoF, lines have 4. In 3D, DoF counting (correctly) predicts there is a finite number of lines which cut four given skew lines: the unknown line has 4 DoF and there are 4 constraints. Similarly, there is a finite number of lines tangent to 4 given spheres.

Decomposition methods are essential, since they permit to solve big systems of geometric constraints, which can not be otherwise.

3.2 Decomposition modulo scaling or homography

Decomposition methods are complex and do not always take into account non geometric unknowns or geometric unknowns such as radii (which are independent on the cartesian frame, contrarily to unknowns). Decomposition methods should be simpler, more general, and decompose not only in subparts well constrained modulo displacements, but also modulo scaling [Schramm and Schreck 2003]: so we can compute an angle, or a distance ratio, in one part, and propagate this information elsewhere, and modulo homography: so we can compute cross ratios in one part and propagate elsewhere.

3.3 Almost decomposition

Hoffmann, Gao and Yang [Gao et al. 2004] introduce almost decompositions. They remark that a lot of irreducible systems in 3D are easier to solve when one of the unknowns is considered as a parameter, and when the system is solved for all (or some sampling) values of this parameter. In this artificial but simple example:

$$\begin{aligned}
f_1(x_1,x_2,x_3,x_4 = u) &= 0 \\
f_2(x_1,x_2,x_3,x_4 = u) &= 0 \\
f_3(x_1,x_2,x_3,x_4 = u) &= 0 \\
f_4(x_1,x_2,x_3,x_4,x_5,x_6,x_7) &= 0 \\
f_5(x_1,x_2,x_3,x_4,x_5,x_6,x_7) &= 0 \\
f_6(x_1,x_2,x_3,x_4,x_5,x_6,x_7) &= 0 \\
f_7(x_1,x_2,x_3,x_4,x_5,x_6,x_7) &= 0
\end{aligned}$$

x_4 is considered as a parameter u with a given value (we call it a key unknown for convenience). The subsystem $S_u : f_1(x_1,x_2,x_3,u) = f_2(x_1,x_2,x_3,u) = f_3(x_1,x_2,x_3,u) = 0$ is solved for all values of u, or in a more realistic way for some sampling of u. Then the rest of the system $T_u : f_4(x) = f_5(x) = f_6(x) = 0$ is solved, forgetting temporarily one equation, say f_7. f_7 is then evaluated at all sampling points on the solution curve of $f_1(x) = \dots f_6(x) = 0$. When f_7 almost vanishes, the possible root is polished with some Newton iterations. For the class of basic 3D configurations studied by Hoffmann, Gao and Yang [Gao et al. 2004], one key unknown is sufficient most of the time, but some rare more difficult problems need two key unknowns. One may imagine several variants of this approach, for instance the use of marching curve methods to follow the curve parameterized with u, or methods to automatically produce the best almost decomposition for irreducible systems: the best is the one which minimizes the number of key unknowns.

Curve tracing [Michelucci and Faudot 2005] can also be used to explore a finite set of solutions when no geometric symbolic solution is available (which is often the case in 3D). If the solution proposed by the solver does not fit the user needs, the idea is to forget one constraint and to trace the corresponding curve. In this case the roots are the vertices of a graph the edges of which correspond to the curves where a constraint has been forgotten. If we have d equations and d unknowns then each vertex is of degree d. One difficulty is that this graph can be disconnected, and there is no guarantee to reach every vertex starting from a given solution.

3.4 Some challenging problems

Some challenging problems resist this last attack of almost decomposition. Consider the graph of the regular icosahedron (20 triangles, 30 edges, 12 vertices). Labelling edges with lengths gives a well constrained system with 30 distance constraints between 12 points in 3D (the regular pentagons of the icosahedron are not constrained to stay coplanar). This kind of problems, with distance constraints only, is called the molecule [Hendrickson 1992; Porta et al. 2003; Laurent 2001] problem because of its applications in chemistry: find the configuration of a molecule given some distances between its atoms. This last system has Bézout number $2^{30} \approx 10^9$.

A seemingly more difficult problem uses the graph of the regular dodecahedron (12 pentagonal faces, 20 vertices, 30 edges). Label edges with lengths; this time, also impose to each of the 12 pentagonal faces to stay planar, for the problem to be well constrained. The dodecahedron problem is not a molecule one, because of the coplanarity constraints. In the same family, the familiar cube gives a well constrained problem, with 8 unknown points, 12 distances, and 6 coplanarity relations.

Considering the regular octahedron gives a simpler molecule problem, with 6 unknown points and 12 distances (no coplanarity condition). This problem was already solved by Durand and Hoffmann [Durand 1998; Durand and Hoffmann 2000] with homotopy. Another solution is to use Cayley-Menger relations [Yang 2002; Porta et al. 2003; Michelucci and Foufou 2004].

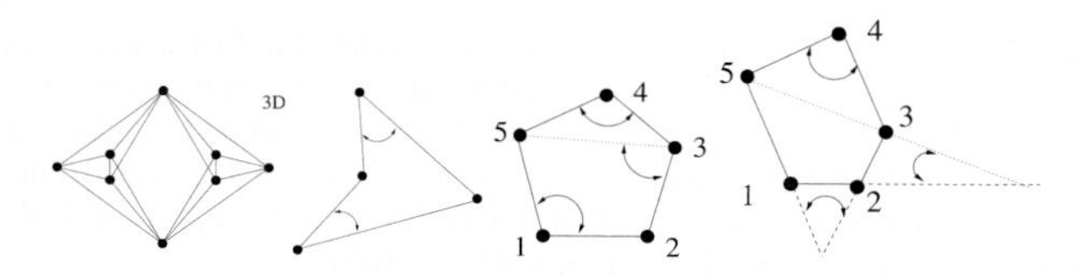

Figure 4: The double banana, and three other 3D configurations due to Auxkin Ortuzar, where DoF counting fails. No four points are coplanar.

3.5 Pitfalls of graph based methods

A pitfall of DoF counting is that geometric constraints can be dependent in subtle ways. In 2D, the simplest counter example to DoF counting is given by the 3 angles of a triangle: they can not be fixed independently (note they can in spherical geometry). Fig. 5 shows a more complex 2D counter example. In 3D, a simple counter example is: point A and B lie on line L, line L lie on plane H, point A lie on plane H; the last constraint is implied by the others. Fig. 4 shows other counter examples which make fail DoF counting in 3D. It is possible to use some ad hoc tests in graph based methods to account for some of these configurations. However every incidence theorem (Desargues, Pappus, Pascal, Beltrami, Cox... see Fig. 6) provide dependent constraints: just use its hypothesis and conclusion (or its negation) as constraints; moreover no genericity assumption (used in Rigidity theory) is violated since incidence constraints do not use parameters. Thus detecting a dependence is as hard as detecting or proving geometric theorems.

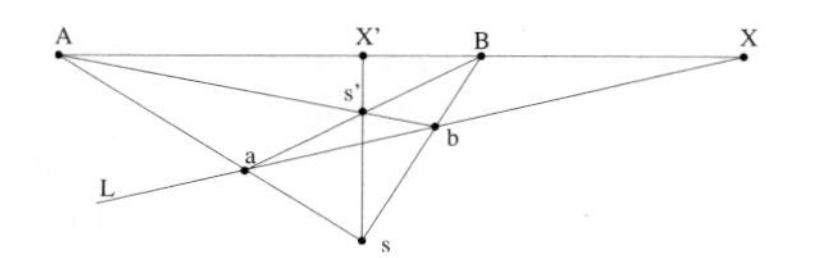

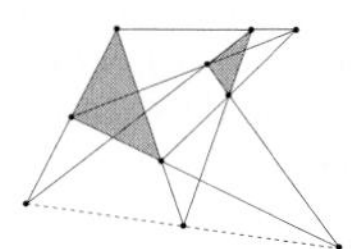

Figure 5: *Left*: be given 3 aligned points A, B, X; for any point s outside AB, for any L through X outside s, define: $a = L \cap As$, $b = L \cap Bs$, $s' = Ab \cap aB$, $X' = ss' \cap AB$; then X' is independent of s and L. *Right*: Desargues theorem: if two triangles (in gray) are perspective, homologous sides cut in three collinear points.

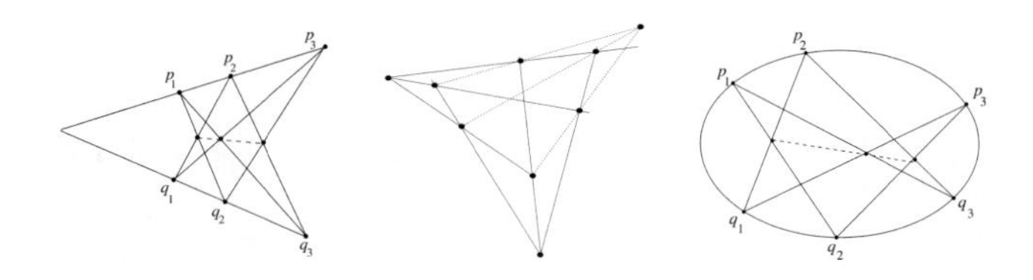

Figure 6: Pappus, its dual, Pascal theorems.

DoF counting is mathematically sound only in a very restricted case, the 2D molecule problem, *i.e.* when all constraints are generic distance constraints between 2D points (thus points can not be collinear): it is Laman theorem [J. Graver 1993]. For the 3D molecule problem, no characterization is known; Fig. 4 leftmost shows the most famous counterexample to DoF counting: the double banana, which uses only distance constraints. Even in the simple case of distance constraints, a combinatorial (*i.e.* in terms of graph or matroids) characterization of well constrainedness seems out of reach.

With the still increasing size of constraints systems, the probability for a subtle dependence increases as well. J-C. Léon (personal communication), who uses geometric constraints to define constrained surfaces or curves, reports this typical behavior: the solver detects no over-constrainedness but fails to find a solution; the failure persists when the user tries to modify the value of parameters (distances, angles) – which is terribly frustrating. This independence to parameter values suggests that the dependence is due to some incidence theorems of projective geometry (such as Pappus, Desargues, Pascal, Beltrami, etc). For conciseness, the other hypothesis: some triangular (or tetrahedral [Serré 2000]) inequality is violated, is not detailed here.

Detecting such dependences -or solving in spite of them when it is a consistent dependence- is a key issue for GCS. Clearly, no graph based method can detect all such dependences. It gives strong motivation for investigating other methods.

3.6 Linear Algebra performs qualitative study

Today decomposition is graph based most of the time. Linear algebra seems a promising alternative. For conciseness, the idea is illustrated for 2D systems of distance constraints only between n points. Assume also the distances are algebraically independent (thus no collinear points), and that points are represented by their cartesian coordinates: $X = (x_1, y_1, \dots x_n, y_n)$. For clarity, we say that $p = (x, y)$ is a "point", and X is a "configuration". After Rigidity theory [J. Graver 1993; Lamure and Michelucci 1998], it is well known that it suffices to numerically study the jacobian at some random configuration $X \in \mathbb{R}^{2n}$. It is the essence of the so called numerical probabilistic method (NPM).

By convention, the k th line of the jacobian J is the derivative of the k th equation of the system. Vectors m such that $Jm^t = 0$ are called infinitesimal motions. The notation $\dot{X} = (\dot{x_1}, \dot{y_1}, \dots \dot{x_n}, \dot{y_n})$ is also used to denote the infinitesimal motion at configuration X.

First, if the rank of the jacobian (at the random, generic configuration) is equal to the number of equations, equations are independent; otherwise it is possible to extract a base of the equations. Second, the system is well-constrained (modulo displacement) if its jacobian has corank 3: actually it is even possible to give a base of the kernel of the jacobian (the kernel is the set of infinitesimal motions). This base is t_x, t_y, r, where $t_x = (1, 0, 1, 0, \dots)$ is a translation in x,

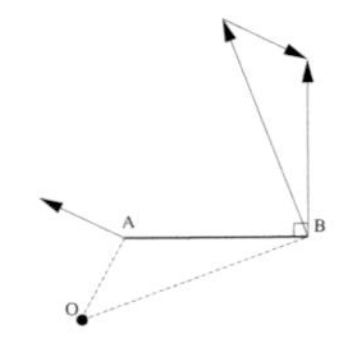

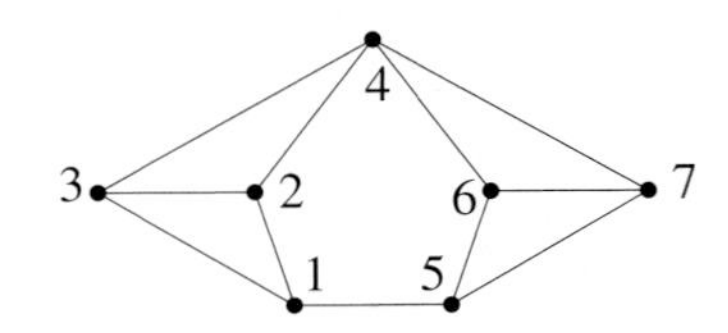

Figure 7: *Left*: the arrows illustrate the infinitesimal rotation around O of points A and B. For a displacement like this rotation, $\dot{A}-\dot{B}$ is orthogonal to AB for all couples A,B. *Right*: this system is well-constrained. Removing the bar (the constraint distance) $1,5$ breaks the system into two well-constrained parts (the left and the right of point 4).

$t_y=(0,1,0,1,\ldots)$ is a translation in y, and r is an instantaneous rotation around the origine: $r=(-y_1,x_1,-y_2,x_2,\ldots-y_n,x_n)$. These 3 infinitesimal motions are displacements, also called isometries; they do not modify the relative location of points, contrarily to deformations (also called flexions).

An infinitesimal motion m is a displacement iff for all couple of points A,B, the difference $\dot{A}-\dot{B}$ between A and B motions is orthogonal to the vector $\overrightarrow{AB}$. Fig. 7 illustrates that for the rotation r.

For convenience, define $d_{i,j}$ as the vector $(\dot{x_1},\dot{y_1},\ldots\dot{x_n},\dot{y_n})$ where $\dot{x_i}=x_i-x_j$, $\dot{x_j}=x_j-x_i$, $\dot{y_i}=y_i-y_j$, $\dot{y_j}=y_j-y_i$ and $\dot{x_k}=\dot{y_k}=0$ for $k\neq i,j$; actually, $d_{i,j}$ is half the derivative of the distance equation $(x_i-x_j)^2+(y_i-y_j)^2-D_{ij}^2=0$. Obviously $d_{i,j}=-d_{j,i}$. It is easy to check that, consistently, t_x, t_y and r are orthogonal to all $d_{i,j}, 1\leq i<j\leq n$: they indeed are displacements, not deformations.

All that is well known, after Rigidity theory [J. Graver 1993]. What seems less known is that linear algebra makes also possible to decompose a well-constrained system into well-constrained subparts.

3.7 The NPM decomposes

Consider for instance the well-constrained system in Fig. 7 Right, and remove the constraint distance (the edge) $1,5$. It increases the corank by 1, adding an infinitesimal flexion (a deformation); a possible base for the kernel is t_x,t_y,r and $f=(0,0,0,0,0,0,0,0,y_4-y_5,x_5-x_4,y_4-y_6,x_6-x_4,y_4-y_7,x_7-x_4)$ *i.e.* an instantaneous rotation of $4,5,6,7$ around 4, or $g=(y_4-y_1,x_1-x_4,y_4-y_2,x_2-x_4,y_4-y_3,x_3-x_4,0,0,0,0,0,0,0,0)$ *i.e.* an instantaneous rotation of $1,2,3,4$ around 4, or any linear combination m of f, g, t_x,t_y,r (outside the range of t_x,t_y,r, to be pedantic). Of course f and g especially make sense for us, but any deformation m is suitable.

The deletion of edge $1,5$ leaves the part 1,2,3,4 well-constrained: it is visually obvious, and confirmed by the fact that $d_{i,j}, 1\leq i<j\leq 4$ is orthogonal to m. Idem for the part 4,5,6,7, because $d_{i,j}, 4\leq i<j\leq 7$ is orthogonal to m. But no $d_{i,j}$ with $i<4<j$ is orthogonal to m. This gives a polynomial time procedure to find maximal (for inclusion) well-constrained parts in a flexible system, and a polynomial time procedure to decompose well-constrained systems into well-constrained subsystems: remove a constraint and find remaining maximal (for inclusion) well-constrained parts, as in the previous example.

This idea can be easily extended to 3D distance constraints, with some minor changes: the corank is 6 instead of 3. Note this method detects the bad constrainedness of the classical double banana, contrarily to graph based methods which extend the Laman condition.

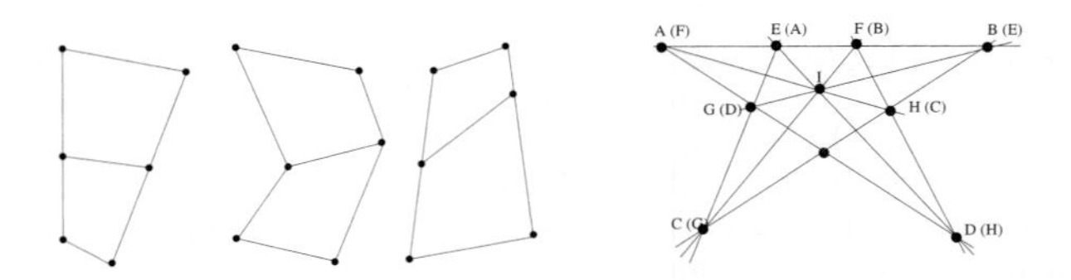

Figure 8: *From left to right*: the unknown solution configuration; a random configuration, not fulfiling incidence constraints; a witness configuration; an irrational configuration with an underlying regular pentagon (or an homography of).

What if other kinds of constraints are used, not only distance constraints? From a combinatorial point of view, the vertices in Fig. 7 can represent points, but also lines (which have also 2 DoFs, like points, in 2D). Thus as far as decomposing an well-constrained graph into well-constrained subparts is concerned, we can consider vertices of the graph as points, and constraints/edges as distance constraints. This first answer is not always satisfactory, for instance when vertices have distinct DoF (in 3D, points and planes have 3 DoF, but lines have 4), or when constraints involve more than 2 geometric objects.

In fact this method has been extended to other kind of constraints [Foufou et al. 2005]. The only serious difficulty occurs when the assumption of the genericity of the relative location of points is contradicted by some explicit (or induced) projective constraints (collinearity or coplanarity constraints). Of course graph based decomposition methods have the same limitation.

3.8 The witness configuration principle

Clearly, the NMP give incorrect results because it studies the jacobian at a random, generic, configuration which does not fulfil these projective constraints. A solution straightforwardly follows: compute a "witness configuration" and study it with the NPM; a witness configuration [Foufou et al. 2005; Michelucci and Foufou 2006] does not satisfy the metric constraints (*i.e.* it has typically lengths or angles different of the searched configuration), but it fulfils the specified projective constraints (see Fig. 9), and also, by construction, the projective constraints (collinearities, coplanarities) due to geometric theorems of projective geometry, *e.g.* Pascal, Pappus, Desargues theorems. First experiments validate the witness configuration method [Foufou et al. 2005]: it works for all counter examples to DoF in this paper (for instance Fig. 4 or 5), and it is even able to detect and stochastically prove incidence theorems which confuse DoF counting (see below). In other words no confusing witness configuration has been found up to now.

3.9 Computing a witness configuration

Most of the time, the sketch is a witness configuration. Otherwise, if the distance and angle parameters are generic (no right angle, for instance), remove all metric constraints and solve the remaining (very under-constrained system); the latter contains only projective constraints, *i.e.* incidence constraints. Even for the challenging problems: icosahedron, dodecahedron, cube, it is trivial to find a witness polyhedron – the latter can be concave or self intersecting.

Finally, if distance and angle values are not generic (*e.g.* right angles are used), the simplest strategy is to consider parameters as unknowns (systems are most of the time of the form: $F(U,X)=0$ where U is a vector of parameters: lengths, angle cosines, etc; their values are known just before resolution), then to solve the very under-constrained resulting system: it is hoped it is easily solvable. Once a solution has been found, it gives a witness configuration which is studied and decomposed with the NPM.

This section has given strong motivations to study the decomposition and resolution of under-constrained systems, and of systems of incidence constraints.

3.10 Incidence constraints

The previous section has already given some motivations to study incidence constraints, but these constraints also arise in photogrammetry, in computer vision, in automatic correction of hand made drawings. We hope the systems of incidence constraints met in our applications to be trivial or almost trivial (defined below), however incidence constraints can be arbitrarily difficult even in 2D.

In 2D, a system of incidence constraints between points and lines reduce to a special 3D molecule problem [Hendrickson 1992; Porta et al. 2003; Laurent 2001]: represent unknown points and lines by unit 3D vectors; the incidence $p \in L$ means that the corresponding vertices on the unit sphere have distance $\sqrt{2}$. To avoid degeneracies (either all points are equal, or all lines are equal), one can impose to four generic points to lie on some arbitrary square on the unit sphere.

3.10.1 Trivial and almost trivial incidence systems

In 2D, a system of incidence constraints (point-line incidences) is trivial iff it contains only removable points and lines. A point p is removable when it is constrained to lie on two lines l_1 and l_2 (or less): then its definition is stored in some data structure (either $p = l_1 \cap l_2$, or $p \in l_1$ is any point on line l_1, or p is any point), it is erased from the incidence system, the rest of the system is solved, then the removed point is added using its definition. Symmetrically (or dually) for a line, when it is constrained to pass through two points (or less). Erasing a point or a line may make removable another point or line. If all points and lines are removed, the graph is trivial. Trivial systems are easily solved, using the definitions of removed elements in reverse order.

The extension to 3D is straightforward. This method finds a witness for every Eulerian 3D polyhedra (a polyhedron is Eulerian iff it fulfils Euler formula). It is easily proved that every Eulerian 3D polyhedron contain a removable vertex or a removable face, and thus is trivial: assume there is a contradicting polyhedron, with V vertices, E edges and F faces. Let $v_1, v_2 \dots v_V$ be the vertex degrees, all greater than 3, and $f_1, f_2 \dots f_F$ the number of vertices of the F faces, all greater than 3 as well; it is well known that $\sum_1^V v_i = 2E = \sum_1^F f_j$, thus $E \geq 2V$ and $E \geq 2F$. By Euler' formula: $V - E + F = 2$. Thus $E + 2 = V + F \geq 2V + 2$ and $E + 2 = V + F \geq 2F + 2$. Add. We get $2E + 4 = 2V + 2F \geq 4 + 2V + 2F$: a contradiction. QED. Unfortunately, this simple method no more applies with non Eulerian polyhedra, say a faceted torus with quadrilateral faces and where every vertex has degree 4 (this last polyhedron has a rational witness too).

Another construction of a witness for Eulerian polyhedra first computes a 2D barycentric embedding (also called a Tutte embedding) of its vertices and edges: an arbitrary face is mapped to a convex polygon and other vertices are barycenters of their neighbors – it suffices to solve a linear system. Maxwell and Cremona already knew that such a 2D embedding is the projection of a 3D convex polyhedron; for instance, the three pairwise intersection edges of the three faces of a truncated tetrahedron concur. It is then easy to lift the Tutte embedding to a 3D convex polyhedron, using propagation and continuity between contiguous faces. In passing, this construction proves Steinitz theorem: all 3D convex polyhedra are realizable with rational coordinates only, and thus with integer coordinates only; this property is wrong for 4D convex polyhedra [Richter-Gebert 1996].

Configurations in incidence theorems are typically almost trivial (the word is chosen by analogy with almost decomposition). A system is almost trivial iff, removing an incidence, the obtained system is trivial: Desargues, Pappus, hexamy[1] configurations are almost trivial.

Almost triviality permits the witness configuration method to detect and prove incidence theorems in a probabilistic way: erase an incidence constraint to make the system trivial; for Pappus, Desargues, hexamy configurations to quote a few, due to the symmetry of the system, every incidence is convenient; solve the trivial system.

- If the obtained configuration fulfils the erased incidence constraint, then this incidence is with high probability a consequence of the other incidences: a theorem has been detected and (probabilistically) proved. A prototype [Foufou et al. 2005], performing computations in a finite field $\mathbb{Z}/p\mathbb{Z}$ (p a prime, near 10^9) for speed and exactness, probabilistically proves this way all theorems cited so far and some others, such as the Beltrami theorem[2] in 3D in a fraction of a second. This shows that using some computer algebra is possible and relevant.
- If the obtained configuration does not fulfil the erased incidence constraint, this constraint is not a consequence of the others. This case occurs with the pentagonal configuration in Fig. 8; the later is not realizable in $\mathbb{Q}$: indeed a regular pentagon (or an homography of) is needed. This configuration is not relevant for CAD-CAM (actually, we know none).

3.10.2 Universality of point line incidences

However, incidence constraints in 2D (and a fortiori in 3D) can be arbitrarily difficult; it is due to the following theorem which is a restatement[3] of the fundamental theorem of projective geometry, known since von Staudt and Hilbert [Bonin 2002; Coxeter 1987; Hilbert 1971]:

Theorem 1 (Universality theorem) *All algebraic systems of equations with integer coefficients and unknowns in a field* $\mathbb{K}$ *(typically* $\mathbb{K} = \mathbb{R}$ *or* $\mathbb{C}$*) reduce to a system of point-line incidence constraints in the projective plane* $\mathbb{P}(\mathbb{K})$*, with the same bit size.*

The proof relies on the possibility to represent numbers by points on a special arbitrary line, and on the geometric construction (with ruler only) of the point representing the sum or the product of two numbers (Fig. 9), from their point representation [Bonin 2002]. Some consequences are:

- Alone, point-line incidences in the projective plane are sufficient to express all geometric constraints of today GCS.

[1] An hexamy is an hexagon the opposite sides of which cut in three collinear points; every permutation of an hexamy is also an hexamy; it is a desguise of Pascal theorem.

[2] Coxeter [Coxeter 1999; Coxeter 1987] credits Gallucci for this theorem, in his books.

[3] D. Michelucci and P. Schreck. Incidence constraints: a combinatorial approach. Submitted to the special issue of IJCGA on Geometric Constraints.

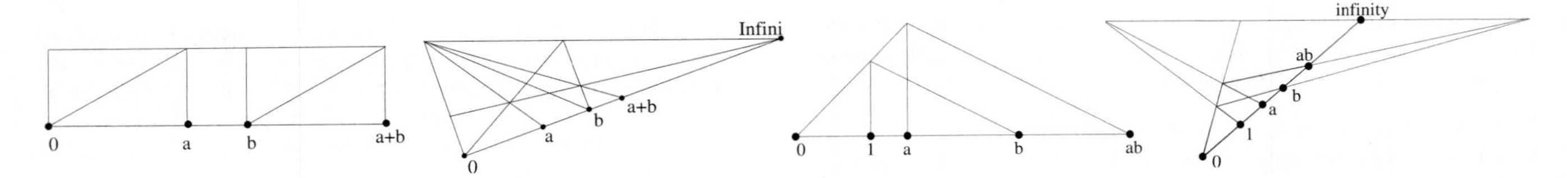

Figure 9: *left*: affine and projective constructions of $a+b$; *right*: affine and projective constructions of $a \times b$

- Programs solving point line incidence constraints (*e.g.* solving the 3D molecule problem [Hendrickson 1992; Laurent 2001; Porta et al. 2003]) can solve all systems of geometric constraints.
- Programs detecting or proving incidence theorems in 2D (as the hexamy prover [Michelucci and Schreck 2004]) address all algebraic systems. Fascinating.
- Algebra reduces to combinatorics: the bipartite graph of the point line incidences contains *all* the information of the algebraic system: no need for edge weights, no genericity assumption (contrarily to Rigidity theory).
- This bipartite graph is a fundamental data structure. What are its properties? its forbidden minors? Which link between its graph properties and properties of the algebraic system?
- Incidence constraints are definitively not a toy problem.

Practical consequences are unclear for the moment: for instance, does it make sense to reduce algebraic systems to a (highly degenerate) 3D molecule problem? Probably not.

4 Solving with a continuum of solutions

Current solvers assume that the system to be solved has a finite number of solutions, and get into troubles or fail when there is a continuum of solutions.

Arguably, computer algebra [Chou 1988; Chou et al. 1987], and geometric solvers (typically ruler and compass) already deal with under constrainedness; both are able to triangularize in some way given under-constrained systems $F(X)=0$; for instance, several elimination methods from computer algebra are (at least theoretically) able to partition the set X of unknowns into $T \cup Y$, where T is a set of parameters, and to compute a triangularized system of equations: $g_1(T,y_1)=g_2(T,y_1,y_2)=\dots g_n(T,y_1,y_2,\dots y_n)=0$ which define the unknowns in Y. In the 2D case, and when a ruler and compass construction is possible, some geometric solvers are able to produce a construction program (also named: straight line program, DAG, etc): $y_1=h_1(T), y_2=h_2(T,y_1),\dots y_n=h_n(T,y_1\dots y_{n-1})$, where h_i are multi-valued functions for computing the intersection between two cercles, or between a line and a cercle, etc; dynamic geometry softwares [Bellemain 1992; Kortenkamp 1999; Dufourd et al. 1997; Dufourd et al. 1998] have popularized this last approach, which unfortunately does not scale well in 3D.

Another approach, typically graph-based, considers that under-constrainedness are due to a mistake from the user or to an incomplete specification; they try to detect and correct these mistakes, or to complete the system to make it well-constrained – and as simple to solve as possible [Joan-Arinyo et al. 2003; Gao and Zhang 2003; Zhang and Gao 2006].

Some systems are intrinsically under-constrained: the specified set is continuous. This happens when designing mechanisms or articulated bodies, when designing constrained curves or surfaces (for instance for blends), when using an almost decomposition, when searching a witness. Thus it makes sense to design more robust solvers, able to deal with a continuum of solutions. Such a solver should detect on the fly that there is a continuum of solutions, should compute the dimension of the solution set (0 for a finite solution set, 1 for a curve, 2 for a surface, etc) and should be able to segment solution curves and to triangulate solution surfaces, etc. Methods for computing the dimension of a solution set already exist in computer graphics (and elsewhere); roughly, cover the solution set with a set of boxes (as in Fig. 1) with size length ε; if halving ε multiplies the number of boxes by about 1, 2, 4, 8, etc, induce that the solution set has dimension 0, 1, 2, 3, etc; this is the Bouligand dimension of fractals [Mandelbrot 1982; Barnsley 1998]. Instead of boxes for the cover, it is possible to use balls or simplices. This ultimate solver will unify the treatment of parameterized surfaces, implicit surfaces, blends, medial axis, and geometric constraints in geometric modeling. C. Hoffmann calls that the "dimensionality paradigm".

Fig. 10 illustrate such an ultimate solver with examples, mainly 2D for clarity. For the first picture, the input is the system:

$$\begin{cases} (x-x_c)^2+(y-y_c)^2=r^2 \\ (x_1-x_c)^2+(y_1-y_c)^2=r^2 \\ (x_2-x_c)^2+(y_2-y_c)^2=r^2 \\ (x_3-x_c)^2+(y_3-y_c)^2=r^2 \end{cases}$$

with x_n, y_n the coordinates of the triangle vertices and x, y, x_c, y_c, r the unknowns.

The second picture represents two circles with the radii defined by an equation; the input of the solver is the system:

$$\begin{cases} x^2+y^2=r^2 \\ (r-1)(r-2)=0 \end{cases}$$

The third one shows the section of a Klein's bottle; the input of the solver is:

$$\begin{cases} (x^2+y^2+z^2+2y-1)((x^2+y^2+z^2-2y-1)^2-8z^2)+ \\ \quad 16xz(x^2+y^2+z^2-2y-1)=0 \\ x-z=1 \end{cases}$$

The latter is the intersection curve between an extruded folium and a sphere; the input of the solver is the system:

$$\begin{cases} x^2+y^2+z^2=1 \\ x^3+y^3-3xy=0 \end{cases}$$

The two last pictures illustrate also an adaptive subdivision in accordance with the curvature of the solution set inside a box and a detection of the boxes containing singular points. In these examples, the Bouligand dimension is used also to get rid of terminal boxes (at the lowest subdivision depth) without solutions.

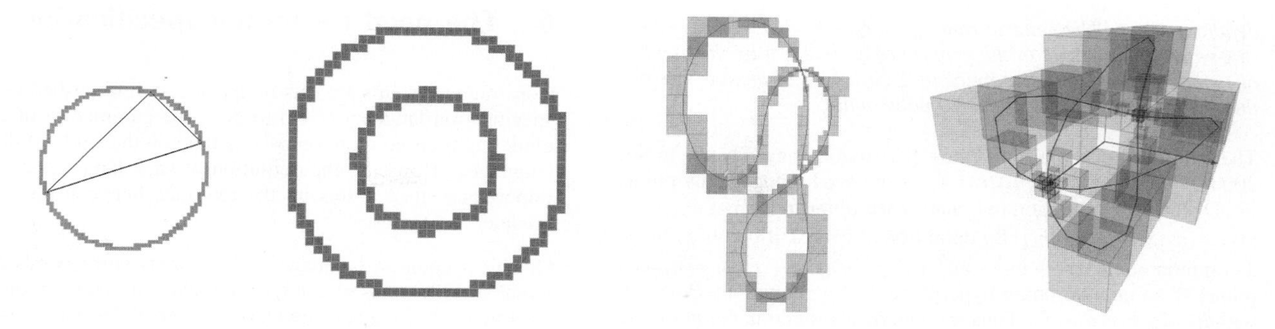

Figure 10: Some preliminary results of a solver based on centered interval arithmetic and Bouligand dimension; *left*: a triangle's circumscribed circle; *middle-left*: two circles with "unknown" radius; *middle-right*: intersection between a plan and the Klein's bottle; *right*: intersection between an extruded folium and a sphere.

5 Coordinates-free constraints

Recently several teams [Yang 2002; Lesage et al. 2000; Serré et al. 1999; Serré et al. 2002; Serré et al. 2003; Michelucci and Foufou 2004] propose coordinate-free formulations, which are sometimes advantageous. For instance, the Cayley Menger determinant links the distances between $d+2$ points in dimension d and gives, for the octahedron problem, a very simple system solvable with Computer Algebra. These intrinsic relations have been extended to other configurations, *e.g.* with points and planes in 3D, points and lines in 2D. An intrinsic relation, due to Neil White, is given in Sturmfels's book [Sturmfels 1993], th. 3.4.7: it is the condition for five skew lines in $3D$ space to have a common transversal line. Philippe Serré, in his PhD thesis [Serré 2000], gives the relation involving distances between two lines AB and CD and between points A, B, C, D. However, for 3D configurations involving not only lines but also points or planes, intrinsic formulations (e.g. extending Cayley-Menger formulations) are missing most of the time. Even the intrinsic condition for a set of points to lie on some algebraic curve or surface with given degree was unknown (it is given just below). Next sections suggest methods to find such relations. These issues are foreseeable topics for GCS.

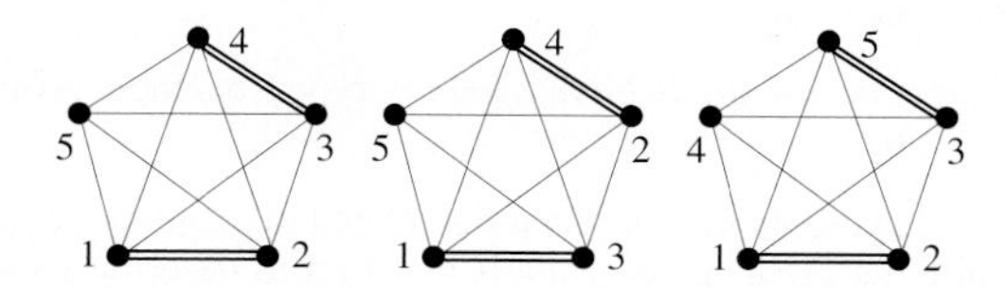

Figure 11: Isomorphic subgraphs of the same class monomials.

5.1 Finding new relations

To find relations linking invariants (distances, cosines, scalar products, signed areas or volumes *i.e.* determinants) for a given configuration of geometric elements, it suffices in theory to use a Grobner package which eliminates variables representing coordinates in some set of equations, for instance equations: $(x_i-x_j)^2+(y_i-y_j)^2+(z_i-z_j)^2-d_{ij}^2=0, i\in[1;4], j\in[i+1;5]$, to find the Cayley-Menger equation relating distances between 5 points in $3D$. In practice, computer algebra is not powerful enough. The polynomial conditions can be computed by interpolation: for instance, to guess the Cayley-Menger equation in $3D$, one can proceed in three steps: first, generate N random configurations of 5 points $(x_i,y_i,z_i)\in\mathbb{Z}^3$, second compute square distances $d_{ij}^{(k)}, i\in[1;4]$ and $j\in[i+1;5]$ for each configuration $k\in[1;N]$; this gives N points with 15 coordinates; third, all these N points lie on the zero-set of an unknown polynomial in the variables d_{ij}: search for this polynomial by trying increasing degrees.

This polynomial has an exponential number of monomials, thus an exponential number of unknown coefficients. Due to symmetry, some monomials have the same coefficients; they are said to lie in the same class. For instance, monomials $d_{12}^2d_{34}^2$, $d_{13}^2d_{24}^2$, etc lie in the same class; monomials in the same class correspond to isomorphic edge weighted subgraphs of K_5, the complete graph with 5 vertices and with edges weighted by the degree of the corresponding monomial (Fig. 11). To be feasible this interpolating method must exploit this symmetry. The fast generation of these classes (and of one instance per class) is an interesting and non trivial combinatorial problem by itself, related to the Reed-Polya counting theory. To validate this approach, we implemented a simple algorithm, which successfully computes Cayley-Menger relations, and distance relations for six 2D (ten 3D) points to lie on the same conic (quadric). A lesson of this prototype is that another good reason to exploit symmetry is to limit the size of the output interpolating polynomial.

5.2 Kernel functions provide intrinsic formulations

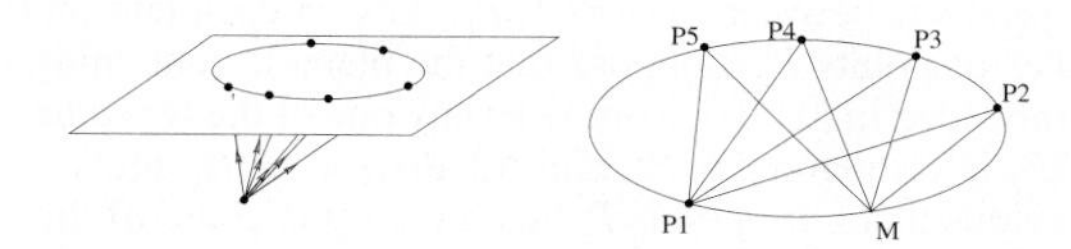

Figure 12: Vectorial condition for points to lie on a common conic or algebraic curve with degree d (two cases).

Given a set of non null vectors $v_i, i=1\dots n$ having a common origin Ω, the set of lines l_i supported by these vectors and a plane π not passing through Ω (Fig. 12), what is the condition on the scalar products between vectors v_i for the intersection points between lines l_i and plane π to lie on the same conic? This section shows that the matrix M with $M_{i,j}=(v_i\cdot v_j)^2=M_{j,i}$ must have rank 5 or less. If the $n\geq 6$ intersection points do not lie on the same conic, but are generic, the matrix M has rank 6 (assuming the v_i lie in 3D space). More generally:

Theorem 2 *The intersection points between a plane π and the lines defined by supporting vectors v_i through a common origin Ω outside π lie on a degree d curve iff the matrix $M^{(d)}$, where $M_{i,j}^{(d)}=(v_i\cdot v_j)^d=M_{j,i}^{(d)}$, has rank $r_d=d(d+3)/2$ or less (r_d for* *deficient rank). The generic rank $g_d = r_d + 1 = (d+1)(d+2)/2$ (the rank of the matrix in the generic case) is given by the number of monomials in the polynomial in 2 variables of degree d, since this curve is the zero set of such a polynomial.*

The proof uses kernel functions [Cristianini and Shawe-Taylor 2000]. Let $p_i = (x_i, y_i, h_i), i = 1 \dots 6$ be six homogeneous points in 2D and ϕ_2 the function that maps each point p_i to $P_i = \phi_2(p_i) = (x_i^2, y_i^2, h_i^2, x_i y_i, x_i h_i, y_i h_i)$. By definition of conics, if points p_i lie on a common conic $ax_i^2 + by_i^2 + ch_i^2 + dx_i y_i + ex_i h_i + fy_i h_i = 0$, then points P_i lie on a common hyperplane, having equation: $P_i \cdot h = 0$ with $h = (a,b,c,d,e,f)$. Thus six generic (or random, *i.e.* not lying on a common conic) 2D points p_i give six lifted points P_i with rank 6, and six 2D points p_i lying on a common conic give six lifted points P_i with rank $r_2 = 5$ or less.

If m vectors $P_1, \dots P_m$ have rank r, their Gram matrix $G_{ij} = P_i \cdot P_j = G_{ji}$ has also rank r. To compute $P_i \cdot P_j$, the naive method compute $P_i = \phi_2(p_i)$, and $P_j = \phi_2(p_j)$, then $P_i \cdot P_j$. Kernel functions avoid the computation of $\phi_2(p_i)$. A kernel function K is such that $K(p_i, p_j) = \phi_2(p_i) \cdot \phi_2(p_j)$.

A first example of kernel function considers given $p = (x, y, h)$ and homogeneous $\phi_2(p) = (x^2, y^2, h^2, \sqrt{2}xy, \sqrt{2}xh, \sqrt{2}yh)$. The cosmetic $\sqrt{2}$ constant does not modify rank but simplifies computations: $K(p, p') = \phi_2(p) \cdot \phi_2(p') = \phi_2(p) \cdot \phi_2(p') = \dots = (p \cdot p')^2$ as the reader will check. More generally, for an homogeneous kernel polynomial of degree d, $K(p, p') = (p \cdot p')^d$: it suffices to adjust the cosmetic constants. Thus the Gram matrix for this homogeneous lift with degree d is: $G_{i,j} = (p_i \cdot p_j)^d$. The proof of the previous theorem follows straightforwardly.

A second example considers a non homogeneous lifting polynomial. Let $p = (x, y)$ and $\phi(p) = (x^2, y^2, \sqrt{2}xy, \sqrt{2}x, \sqrt{2}y, 1)$. As above, the $\sqrt{2}$ does not modify rank but simplifies computations: $K(p, p') = \phi(x, y) \cdot \phi(x', y') = \dots = (p \cdot p' + 1)^2$ as the reader will check. More generally, for a non homogeneous lifting polynomial of degree d, $K(p, p') = (p \cdot p' + 1)^d$. Thus the Gram matrix for this lift with degree d is $G_{i,j} = (p_i \cdot p_j + 1)^d$. We use the latter to answer the question: what is the coordinate-free condition for six 2D points $P_i, i = 0 \dots 5$ to lie on a common quadric, or on a common algebraic curve with degree d? We search a condition involving scalar products between vectors $P_0 P_j$, thus independent on the coordinates of points P_i. Suppose that the plane π containing points P_i is embedded in 3D space, let Ω be any one of the two points such that ΩP_0 is orthogonal to π, and the distance ΩP_0 equals 1. Use the previous theorem: points P_i lie on the same conic iff the matrix M, where $M_{i,j} = (\overrightarrow{\Omega P_i} \cdot \overrightarrow{\Omega P_j})^2$, has rank five or less, and the P_i lie on the same algebraic curve with degree d iff the matrix M, where $M_{i,j} = (\overrightarrow{\Omega P_i} \cdot \overrightarrow{\Omega P_j})^d$ has deficient rank $r_d = d(d+3)/2$. We remove Ω:

$$\begin{aligned} \overrightarrow{\Omega P_i} \cdot \overrightarrow{\Omega P_j} &= (\overrightarrow{\Omega P_0} + \overrightarrow{P_0 P_i}) \cdot (\overrightarrow{\Omega P_0} + \overrightarrow{P_0 P_j}) \\ &= \overrightarrow{\Omega P_0} \cdot \overrightarrow{\Omega P_0} + \overrightarrow{\Omega P_0} \cdot \overrightarrow{P_0 P_j} + \overrightarrow{P_0 P_i} \cdot \overrightarrow{\Omega P_0} + \overrightarrow{P_0 P_i} \cdot \overrightarrow{P_0 P_j} \\ &= 1 + 0 + 0 + \overrightarrow{P_0 P_i} \cdot \overrightarrow{P_0 P_j} \end{aligned}$$

Theorem 3 *Coplanar points P_i lie on the same algebraic curve with degree d iff the matrix M has deficient rank (*i.e. *$r_d = d(d+3)/2$) or less, where $M_{i,j} = (1 + \overrightarrow{P_0 P_i} \cdot \overrightarrow{P_0 P_j})^d$.*

These theorems nicely extends to surfaces and beyond. All relations involving scalar products can be translated into relations involving distances only, using: $\overrightarrow{u} \cdot \overrightarrow{v} = (\overrightarrow{u}^2 + \overrightarrow{v}^2 - (\overrightarrow{v} - \overrightarrow{u})^2)/2$.

6 The need for formal specification

Constraints systems are sets of specification described using some specification languages. Up to now, the community of constraint modeling focused more on solvers than on the study of description languages. However, the definition of such languages is of major importance since it sets up the interface between the solver, the modeler and the user.

On this account, a language of constraints corresponds to the external specifications of a solver: it makes the skeleton of the reference manual of a given solver, or, conversely, it defines the technical specifications for the solver to be realized. On the other hand, a language of constraints has to be clearly and fully described in order to be able to define the conversion from a proprietary architecture to an exchange format (which is itself described by such a language). CAD softwares are currently offering several exchange formats, but, in our sense, they are very poorly related to the domain of geometric constraints and they are unusable, for instance, for sharing benchmarks [Aut 2005].

It seems to us that a promising track in this domain consists in considering the *meta*-level. More clearly, we argue that we need a standard for the description of languages of geometric constraints rather than (or in addition to) specific exchange formats. This is the point of view adopted by the STEP consortium which is, as far as we know, not concerned by geometric constraints [4]. Besides, a meta-level approach allows to consider a *geometric universe* as a *parameter* of an extensible solver.

A first attempt in this direction was presented in [Wintz et al. 2005]. This work borrows the ideas and the terminology of the algebraic specification theory [Wirsing 1990; Goguen 1987]: a constraint system is syntactically defined by a triple (C, X, A) where X and A are some symbols respectively referring to unknowns and parameters, and C is a set of predicative terms built on a heterogeneous signature Σ. Recall that a heterogeneous signature is a triple $< S, F, P >$ where

- S is a set of *sorts* which are symbols referring to types,
- F is a set of *functional symbols* typed by their profile $f : s_1 \dots s_m \longrightarrow s$,
- P is a set of *predicative symbols* typed by their profile $p : s_1 \dots s_k$

Functional symbols express the tools related to geometric construction while the predicative symbols are used to describe geometric constraints. The originality of the approach described in [Wintz et al. 2005] consists in the possibility of describing the semantic —or more precisely, several semantics like visualization, algebra, logic— within a single framework allowing to consider as many tool sets as provided semantic fields.

Since the main tools are based on syntactic analyzers, the support language considered is XML which is flexible enough to allow the description of geometric universes and which comes with a lot of facilities concerning the syntactic analysis.

The advantages of using formal specifications can be summarized in the two following points: (i) Clarifying and expliciting the semantics, which helps to avoid the misunderstandings that commonly occur between all the partners during data exchanges; and (ii) There are more and more software tools: parsers, but also provers, code generators, compilers, which are able to use these

[4] although some searchers are working on such a task (personal communication of Dr. Mike Pratt)

explicited semantics; these tools make it easier to ensure the reliability and the consistence between distinct pieces of software, to extend the software and to document it.

There is, of course, a lot of works to do in this domain, let us enumerate some crucial points:

- the definition of tools able to describe and handle the translation between two languages of constraints (for instance using the notion of signature morphism).
- the automatic generation of tools from a given language of constraints.
- the possibility to take into account robustness consideration in the framework (see, for instance, [Schreck 2001])
- ideally, it is possible to imagine that such languages would be able to fully describe geometric solvers from the input of the constraints to the expression and visualization of the solutions

Tackling the problem from another point of view, the user, that is the designer, should be allowed to enter his proper solutions of a constraints system or his proper geometric constructions within the application. This should be made easier by considering a precise language of constraints. This family of tools come with the ability of compiling constructions and doing parametric design (see [Hoffmann and Joan-Arinyo 2002]). This problem is naturally very close to the generative modeling problem and the well known notion of features.

We think that the fields of geometric constraints solving and features modeling are mature enough for attempting to join them together (see for instance [Sitharam et al. 2006]). Indeed, the geometric constraints solving field addresses routinely 3D problems and takes more and more semantic aspects into consideration. This should give some hints to handle the problematic of under-constrained or over-constrained constraint systems. Indeed, up to now, researchers in GCS field have considered this problem from a quite combinatorial point of view (see [Joan-Arinyo et al. 2003; Hoffmann et al. 2004]); maybe the user intentions should deserve more considerations.

7 Conclusion

This article posed several important problems for GCS and proposed several research tracks: the use of the simplicial Bernstein base to reduce the wrapping effect, the computation of the dimension of the solution set, the pitfalls of graph based decomposition methods, the alternative provided by linear algebra, the witness configuration method which overcomes the limitations of DoF counting and which is even able to probabilistically detect and prove incidence theorems (Desargues, Pappus, Beltrami, hexamy, harmonic conjugate, etc and their duals), the study of incidence constraints, the search for intrinsic (coordinate-free) formulations. Maybe the more surprising conclusion concerns the importance of incidence constraints.

Acknowledgements to our colleagues from the French Group working on geometric constraints and declarative modelling, partly funded by CNRS in 2003–2005.

References

AIT-AOUDIA, S., JEGOU, R., AND MICHELUCCI, D. 1993. Reduction of constraint systems. In *Compugraphic*, 83–92.

AUTODESK. 2005. *DXF reference*, July.

BARNSLEY, M. 1998. *Fractal everywhere*. Academic Press.

BELLEMAIN, F. 1992. *Conception, réalisation et expérimentation d'un logiciel d'aide à l'enseignement de la géométrie : Cabri-géomètre*. PhD thesis, Université Joseph Fourier - Grenoble 1.

BONIN, J. 2002. *Introduction to matroid theory*. The George Washington University.

CHOU, S.-C., SCHELTER, W., AND YANG, J.-G. 1987. Characteristic sets and grobner bases in geometry theorem proving. In *Computer-Aided geometric reasoning, INRIA Workshop, Vol. I*, 29–56.

CHOU, S.-C. 1988. *Mechanical Geometry theorem Proving*. D. Reidel Publishing Company.

COXETER, H. 1987. *Projective geometry*. Springer-Verlag, Heidelberg.

COXETER, H. 1999. *The beauty of geometry. 12 essays*. Dover publications.

CRISTIANINI, N., AND SHAWE-TAYLOR, J. 2000. *An introduction to support vector machines*. Cambridge University Press.

DUFOURD, J.-F., MATHIS, P., AND SCHRECK, P. 1997. Formal resolution of geometrical constraint systems by assembling. *Proceedings of the 4th ACM Solid Modeling conf.*, 271–284.

DUFOURD, J.-F., MATHIS, P., AND SCHRECK, P. 1998. Geometric construction by assembling solved subfigures. *Artificial Intelligence Journal 99(1)*, 73–119.

DURAND, C., AND HOFFMANN, C. 2000. A systematic framework for solving geometric constraints analytically. *Journal on Symbolic Computation 30*, 493–520.

DURAND, C. B. 1998. *Symbolic and Numerical Techniques for Constraint Solving*. PhD thesis, Purdue University.

FARIN, G. 1988. *Curves and Surfaces for Computer Aided Geometric Design — a Practical Guide*. Academic Press, Boston, MA, ch. Bézier Triangles, 321–351.

FOUFOU, S., MICHELUCCI, D., AND JURZAK, J.-P. 2005. Numerical decomposition of constraints. In *SPM '05: Proceedings of the 2005 ACM symposium on Solid and physical modeling*, ACM Press, New York, USA, 143–151.

GAO, X.-S., AND ZHANG, G. 2003. Geometric constraint solving via c-tree decomposition. In *ACM Solid Modelling*, ACM Press, New York, 45–55.

GAO, X.-S., HOFFMANN, C., AND YANG, W. 2004. Solving spatial basic geometric constraint configurations with locus intersection. *Computer Aided Design 36*, 2, 111–122.

GARLOFF, J., AND SMITH, A. 2001. Solution of systems of polynomial equations by using bernstein expansion. *Symbolic Algebraic Methods and Verification Methods*, 87–97.

GARLOFF, J., AND SMITH, A. P. 2001. Investigation of a subdivision based algorithm for solving systems of polynomial equations. *Journal of nonlinear analysis : Series A Theory and Methods 47*, 1, 167–178.

GOGUEN, J. 1987. Modular specification of some basic geometric constructions. In *Artificial Intelligence*, vol. 37 of *Special Issue on Computational Computer Geometry*, 123–153.

HENDRICKSON, B. 1992. Conditions for unique realizations. *SIAM J. Computing 21*, 1, 65–84.

HILBERT, D. 1971. *Les fondements de la géométrie, a french translation of Grunlagen de Geometrie, with discussions by P. Rossier*. Dunod.

HOFFMANN, C., AND JOAN-ARINYO, R. 2002. *Handbook of Computer Aided Geometric Design*. North-Holland, Amsterdam, ch. Parametric Modeling, 519–542.

HOFFMANN, C., LOMONOSOV, A., AND SITHARAM, M. 2001. Decomposition plans for feometric constraint systems, part i : Performance measures for cad. *J. Symbolic Computation 31*, 367–408.

HOFFMANN, C., SITHARAM, M., AND YUAN, B. 2004. Making constraint solvers more usable: overconstraint problem. *Computer-Aided Design 36*, 2, 377–399.

HU, C.-Y., PATRIKALAKIS, N., AND YE, X. 1996. Robust Interval Solid Modelling. Part 1: Representations. Part 2: Boundary Evaluation. *CAD 28*, 10, 807–817, 819–830.

J. GRAVER, B. SERVATIUS, H. S. 1993. *Combinatorial Rigidity. Graduate Studies in Mathematics*. American Mathematical Society.

JOAN-ARINYO, R., SOTO-RIERA, A., VILA-MARTA, S., AND VILAPLANA, J. 2003. Transforming an unerconstrained geometric constraint problem into a wellconstrained one. In *Eight Symposium on Solid Modeling and Applications*, ACM Press, Seattle (WA) USA, G. Elber and V.Shapiro, Eds., 33–44.

KORTENKAMP, U. 1999. *Foundations of Dynamic Geometry*. PhD thesis, Swiss Federal Institue of Technology, Zurich.

LAMURE, H., AND MICHELUCCI, D. 1998. Qualitative study of geometric constraints. In *Geometric Constraint Solving and Applications*, Springer-Verlag.

LAURENT, M. 2001. Matrix completion problems. *The Encyclopedia of Optimization 3*, Interior - M, 221–229.

LESAGE, D., LÉON, J.-C., AND SERRÉ, P. 2000. A declarative approach to a 2d variational modeler. In *IDMME'00*.

MANDELBROT, B. 1982. *The Fractal Geometry of Nature*. W.H. Freeman and Company, New York.

MICHELUCCI, D., AND FAUDOT, D. 2005. A reliable curves tracing method. *IJCSNS 5*, 10.

MICHELUCCI, D., AND FOUFOU, S. 2004. Using Cayley Menger determinants. In *Proceedings of the 2004 ACM symposium on Solid modeling*, 285–290.

MICHELUCCI, D., AND FOUFOU, S. 2006. Geometric constraint solving: the witness configuration method. *To appear in Computer Aided Design, Elsevier*.

MICHELUCCI, D., AND SCHRECK, P. 2004. Detecting induced incidences in the projective plane. In *isiCAD Workshop*.

MORIN, G. 2001. *Analytic functions in Computer Aided Geometric Design*. PhD thesis, Rice university, Houston, Texas.

MOURRAIN, B., ROUILLIER, F., AND ROY, M.-F. 2004. Bernstein's basis and real root isolation. Tech. Rep. 5149, INRIA Rocquencourt.

NATARAJ, P. S. V., AND KOTECHA, K. 2004. Global optimization with higher order inclusion function forms part 1: A combined taylor-bernstein form. *Reliable Computing 10*, 1, 27–44.

NEUMAIER, A. Cambridge, 2001. *Introduction to Numerical Analysis*. Cambridge Univ. Press.

OWEN, J. 1991. Algebraic solution for geometry from dimensional constraints. In *Proc. of the Symp. on Solid Modeling Foundations and CAD/CAM Applications*, 397–407.

PORTA, J. M., THOMAS, F., ROS, L., AND TORRAS, C. 2003. A branch and prune algorithm for solving systems of distance constraints. In *Proceedings of the 2003 IEEE International Conference on Robotics & Automation*.

RICHTER-GEBERT, J. 1996. *Realization Spaces of Polytopes*. Lecture Notes in Mathematics 1643, Springer.

SCHRAMM, E., AND SCHRECK, P. 2003. Solving geometric constraints invariant modulo the similarity group. In *International Workshop on Computer Graphics and Geometric Modeling, CGGM'2003*, Springer-Verlag, Montréal, LNCS Series.

SCHRECK, P. 2001. Robustness in cad geometric constructions. In *IEEE Computer Society Press*, P. of the 5th International Conference IV in London, Ed., 111–116.

SERRÉ, P., CLÉMENT, A., AND RIVIÈRE, A. 1999. *Global consistency of dimensioning and tolerancing*. ISBN 0-7923-5654-3. Kluwer Academic Publishers, March, 1–26.

SERRÉ, P., CLÉMENT, A., AND RIVIÈRE, A. 2002. Formal definition of tolerancing in CAD and metrology. In *Integrated Design an Manufacturing in Mechanical Engineering. Proc. Third IDMME Conference, Montreal, Canada, May 2000*, Kluwer Academic Publishers, 211–218.

SERRÉ, P., CLÉMENT, A., AND RIVIÈRE, A. 2003. Analysis of functional geometrical specification. In *Geometric Product Specification and Verification : Integration of functionality*. Kluwer Academic Publishers, 115–125.

SERRÉ, P. 2000. *Cohérence de la spécification d'un objet de l'espace euclidien à n dimensions*. PhD thesis, Ecole Centrale de Paris.

SHERBROOKE, E. C., AND PATRIKALAKIS, N. 1993. Computation of the solutions of nonlinear polynomial systems. *Computer Aided Geometric Design 10*, 5, 379–405.

SITHARAM, M., OUNG, J.-J., ZHOU, Y., AND ARBREE, A. 2006. Geometric constraints within feature hierarchies. *Computer-Aided Design 38*, 2, 22–38.

STURMFELS, B. 1993. *Algorithms in Invariant Theory*. Springer.

WINTZ, J., MATHIS, P., AND SCHRECK, P. 2005. A metalanguage for geometric constraints description. In *CAD Conference (presentation only, available at http://axis.u-strasbg.fr/~schreck/Publis/gcml.pdf)*.

WIRSING, M. 1990. *Handbook of Theoretical Computer Science*. Elsevier Science, ch. Algebraic specification, 677–780.

YANG, L. 2002. Distance coordinates used in geometric constraint solving. In *Proc. 4th Intl. Workshop on Automated Deduction in Geometry*.

ZHANG, G., AND GAO, X.-S. 2006. Spatial geometric constraint solving based on k-connected graph decomposition. *To appear in Symposium on Applied Computing*.

A Higher Dimensional Formulation for Robust and Interactive Distance Queries

Joon-Kyung Seong David E Johnson Elaine Cohen

School of Computing, University of Utah

Abstract

We present an efficient and robust algorithm for computing the minimum distance between a point and freeform curve or surface by lifting the problem into a higher dimension. This higher dimensional formulation solves for all query points in the domain simultaneously, therefore providing opportunities to speed computation by applying coherency techniques. In this framework, minimum distance between a point and planar curve is solved using a single polynomial equation in three variables (two variables for a position of the point and one for the curve). This formulation yields two-manifold surfaces as a zero-set in a 3D parameter space. Given a particular query point, the solution space's remaining degrees-of-freedom are fixed and we can numerically compute the minimum distance in a very efficient way. We further recast the problem of analyzing the topological structure of the solution space to that of solving two polynomial equations in three variables. This topological information provides an elegant way to efficiently find a global minimum distance solution for spatially coherent queries. Additionally, we extend this approach to a 3D case. We formulate the problem for the surface case using two polynomial equations in five variables. The effectiveness of our approach is demonstrated with several experimental results.

CR Categories: I.3.5 [Computer Graphics]: Computational Geometry and Object Modeling—Geometric algorithms, languages, and systems; I.3.5 [Computer Graphics]: Computational Geometry and Object Modeling—Splines

Keywords: minimum distance, dimensionality lifting, spline models, problem reduction scheme

1 Introduction

Minimum distance queries on computer models are one of the most important geometric operations in simulation [Baraff 1990], haptics [II et al. 1997], robotics [Quinlan 1994], registration [Pottmann et al. 2003], and distance volume computation [Breen et al. 1998]. Often, efficiency and robustness are important for these applications, yet prior formulations have not been able to combine the efficiency of numerical solutions with robustness and global convergence. In this article, we develop an algorithm for reliable and efficient minimum distance queries from a point to a spline model by casting the problem into a higher-dimensional space parameterized not only by the curve or surface, but also by potential positions of the query point.

SPM 2006, Cardiff, Wales, United Kingdom, 06–08 June 2006.
© 2006 ACM 1-59593-358-1/06/0006 $5.00

The minimum distance between a point and a spline model can be computed either symbolically or numerically. Symbolic approaches are robust and can find a global minimum solution but are relatively slow. Because the underlying equations change when the query point changes, symbolic approaches have not been able to exploit spatially coherent queries to accelerate their solution. On the other hand, numerical methods depend on spatial coherency to yield rapid solutions to distance queries, but often can be unreliable since they typically search only for local solutions and miss global solutions that may "jump" from one branch of the model to another. In this paper, a symbolic-numeric hybrid approach is presented to exploit both advantages.

Our higher-dimensional approach symbolically represents the distance between a spline model and all query points in a bounded domain. Given a planar curve $C(t)$ and a parameterization of all possible points in the plane $P(x,y)$, minimum distance queries between them are solved using a single polynomial equation. Using this single equation in three variables, the two-dimensional solution space is constructed as a zero-set in the xyt-parameter space. A parameter point (x,y,t) located on the solution space is mapped to a distance extrema for $C(t)$ in the real space. Thus, a search for the distance extrema can be reduced to finding a solution point on the two-dimensional zero-sets in the parameter space.

After performing one minimum distance query using a symbolic approach, repeated queries can be updated using fast numerical methods. Since we have a two-dimensional solution space implicitly defined by a single polynomial equation, spatially coherent queries can be solved by numerically tracing the solution manifolds. In the parameter space, this marching technique computes solution points for the next query point based on the previous ones, which results in a very fast computation of minimum distances.

Local searches using the numerical tracing method may not guarantee that they find all the solutions, one of which may be the global minimum. A new component of the solution space may appear or an existing one disappear when the point encounters a global changes in the higher-dimensional topological structure. We recast the problem of analyzing global topology of the solution space to that of solving two polynomial equations in the parameter space – this part of the algorithm is further accelerated through preprocessing of the higher-dimensional space. This topology information supports the local numerical method to efficiently find a global solution for spatially coherent queries.

The approach developed for curves in the plane can be extended to freeform surfaces in a 3D space. For the surface case, a higher-dimensional solution space is implicitly defined by two polynomial equations in five variables. Experimental results show the robustness of the approach and the speed advantage for coherent queries. We are able to achieve simultaneous tracking of multiple extrema for global minimum distance thousands of times per second for curves and hundreds of times per second for surfaces.

The rest of this paper is organized as follows. In Section 2, some related works are discussed. Section 3 presents an algorithm for computing minimum distance between a point and planar curve, an approach that is based on the dimensionality lifting scheme in the parameter space. In Section 4, the minimum distance algorithm is extended to freeform surfaces in a 3D space. Some examples are presented in Section 5, and finally, in Section 6, this paper is concluded.

2 Related Works

Distance query solution methods tend to vary depending on the type geometric primitive being used. For discrete geometry, such as triangular models, the predominant approaches accelerate queries using bounding volume hierarchies. In this case, research has focused on more efficient bounding primitives. For continuous geometry, such as spline surfaces, the distance equations can be directly formulated and solution approaches have used numerical or symbolic techniques to find distance minima.

2.1 Polygonal Models

An early hierarchical approach to distance queries used bounding spheres to prune away portions of the model further away than an upper bound on distance [Quinlan 1994]. In [Larsen et al. 2000], swept spheres volumes proved more efficient at pruning than spheres for near contact cases. More recently, [Ehmann and Lin 2001] has used temporal coherence on hierarchies of convex surface patches to accelerate distance queries.

These techniques all used Euclidean space bounds to find a global minimum. Hierarchical bounds on surface normals were used in [Johnson and Cohen 2001] to efficiently find local minima in distance for polygonal models as well as spline models [Johnson and Cohen 2005].

2.2 Spline Models

The basic approach of formulating equations that reflect the extrema of a distance function have been well-known and available in textbooks for some time [Mortensen 1985]. Research issues have been in improving robustness and speed of convergence. Baraff used distance measures on convex smooth models as an efficient collision test [Baraff 1990]. Snyder used an interval Newton's method to robustly update model penetration [Snyder 1993b]. A combination of numerical methods solved for the closest point on a space curve in [Wang et al. 2002].

Subdivision of constraint spaces have been used as more robust, yet slower, alternatives to numerical methods. Interval methods searched the four dimensional parameter space representing the distance between two parametric surfaces in [Snyder 1993a]. Elber [Elber 1992] represented the distance equation and its derivatives as NURBS surfaces and used subdivision along with numerical methods to search for solutions.

2.3 Distance Transform

The distance transform is a mapping between point position and a minimum distance to a model. Often, this transform is represented as a distance volume, which is a discrete sampling of point positions with their minimum distances to a model stored in a volumetric array. Distance volumes have the advantage of being extremely fast, since finding a distance is just looking up the appropriate value in a table [Museth et al. 2005]. They have the disadvantage of requiring large storage and of being a fixed, usually crude, resolution. Our approach is analogous to a continuous version of a distance volume, since we can quickly find a minimum distance for any point position in a bounded domain. Distance field computations are also explored in many papers [Frisken et al. 2000; Sigg et al. 2003; Sud et al. 2004]. We present a continuous representation of points and parameters that satisfy a local distance extremum solution, while distance fields are discrete representations of the global minimum distance.

2.4 Problem Reduction to Parameter Space

A variety of geometric problems involving freeform curves or surfaces can be reduced to the single question of finding the zero-set of a system of non-linear polynomial equations in the parameter space of the original curves or surfaces [Elber et al. 2001; Seong et al. 2004; Seong et al. 2005]. Techniques for solving a set of polynomial equations are developed and applied to various geometric problems as a primitive tool [Sherbrooke and Patrikalakis 1993; Elber and Kim 2005; Dokken 1985; Grandine et al. 2000]. The minimum distance algorithm employed in this paper also operates on the same premise as taken in [Kim and Elber 2000; Patrikalakis and Maekawa 2002].

3 Distance from a Point to Planar Curve

In this section, we present a dimensionality lifting approach for computing minimum distances between a point and planar curve. Considering all the possible position of the point, we lift the problem into a 3D parameter space and construct a solution space for the minimum distance problem. In the higher dimension, we find that the solution is simplified and it becomes easy to analyze the topological structure of the solution space. This topology information makes it possible to search the global solution in a very efficient way.

3.1 Computing a High Dimensional Solution Space

Given a planar curve $C(t)$ and a point in the plane, $\mathbf{P}$, the squared distance

$$D^2(t) = \langle C(t) - \mathbf{P}, C(t) - \mathbf{P} \rangle$$

is represented as a B-spline scalar function. The minimum of $D^2(t)$ can be found by computing all its extrema and choosing the smallest. Assuming that $C(t)$ is C^1-continuous, extrema of $D^2(t)$ occur where its derivative is zero, which in turn we seek the zeros of the scaled extremal equation $E(t)$,

$$E(t) = \left\langle C(t) - \mathbf{P}, C'(t) \right\rangle .$$

Considering all the possible position in the plane, the point $\mathbf{P}$ can be parameterized by $P(x,y)$. Then, a foot-point of $P(x,y)$ onto the curve $C(t)$ can be defined by lifting a distance extremal equation $E(t)$ into a 3D space,

Definition 1 *A foot-point of $P(x,y)$ onto the planar curve $C(t)$ satisfies the following polynomial equation:*

$$\mathscr{F}(x,y,t) = \left\langle C(t) - P(x,y), C'(t) \right\rangle = 0. \quad (1)$$

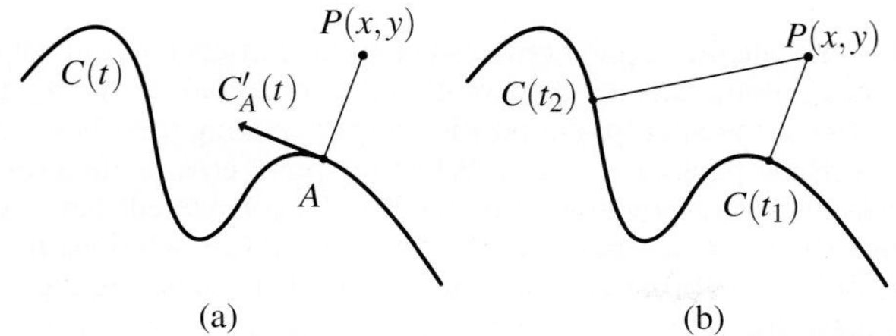

Figure 1: (a) A curve point A is a foot-point of $P(x,y)$ since $C'_A(t)$ is orthogonal to the directional vector $A - P(x,y)$. Two foot-points of $P(x,y)$, $C(t_1)$ and $C(t_2)$, are shown in (b).

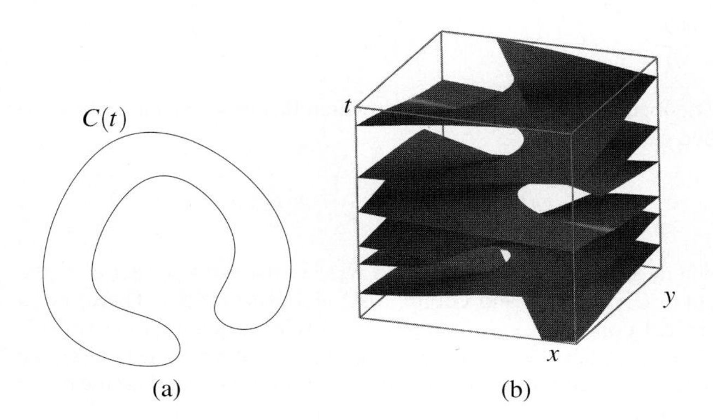

Figure 2: Given a planar curve $C(t)$ in (a), a zero-set surface of Equation (1) is represented using a red-colored surface in the xyt-parameter space (b).

In Figure 1(a), a curve point A is a foot-point of $P(x,y)$ since $C'_A(t)$, the tangent vector evaluated at the point A, is orthogonal to the directional vector $(A - P(x,y))$. A general freeform curve may have a set of foot-points, $C(t_i), i = 0, 1, \cdots, n$, from the point $P(x,y)$ (Figure 1(b) shows two of them). Then, the problem for computing a minimum distance from the point $P(x,y)$ to the planar curve $C(t)$ can be posed as

$$\text{find } \min \|P(x,y) - C(t_i)\|^2,$$

where $C(t_i), i = 0, 1, \cdots, n$, are foot-points of $P(x,y)$ onto the curve $C(t)$.

The solution space to this minimum distance problem is constructed in the higher dimensional parameter space. Denoted by $\mathscr{M}$, a zero-set of Equation (1) is a two-manifold surface in the xyt-parameter space. Given a C^1-continuous curve $C(t)$, $\mathscr{M}$ is continuous and closed in the domain. The surface $\mathscr{M}$, in the parameter space, implicitly represent the solution space for the problem of minimum distance queries. A parameter point (x,y,t) located on $\mathscr{M}$ corresponds to a foot-point $C(t)$ from $P(x,y)$ in the real world. Figure 2(a) shows a planar curve $C(t)$. Assuming that the plane is parameterized by $P(x,y)$, Figure 2(b) represents a zero-set of Equation (1) using a red-colored surface in the xyt-parameter space. Please note that the solution space $\mathscr{M}$ is constructed implicitly as a zero-set of the single polynomial equation in three variables, while the polynomial equation itself is explicitly represented using a NURBS function.

3.2 Solving for Single Queries

Given a specific position of the point $P(x_0,y_0)$, finding a minimum distance to the planar curve $C(t)$ is considered. Having a solution

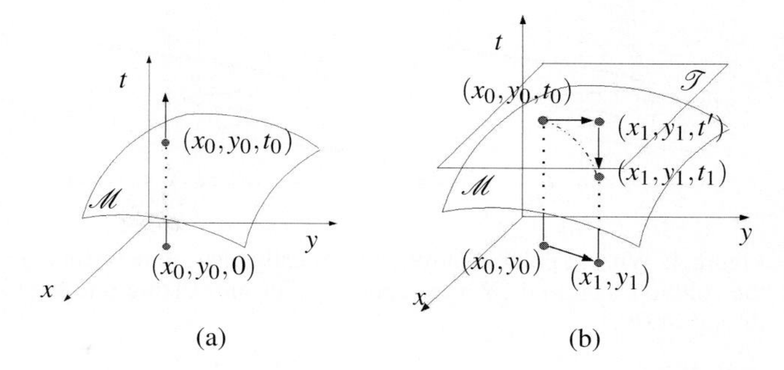

Figure 3: (a) Given a zero-set $\mathscr{M}$ of Equation (1), its intersection points with a ray staring at $(x_0,y_0,0)$ and directing positive t-direction correspond to foot-points of $P(x_0,y_0)$ onto the curve $C(t)$. When the point dynamically moves to the next position, an appropriate numerical marching process is shown in (b).

space $\mathscr{M}$ in the parameter space, one can imagine intersections of $\mathscr{M}$ with a ray of positive t-direction starting at $(x_0,y_0,0)$. A set of such intersection points, $\{(x_0,y_0,t_i)\}, i = 0, 1, \cdots, n$, between the ray and the solution space $\mathscr{M}$ corresponds to foot-points, $C(t_i), i = 0, 1, \cdots, n$, from $P(x_0,y_0)$. Figure 3(a) shows the solution space $\mathscr{M}$ and its intersection with the ray in the xyt-parameter space. A problem of computing a minimum distance between a point and planar curve can then be reduced to that of intersecting such a ray with the solution space $\mathscr{M}$.

The abstract idea of intersecting the ray with the zero-set $\mathscr{M}$ is computed by symbolically solving zeros of Equation (1) at a given parameter (x_0,y_0) without an explicit representation of $\mathscr{M}$. Since $\mathscr{F}(x,y,t)$ is a piecewise rational function, the zero-set can be constructed by exploiting the convex hull and subdivision properties of NURBS, yielding a highly robust divide-and-conquer zero-set computation that is reasonably efficient (see [Elber and Kim 2005] for details). The subdivision process continues until a given maximum depth of subdivision or some other termination criteria is reached. At the end of the subdivision step, a leaf node of the subdivision tree contains a single solution point and a set of these discrete points are improved using multivariate Newton-Rapson method into a highly precise solutions.

3.3 Solution to Consecutive Queries

Without spatial coherency, solving a minimum distance query requires an evaluation and subdivision of the solution space $\mathscr{M}$ to be performed for every new position of the point. But, a point may move continuously in the plane and the distance query can be reformulated for a consecutive position of the point. In such a case, we present an efficient numerical marching method, which traces the solution space $\mathscr{M}$ in the parameter space. Let's assume that a point $P(x_0,y_0)$ in the plane moves to the next point $P(x_1,y_1)$ and that two points are close enough to each other. Then, in the xyt-parameter space, the corresponding solution point (x_0,y_0,t_0) needs to be updated to the next point (x_1,y_1,t_1), while the new point should be on the zero-set manifold $\mathscr{M}$. We compute new solution point using an iterative marching method. The point (x_0,y_0,t_0) first proceeds to the point (x_1,y_1,t') located on the tangent plane, $\mathscr{T}$, of the zero-set at (x_0,y_0,t_0):

$$(x_1,y_1,t') = (x_0,y_0,t_0) + \Delta P - \nabla\mathscr{F}\langle\nabla\mathscr{F},\Delta P\rangle,$$

where $\Delta P = (x_1,y_1,0) - (x_0,y_0,0)$. Here, we use the normal vector of the zero-set surface $\mathscr{M}$, $\nabla\mathscr{F}$, that is normalized to a unit vector. We now project the point (x_1,y_1,t') back onto the manifold $\mathscr{M}$

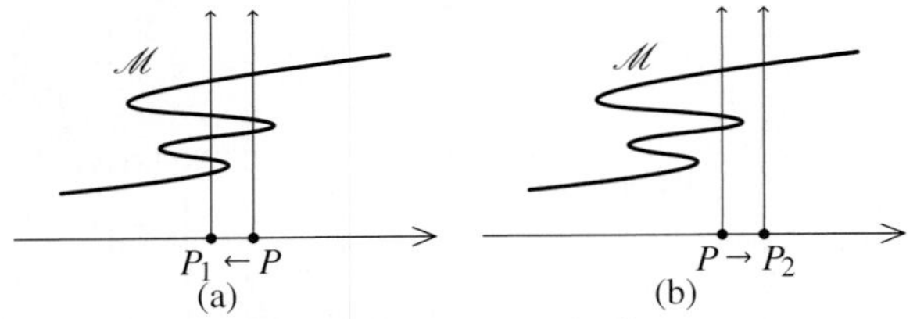

Figure 4: When a point P moves dynamically, new components of the solution manifold $\mathcal{M}$ may appear (a) or an existing one may disappear (b).

using a high-dimensional Newton's method. We do this process iteratively until the point (x_1, y_1, t') is placed on the solution space $\mathcal{M}$. Figure 3(b) shows the zero-set manifold $\mathcal{M}$ and the marching process over $\mathcal{M}$. Experimental results show that two or three iterations are enough for the convergence of the numerical projection operation.

3.4 Symbolic Analysis of the Topology

A continuous change of the position $P(x,y)$ may require a global analysis of the topological structure of the solution space. For consecutive distance queries, a numerical marching method may fail since it considers only the local properties of the solution space. Figure 4 shows a one-dimensional analogy to such a case. In Figure 4(a), when the point P moves to P_1, a new component of the zero-set $\mathcal{M}$ appears. Similarly, a proceeding toward P_2 causes an event such that an existing component of the zero-set disappears (see Figure 4(b)). These critical events need to be properly handled for a global solution to a distance query. Considering the case that a new component appears (Figure 4(a)), a new part of the planar curve may contribute to a minimum distance. Local tracing algorithms have difficulties in searching for a global solution since the global solution jumps from one part of the curve to another. On the other hand, the projection of the new solution point onto the zero-set manifold may not converge if it encounters the second case of the critical events (Figure 4(b)), as the existing component disappears.

In this section, the problem for computing critical positions where such global changes occur is reduced to that of solving two polynomial equations in three variables. As presented in Section 3.1, the zero-set surface of Equation (1) determines the solution space for querying minimum distances from all the possible positions of the point to planar curve. Since the point moves continuously in the xy-plane, critical events occur at points where the t-component of the normal vector of $\mathcal{M}$ vanishes. We, therefore, reformulate the problem for computing topological changes to that of solving the following two polynomial equations:

$$\begin{aligned} \mathcal{F}(x,y,t) &= 0, \\ \mathcal{G}(x,y,t) &= \frac{\partial \mathcal{F}}{\partial t}(x,y,t) = 0. \end{aligned} \quad (2)$$

Having two equations in three variables, the simultaneous zeros of Equations (1) and (2) are one-dimensional curves in the xyt-parameter space. By projecting them onto the xy-plane, a set of critical curves that causes the topological change can be obtained. The plane is then decomposed into several small sub-regions having boundaries at the critical curves. Inside the connected sub-region, the topology (i.e. the number of foot-points) does not change. Given the set of critical curves, one can easily detect such events that bring a global change to the topology of the solution space. We check whether a point crosses one of the critical curves by simply subdividing each of the two B-spline functions $\mathcal{F}(x,y,t)$ and $\mathcal{G}(x,y,t)$ at the appropriate position and evaluating their bounding boxes in the parameter space. When the point crosses the critical curves, the corresponding event needs to be considered: new foot-points should be inserted into the set of candidate solutions in the case of Figure 4(a) or existing one is deleted encountering the case of Figure 4(b).

A geometric interpretation of the critical curves shows an interesting relationship between those curves and the *curve evolute* of the given planar curve. Equation (2) expands to

$$\frac{\partial \mathcal{F}}{\partial t}(x,y,t) = \langle C'(t), C'(t)\rangle + \langle C(t) - P(x,y), C''(t)\rangle = 0.$$

The locus of the points $P(x,y)$ at which the above equation is satisfied can be written as

$$P(x,y) = C(t) + \frac{1}{\kappa(t)} N(t),$$

where $\kappa(t)$ is the curvature and $N(t)$ is the normal vector of the curve $C(t)$ [Bruce and Giblin 1992; Johnson 2005]. Therefore, a critical point $P(x,y)$, computed by solving Equations (1) and (2) simultaneously, lies at the center of the osculating circle at $C(t)$, the locus of which form a possibly discontinuous curve called the *curve evolute*.

Figure 5(a) shows a planar curve and its curve evolute. In Figure 5(a), the curve evolute is represented in gray-colored lines, which is computed symbolically using the formula, $C(t) + \frac{1}{\kappa}N(t)$. The evolute is possibly discontinuous at inflection points of the original curve, resulting in multiple curve sections (see Figure 5(a)). The zero-set of Equation (1) is shown as a red-colored surface and the critical curves are shown in green-colored lines in Figure 5(b). As one can see from Figure 5(c), the projection of critical curves onto the xy-plane matches with the curve evolute of Figure 5(a).

We now consider the combination of the global topology information with the previous numerical marching method for consecutive distance queries. Assuming $\mathcal{F}(x,y,t)$ in Equation (1) as a single variate function with t variable, its Jacobian becomes a matrix containing a single component $\frac{\partial \mathcal{F}}{\partial t}(x,y,t)$. For the case of finding roots of Equation (1), an iteration of Newton's method becomes degenerate when the determinant of its Jacobian vanishes:

$$\det(\mathbf{J}) = \det([\mathcal{F}_t]) = \frac{\partial \mathcal{F}}{\partial t}(x,y,t) = 0.$$

This shows that the meaning of Equation (2) can be reviewed from the Jacobian of Equation (1), which determines the degeneracy condition of the Newton's method. Thus, the critical curves provide an elegant way not only to efficiently find the global solution but also to support the numerical method even in its degenerate cases.

To get an efficient system for the minimum distance queries, a hierarchical subdivision of the plane is implemented. We first precompute critical points by symbolically solving Equations (1) and (2) with a rough tolerance. Four or five level of subdivision is used along each parametric axes for the hierarchical subdivision in the examples of Section 5. Complex geometry of the original curve may require high level of subdivision. Each sub-region of the subdivided plane contains a list of critical points which are located inside the region. More precisely speaking, they maintain a set of bounding boxes, at which the critical point resides, of the size of the given tolerance used from the symbolic computation. Figure 5(d) shows a planar curve and the critical points with their bounding boxes. Since it is relatively expensive to test whether the point crosses one

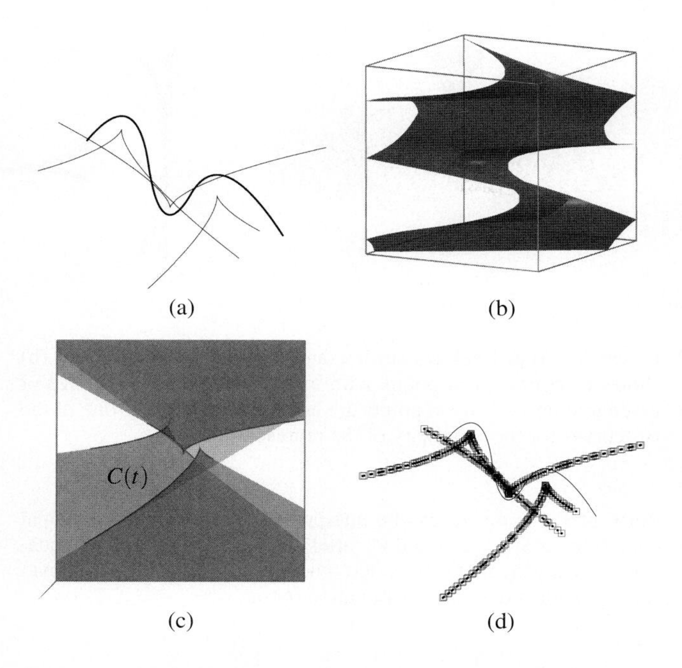

Figure 5: (a) A planar curve is shown in bold lines togethered with its curve evolute. (b) A zero-set of Equation (1) is represented in red-colored surface and critical curves are shown in green and bold lines. (c) The projection of the critical curves onto the xy-plane matches with the curve evolute of $C(t)$. A set of bounding boxes are shown in (d) with their center at the corresponding critical point.

of the critical curves, the system checks the crossing only when the point moves into one of the bounding boxes of the critical points. Hierarchical subdivision of the plane prunes away the movement of the point using a simple bounding box test. Experimental results show that this grid-based approach makes the performance of querying system about ten times faster.

4 Minimum Distance Queries Between a Point and Freeform Surface

The algorithm for computing minimum distances between a point and planar curve can be extended to the 3D case: distance queries from a point to freeform surface. For the surface case, a high-dimensional solution space is constructed by two polynomial equations in five variables. We therefore solve and update the distance query by tracing the solution space numerically in five-dimensional parameter space. Furthermore, we reduce the problem for analyzing the global topology of the solution space to that of solving three polynomial equations in five variables. Based on the topology information, we get a reliable tracing algorithm to find global minima.

4.1 Computing a High Dimensional Solution Space

Similarly to the curve case, the squared distance from a point in a 3D space, $\mathbf{P}$, to a freeform surface $S(u,v)$

$$D^2(u,v) = \langle S(u,v) - \mathbf{P}, S(u,v) - \mathbf{P} \rangle$$

is represented by B-spline scalar function. The minimum of $D^2(u,v)$ can be found by computing all its extrema and choosing the smallest. Assuming that $S(u,v)$ is C^1-continuous, extrema of $D^2(u,v)$ occur where its partial derivatives with respect to u and v become zero simultaneously. Using scaled extremal equations $E_1(u,v)$ and $E_2(u,v)$

$$E_1(u,v) = \langle S(u,v) - \mathbf{P}, S_u(u,v) \rangle$$
$$E_2(u,v) = \langle S(u,v) - \mathbf{P}, S_v(u,v) \rangle,$$

where $S_{u/v}$ is u/v-partial derivative of $S(u,v)$, we find their simultaneous zeros.

Given a parameterization of a point in a 3-D space, $P(x,y,z)$, we define a foot-point of $P(x,y,z)$ onto the surface $S(u,v)$ by lifting extremal distance equations $E_1(u,v)$ and $E_2(u,v)$ into a five variate functions,

Definition 2 *A foot-point of the point $P(x,y,z)$ onto freeform surface $S(u,v)$ satisfies the following two polynomial equations:*

$$\mathscr{H}(x,y,z,u,v) = \left\langle S(u,v) - P(x,y,z), \frac{\partial S}{\partial u}(u,v) \right\rangle = 0, \quad (3)$$
$$\mathscr{I}(x,y,z,u,v) = \left\langle S(u,v) - P(x,y,z), \frac{\partial S}{\partial v}(u,v) \right\rangle = 0. \quad (4)$$

A general freeform surface may have a set of foot-points, $S(u_i,v_i), i = 0,1,\cdots,n$, from the point $P(x,y,z)$. Then, the problem for computing a minimum distance from $P(x,y,z)$ to $S(u,v)$ can be posed as

$$\text{find } \min \|P(x,y,z) - S(u_i,v_i)\|^2,$$

where $S(u_i,v_i), i = 0,1,\cdots,n$, are foot-points of $P(x,y,z)$ onto the surface $S(u,v)$.

The solution space to the minimum distance problem for the surface case is now constructed using Equations (3) and (4). Having two equations in five variables, one gets three-dimensional manifolds, $\mathscr{Z}$, as their simultaneous zero-set in five-dimensional parameter space. A parameter point (x,y,z,u,v) located on $\mathscr{Z}$ corresponds to a foot-point $S(u,v)$ from $P(x,y,z)$ in the real world. Thus, two polynomial equations implicitly define the solution space $\mathscr{Z}$ in five-dimensional parameter space.

4.2 Computing Minimum Distances

As the solution space to the Equations (3) and (4) has three degrees-of-freedom, specifying a point (x,y,z) yields discrete set of zero-dimensional solution points. For a specific point $P(x_0,y_0,z_0)$, one can imagine a hyperplane in the $xyzuv$-parameter space. Then, intersection points between the hyperplane and the three-dimensional solution space are mapped to a set of foot-points and one can choose one of them having the smallest distance to $P(x_0,y_0,z_0)$. Similarly to the curve case, the intersection points are computed using a multivariate constraint solver [Elber and Kim 2005].

Solving for a single query requires an evaluation and subdivision of B-spline functions for every new query. However, a spatially coherent movement of a space point $P(x,y,z)$ yields an efficient numerical algorithm by utilizing a higher dimensional solution space, $\mathscr{Z}$. Assuming that a set of solution points, $\{(x_0,y_0,z_0,u_i,v_i)\}, i = 0,1,2,\cdots,n$, is given at current time step, a minimum distance query for the next position is solved numerically by extending the tracing algorithm presented in Section 3.3 to that of five dimensional space. When the point $P(x_0,y_0,z_0)$ moves to $P(x_1,y_1,z_1)$, we numerically march each solution points $\{(x_0, y_0, z_0, u_i, v_i)\}$ in the parameter space to the next point located on the solution space $\mathscr{Z}$. For the simplicity of the explanation, we consider a single solution point $(x_0, y_0, z_0, u_0, v_0)$ in the parameter space. Then, the point $(x_0, y_0, z_0, u_0, v_0)$ proceeds to the point (x_1, y_1, z_1, u', v') located on the hyperplane which is tangent to $\mathscr{Z}$ at $(x_0, y_0, z_0, u_0, v_0)$,

$$(x_1, y_1, z_1, u', v') = (x_0, y_0, z_0, u_0, v_0) + \Delta P - \nabla\mathscr{H}\langle\nabla\mathscr{H}, \Delta P\rangle - \nabla\mathscr{I}\langle\nabla\mathscr{I}, \Delta P\rangle.$$

Here, $\Delta P = (x_1, y_1, z_1, 0, 0) - (x_0, y_0, z_0, 0, 0)$ and we use two gradient vectors, $\nabla\mathscr{H}$ and $\nabla\mathscr{I}$, that are orthonormal to each other. Finally, the point (x_1, y_1, z_1, u', v') is projected onto the manifold using a high dimensional Newton's method. Note, however, that we project the point onto the manifold while keeping the *xyz* components of the point unchanged. We do this process iteratively until the point (x_1, y_1, z_1, u', v') is placed on the solution space $\mathscr{Z}$.

4.3 Analysis of the Topology

As a logical extension from the curve case, we provide a global analysis on the topology of the solution manifold for the surface case and show how this topology information helps the numeric marching method especially in degenerate cases where the numerical method fails. A geometric interpretation of the critical points also shows its relationship with the *focal set* of the surface.

Similar to the curve case, the global structure of the solution space changes when the space point $P(x, y, z)$ crosses critical points. Since the solution space $\mathscr{Z}$ is determined by two five-variate Equations (3) and (4), a condition for the critical point becomes

$$\mathscr{K}(x, y, z, u, v) = \det\begin{bmatrix} \frac{\partial\mathscr{H}}{\partial u} & \frac{\partial\mathscr{H}}{\partial v} \\ \frac{\partial\mathscr{I}}{\partial u} & \frac{\partial\mathscr{I}}{\partial v} \end{bmatrix} = 0. \quad (5)$$

Having three Equations (3), (4) and (5) in five variables, one gets two-manifold surfaces as their simultaneous zero-set. By projecting them onto the *xyz*-space, a set of critical surfaces that causes the topological change can be obtained. Figure 6(a) shows a freeform surface and the set of critical points.

A hierarchical subdivision of the *xyz*-space to efficiently maintain a set of critical points is constructed like the curve case. First, critical points are precomputed by symbolically solving Equations (3), (4) and (5) with a rough tolerance. Since the symbolic solver computes critical points by recursively subdividing the parameter space, we can get a hierarchical subdivision of the *xyz*-space without any other costs, which contains a critical point at the end of the subdivision stage. Figure 6(b) shows a set of critical points with its bounding box of the size of the given tolerance. In Figure 6(b), critical points are evaluated only along one of the iso-curves for the simplicity of the representation.

A geometric meaning of critical points shows that intrinsic properties of a surface, such as the principal curvatures of a surface, characterize the topological structure of the solution space, $\mathscr{Z}$. Since critical points satisfy Equations (3) and (4), we can restrict the points to be on the normal line of a surface. Then, the directional vector $(S(u, v) - P(x, y, z))$ can be replaced by $\ell N(u, v)$, where $N(u, v)$ is a normal vector of $S(u, v)$ and ℓ is some real variable. Plugging in this to Equation (5) results in

$$\det\begin{bmatrix} \langle\ell N(u,v), S_{uu}\rangle + S_u^2 & \langle\ell N(u,v), S_{uv}\rangle + \langle S_u, S_v\rangle \\ \langle\ell N(u,v), S_{uv}\rangle + \langle S_u, S_v\rangle & \langle\ell N(u,v), S_{vv}\rangle + S_v^2 \end{bmatrix} = 0. \quad (6)$$

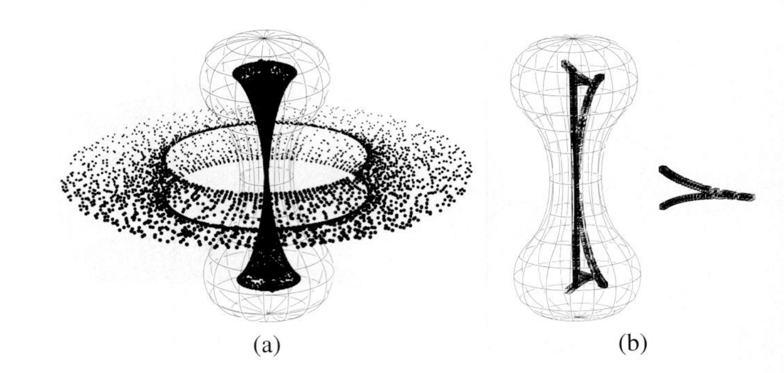

Figure 6: (a) A freeform surface and a set of critical points. (b) shows a set of critical points with its bounding box of the size of given tolerance. Critical points are evaluated only along one of the iso-curves for the simplicity of the representation.

Zeros of Equation (6) can be interpreted in terms of the principal curvatures at $S(u, v)$, κ_1 and κ_2, after a simple substitution of Equation (6) using the first and second fundamental forms. These derivations provide a rewriting of Equation (6) as

$$\ell^2\kappa_1\kappa_2 + \ell(\kappa_1 + \kappa_2) + 1 = 0,$$

and the zeros are just

$$\ell = -\frac{1}{\kappa_1} \quad \textit{and} \quad \ell = -\frac{1}{\kappa_2}.$$

Therefore, the critical point $P(x, y, z)$ is a distance along the normal equal to one of the principal radii of curvature of the closest point on the surface $S(u, v)$. Compare this geometric interpretation of the critical point to that of the curve case.

5 Experimental Results

Several examples of computing minimum distances between a point and planar curve or freeform surface are now presented. First, some examples for minimum distance queries to planar curve are shown. Figure 7 presents an example of a curve with a trajectory of the moving point. In Figure 7(a), the planar curve is shown in bold lines with a trajectory curve in gray. For each sampled position of the moving point, the corresponding curve point which gives the minimum distance to the point is connected using a line segment. Figure 7(b) shows a higher dimensional solution space using a red-colored surface in the *xyt*-parameter space. Minimum distance points in real Euclidean space have their corresponding points in the higher dimensional parameter space. In Figure 7(b), a yellow-colored sphere represents a parameter point corresponding to each of the minimum distance query shown in Figure 7(a). One more example is presented in Figure 8. The planar curves presented in these experimental examples are represented by cubic NURBS having about 10-20 control points. In these experimental examples, a tolerance of 0.05 (planar curves' dimensions span about a unit length) is used for the preparation of the critical points. And about 90% of distance queries are pruned away during the test of crossing the critical curves.

Figure 9 presents two more complex examples of minimum distance queries for planar curves. Here, planar curves are shown in bold lines and a trajectory curve for the moving point in gray. 200 positions of the trajectory curve are sampled for the minimum distance queries. With the aid of the higher dimensional solution

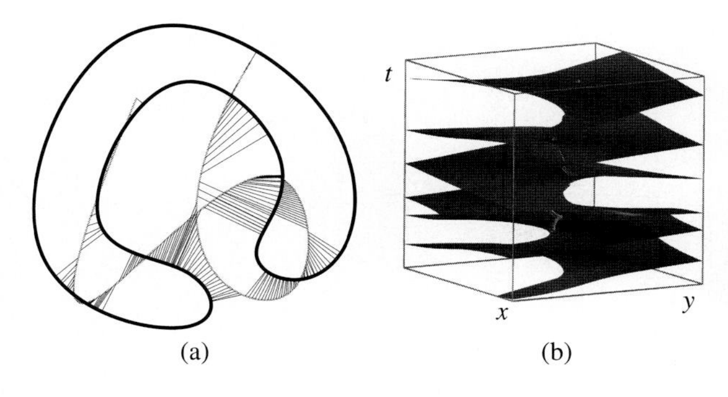

Figure 7: (a) A planar curve is shown in bold lines togethered with a trajectory curve for the moving point. For a sampled point of the trajectory curve, a line segment connects its corresponding curve point which gives a minimum distance to the curve. (b) A higher dimensional solution space is represented in red-colored surface and solution point in the parameter space, which is corresponding to the minimum distance, is shown in yellow sphere.

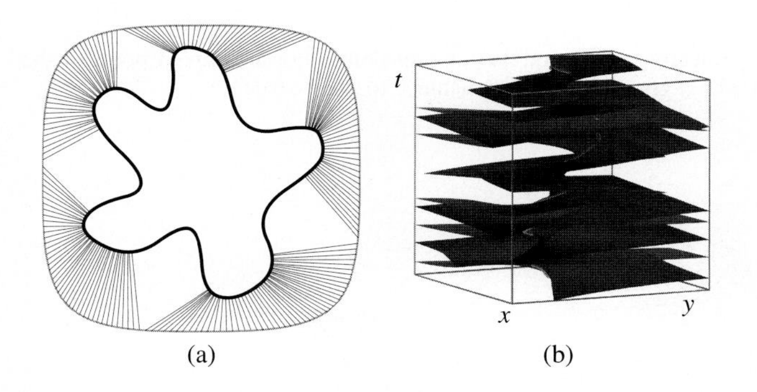

Figure 8: (a) A planar curve is shown in bold lines togethered with a trajectory curve for the moving point. For a sampled point of the trajectory curve, a line segment connects its corresponding curve point which gives a minimum distance to the curve. (b) A higher dimensional solution space is represented in red-colored surface and solution point in the parameter space, which is corresponding to the minimum distance, is shown in yellow sphere.

space, global solutions to the minimum distance queries are computed and shown in Figure 9 using line segments. Having the hierarchical subdivision of the plane and early pruning technique, the computation time for these results for the curve case are about the same, 10000 distance queries took about 1.2 to 1.4 seconds. Please note that we take all the solution points of local minima being tracked to yield a global solution. Thus, a single query contains multiple distance extrema computations. Testing was done on a Pentium IV 2GHz desktop machine.

Continuing to examples of computing minimum distance for freeform surfaces, Figure 10 shows two examples. In Figure 10(a), the same surface as in Figure 6 is used for computing a minimum distance. As one can see from Figure 6(a), the minimum distance point jumps to a different part of the surface when the moving point crosses the critical points. With the aid of global analysis on the topology of the solution space, our approach properly traces the global solution. The freeform surfaces presented in these experimental examples are represented by bicubic NURBS having about 50-70 control points. A symbolic computation of the critical points used a tolerance of 0.1 (freeform surfaces' dimensions span about a unit length) and the symbolic computation time for these examples are within a minute including a construction of the hierarchical bounding boxes, which were used in early pruning of distance queries. On a 2GHz Pentium IV machine, about 300 to 500 queries were computed in a second.

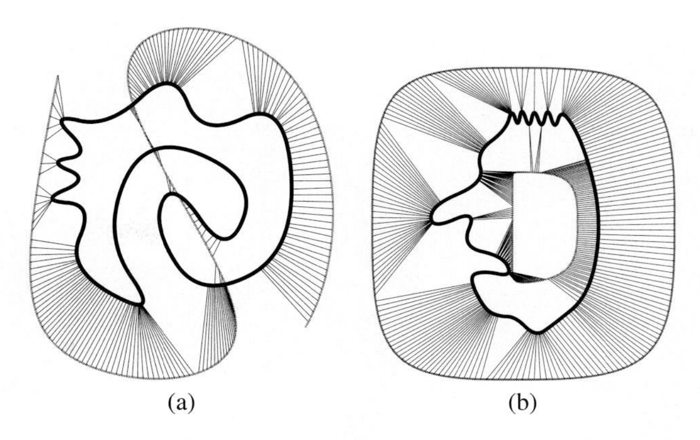

Figure 9: A planar curve is shown in bold lines togethered with a trajectory curve for the moving point. For a sampled point of the trajectory curve, a line segment connects its corresponding curve point which gives a minimum distance to the curve.

Figure 11 shows two more complex examples. Here, freeform surfaces are shown in bold lines and the trajectory curve for the moving point in gray. 200 positions of the trajectory curve are sampled for the minimum distance queries. In Figure 11(b), our minimum distance search is applied to a collection of surfaces that form a teapot. It is quite straightforward to extend our approach to such a scene that has multiple objects. Obviously, distance computations between a point and boundary curves of each surface patch may be taken into consideration to handle multiple objects, while that doesn't need to be invoked in the example of Figure 11(b).

6 Conclusion

A robust and efficient algorithm for computing minimum distances between a point and planar curve or freeform surface has been presented. The presented approach is based on the dimensionality lifting of the problem into a higher-dimensional parameter space. This higher dimensional formulation solves for all query points in the domain simultaneously, which provides opportunities to speed computation by applying coherency techniques. For the curve case, a higher dimensional solution space is defined by a single equation in three variables. This formulation yields two-manifold surfaces in the parameter space, from which a minimum distance query is solved numerically in a very efficient way. Global convergence is assured by detecting changes in the higher-dimensional topological structure. We reduced this problem of analyzing the topological structure of the solution space to that of solving two polynomial equations in three variables. The topology information supports local searches in the numerical tracing algorithm to guarantee that it finds all the solutions and finally gives a global minima. This symbolic computation part of the algorithm is accelerated through preprocessing of the higher-dimensional space. The approach developed for curves in the plane has been extended to freeform surfaces in a 3D space. For the surface case, a higher-dimensional solution space is implicitly defined by two polynomial equations in five variables.

A variety of hierarchical space subdivision techniques can be applied to the current system for computing minimum distances. An efficient integration of the hierarchical technique is expected to

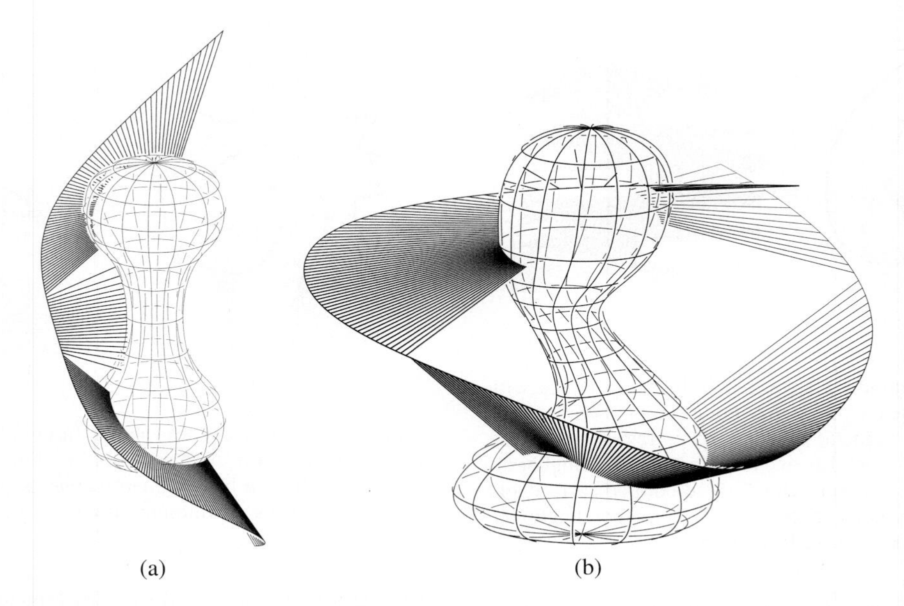

(a) (b)

Figure 10: A freeform surface is shown in bold lines togethered with a trajectory curve for the moving point. For a sampled point of the trajectory curve, a line segment connects its corresponding surface point which gives a minimum distance to the surface.

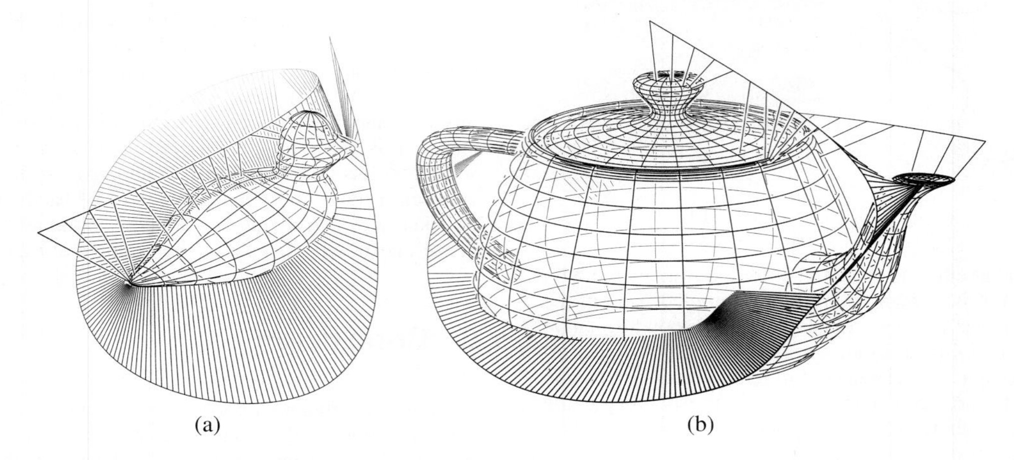

(a) (b)

Figure 11: A freeform surface is shown in bold lines togethered with a trajectory curve for the moving point. For a sampled point of the trajectory curve, a line segment connects its corresponding surface point which gives a minimum distance to the surface.

yield a distance querying algorithm of higher performance. The presented approach may easily be applicable to a trimmed model. Solution points for the local distance extrema only need to be tested whether they are located inside the valid parameter domain or not in the case of trimmed models. We are also working to extend the presented approach to the computation of minimum distances for the curve-curve or surface-surface case. To this end, we need to deal with even higher-dimensional solution spaces.

Acknowledgments

All the algorithms and figures presented in this paper were implemented and generated using the IRIT solid modeling system [Elber 2000] developed at the Technion, Israel. This work was supported in part by ARO (DAAD19-01-1-0013) and NSF (IIS0218809, CCR0310705). All opinions, findings, conclusions or recommendations expressed in this document are those of the author and do not necessarily reflect the views of the sponsoring agencies.

References

BARAFF, D. 1990. Curved surfaces and coherence for non-penatrating rigid body simulation. *Computer Graphics 24*, 4, 19–28.

BREEN, D., MAUCH, S., AND WHITAKER, R. 1998. 3d scan conversion of csg models into distance volumes. In *Proceedings of the 1998 Symposium on Volume Visualization*, ACM SIGGRAPH, 7–14.

BRUCE, J., AND GIBLIN, P. 1992. *Curves and Singularities*. Cambridge University Press.

DOKKEN, T. 1985. Finding intersections of b-spline represented geometries using recursive subdivision techniques. *Computer Aided Geometric Design 2*, 1, 189–195.

EHMANN, S. A., AND LIN, M. C. 2001. Accurate and fast proximity queries between polyhedra using convex surface decomposition. *Eurographics (EG) 2001 Proceedings 20*, 500–510.

ELBER, G., AND KIM, M. S. 2005. Geometric constraint solver using multivariate rational spline functions. In *Proc. of International Conference on Shape Modeling and Applications*, MIT, USA, 216–225.

ELBER, G., KIM, M. S., AND HEO, H. 2001. The convex hull of rational plane curves. *Graphical Models 63*, 151–162.

ELBER, G. 1992. *Free Form Surface Analysis using a Hybrid of Symbolic and Numeric Computation*. PhD thesis, Department of Computer Science, University of Utah.

ELBER, G., 2000. Irit 9.0 user's manual. http://www.cs.technion.ac.il/~irit, October.

FRISKEN, S., PERRY, R., ROCKWOOD, A., AND JONES, T. 2000. Adaptively sampled distance fields: A general representation of shape for computer graphics. In *Proc. of SIGGRAPH 2000*, 249–254.

GRANDINE, T., CRACIUN, B., HEITMANN, N., INGALLS, B., GIA, Q., OU, M., AND TSAI, Y. 2000. The bivariate contouring problem. *IMA Bebruary 2000 Preprint Series*.

II, T. T., JOHNSON, D., AND COHEN, E. 1997. Direct haptic rendering of sculptured models. In *Proc. of 1997 Symposium on Interactive 3D Graphics*, 167–176.

JOHNSON, D., AND COHEN, E. 2001. Spatialized normal cone hierarchies. In *ACM SIGGRAPH Symposium on Interactive 3D Graphics (I3D)*, 129–134.

JOHNSON, D., AND COHEN, E. 2005. Distance extrema for spline models using tangent cones. In *Graphics Interface*.

JOHNSON, D. 2005. Minimum distance queries for haptic rendering. *PhD Thesis*.

KIM, M. S., AND ELBER, G. 2000. Problem reduction to parameter space. In *The Mathematics of Surface IX (Proc. of the Ninth IMA Conference)*, London, 82–98.

LARSEN, E., GOTTSCHALK, S., LIN, M., AND MANOCHA, D. 2000. Fast distance queries with rectangular swept sphere volumes. In *IEEE International Conference on Robotics and Automation (ICRA)*, 24–48.

MORTENSEN, M. 1985. *Geometric Modeling*. John Wiley & Sons, New York.

MUSETH, K., BREEN, D., WHITAKER, R., MAUCH, S., AND JOHNSON, D. 2005. Algorithms for interactive editing of level set models. *Computer Graphics Forum 24*, 4, 1–22.

PATRIKALAKIS, N., AND MAEKAWA, T. 2002. *Shape Interrogation for Computer Aided Design and Manufacturing*. Springer Verlag.

POTTMANN, H., LEOPOLDSEDER, S., AND ZHAO, H. 2003. The d^2-tree: A hierarchical representation of the squared distance function. In *Technical Report No. 101*, Institut fur Geometrie, TU Wien.

QUINLAN, S. 1994. Efficient distance computation between non-cnovex objects. In *IEEE Int. Conference on Robotics and Automation*, 3324–3329.

SEONG, J., ELBER, G., JOHNSTONE, J., AND KIM, M. 2004. The convex hull of freeform surfaces. *Computing 72*, 1, 171–183.

SEONG, J., KIM, K., KIM, M., ELBER, G., AND MARTIN, R. 2005. Intersecting a freeform surfaces with a general swept surface. *Computer-Aided Design 37*, 5, 473–483.

SHERBROOKE, E., AND PATRIKALAKIS, N. 1993. Computation of the solutions of nonlinear polynomial systems. *ComputerAided Geometric Design 10*, 5, 379–405.

SIGG, C., PEIKERT, R., AND GROSS, M. 2003. Signed distance transform using graphics hardware. In *Proc of IEEE VIS*.

SNYDER, J. 1993. Interval analysis for computer graphics. *Computer Graphics 26*, 2, 121–130.

SNYDER, J. 1993. Interval methods for multi-point colilisions between time-dependent curved surfaces. *Computer Graphics 27*, 2, 321–334.

SUD, A., OTADUY, M., AND MANOCHA, D. 2004. Difi: Fast 3d distance field computation using graphics hardware. *Computer Graphics Forum 23*, 3, 557–566.

WANG, H., KEARNEY, J., AND ATKINSON, K. 2002. Robust and efficient computation of the closest point on a spline curve. In *Proceedings of the 5th International Conference on Curves and Surfaces*, San Malo, France, 397–406.

Duplicate-Skins for Compatible Mesh Modelling

Yu Wang*
The Hong Kong University of Science and Technology

Charlie C.L. Wang†
The Chinese University of Hong Kong

Matthew M.F. Yuen‡
The Hong Kong University of Science and Technology

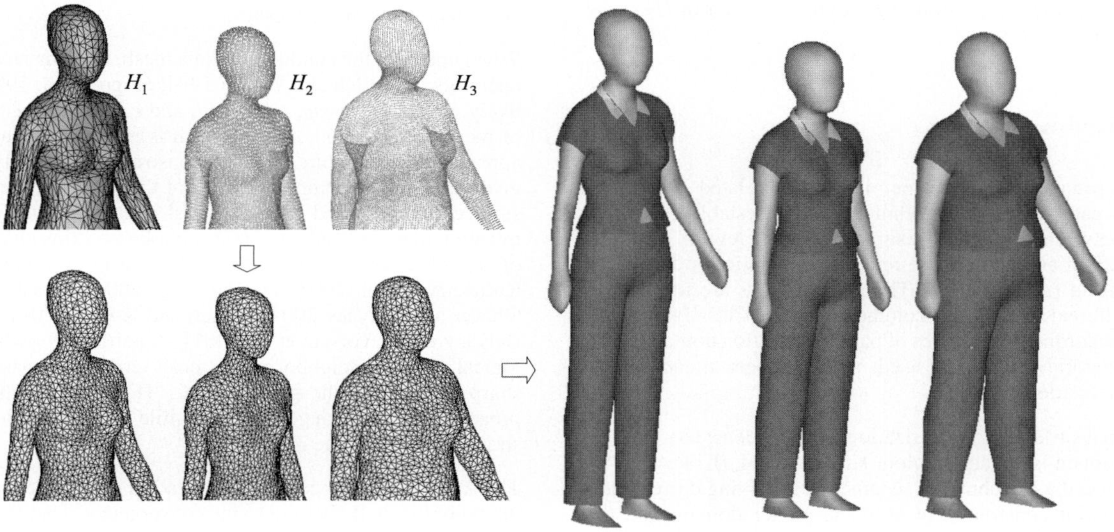

Figure 1: A design automation application — after modelling the compatible meshes by our duplicate-skins algorithm on three human models given in various representations (where H_1 is a two-manifold mesh, H_2 is a polygon soup with many holes, and the shape of H_3 is represented by a point cloud), the clothes designed around H_1 can be automatically "graded" to fit the body shape of H_2 and H_3.

Abstract

As compatible meshes play important roles in many computer-aided design applications, we present a new approach for modelling compatible meshes. Our compatible mesh modelling method is derived from the skin algorithm [Markosian et al. 1999] which conducts an active particle-based mesh surface to approximate the given models serving as skeletons. To construct compatible meshes, we developed a duplicate-skins algorithm to simultaneously grow two skins with identical connectivity over two skeleton models; therefore, the resultant skin meshes are compatible. Our duplicate-skins algorithm has less topological constraints on the input models: multiple polygonal models, models with ill-topology meshes, or even point clouds could all be employed as skeletons to model compatible meshes. Based on the results of our duplicate-skins algorithm, the modelling method of n-Ary compatible meshes is also developed in this paper.

CR Categories: I.3.5 [Computational Geometry and Object Modeling]: Boundary representations—Curve, surface, solid, and object representations; J.6 [COMPUTER-AIDED ENGINEERING]: Computer-aided design (CAD)—Computer-aided design (CAD)

Keywords: Compatible meshes, skin algorithm, free-form modelling, deformation, design automation

*e-mail: mewangyu@ust.hk
†e-mail: cwang@acae.cuhk.edu.hk
‡e-mail: meymf@ust.hk

1 Introduction

Representing the models in compatible meshes is a fundamental problem for a large class of applications, such as mesh metamorphosis [Alexa 2002; Kanai et al. 2000; Lee et al. 1999], n-way shape blending/editing [Biermann et al. 2002; Kraevoy and Sheffer 2004; Praun et al. 2001], detail and texture transferring [Kraevoy and Sheffer 2004], parametric design of free-form models [Wang 2005; Seo and Magnenat-Thalmann 2004; Allen et al. 2003; Praun et al. 2001; Marschner et al. 2000], and design automation [Wang et al. 2005]. Compatible meshes, i.e. meshes with an identical connectivity, support bijective mapping between two or more models which establish immediate point correspondences between models. Therefore, each vertex in one mesh has a unique corresponding vertex in every other mesh. The research presented in this paper develops a new algorithm to construct compatible meshes between given models. For two given models M_1 and M_2, our duplicate-skins algorithm manipulates two skin meshes with consistent connectivity to approximate the geometry of M_1 and M_2. Note that the model here means the geometry represented by various representations (e.g., a polygonal mesh or a point cloud — see H_1, H_2 and H_3 in Fig.1). The correspondences of semantic features on the given models M_1

SPM 2006, Cardiff, Wales, United Kingdom, 06–08 June 2006.
© 2006 ACM 1-59593-358-1/06/0006 $5.00

and M_2 are specified by users or by a feature recognition algorithm. Our duplicate-skins algorithm can construct identical entities on the resultant meshes for corresponding pairs of position markers.

Compatible meshes are usually requested on the models with similar features (i.e., we seldom have the need to build a compatible mesh between a tori and a cube); also, we assumed that the corresponding features have been correctly specified. It is meaningless to map the leg of a human H_1 to the head of another human body H_2 and correlate the bellybutton of H_1 to the shoulder of H_2 at the same moment.

1.1 Related work

The work presented in this paper is closely related to the so-called cross-parameterization technique, which established bijective maps between models. Alexa gave a good review of cross-parameterization and compatible remeshing techniques developed for morphing in [Alexa 2002]. The general ways are to parameterize the different models to a common domain. Classifying these techniques according to the types of parameterization domain, there are three categories commonly used: planar, spherical and simplicial parameterization.

The traditional surface parameterization problem considers the case where the domain is a planer region. Kraevoy et al. [Kraevoy et al. 2004] introduced a Matchmaker scheme for satisfying corresponding feature point constraints in both the planer domain and the model's surface. When cross-parameterization is used for geometry processing, it is sometimes possible to limit the computation to disk-like parts of the surfaces [Biermann et al. 2002; Desbrun et al. 2002]. After the entire surface is cut to disk-like parts, each part is parameterized independently. In some techniques, the surface is cut into a single chart [Sheffer and Hart 2002; Sorkine et al. 2002], while in others, it is cut into an atlas of parts (e.g., [Julius et al. 2005; Sander et al. 2003; Levy et al. 2002; Sander et al. 2002]). In either case, the cuts break the continuity of the parameterization, and make it difficult to use a planar parameterization approach to construct a low distortion bijective mapping between two different models.

Another popular choice is spherical parameterization, which uses a sphere as the base domain [Alexa 2002; Gotsman et al. 2003; Praun and Hoppe 2003]. An important limitation of spherical parameterization is that it can only deal with a closed and genus zero surface. One more general approach is to let the domain be a coarse base mesh, called simplicial parameterization. The surface is partitioned into matching patches with a consistent inter-patch connectivity [Praun et al. 2001; Kraevoy and Sheffer 2004; Schreiner et al. 2004]. The challenge in this way is that it is difficult to globally optimize the parameterization. In addition, all of these techniques requires the meshes on given models are valid and two-manifold. Our duplicate-skins algorithm does not have this constraint so that the range of models to be processed is broadened. Although, several researches in literature (e.g., [Nguyen et al. 2005; Ju 2004]) mentioned techniques that can repair the meshes with ill-topology, an operation with less constraints is always welcome.

Avoiding explicit parameterization, [Allen et al. 2003] employed a mesh surface as a template and the connectivity of this template is fixed to approximate the geometry of the input point cloud. They formulated an optimization problem in which the degrees of freedom are an affine transformation at each template vertex. However, their solution is limited to very specified inputs and this can introduce severe approximation errors when the input models have a significantly different geometry.

To explicitly and rapidly construct compatible meshes with different geometries, the skin algorithm presented in [Markosian et al. 1999] is adopted to develop our duplicate-skins algorithm. The original purpose of Markosian et al. is to rapidly design a rough free-form shape via direct interactivities. A user could interactively sculpt a free-from surface (skin) that approximates the underlying given models. The mesh connectivity of skin is updated in the iterations of skin evolution. Inspired by their work, we developed a new algorithm, *duplicate-skins*, to grow over various geometries and obtain compatible meshes.

When updating the conductivity of a mesh, three mesh optimization operators in [Welch and Witkin 1994; Hoppe et al. 1993] are iteratively used: *edge swap*, *edge split*, and *edge collapse*. In the sense of mesh optimization, our approach is related to many remeshing approaches in literature, which reconstruct high-quality meshes for given surfaces. A complete review of the remeshing techniques for surfaces can be found in [Alliez et al. 2005]. As the dynamic optimization manner is adopted in our duplicate-skins algorithm, some of our ideas are borrowed from [Surazhsky and Gotsman 2003; Kartasheva et al. 2003; Ohtake et al. 2003; Vorsatz et al. 2003; Ohtake and Belyaev 2002; Botsch and Kobbelt 2001; Ohtake and Belyaev 2001; Vorsatz et al. 2001] — particularly when the topological changes neighboring to mesh entities are associated with sharp features on the given model. However, all the above approaches consider single meshes while our approach extends the strategies to duplicate meshes.

The same as other dynamic optimization approaches, a good starting point is usually helpful to the convergency. Therefore, in our algorithm, the *radial basis function* (RBF) based shape interpolation techniques are employed to create desirable initial skin meshes to fit the underlying skeleton models. An RBF offers a compact functional description of a set of surface data. Interpolation and extrapolation are inherent in the functional representation. The benefits of modeling surfaces with RBFs have been recognized in [Yngve and Turk 2002; Cohen-Or et al. 1998; Carr et al. 1997; Savchenko et al. 1995]. The radial basis functions associated with a surface can be evaluated at any location to produce a mesh at the desired resolution. [Carr et al. 2001] suggested a RBF-based approach, which can be used for reconstructing the incomplete scan data. Similar to [Cohen-Or et al. 1998], we adopted the global RBF in our approach to deform initial skin meshes into desirable shapes and orientations.

1.2 Definition of Terms

Surface representation based on polygonal meshes has become a standard in many geometric modelling applications. The mesh representation is flexible and general with respect to shape and topology as well as conducive to efficient algorithm processing on meshes. We use the now widespread terminology of mesh from [Spanier 1966]. A triangular mesh is described by a pair (K,V), where $V=(v_1,\ldots,v_n)$ describes the geometric position of the vertices in $\Re^d$ (typically $d=3$) and K is a simplicial complex representing the connectivity of vertices, edges, and faces. The abstract complex K describes vertices, edges and faces as $0,1,2$ simplicies, that is, edges are pairs $\{i,j\}$, and faces are triples $\{i,j,k\}$ of vertices. The neighborhood ring of a vertex $\{i\}$ is the set of adjacent vertices $N(i)=\{j|\{i,j\}\in K\}$ and its *star* is the set of incident simplices $star(i)=\bigcup_{i\in\varsigma,\varsigma\in K}\varsigma$.

Skeletons are the given models and regarded as a geometric reference, they may be closed mesh surfaces, surfaces with boundaries, non-manifold surfaces, polylines or even isolated points, as shown in Fig.2. In the duplicate-skins algorithm, a pair of skeletons are assigned as *source* and *target* skeletons respectively. Our method does not require these two skeletons share the same number of vertices or triangles, or have identical connectivity. *Sharp edges* on the given skeletons are edges with relatively large curvatures.

Skin is the mesh growing over a skeleton. We refer to the vertices of the skin as *particles*. For a skin, a *target edge length* is defined which is the expected skin edge length. We take the target edge length as the criterion of skin connectivity modification. The target edge length is measured in terms of the average edge length of skeletons, i.e. $L_{tag} = ratio \times L_{avg}$. Each particle p should track to one position locally closest to p on the relative skeleton. This position is called the *tracking position*. We refer to the face containing the tracking position as the *tracking face*. Our duplicate-skins algorithm simultaneously constructs two skins respectively for the *source* and *target* skeletons. These two skins should be guaranteed compatible and we call these two skins as *duplicate-skins*.

Like most of the technologies used to build the correspondence between meshes [Allen et al. 2003; Sumner and Popović 2004; Kraevoy and Sheffer 2004; Schreiner et al. 2004], a small set of *position markers* are necessary to be specified on the source and target skeletons. These markers are enforced as mapping constraints, taking two head models as an example, the markers constrains the correspondence of ear, nose and eyes as well as other facial elements/features.

1.3 Contribution

We propose a new method for compatible mesh modelling — a duplicate-skins algorithm, which simultaneously grows two skins with identical connectivity over two skeleton models while satisfying the feature correspondences. Compared to other recent approaches for the same purpose, the method presented in this paper shows almost no topological constraint on the models to be approximated (i.e., the input models can be in various geometry representations).

Based on the results of our duplicate-skins algorithm, n-Ary compatible meshes also can be easily determined. As the feature vertices (i.e., position markers) are correlated to particles on the skin meshes, the feature correspondences among all n skeleton models are satisfied.

Sharp edges are well preserved on the resultant compatible meshes from our algorithm, which is important for many applications. A new sharp feature tracking method is developed to guarantee that the sharpness-preserved results can be given on the compatible meshes.

Thanks to the connectivity optimization in the duplicate-skins algorithm, the resultant compatible meshes are relatively regular, so that they can serve as good inputs for the downstream geometry processing applications where the irregularity usually leads to unsatisfactory results.

The rest of the paper is organized as follows. Section 2 gives a more detailed description of the original skin algorithm [Markosian et al. 1999]. Section 3 details our duplicate-skins algorithm. Section 4 presents the method to construct n-Ary compatible meshes. In section 5, several applications of the compatible meshes are demonstrated — the applications fall into two categories: free-form modelling and design automation for customized free-form products. Finally, section 6 discusses the limitations of our algorithm and suggests future research directions.

2 Skin Algorithm

We give a more detailed description of the original skin algorithm [Markosian et al. 1999] as below which works as the basis of our duplicate-skins algorithm. With a skeleton mesh M as the input, the skin algorithm governs a skin mesh S growing over M to approximate its shape with a smooth mesh whose connectivity is more regular. The algorithm consists of five steps: 1) the construction of the first skin S; 2) search for the tracking position and face for each particle on S; 3) reposition particles; 4) modify the connectivity of S; 5) update the tracking position and face for particles on S. The 3rd to 5th steps are iteratively applied on S until the movement of all particles on S is less than a small value ε and no further modification of mesh connectivity is needed.

There is no limitation to the topology of the first skin S. The only requirement is that S should show the shape that can vary into the source/target model by an elastic deformation. For example, for genus-0 models, a mesh with a box shape or a spherical shape bounding them is a good initial skin. However, for genus-1 models, we must introduce tori-like initial skins.

Before entering the iteration of skin evolution, the closest point and face of each particle on the skeleton S must be found to serve as the tracking point and face. The global searching strategy is applied to accurately obtain the first tracking. In the later iteration steps, to speed up, a local search strategy replaces the global one.

To gradually attract skin toward the skeleton, the movement of each particle is measured from its tracking position, its current position and its neighbors on the skin. The new location $v_{new}(p)$ of the particle p is computed by

$$v_{new}(p) = \alpha v_0(p) + \beta v_c(p) + \gamma v_t(p) \tag{1}$$

where $v_0(p)$ is the current location of the particle, $v_c(p)$ is the center of p's 1-ring neighbors, and $v_t(p)$ is the target position that computed by

$$v_t(p) = v_0(p) + w_m(d_s - r_s)v_{trk}(p) \tag{2}$$

where $v_{trk}(p)$ is the unit tracing direction to the closest point on skeleton. d_s is the distance from p to its tracking position, r_s is the user specified offset between the skeleton model and the final skin, and w_m is the moving ration in the range $[0,1]$ which controls the amount of movement. α, β and γ are positive coefficients, and $\alpha+\beta+\gamma = 1$. β controls the smoothness of the skin. The trade-off for selecting β determines the behavior of the skin, e.g. a large β leads to the over smoothed skin; on the other hand, if a very small β is adopted, uneven particle distribution and sliver triangles could be produced. γ is calculated by $\gamma = 1 - \alpha - \beta$. From our experiments, we choose $\alpha = 0.3$. A non-linear and attenuating function is used to evaluate the value of β where the iteration step is the function variable.

In the connectivity modification step of the skin algorithm, three operators: *edge swap*, *edge split*, and *edge collapse* from [Welch and Witkin 1994; Hoppe et al. 1993] are iteratively applied to remove extreme long and short edges, and at the same time increase the minimal angle in triangles. Through this, the shape quality of the triangles on the skin mesh is optimized.

As claimed by [Markosian et al. 1999], the skin algorithm can work with the skeleton models in the form of closed mesh surfaces, surfaces with boundaries (more generally saying - non-manifold models), polylines or even isolated points. Examples are given in Fig.2. Benefitting from this characteristic of the basic skin algorithm, our duplicate-skins algorithm can model compatible meshes approximating various skeleton models. Details are addressed below.

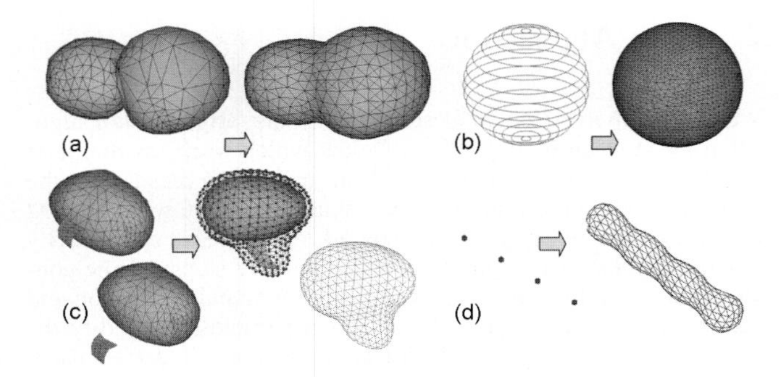

Figure 2: Various models could be employed as skeletons: (a) a closed mesh surface, (b) a wire-frame, (c) a non-manifold structure assembled from several mesh patches, and (d) isolated points.

3 Duplicate-Skins

To seek a method to construct compatible meshes for various free-form models, we propose a new technique and name it *duplicate-skins* which is derived from the original skin algorithm. The most important difference is that we can now simultaneously manipulate two skins in various geometries but with identical mesh connectivity. Benefitting from the desirable high-quality of skin meshes, the resultant compatible meshes give a bijective mapping between them that is guaranteed to be smooth and continuous. The following addresses the details of our duplicate-skins algorithm for generating compatible meshes between a pair of skeletons.

3.1 Algorithm overview

The input to the algorithm consists of two skeletons $M_0 = (K_0, V_0)$ (source) and $M_1 = (K_1, V_1)$ (target) and two sets of position markers (including $P_0 = \{(x_i, y_i, z_i), i = 1, \ldots, n\}$ defined on M_0 and its correspondence $P_1 = \{(x'_i, y'_i, z'_i), i = 1, \ldots, n\}$ defined on M_1). A point $(x_i, y_i, z_i) \in P_0$ should be mapped to the point $(x'_i, y'_i, z'_i) \in P_1$. The duplicate-skins algorithm is then outlined as below, after which the key phases of the algorithm are detailed successively.

1: Construct the first skin S_0 for M_0;
2: Construct the first skin S_1 for M_1 by copying S_0 to S_1 and deform S_1 to be around M_1 by the markers P_0 and P_1;
3: Initialize the tracking position to every particle globally;
4: Determine the particles that should track to the position markers;
5: **repeat**
6: Reposition the particles in tandem with S_0 and S_1;
7: Modify the mesh connectivity;
8: Update the tracking position of a particle if it does not track to a marker;
9: **until** no change occurs;
10: **if** M_0 or M_1 has sharp features **then**
11: Find corresponding sharp-tracking-edges on skins S_0 and S_1;
12: **repeat**
13: Reposition the particles in tandem with S_0 and S_1;
14: Modify the mesh connectivity;
15: Update the tracking position of a particle if it tracks neither a position marker nor a sharp edge;
16: **until** no change occurs;
17: **end if**

To let the resultant meshes have the correct correspondence on the position markers, if a particle on S_0 tracks to a position marker $\tau = (x_m, y_m, z_m) \in P_0$, the identical particles (in terms of topology) on S_1 should track to its corresponding marker (i.e., $\tau' = (x'_m, y'_m, z'_m) \in P_1$). For searching the tracking positions at the very beginning, space subdivision techniques (e.g., Octree or *k*D-tree) could be used to speed up the algorithm. We employ an Octree with a fixed depth in our implementation. Different from the particles tracking to markers, the other particles can freely track to either a vertex or a surface point (i.e., an interior point on a triangle) on skeletons. However, even if the space partition strategy is employed, it is still inefficient to conduct a global closest point search in every iteration step. In [Markosian et al. 1999], a local update strategy is conducted: for a particle p, the new tracking point is only searched on a limited number of faces – the set of entities to be searched, Γ, includes the current tracking face f of p and the faces containing any vertex of f on the skeleton; and in order to get out of a local minimum, for any particle $p' \in N(p)$, the tracking face f' of p' and the faces $f_i \in star(j)$ for $j \in f'$ are all added into Γ. The above strategy relies on the local connectivity on skeletons, thus fails on models with ill-meshes or point cloud models. To overcome the limitation, we change the strategic rules for locally updating tracking points to the follows:

- for a particle p, if an Octree node Υ_d contains the current tracking point of p, its new tracking point is searched among the faces/vertices held by Υ_d and the spatial neighboring nodes of Υ_d;
- for the purpose of jumping out of the local minimum, again the Octree nodes containing the tracking points of the particles $p' \in N(p)$ are added into the range of searching.

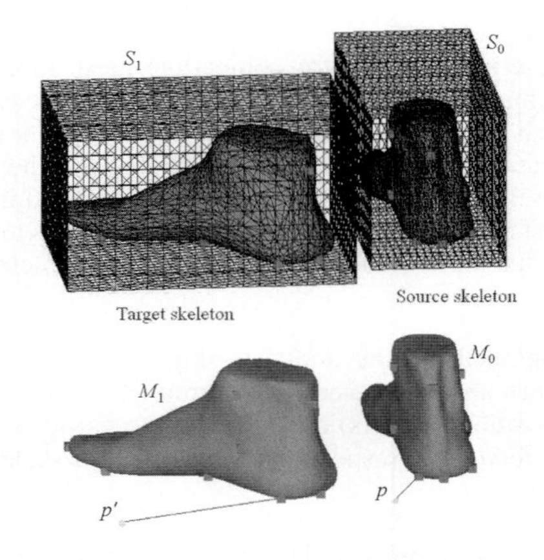

Figure 3: Illustration to explain why the RBF-deformation is needed for constructing initial skins. On the source skin and skeleton (right), the green particle p is the point on the skin S_0 closest to the marker at heel on M_0 (they are linked by the tracking vector in blue); however, the corresponding particle of p, $p' \in S_1$, is not the closest point to the corresponding marker on M_1, so that the dispersion of tracking directions occurs around p' since the particles near to p' all track to their closest points on S_1 but p' does not. Note that the topologies of S_0 and S_1 are always identical.

3.2 Initial Skins

To describe how we generate the initial skins in more detail, let us use the example shown in Fig.3. For two feet models, as the source and target skeletons (i.e., M_0 and M_1 respectively), are placed in different positions and orientations. As explained previously, the easiest way to generate initial skins for genus-0 models is to construct skin meshes as the bounding boxes orthogonal to *x,y,z*-axes.

However, problems may be encountered when position markers are specified on skeletons. For example, see Fig.3, after searching through the skin for the source skeleton, the green particle p on skin S_0 is closest to the marker point (in red) at the heel. Since the particles tracking the markers on the source and target skeletons should be identical, the corresponding particle of p on the target skin - $p' \in S_1$ is *not* the closest point to the heel marker. Figure 3 gives the tracking directions of p and p' in blue color. However, all the particles around p' will still track to their closest points on M_1; in other words, the tracking directions are dispersed, which easily leads to poor or even invalid meshes (e.g., face flipped). Therefore, to eliminate the occurrence of the above situation, the construction problem of S_1 after obtaining S_0 is reformulated as follows.

Problem: Given a set of position markers P_0 defined on M_0 and P_1 with respect to M_1 and a surface S_0, find a surface S_1 which is transformed from S_0 and the deformation from S_0 to S_1 is equivalent to the deformation from P_0 to P_1.

The above problem is solved by defining a deformation function $\Psi(\ldots)$ letting $P_1 = \Psi(P_0)$ so that $S_1 = \Psi(S_0)$ is easily obtained. The radial basis function (RBF) is the most suitable candidate for this deformation function [Botsch and Kobbelt 2005; Turk and O'Brien 2002]. In general, a RBF is represented in the piecewise form

$$\Psi(x) = p(x) + \sum_i^n \lambda_i \phi(\|x - \tau_i\|) \tag{3}$$

where $p(x)$ is a linear polynomial that accounts for the rigid transformation, the coefficients λ_i are real numbers to be determined and $\|\cdot\|$ is the Euclidean norm on $\Re^3$. To achieve a global deformation, the basis function $\phi(t)$ is chosen as $\phi(t) = t^3$ (the triharmonic spline as [Yngve and Turk 2002]). The coefficients λ_i and the coefficients of $p(x)$ can be easily determined by letting $\Psi(\tau_i) \equiv \tau_i'$ for all pairs of $\tau_i \in P_0$ and $\tau_i' \in P_1$ plus the compatibility conditions $\sum_i^n \lambda_i = \sum_i^n \lambda_i \tau_i = 0$. The formulated linear equation system has been proven to be positive definite unless all the points in P_0 and P_1 are coplanar.

The use of RBF guarantees smooth geometric deformation. Thus, the geometry of target skin S_1 can be determined by smoothly blending the positions of the vertices on the source skin S_0 as

$$(p' \in S_1) = \Psi((p \in S_0)). \tag{4}$$

The result of the foot example by this deformation is shown in Fig.4(a), where the initial skin S_1 follows the orientation of skeleton M_1. Also, compared to Fig.3, the green particle on S_1 is much closer to the marker at heel. This greatly reduces the chance to generate self-overlapped skins. However, uneven triangulation could still happen (see Fig.4(b)), since the somewhat conflicting tracking directions still exist among the particle which are enforced to track markers and its neighbors. Therefore, to completely eliminate the conflicting tracking directions, an alternative way with pre-skinning is proposed to construct the initial duplicate skins. Firstly, we perform several runs of the single skin algorithm for the source skeletons. The result skin is applied as S_0 and then this skin is deformed to fit the target skeleton by Eq.(3) and (4), so that S_1 is created. Figure 4(c) shows the result from this change. The source and target skeletons are respectively approximated by their corresponding skins. In this way, the green particle is almost the closest particle to its corresponding markers through the whole skin S_1. Consequently, a significant feature is introduced here: our duplicate-skins algorithm is independent of the placement of input skeletons.

The above RBF-deformation based method also enables our method to work for those genus-k models ($k \neq 0$). If enough number of position markers are well defined around each handle, we can duplicate a mesh from the source skeleton M_0 to serve as S_0 and employ

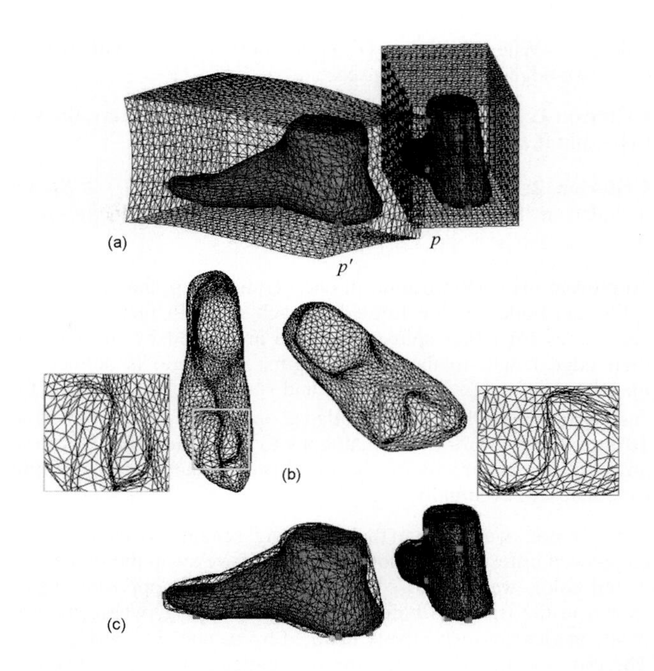

Figure 4: Illustration of the method with pre-skinning to construct initial duplicate skins. (a) Initial duplicate skins in box shape; after S_0 have been obtained, the S_1 is obtained by RBF-deformation. (b) Beginning with the initial duplicate skins in (a), the result of the duplicate-skins algorithm is shown in very uneven triangulations (see two close-up views) — this is because that some faraway particles are enforced to track markers on the target skeleton. (c) Construct initial duplicate skins by the method with pre-skinning, where the shape of S_0 after several (5 to 8) runs of the single-skin algorithm is adopted to create S_1 instead of the box shape in (a).

the above RBF-deformation function to create S_1 around M_1. In our experience, four to eight pairs of uniformly distributed markers for each handle will be enough. For instance, as the example shown in Fig.10, red points are the position markers. The initial skins consist of one tori mesh and another mesh deformed from tori by RBF. Therefore, starting from the genus-1 initial skins, our duplicate skins are iteratively evolved to take the form of the skeleton shape with mesh optimized.

3.3 Optimize connectivity on duplicate skins

In our duplicate-skins algorithm, two skins are adopted to approximate their relative skeletons. In the course of skin evolving, the mesh topology updating is identical on duplicate skins, even though they interpolate different underlying geometries. That is the primary ingredient for the generation of compatible skins. Obviously, when we apply the edge-based optimization operators, the measurements on two skins should both be under consideration.

To evaluate the criteria of edge splitting, the edge lengths on S_0 and S_1 are both taken into account. We perform the edge split if either edge $\{i, j\} \in S_0$ or its corresponding edge $\{i', j'\} \in S_1$ satisfies the splitting condition. Note that the duplicate skins share one target edge length L_{tag}. For two skeletons with sizes that differ significantly, we scale M_1 by the ratio $\rho = DL_0/DL_1$ before applying our algorithm where DL_0 and DL_1 are the diagonal lengths of the bounding boxes of M_0 and M_1. After computing the compatible meshes, we scale M_1 and S_1 back to the original dimension by the ratio ρ^{-1}. When $\{i,j\}$ and $\{i',j'\}$ are both less than half of L_{tag}, these two edges can be collapsed.

Criterion 1: A pair of edges $\{i,j\} \in S_0$ and $\{i',j'\} \in S_1$ are allowed to be split if $\|v_iv_j\| > 1.5L_{tag}$ or $\|v'_iv'_j\| > 1.5L_{tag}$.

Criterion 2: A pair of edges $\{i,j\} \in S_0$ and $\{i',j'\} \in S_1$ are included in the edge-collapse candidates if and only if $\|v_iv_j\| < 0.5L_{tag}$ and $\|v'_iv'_j\| < 0.5L_{tag}$.

To prevent over-optimization in one iteration step, the numbers of splits and collapses are limited in each run. In fact, the available edges for either split or collapse are sorted by the ratio of their edge length to the target length. The ratio is defined by $\max(\|v_iv_j\|, \|v'_iv'_j\|)/L_{tag}$ for split and $(\|v_iv_j\| + \|v'_iv'_j\|)/2L_{tag}$ for collapse. We do the split or collapse operations on only the first 10% of edges in priority. Analogous to the basic skin algorithm, any edge that is too long (or too short) is still guaranteed to be split (or collapsed) eventually.

Furthermore, we propose the following scheme to estimate the edge-swap criterion. As shown in Fig.5, if we swap the two edges in red color, we need to compare the maximum opposite angles shown in the triangles before and after swapping, where the opposite angles are respectively denoted by α_i and β_i $(i = 1,2,3,4)$. The criterion for edge-swapping on duplicate meshes is given as follows.

Criterion 3: Defining $\alpha_{\max} = \max\{\alpha_i\}$ and $\beta_{\max} = \max\{\beta_i\}$, if and only if $\alpha_{\max} > \beta_{\max}$, the pair of edges $\{i,j\} \in S_0$ and $\{i',j'\} \in S_1$ are considered to be swapped.

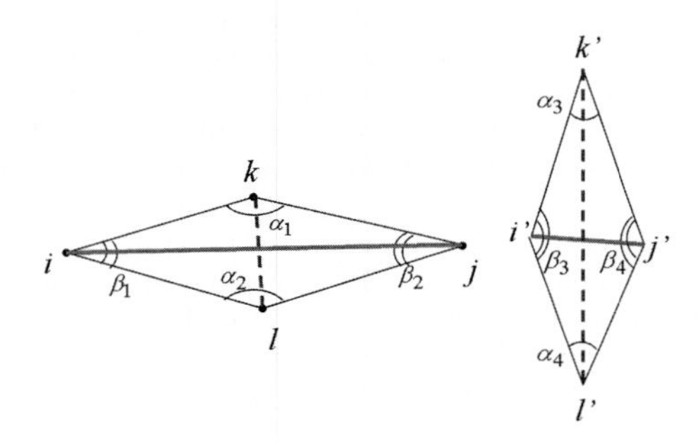

Figure 5: Edge-swap for duplicate-skins: left part – the edge $\{i,j\}$ and its adjacent triangles on S_0, right part – the edge $\{i',j'\}$ and its adjacent triangles on the target skin S_1.

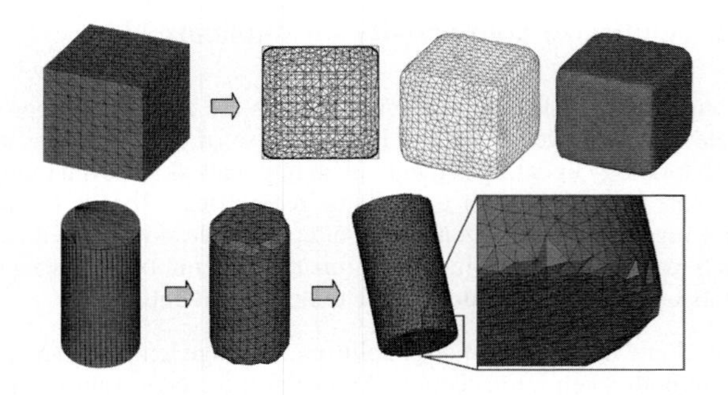

Figure 6: Aliasing errors are introduced by the skin algorithm: top row, for the box skeleton given in the most left, the resultant mesh from the skin algorithm has a high quality but degenerates in the sharp edges and corners; bottom row, for the given cylinder with a coarse and irregular mesh, even if a denser mesh is employed, the original skin algorithm can hardly recover the aliased sharp curves.

3.4 Sharp edge recovering

The original skin algorithm cannot give a correct construction at sharp features of a skeleton. If a skeleton has sharp geometric features (i.e., creases where the surface does not have continuous tangent planes) on it, the skin always aliases the sharpness with approximation artifacts. As pointed out in [Kobbelt et al. 2001], increasing the sampling rate of surfaces will not cause the skin to converge to the sharp edges and corners if no special treatment is given. Our tests also prove this (see Fig.6). In this section, we propose an efficient scheme that recovers sharp features on the resultant skins. Note that this only works for the skins interpolating skeletons (i.e., $r_s = 0$ in Eq.(2)).

On the skeleton meshes, all the endpoints of sharp edges are defined as *sharp vertices* and all the triangle faces belonging to the stars of sharp vertices are called *sharp faces*. After the sharp edges have been identified on the skeleton meshes, several edges of the skin, named as *sharp-tracking-edges*, are enforced to align to these features. Also, to prevent breaking the sharpness on the resultant skins, the meshes around sharp tracking edges should be specially treated during connectivity optimization. As mentioned in the algorithm overview, the sharp edge recovering procedure is regarded as a type of post-processing. When applying the algorithm, the skins have almost interpolated the given skeletons, so only several runs are needed and can be completed in a short time. The most important phase is to determine the sharp-tracking-edges on S_0 and S_1 which can be decomposed into three steps:

- Find the sharp edges, vertices and faces on skeletons;
- Determine the particles tracking to sharp vertices;
- Compute the shortest path on the skins between each pair of particles which track to the two endpoints of a sharp edge, where the path passes along the skin edges.

For extracting the sharp edges on skeletons, we can either manually pick the edges or automatically detect them by the discrete sharp operator referring to the discrete mean curvature at the mesh edges. For an edge e with dihedral angle θ_e, $H_e = 2\|e\|\cos\frac{\theta_e}{2}$ (ref. [Hildebrandt and Polthier 2004]) is given as the mean curvature on e. If H_e exceeds a threshold, the edge is labelled as a sharp edge.

Next, we need to find the corresponding sharp-tracking-edges on the skins. More specifically, a list C_e of edges are enforced to track each of the edges e with '*sharp*' label. To successfully recover sharp features on skeletons, the curve formed by C_e must be single-wide. Thus, the particles tracked to sharp vertices in the bijective manner (i.e., without repeating) are found first. The shortest path linking the two particles tracking to a pair of endpoints are then determined by the Dijkstra's algorithm. To speed up the searching, we filtered out most of the edges on the skin – only the edges with their endpoint tracking to sharp vertices/edges/faces are regarded as legal paths. Also, the edges that have been labelled as sharp-tracking-edges in previous searches are prevented from the searching (by assigning their length to ∞). As a result, for every list of sharp-tracking-edges, C_e, its starting and ending particles track to the endpoints of e, and the interior particles of C_e are restricted to the inertia points on e by the proportion of lengths. In other words, the particles in C_es are tracked to the sharp edges exactly. Therefore, the sharp edges are preserved list by list. See the models in Fig.7, the red edges on a skin are the sharp-tracking-edges on the skin while the blue ones are sharp edges on the skeleton.

Finally, as pointed out in the algorithm overview, when we iteratively optimize the connectivity on a skin mesh, several configurations need to be discussed if any entity related to sharp features is under consideration.

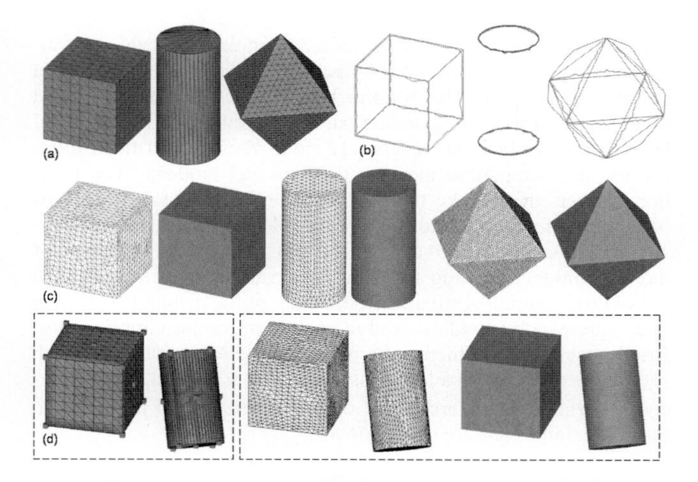

Figure 7: Shape edge recovering: (a) given skeleton meshes, (b) the sharp edges (in blue) on skeletons and their corresponding sharp-tracking-edges (in red) on skins, (c) the sharp edge recovering results on individual skeletons, and (d) the result of duplicate-skins with sharp edges recovered.

Criterion 4: If an edge on skin is a sharp-tracking-edge, this edge is prevented from swapping.

Criterion 5: For the convenience of implementation, if an edge e is a sharp-tracking-edge, we prevent edge-split on it; otherwise, the list C_e holding e needs to be updated and the tracking position of the newly inserted vertex needs to be searched.

Criterion 6: If both two particles of an edge are endpoints of sharp-tracking-edges, collapse on this edge is not allowed.

The configuration considered by the above criterion could be either of the two cases shown in Fig.8(a), where the first one eliminates the sharp-tracking-edge and the second one merges two diverse sharp-tracking-edges into one.

Criterion 7: If an edge $\{i, j\}$ satisfies the collapse-length criterion and only one of its endpoints $\{i\}$ tracks to sharp features, this edge could be collapsed.

The degenerated vertex $\{h\}$ from $\{i, j\}$ and the edges $\in star(i) \bigcup star(j)$ must be carefully processed. The position of $\{h\}$ is not located at the middle of $\{i, j\}$ after collapsing, but should inherit the position of the particle tracking to the sharp features (i.e., the position of $\{i\}$ in Fig.8(b)). In addition, if either $\{k, i\}$ or $\{k, j\}$ is a sharp-tracking-edge, the edge $\{k, h\}$ degenerated from the triangle $\{i, k, j\}$ must be assigned as a sharp-tracking-edge as illustrated in Fig.8(c). Similarly, the edge $\{l, h\}$ should be a sharp-tracking-edge, if either $\{l, i\}$ or $\{l, j\}$ is sharp tracking edge.

Criterion 8: For a pair of edges $\{i, j\} \in S_0$ and $\{i', j'\} \in S_1$, if either of them satisfy one of the above four criteria, the topology update of the edges should follow the above rules.

Criterion 9: During the iteration of sharp edge recovering, we update the tracking position and tracking face of the particles, only if they do not track to any sharp vertex/edge.

As mentioned in the outline of our algorithm, the loop for interpolating sharp features works as a post-processing procedure; hence, the iteration should have a small number of steps and this additional loop does not degenerate the efficiency of our algorithm.

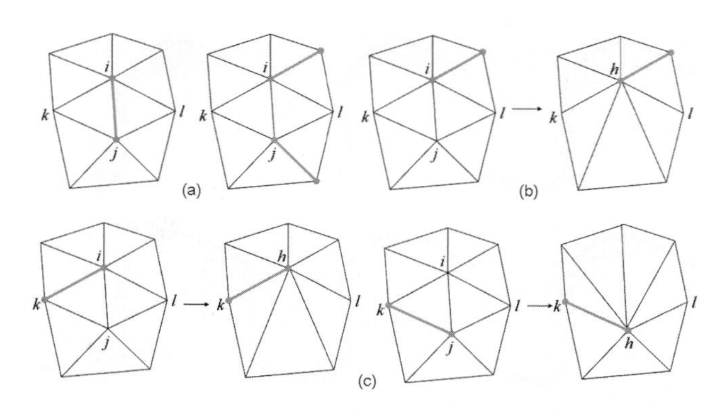

Figure 8: Cases that should be specially treated in the sharp-edge-preserved mesh optimization, where red edges denote the sharp-tracking-edges and red points are the particles tracking to sharp features. (a) Two cases that the edge $\{i, j\}$ cannot be collapsed. (b) If the particle $\{i\}$ tracks to sharp features, the degenerated particle $\{h\}$ maintains the position and tracking information of $\{i\}$. (c) The configurations for collapsing $\{i, j\}$ if either $\{k, i\}$ or $\{k, j\}$ is a sharp-tracking-edge.

4 n-Ary Compatible Meshes

The duplicate-skins algorithm above can successfully generate the compatible meshes S_0 and S_1 to approximate the given two skeleton models M_0 and M_1. However, for some applications (e.g., n-way blending, sample based parametric design of freeform models, etc.), the compatible meshes are requested for more than two skeleton models. The n-Ary compatible meshes can be generated through the vertex transformations on the results from our duplicate-skins algorithm.

Suppose that the compatible meshes are requested on n skeleton models M_i ($i = 0, \ldots, n-1$), the duplicate-skins algorithm is first applied $n-1$ times on the pairs of skeletons — M_0 and M_j ($j = 1, \ldots, n-1$). Thus, $n-1$ pairs of skins are obtained; for the convenience of description they are denoted by $S_{0(j)}$ and $S_{(j)}$ respectively. Letting $S_0 = S_{0(1)}$ and $S_1 = S_{(1)}$, we conduct the following algorithm to determine the meshes S_j ($j > 1$) compatible to S_0 on M_j.

1: Duplicate a skin mesh S_j with S_0;
2: **for all** vertex $p \in S_j$ **do**
3: The closest point p_c of p on $S_{0(j)}$ is found;
4: p_c must be inside a triangle $T \in S_{0(j)}$, the barycentric coordinates of p_c on T - $(\alpha_c, \beta_c, \gamma_c)$ is then recorded;
5: By the pair of skins $S_{0(j)}$ and $S_{(j)}$, T could easily find its corresponding triangle T' on $S_{(j)}$;
6: Applying the braycentric coordinate $(\alpha_c, \beta_c, \gamma_c)$ on T', the new position of p on $S_{(j)}$ can be computed;
7: Move p to the new position;
8: **end for**

Repeating the vertex transformation, the new shape of S_j approximating M_j is easily determined. It is easy to find that all steps of the above algorithm can be finished in a short time except the closest point search step. For this step, the space partition strategy which has been previously employed in the tracking point search is used again to accelerate the process.

About the approximation error. Although easy to implement, the above algorithm for n-Ary compatible meshes enlarge the L^2 approximate error (see Fig.9a). This enlargement is led by the closest point projection, where $S_{(j)}$ is employed to approximate M_j. The

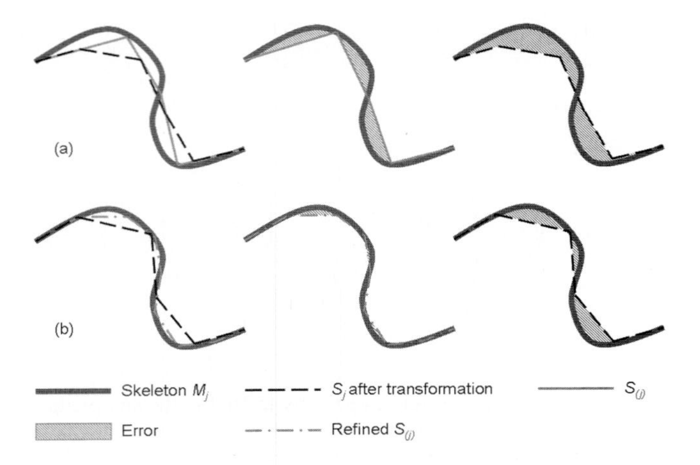

Figure 9: L^2 approximate error analysis: (a) L^2 is enlarged on the compatible meshes generated using vertex transformation (comparing the approximation error on $S_{(j)}$ and S_j), and (b) the approximation error (both $S_{(j)}$ and S_j) can be reduced by refining $S_{(j)}$.

Table 1: Computational Statistics

Examples	Fig.10	Fig.11	Fig.12
Number of Triangles (skeleton I)	3,356	3,270	4,638
Number of Vertices (skeleton I)	1,128	1,648	1,821
Number of Triangles (skeleton II)	616	2,228	8,030
Number of Vertices (skeleton II)	308	1,165	4,017
Number of Triangles (skin)	2,634	13,654	19,880
Number of Vertices (skin)	1,317	6,829	9,942
Comp. Time (in sec.)	~10	~83	~40

error can be reduced by refining $S_{(j)}$ while keeping the same connectivity on S_j. By the refinement, the approximation error shown on $S_{(j)}$ is decreased so that the error given on S_j is also reduced (comparing the errors shown in different rows of Fig.9). We can also apply the repositioning step of our duplicate-skins on S_j for several runs to decrease the approximation error.

5 Results and Applications

All the examples shown in this paper are tested on a standard PC with AMD 1.6 GHz mobile CPU and 480MB RAM. For the examples shown in Fig.10, 11 and 12, the computational statistics are listed in Table 1.

5.1 Free-form modelling

Our first example is to apply our duplicate-skins algorithm on genus-1 models (a torus and a mechanical-part with sharp edges). As seen in Fig.10, twelve pairs of manually specified position markers (red points) govern our algorithm and establish correct correspondences on the resultant compatible meshes. Sharp edges are well recovered. Then, the compatibility of the meshes are employed to partially deform the mechanical part into the torus. Our duplicate-skins algorithm is applied to the design of a toy bear in the example shown in Fig.11. The original bear model consists of 1 hemisphere, 3 spheres and 3 cylinders. Although the bear model is not a single manifold surface, benefited from the duplicate-skins algorithm, we can still generate compatible meshes on a rabbit model and the bear model. The body of bear was selected and reshaped to the shape of its corresponding part on the rabbit, so that the final bear model is obtained. Figure 12 gives the original head models and the compatible meshes for two head models. The position markers in red take the role of defining the correspondences of semantic features. After the compatible meshes have been constructed, it is very easy to change the female's nose by the shape of the male model's (see Fig.13). The shape variation performed in the above three examples are all with the help of the displacement-map technique which is widely used in computer graphics applications. Briefly, the detail geometry of a mesh surface M (or part of a mesh surface) is encoded on a low-pass filtered $\overline{M}$ of M, and then the encoded surface details are added onto another filtered mesh $\overline{M'}$ so that the details of M are shown on M'.

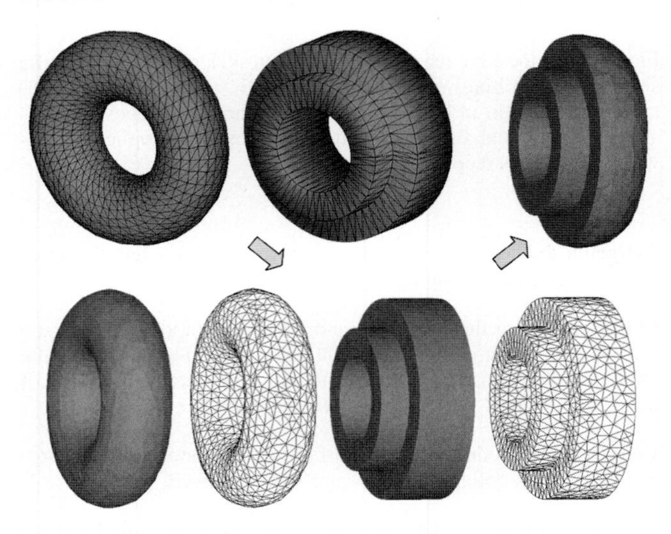

Figure 10: Tori-MechPart example: our duplicate-skins algorithm can generate compatible meshes on genus-1 models with sharp-edges well recovered.

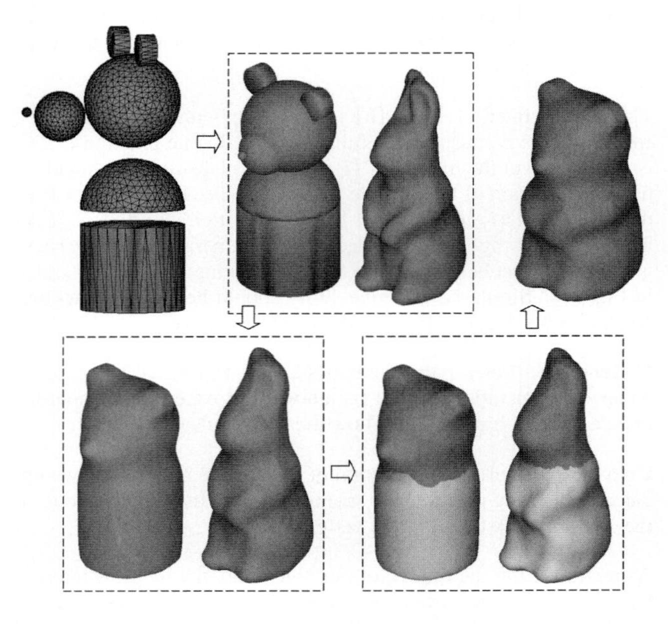

Figure 11: Toy bear design with our duplicate-skins algorithm.

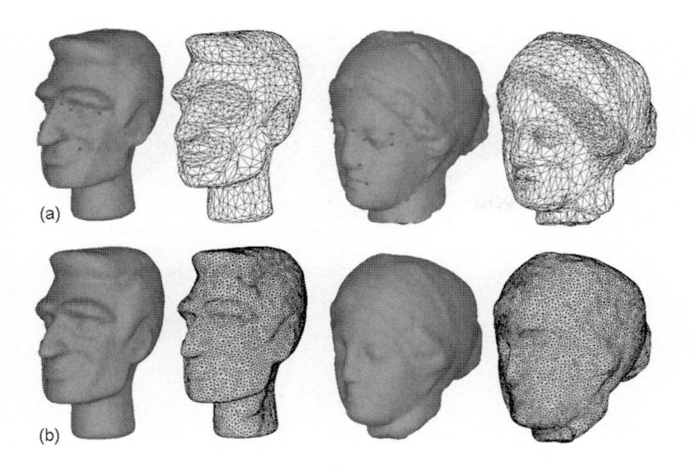

Figure 12: Compatible meshes generated on head models: (a) given heads models and the position markers defined on it (red ones), and (b) resultant compatible meshes.

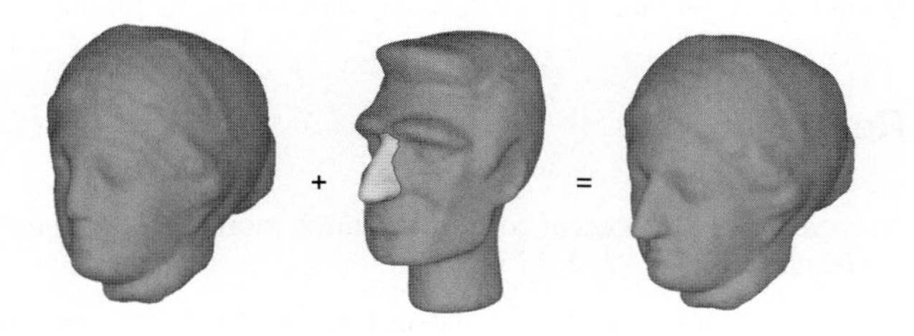

Figure 13: Cut-and-Paste modelling for changing the nose on a head model.

Figure 14 demonstrates the compatible meshes on 3 head models so that the interpolation triangle among three head models could be determined. The n-Ary compatible meshes are usually employed in the application of the parametric design of free-form models (e.g., the parametric design of human models [Wang 2005]). As shown in Fig.15, after obtaining the input parameters from users, a set of human models are selected from the human model database, where all human models stored in the database come with surfaces have compatible meshes. By a numerical optimization scheme, we can determine the synthesis weights of this set of human models so that the result synthesized models give the user specified parameters, where the synthesis is eventually a weighted blending procedure.

5.2 Design automation for customized products

The application of the resultant compatible meshes is not limited to the variation of mesh surfaces themselves. The compatible meshes are also very important to the design automation problem in several industries (e.g., apparel industry, shoe industry, jewelling industry, eye-wear industry, etc.). In all these industries, there exists a common quest for design automation: after carefully designing a product's geometry around a model in standard size and shape, it is desired to automatically transform the geometry of the product to other models with customized shapes while maintaining the originally spatial relationship between the product and the model (i.e., preserving the fitness). For example, Fig.1 shows this application in the apparel industry. After constructing the compatible meshes from various input models (two-manifold mesh model, a polygon soup with holes, or a point cloud), we can apply the t-FFD or p-FFD technique [Kobayashi and Ootsubo 2003; Wang et al. 2005] to

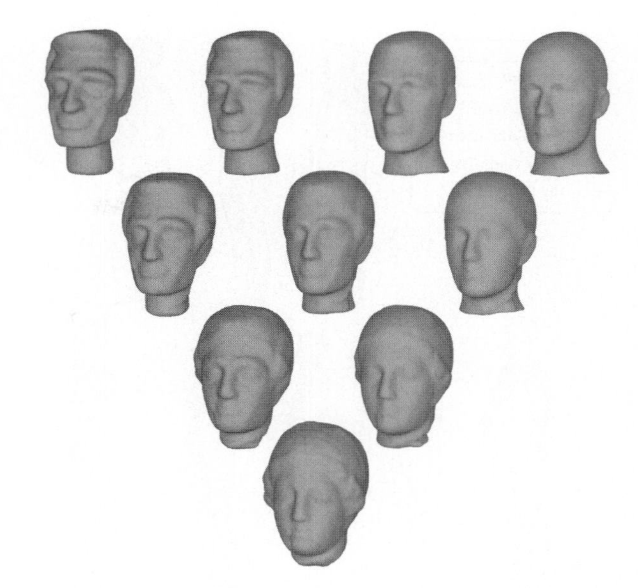

Figure 14: Shape interpolation among three head models.

encode the coordinate of each vertex on clothes by the compatible mesh H_1. Then, based on the correspondences between triangles among S_1, S_2, and S_3, we can easily fit the clothes to the shape around S_2 and S_3. Figure 16 shows a similar design automation application in the shoe industry, where the foot models have been previously shown in Fig.4.

6 Limitations and Discussion

This paper presents an approach for modelling compatible meshes on give models. Our duplicate-skins algorithm works on the models represented by polygonal meshes, a polygon soup, or a point cloud. The algorithm drives two active particle-based mesh surfaces (i.e., skins) with identical connectivity to approximate the given models that serves as the skeleton. One major limitation of the algorithm is the shrinking effects: when using skins with a relatively large triangle size to model compatible meshes on the skeleton with complex details, the shrinking effect occurs around the details since the sampling rate on skins is low. Also, the sharp edge can hardly preserved if the resolution of skin mesh is lower than the resolution of skeletons. This is in fact the problem addressed by the sampling theory. Increasing the sampling frequency can solve this problem to a certain degree but cannot guarantee the preservation of details since there is no mechanism in our current approach to ensure detail preservation. This is definitely an area we should consider in the future. One possible approach for solving this problem is to employ the compatible meshes generated by our method as the control network of subdivision surfaces, then the detail geometry could be recovered in the later mesh refinement procedure. The approximation error of our approach also needs to be further studies, and a curvature-based mesh refinement could be considered in our framework to reduce the error.

Another limitation of this approach and also all the other cross-parameterization algorithms (e.g., [Kraevoy and Sheffer 2004; Schreiner et al. 2004]) is that the algorithm relies too heavily on the position markers when modelling the genus-k models. For instance in the Tori-MechPart example in Fig.10, if all the markers on the tori M_1 are mapped to the markers all being placed on the left

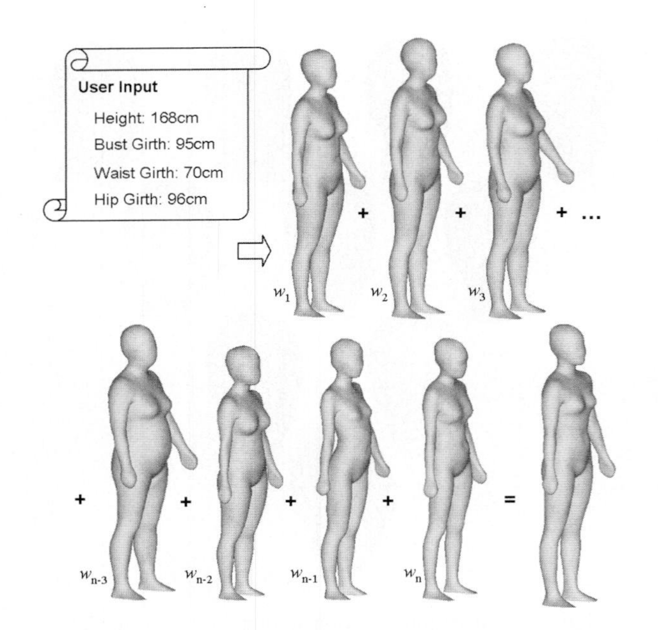

Figure 15: Parametric design of human models, where the human models with compatible meshes are synthesized into a model satisfying the user input parameters.

part of the MechPart M_2 (instead of the similar positioning of markers shown on M_2 in Fig.10), compatible meshes for these genus-1 models will not be correctly constructed. Therefore, an important area of future research is how to develop a method to add markers to ensure correct mapping. Also, the positions of markers effect on the quality of triangles and the result of shape approximation. For example, if two skeletons are much different from each other, the over-strict constraints on position markers may lead a result with great distortion. Or if there is no marker defined in the cavity of a U-shaped skeleton, our current method can hardly pull the skin mesh into the cavity. All these problems are to be investigated in our future research to improve our approach.

The last limitation in our current implementation is that the generation of n-Ary compatible meshes does not preserve sharp-edges. However, since the sharp-edges have already been recovered in our basic duplicate-skins algorithm, it is possible for us to find some methods for preserving the sharp-edges when generating n-Ary compatible meshes. Our first idea is that after we have found the corresponding paths of sharp-edges (similar to what we did in section 3.4), we restrict the mapping of such paths to be along the triangle edges (instead of triangle faces) in the mapping of n-Ary compatible meshes. This will be investigated in our future research.

Although there are several limitations, our approach shows an important advantage comparing to other approaches for the same purpose — there is less topological constraint on input models: multiple polygonal models, models with ill-topology meshes, or even point clouds can be employed as skeletons.

Acknowledgement

The authors from the Hong Kong University of Science and Technology are partially supported by the HKUST6234/02E project. The author from the Chinese University of Hong Kong would like to thank the support from the projects CUHK/2050341.

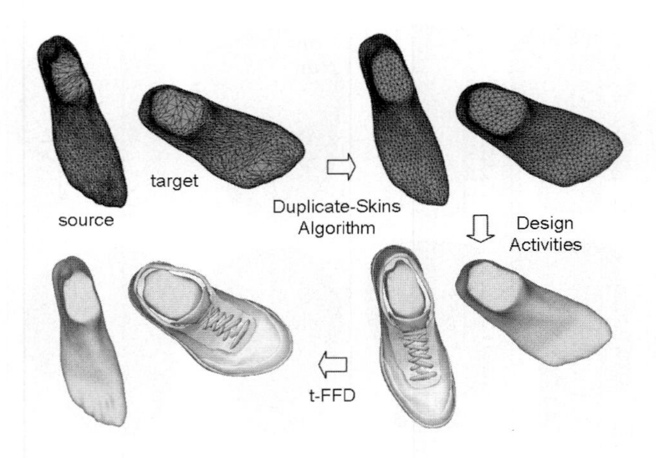

Figure 16: Design automation of shoes: for the given two foot models with position markers (top-left), our duplicate-skins can generate a pair of compatible meshes (top-right); after designing the shoe around the source model, we can automatically re-warp the shoe around the target model by t-FFD using the compatible meshes.

References

ALEXA, M. 2002. Recent advances in mesh morphing. *Computer Graphics Forum 21*, 2, 173–196.

ALLEN, B., CURLESS, B., AND POPOVIĆ, Z. 2003. The space of human body shapes: reconstruction and parameterization from range scans. *Computer Graphics Forum 22*, 3, 612–619.

ALLIEZ, P., UCELLI, G., GOTSMAN, C., AND ATTENE, M., 2005. Recent advances in remeshing of surfaces. Part of the state-of-the-art report of the AIM@SHAPE EU network.

BIERMANN, H., MARTIN, I., BERNARDINI, F., AND ZORIN, D. 2002. Cut-and-paste editing of multiresolution subdivision surfaces. *ACM Transactions on Graphics 21*, 3, 312–321.

BOTSCH, M., AND KOBBELT, L. 2001. Resampling feature and blend regions in polygonal meshes for surface anti-aliasing. In *Proceedings of Eurographics 2001*, 402–410.

BOTSCH, M., AND KOBBELT, L. 2005. Real-time shape editing using radial basis functions. *Computer Graphics Forum 24*, 3, 611–621.

CARR, J., FRIGHT, W., AND BEATSON, R. 1997. Surface interpolation with radial basis functions for medical imaging. *IEEE Transactions on Medical Imaging 16*, 1, 96–107.

CARR, J., BEATSON, R., CHERRIE, J., MITCHELL, T., FRIGHT, W., MCCALLUM, B., AND EVANS, T. 2001. Reconstruction and representation of 3d objects with radial basis functions. In *Proceedings of SIGGRAPH 2001*, 67–76.

COHEN-OR, D., SOLOMOVIC, A., AND LEVIN, D. 1998. Three-dimensional distance field metamorphosis. *ACM Transactions on Graphics 17*, 2, 116–141.

DESBRUN, M., MEYER, M., AND ALLIEZ, P. 2002. Intrinsic parameterizations of surface meshes. *Computer Graphics Forum 21*, 3, 209–218.

GOTSMAN, C., GU, X., AND SHEFFER, A. 2003. Fundamentals of spherical parameterization for 3d meshes. *ACM Transactions on Graphics 22*, 3, 358–363.

HILDEBRANDT, K., AND POLTHIER, K. 2004. Anisotropic filtering of non-linear surface features. *Computer Graphics Forum 23*, 3, 391–400.

HOPPE, H., DEROSE, T., DUCHAMP, T., MCDONALD, J., AND STUDTZLE, W. 1993. Mesh optimization. In *Proceedings of SIGGRAPH 93*, 19–26.

JU, T. 2004. Robust repair of polygonal models. *ACM Transactions on Graphics 23*, 3, 888–895.

JULIUS, D., KRAEVOY, V., AND SHEFFER, A. 2005. D-charts: Quasi-developable mesh segmentation. *Computer Graphics Forum 24*, 3, 981–990.

KANAI, T., SUZUKI, K., AND KIMURA, F. 2000. Metamorphosis of arbitrary triangular meshes. *IEEE Computer Graphics and Applications 20*, 2, 62–75.

KARTASHEVA, E., ADZHIEV, V., PASKO, A., FRYAZINOV, O., AND GASILOV, V. 2003. Discretization of functionally based heterogeneous objects. In *Proceedings of the eighth ACM symposium on Solid modeling and applications*, 145–156.

KOBAYASHI, K., AND OOTSUBO, K. 2003. t-ffd: freeform deformation by using triangular mesh. In *Proceedings of the eighth ACM symposium on Solid modeling and applications*, 226–234.

KOBBELT, L., BOTSCH, M., SCHWANECKE, U., AND SEIDEL, H.-P. 2001. Feature sensitive surface extraction from volume data. In *Proceedings of SIGGRAPH 2001*, 57–66.

KRAEVOY, V., AND SHEFFER, A. 2004. Cross-parameterization and compatible remeshing of 3d models. *ACM Transactions on Graphics 23*, 3, 861–869.

KRAEVOY, V., SHEFFER, A., AND GOTSMAN, C. 2004. Matchmaker: Constructing constrained texture maps. *ACM Transactions on Graphics 22*, 3, 326–333.

LEE, A., DOBKIN, D., SWELDENS, W., AND SCHRÖDER, P. 1999. Multiresoltion mesh morphing. In *Proceedings of SIGGRAPH 99*, 343–350.

LEVY, B., PETITJEAN, S., RAY, N., AND MAILLOT, J. 2002. Least squares conformal maps for automatic texture atlas generation. *ACM Transactions on Graphics 21*, 3, 362–371.

MARKOSIAN, L., COHEN, J., CRULLI, T., AND HUGHES, J. 1999. Skin: a constructive approach to modeling free-form shapes. In *Proceedings of SIGGRAPH 99*, 393–400.

MARSCHNER, S. R., GUENTER, B. K., AND RAGHUPATHY, S. 2000. Modeling and rendering for realistic facial animation. In *Proceedings of the Eurographics Workshop on Rendering Techniques 2000*, 231–242.

NGUYEN, M., YUAN, X., AND CHEN, B. 2005. Geometry completion and detail generation by texture synthesis. *The Visual Computer 21*, 8-10, 669–678.

OHTAKE, Y., AND BELYAEV, A. 2001. Mesh optimization for polygonized isosurfaces. *Computer Graphics Forum 20*, 3, 368–376.

OHTAKE, Y., AND BELYAEV, A. 2002. Dual/primal mesh optimization for polygonized implicit surfaces. In *Proceedings of the seventh ACM symposium on Solid modeling and Applications*, 171–178.

OHTAKE, Y., BELYAEV, A., AND PASKO, A. 2003. Dynamic mesh optimization for polygonized implicit surfaces with sharp features. *The Visual Computer 19*, 2-3, 115–126.

PRAUN, E., AND HOPPE, H. 2003. Spherical parameterization and remeshing. *ACM Transactions on Graphics 22*, 3, 340–349.

PRAUN, E., SWELDENS, W., AND SCHRÖDER, P. 2001. Consistent mesh parameterization. In *Proceedings of SIGGRAPH 01*, 179–184.

SANDER, P., GORTLER, S., SNYDER, J., AND HOPPE, H. 2002. Signal-specialized parameterization. In *Proceedings of Eurographics Workshop on Rendering 2002*, 87–100.

SANDER, P., WOOD, Z., GORTLER, S., SNYDER, J., AND HOPPE, H. 2003. Multi-chart geometry images. In *Proceedings of the 2003 Eurographics/ACM SIGGRAPH symposium on Geometry processing*, 146–155.

SAVCHENKO, V., PASKO, A., OKUNEV, O., AND KUNII, T. 1995. Function representation of solids reconstructed from scattered surface points and contours. *Computer Graphics Forum 14*, 4, 181–188.

SCHREINER, J., PRAKASH, A., PRAUN, E., AND HOPPE, H. 2004. Inter-surface mapping. *ACM Transactions on Graphics 23*, 3, 870–877.

SEO, H., AND MAGNENAT-THALMANN, N. 2004. An example-based approach to human body manipulation. *Graphical Models 66*, 1, 1–23.

SHEFFER, A., AND HART, J. 2002. Seamster: Incospicuous low-distortion texture seam layout. In *Proceedings of IEEE Visualization 2002*, 291–298.

SORKINE, O., COHEN-OR, D., GOLDENTHAL, R., AND LISCHINSKI, D. 2002. Seamster: Incospicuous low-distortion texture seam layout. In *Proceedings of IEEE Visualization 2002*, 355–362.

SPANIER, E. 1966. *Algebraic Topology*. McGraw-Hill, New York.

SUMNER, R., AND POPOVIĆ, J. 2004. Deformation transfer for triangle meshes. *ACM Transactions on Graphics 23*, 3, 399–405.

SURAZHSKY, V., AND GOTSMAN, C. 2003. Explicit surface remeshing. In *Proceedings of the 2003 Eurographics/ACM SIGGRAPH symposium on Geometry processing*, 20–30.

TURK, G., AND O'BRIEN, J. 2002. Modelling with implicit surfaces that interpolate. *ACM Transactions on Graphics 21*, 4, 855–873.

VORSATZ, J., RÖSSL, C., KOBBELT, L., AND SEIDEL, H.-P. 2001. Feature sensitive remeshing. *Computer Graphics Forum 20*, 3, 393–401.

VORSATZ, J., RÖSSL, C., AND SEIDEL, H.-P. 2003. Dynamic remeshing and applications. In *Proceedings: 8th ACM Symposium on Solid Modeling and Applications (SM-03)*, 167–175.

WANG, C., WANG, Y., AND YUEN, M. 2005. Design automation for customized apparel products. *Computer-Aided Design 37*, 7, 675–691.

WANG, C. 2005. Parameterization and parametric design of mannequins. *Computer-Aided Design 37*, 1, 83–98.

WELCH, W., AND WITKIN, A. 1994. Free-form shape design using triangulated surface. In *Proceedings of SIGGRAPH 94*, 247–256.

YNGVE, G., AND TURK, G. 2002. Robust creation of implicit surfaces from polygonal meshes. *IEEE Transactions on Visualization and Computer Graphics 8*, 4, 346–359.

Simultaneous Shape Decomposition and Skeletonization*

Jyh-Ming Lien† John Keyser‡ Nancy M. Amato§

Department of Computer Science, Texas A&M University

Abstract

Shape decomposition and skeletonization share many common properties and applications. However, they are generally treated as independent computations. In this paper, we propose an iterative approach that simultaneously generates a hierarchical shape decomposition and a corresponding set of multi-resolution skeletons. In our method, a skeleton of a model is extracted from the components of its decomposition — that is, both processes and the qualities of their results are interdependent. In particular, if the quality of the extracted skeleton does not meet some user specified criteria, then the model is decomposed into finer components and a new skeleton is extracted from these components. The process of simultaneous shape decomposition and skeletonization iterates until the quality of the skeleton becomes satisfactory. We provide evidence that the proposed framework is efficient and robust under perturbation and deformation. We also demonstrate that our results can readily be used in problems including skeletal deformations and virtual reality navigation.

CR Categories: I.3.5 [COMPUTER GRAPHICS]: Computational Geometry and Object Modeling—Geometric algorithms, languages, and systems

Keywords: skeletonization, multi-resolution skeleton, convex decomposition

1 Introduction

Shape decomposition partitions a model into (visually) meaningful components. Recently shape decomposition has been applied to texture mapping [Sander et al. 2003], shape manipulation [Katz and Tal 2003], shape matching [Mangan and Whitaker 1999; Dey et al. 2003; Funkhouser et al. 2004], and collision detection [Li et al. 2001]. Early work on shape decomposition can be found in pattern recognition and computer vision; see surveys in [Rom and Medioni 1994; Wu and Levine 1997].

A *skeleton* is a lower dimensional object that essentially represents the shape of its target object. Because a skeleton is simpler than the original object, many operations, e.g., shape recognition and deformation, can be performed more efficiently on the skeleton than on the full object. The process of generating such a skeleton is called skeleton extraction or *skeletonization*. Examples of *automatic* skeleton extraction include the Medial Axis Transform (MAT) [Blum 1967] and skeletonization into a *one dimensional poly-line skeleton* (or simply *1D skeleton*) [Capell et al. 2002; Lien and Amato 2006; Katz and Tal 2003].

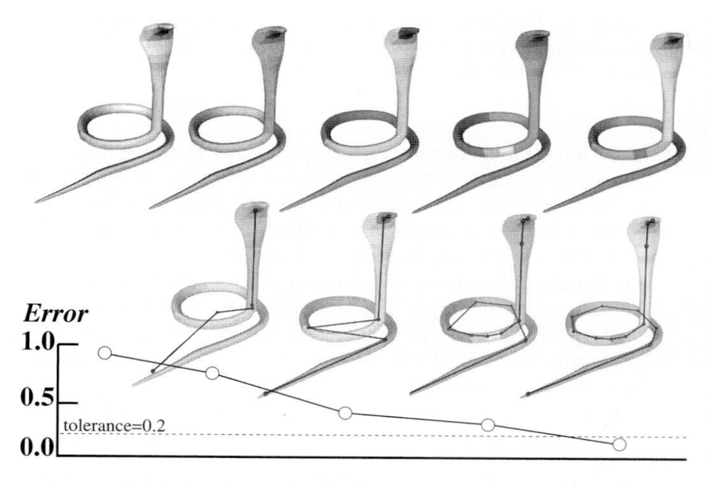

Figure 1: The skeleton (shown in the lower row) evolves with the shape decomposition (shown in the upper row).

Skeletons have been extracted from different sources, such as voxel (image) based data [Zhou and Toga 1999; Palenichka et al. 2002; Bitter et al. 2001], boundary represented models [Chuang et al. 2000; Amenta et al. 2001; Wu et al. 2003], and scattered points [Verroust and Lazarus 1999], and for different purposes, such as shape description [Sheehy et al. 1996; Shinagawa and Kunii 1991], shape approximation [Attali et al. 1994; Wyvill and Handley 2001], similarity estimation [Hilaga et al. 2001], collision detection [Bradshaw and O'Sullivan 2002; Hubbard 1996], biological applications [Amenta et al. 2002], navigation in virtual environments [Li et al. 1999], and animation [Teichmann and Teller 1998; Katz and Tal 2003].

Although it has been noted before that a good shape decomposition can be used to extract a high quality skeleton [Lien and Amato 2006; Katz and Tal 2003] and that a high quality skeleton can be used to produce a good decomposition [Li et al. 2001], this relationship between shape decomposition and skeleton extraction is a relatively unexplored concept, especially in 3D. Instead, when a relationship is noted, the skeletons are usually treated as an intermediate result or a by-product of the shape decomposition.

In this paper, we propose an integrated framework for simultaneous shape decomposition and skeleton extraction that not only acknowledges, but actually exploits the interdependence between these two operations. First, a simple skeleton is extracted from the components of the current decomposition. Then, this extracted skeleton is used to eval-

*This research supported in part by NSF Grants EIA-0103742, ACR-0081510, ACR-0113971, CCR-0113974, ACI-0326350, and by the DOE.

†e-mail:neilien@cs.tamu.edu

‡e-mail:keyser@cs.tamu.edu

§e-mail:amato@cs.tamu.ed

SPM 2006, Cardiff, Wales, United Kingdom, 06–08 June 2006.
© 2006 ACM 1-59593-358-1/06/0006 $5.00

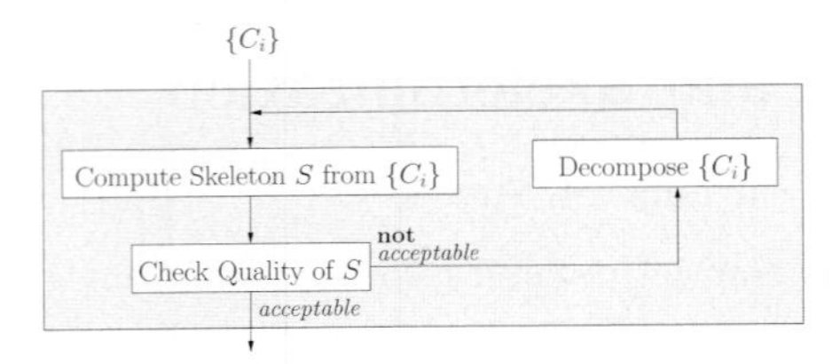

Figure 2: Simultaneous shape decomposition and skeleton extraction. The set $\{C_i\}$ is a decomposition of the input model P and initially $\{C_i\} = \{P\}$.

uate the quality of the decomposition. If the skeleton is satisfactory under some user defined criteria, we report the skeleton and the decomposition as our final results. Otherwise, the components are further decomposed into finer parts using approximate convex decomposition (ACD) [Lien and Amato 2006; Lien and Amato 2004], which decomposes a given component by 'cutting' its most concave features. Figure 2 illustrates this proposed framework and Figure 1 shows an example of the co-evolution process of the shape decomposition and skeleton extraction.

As we will show, our proposed approach has several advantages and makes contributions as listed below.

- This recursive refinement strategy generates multi-resolution skeletons, from coarse to fine levels of detail, which are useful for some applications.
- *Divide-and-conquer* algorithms which operate on the decompositions or skeletons can be more efficient because refinement is applied to the more complex regions but not to areas with less variation.
- The extracted skeleton is invariant under translation, rotation, and uniform scale, and is not very sensitive to boundary noise and skeletal deformations.
- Our approach does not require any pre-processing, e.g., model simplification, or any post-processing, e.g., skeleton pruning, which are required by many of the existing methods, e.g., [Li et al. 2001; Katz and Tal 2003; Wu et al. 2003].
- Our framework is general enough to work for both 2D polygons and 3D polyhedra.

2 Related Work

Both shape decomposition and skeleton extraction have been studied for decades and there is a large amount of previous work. In this review, we concentrate on recent developments most relevant to our work.

Shape decomposition. Inspired by psychological studies, such as recognition by components [Biederman 1987] and the minima rule [Hoffman and Richards 1984; Hoffman and Singh 1997], methods have been proposed to partition models at salient features to produce visually meaningful components. In pattern recognition, Rom and Medioni [Rom and Medioni 1994] partition a model into a set of tubular (generalized cylinder) shapes according to their curvature properties. As a preprocessing step for mesh generation, Sonthi et al. [Lu et al. 1999] identify closed sets (loops) of edges with required convexity and use them to decompose a model into solid parts. However, these methods work best with simple models with sharp internal angles, such as mechanical parts.

Methods that are applicable to models with general shapes also exist. Wu and Levine [Wu and Levine 1997] propose a partitioning method based on a simulated electrical charge distribution on the surface of a model. Mangan and Whitaker [Mangan and Whitaker 1999] and Page et al. [Page et al. 2003] decompose polygonal meshes by applying watershed segmentation with curvature computation. Li et al. [Li et al. 2001] decompose polygonal meshes at critical points along skeletons obtained via model simplification. Dey et al. [Dey et al. 2003] segment a model, in $\mathbb{R}^2$ or $\mathbb{R}^3$, into *stable manifolds*, which are collections of Delaunay complexes of sampled points on the boundary. Katz and Tal [Katz and Tal 2003] cluster mesh facets into fuzzy regions, carefully partition facets in those regions, and successfully produce perceptually clean cuts between decomposed components. A similar approach using a different clustering technique can also be found in [Liu and Zhang 2004]. Interactive methods [Lee et al. 2004; Funkhouser et al. 2004] that identify features via human assistance have also been shown to produce high quality and clean decompositions.

Skeletonization. The Medial Axis (MA), Voronoi diagram, Shock graph and Reeb graph are common skeleton representations. Although the MA can represent a lossless shape descriptor [Blum 1967], it is difficult and expensive to compute accurately in high (> 2) dimensional space [Culver et al. 2004]. Several ideas for approximating the MA have been proposed, e.g., using Voronoi diagram, and its dual Delaunay triangulation [Amenta et al. 2001; Attali and Lachaud 2001; Dey and Zhao 2002], of densely sampled points from the object boundary. Shock graphs [Siddiqi et al. 1998; Cyr and Kimia 2001], another representation of the MA, encode the formation order and, therefore, the importance of each part of the MA. Reeb graphs, a type of 1D skeleton, extracted from various Morse functions, are a powerful tool for shape matching [Verroust and Lazarus 1999; Shinagawa et al. 1991; Attene et al. 2001; Hilaga et al. 2001]. Since Morse functions are defined on mesh vertices, re-meshing [Hilaga et al. 2001; Attene et al. 2001] is usually needed to generate a good (accurate) skeleton.

Several methods have been proposed to extract a skeleton from the components of a decomposition [Lien and Amato 2006; Katz and Tal 2003]. Skeletons can also be constructed by simplifying (contracting) a polygonal mesh to line segments [Li et al. 2001].

Multi-scale and multi-resolution skeletons. Multi-scale skeletons [Rom and Medioni 1993; Ogniewicz and Kubler 1995] consist of a set of skeletons, $S_0, \ldots, S_N$, whose union represents a complete skeleton of the model. S_0 is the most important part of the skeleton, representing global topology, while S_N encodes local features and is sensitive to local changes. Multi-resolution skeletons [Hilaga et al. 2001] consist of a set of skeletons, $S_0, \ldots, S_N$, that encode topology at different levels of detail. S_0 will have the coarsest skeleton and S_N will contain the most detailed information. This representation is desired for some applications. For instance, to extract similar items from a 3D database, a rough skeleton can be used to reject many unlikely models and incrementally refine the skeleton to get better matches. As previously mentioned, one of the features of our framework is that its recursive nature results in the construction of multi-resolution skeletons.

3 Preliminaries

Let P be a polyhedron represented by its boundary ∂P and let H_P be the convex hull of P.

Approximate Convex Decomposition. A set of components $\{C_i\}$ is a *decomposition* of P if their union is P and all C_i are interior disjoint, i.e., $\{C_i\}$ must satisfy:

$$\mathrm{D}(P) = \{C_i \mid \cup_i C_i = P \text{ and } \forall_{i \neq j} C_i^\circ \cap C_j^\circ = \emptyset\}\ , \quad (1)$$

where C° is the open set of C.

A component C is *τ-approximate* convex if C has *concavity* less than or equal to a tunable variable τ. We use concave(C) to denote the concavity measurement of C. Therefore, the *τ-approximate* convex decomposition of P is:

$$\mathrm{ACD}_\tau(P) = \{C_i \in \mathrm{D}(P) \mid \mathrm{concave}(C_i) \leq \tau\}\ . \quad (2)$$

We define the concavity of a vertex x of C as the distance from x to the convex hull surface of C and the concavity of C as the maximum concavity of its vertices, i.e., $\mathrm{concave}(C) = \max_{x \in C}(\mathrm{concave}(x))$. ACD iteratively identifies and resolves concave features with maximum concavity. Figure 3 shows an example of this process. Due to space limitations, we refer readers to [Lien and Amato 2004; Lien and Amato 2006] for details regarding ACD, including how ACD handles models with holes and handles.

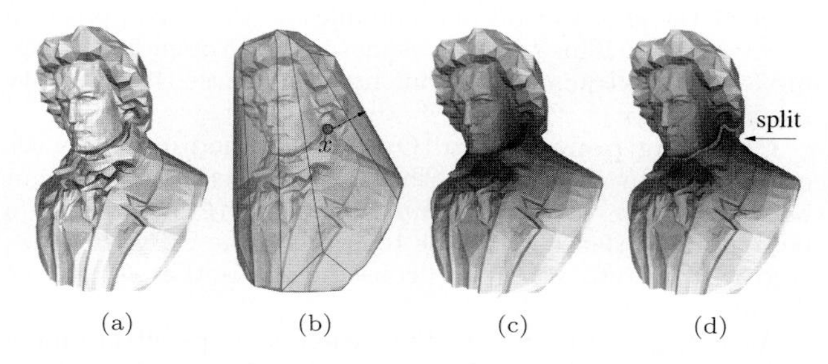

Figure 3: (a) The input model. (b) The convex hull of the input model. The concavity of x is measured as the distance from x to the convex hull surface. (c) The shading of the model represents concavity, i.e., darker areas have higher concavity. (d) The model is decomposed by partitioning at the high concavity region (indicated by an arrow).

The Principal Axis. Let X be a set of points and ℓ be a line. We define $\mathrm{dist}(X, \ell)$, the distance from X to ℓ, as $\sum_X \mathrm{dist}(x, \ell)$, where $x \in X$. Then, the *principal axis* (PA) of a set of points X is a line ℓ such that distance $\sum_X \mathrm{dist}(x, \ell)$ is minimized over all possible lines $\kappa \neq \ell$.

Algorithm 3.1 SSS(P)

```
1: S = Ext_Skeleton(P)
2: if Error(P, S) ≤ τ then
3:    Report S as P's skeleton and report P as a component
4: else
5:    {C_i} = Decompose(P)
6:    For each C ∈ {C_i} do return SSS(C)
```

4 Framework

We propose a framework that simultaneously performs shape decomposition and skeleton extraction. For a given polyhedron P, **S**imultaneous **S**hape decomposition and **S**keleton extraction (SSS) (see Algorithm 3.1) constructs a skeleton for the model from (local) skeletons extracted from each component of a decomposition, evaluates the extracted skeleton components, and continues *refining* the decomposition and the associated skeleton components until the quality of the skeleton for each component is satisfactory, e.g., the error estimation of the skeleton for the respective component is smaller than a tunable threshold τ.

There are three important sub-routines that are required by Algorithm 3.1: *Ext_Skeleton*(P), which extracts a skeleton from a component P, *Error*(P, S), which estimates the quality of the extracted skeleton, and *Decompose*(P), which separates P into sub-components when the extracted skeleton is not acceptable. We discuss methods for skeleton extraction *Ext_Skeleton*(P) in Section 4.1, and methods for quality measurement *Error*(P, S) in Section 4.2. Recall that our choice for the *Decompose*(P) sub-routine is approximate convex decomposition.

4.1 Extracting Skeletons

In this section, we discuss two simple methods to extract a (*local*) skeleton from a component of a decomposition. These local skeletons can be connected to form a global skeleton of the input model. The centroid method is a simple approach that can result in skeletons that do not represent the shape of the object. The second method, based on the principal axis of a component, is slightly more expensive to compute, but leads to improved skeletons in some cases.

Using Centroids. One of the easiest ways to construct a skeleton for a component C (in a decomposition) is to connect the centroids of the openings, called *opening centroids*, on ∂C to the centroid of C. These openings are generated when a component is split into sub-components during the decomposition process,

Several similar methods for extracting skeletons have been proposed [Lien and Amato 2006; Katz and Tal 2003]. Although this approach is simple and generates fairly good results one of the major drawbacks of this type of skeleton is its inability to represent some types of shapes. For example, the skeleton of a cross-like model in Figure 4 extracted using its centroids is only a line segment instead of two crossing line segments. The method described next attempts to address this problem.

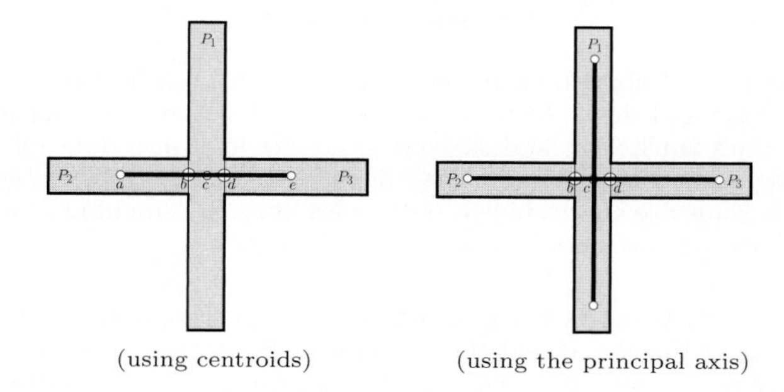

Figure 4: This example shows a problem that arises when skeletonization is based only on the centroids. Points b and d are the centers of the openings and a, c and e are the centers of the components P_1, P_2 and P_3, respectively. This problem can be addressed using the principal axis.

Using the Principal Axis. In this method, we extract a skeleton from a component C (in a decomposition) using the principal axis of the convex hull H_C of C. Instead of connecting the centroids of C's openings to the center of mass of C, we connect these centroids to the principal axis enclosed in H_C. Figure 5 shows an example of skeletons constructed in this manner.

Let $\mathrm{PA}(H_C)$ be a line through the center of mass of H_C. parallel to the principal axis of H_C. Our method connects an opening centroid to one of the k points on $\mathrm{PA}(H_C) \cap H_C$. These k points, denoted by $\mathcal{P}$, evenly subdivide $\mathrm{PA}(H_C) \cap H_C$ into $k+1$ line segments. The selection of the value of k is based on the desired minimum skeleton link length. Let $\mathcal{P}' \subset \mathcal{P}$ be a set of points to which the opening centroids connect. Figure 5 illustrates $\mathcal{P}$ and $\mathcal{P}'$ with circles along $\mathrm{PA}(H_C)$. Then, the final skeleton S of C contains line segments that connect the opening centroids to $\mathcal{P}'$ and line segments that connect the $\mathcal{P}'$.

To minimize the chance of getting a long skeleton with many joints, we match the opening centroids to $\mathcal{P}$ so that the cardinality of $\mathcal{P}'$ and the distances from the opening centroids to $\mathcal{P}'$ are minimized. We solve this optimization matching problem using dynamic programming. Details of how we find the optimal solution are discussed in Appendix A.

In cases where all the points in $\mathcal{P}'$ lie only on one side of the center of mass c of H_C, e.g., $\mathcal{P}'$ in Figure 5(b), line segments that connect to the points in $\mathcal{P}'$ are not enough to represent the entire component. In such cases, the skeleton will connect $\mathcal{P}'$ with the end point of $\mathcal{P}$ on the other side of the center of mass c. Similarly, when $\mathcal{P}'$ contains only c, the skeleton will connect c with the end points of $\mathcal{P}$ on both sides of c, e.g., the skeleton of the component P_1 in Figure 4 (using the principal axis).

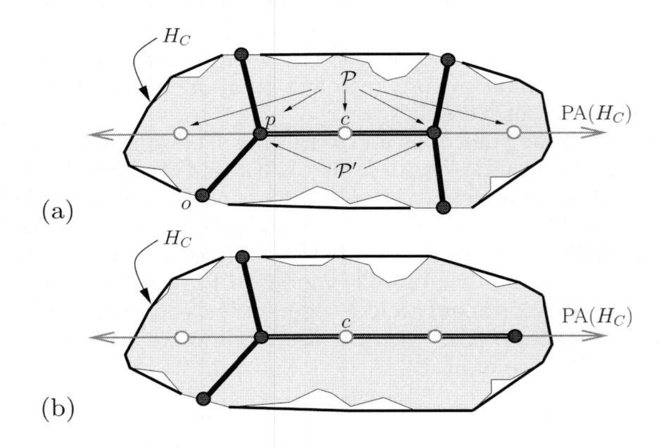

Figure 5: Using the principal axis of the convex hull H_C to extract a skeleton from a component. Skeletons are shown in dark thick lines and skeletal joints are shown in dark circles and c denotes the center of mass of H_C. (a) Opening centroids are connected to both sides of c. (b) Opening centroids are connected to only one side of c.

Figure 6 shows three skeletons: two extracted skeletons using the centroid and the principal axis methods, and one skeleton manually generated by a professional animator. One can see that the skeleton extracted using the principal axis is topologically more similar to the animator generated skeleton than the skeleton generated using the centroid method. In Section 5, we analyze the similarity of these skeletons using graph edit distance.

4.2 Measuring Skeleton Quality

Although several criteria exist for measuring the quality of a skeleton, the general principles we adopt are that the skeleton should reside in the interior of the model and it should encode the "topology" of the model's shape. Thus, using these general criteria, our strategy to compute the quality of a skeleton S is to compare S with its associated component C. In this section, we propose three methods for measuring quality. This first method checks whether S intersects ∂C and the second method checks the topological representation of S w.r.t. C. In the third method, we propose an adaptive measurement based on the volume of the component.

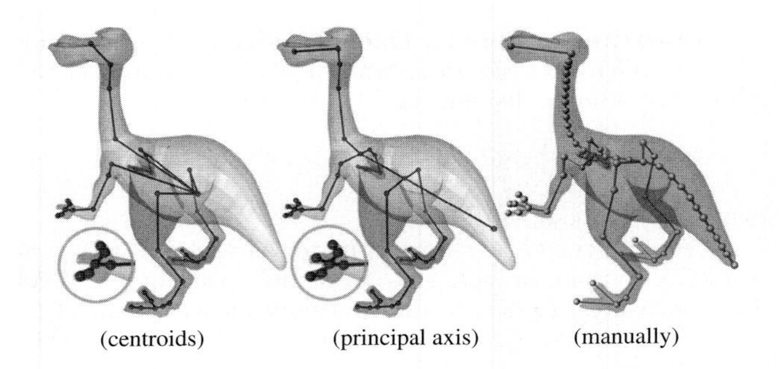

Figure 6: Notice the differences of these skeletons at the torso, the head, and the fingers.

An important property of these three methods is that the error of the skeleton becomes smaller as the decomposition becomes finer. This property is justified in Appendix C. Figure 8 shows extracted skeletons based on these three quality measurements.

Checking penetration. Our first method measures the quality of S by checking whether S intersects the component boundary ∂C. If so, the function $Error(C, S)$ returns a large number (larger than the tolerable value τ). Otherwise, zero will be returned. The consequence is that C will be decomposed if $\partial C \cap S \neq \emptyset$.

As seen in Figure 8, skeletonization using penetration detection stops evolving after a few iterations and does not produce skeletons that represent the dragon or the bird.

Measuring centeredness. In the second method, we measure the offsets of S from the level sets of the geodesic distance map on ∂C. The value for each point in this map is the shortest distance to its closest opening of C. Ideally, a skeleton should pass through all connected components in all level sets. Therefore, this measurement method simply checks the number of times that S does not do so. An example of this measurement is shown in Figure 7.

Let L_C be all the connected components in the level sets of C. We define the error of a skeleton S as:

$$Err(C, S) = \frac{\sum_{l_c \in L_C} f(l_c, S)}{|L_C|}, \tag{3}$$

where $f(l_c, S)$ returns 0 if S intersects component l_c, and 1 otherwise, and $|L_C|$ is the total number of the connected components in C. Details of how we compute the level sets and $f(l_c, S)$ are discussed in Appendix B.

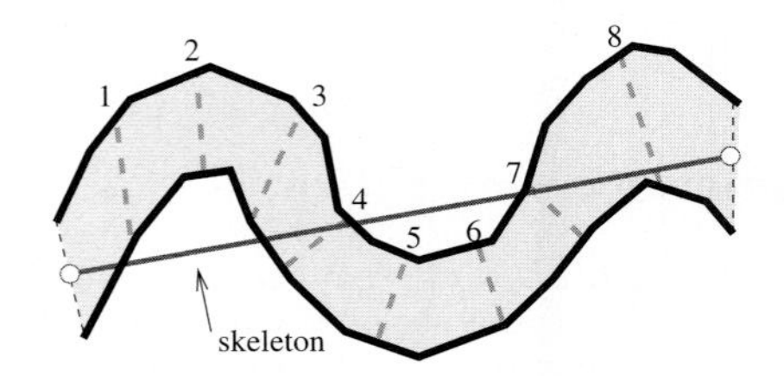

Figure 7: The error measurement for this skeleton, which intersects level sets 4, 7 and 8, is $\frac{5}{8}$.

As seen in Figure 8, skeletonization using the centeredness measurement captures the shape of the dragon and the bird better then simply using penetration detection, but it over segments the tail of the bird and does not produce accurate skeletons in the feet of the dragon or the bird.

Measuring convexity. Our idea for the last quality measurement comes from the observation that in many cases the significance of a feature depends on its volumetric proportion to its "base". For example, a 5 cm stick attached to a ball with 5 cm radius is a more significant feature than a 5 cm stick attached to a ball with 5 km radius. This intuition can be captured by the concept of the *convexity* of a component C defined as $\text{convexity}(C) = \frac{\text{vol}(C)}{\text{vol}(H_C)}$, where $\text{vol}(X)$ is the volume of a set X. Thus, we can define the error measurement as:

$$Err(C, S) = 1 - \text{convexity}(C) \ . \qquad (4)$$

Assume that the skeleton S is a good representation of the convex hull H_C. Then, a smaller difference between H_C and C means that S is a better representation of C. Thus, although the skeleton S is not included in Equation 4, S is implicitly considered in terms of H_C.

As seen in Figure 8, using convexity produces the most realistic skeleton that captures the overall shape of the dragon and the bird and also identifies the detailed features of their feet.

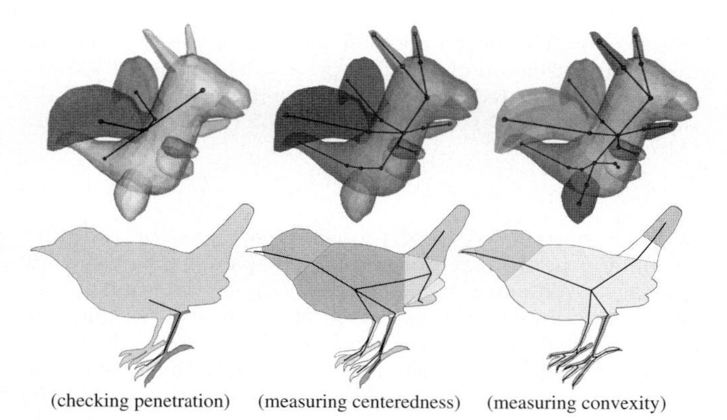

(checking penetration) (measuring centeredness) (measuring convexity)

Figure 8: Final skeletons of a dragon polyhedron and a bird polygon extracted using different quality estimation functions: checking penetration, measuring centeredness, and measuring convexity. The maximum tolerable errors for centeredness and convexity are 0.2 and 0.3, respectively.

4.3 Putting it All Together

Algorithm 4.1 $SSS_{ACD}(P)$

1: Compute a skeleton S from P using the *Principal Axis* of H_P.
2: Estimate the quality of S using *convexity*.
3: **if** S is acceptable **then**
4: Report S as P's skeleton and report P as a component.
5: **else**
6: $\{C_i\} = ACD(P)$.
7: **For** each $C \in \{C_i\}$ **do** return $SSS_{ACD}(C)$

Algorithm 4.1 shows a fleshed-out version of the proposed simultaneous shape decomposition and skeletonization framework. Here we suggest using the principal axis, convexity and approximate convex decomposition for local skeleton extraction, quality measurement and partitioning, respectively. Algorithm 4.1 is used for all the experiments in Section 5. We would like to emphasize that the choice of these methods is made based on our own experience. The framework is not restricted to these selected sub-routines, which can be replaced by other methods to fit particular needs of an application.

5 Implementation and Results

The experiments in this section are used to demonstrate the *efficiency*, the *robustness*, and several *applications* of the proposed method. The method was implemented in C++ and all these experiments are performed on a Pentium 2.0 GHz CPU with 512 Mb RAM. Seventeen decompositions and their associated skeletons are shown in Figures 8 to 13 and in Tables 1 and 2.

Efficiency. A summary of the studied models, which include several game characters, a high genus model, and two scanned models, and the skeletonization and decomposition time of these models is shown in Table 1. Table 1 shows that the processing time of SSS depends on both the size of the model and on the complexity of the shape. For example, even though the model in Figure 9 has the fewest triangles, its large genus (18) increases the processing time. In general, our proposed SSS method can handle models with thousands of triangles in less than a half a minute and scales well for models with tens or hundreds of thousands of triangles.

We further show that SSS is efficient by comparing our results to two recently proposed shape decomposition and skeletonization methods that have been shown to produce very promising results; see Figures 10 and 11, respectively. In both experiments, SSS generates results similar to those results reported previously but SSS can produce the shape decomposition and the skeletons about 30 times and 5 times faster than those methods reported in [Katz and Tal 2003] and [Wu et al. 2003], respectively. We note that there are no well-accepted criteria to compare the quality of these decompositions and skeletons quantitatively, and therefore we do not intend to claim that our results are necessarily better.

Robustness. In this set of experiments, we show that SSS is robust under perturbation and deformation, meaning that the shape decompositions and skeletons remain approximately the same after the input models are perturbed and deformed. The results are shown in Table 2.

Although there are no well accepted criteria to measure the differences among decompositions, we can measure the similarity of these skeletons, e.g., using graph edit distance [Bunke and Kandel 2000] which computes the cost of operations (i.e., inserting/removing vertices or edges) needed to convert one graph to another. In this paper, we simply associate one unit of cost with each operation.

We measure two types of distances, denoted as D_O and D_O^2. D_O is the graph edit distance from a skeleton to the skeleton extracted from the original mesh. Because removing or inserting a degree-two node does not change the topology of a graph, we are also interested in the distance, denoted as D_O^2, that does not count operations that create and remove degree-two nodes. Table 2 shows that D_O remains small for both perturbed and deformed models and D_O^2 is zero in all cases.

Applications. The extracted skeleton can be readily used to create animations. We demonstrate this advantage by re-targeting motion captured data to the skeletons extracted using our method. In Figure 12, we show a sequence

Table 1: Experimental Results

Model	Figure 9	Figure 12	Table 2	Figure 6	Figure 11	Figure 12	Table 2	Figure 1	Table 2
Size	1,984	3,392	5,660	6,564	8,276	11,180	39,694	48,312	243,442
Time	15.6	2.6	1.7	1.5	8.8	3.4	19.4	30.1	73.3

Size is measured as the number of the triangles of each model and the processing time is measured in seconds.

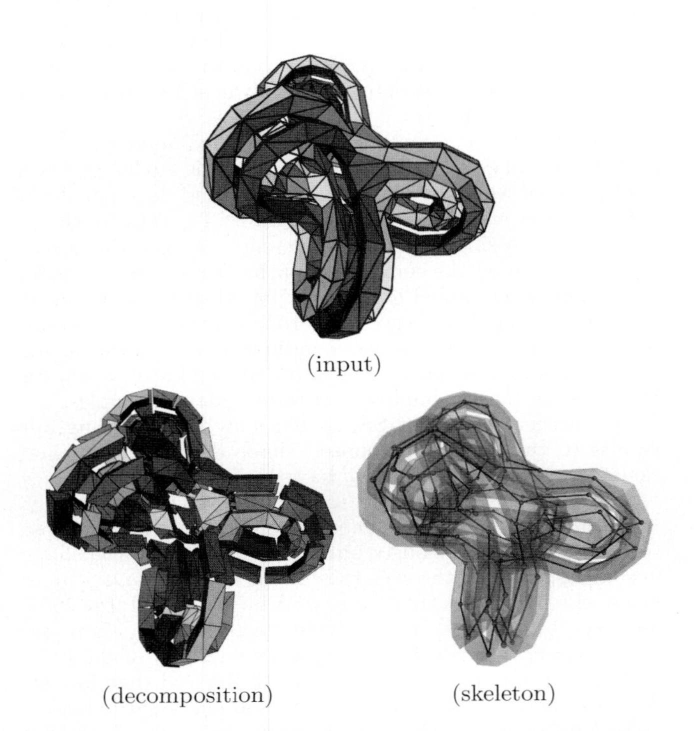

Figure 9: This figure shows the decomposition and the skeleton of a model with 18 handles.

of images obtained from a skeleton-based boxing animation of a baby and a robot using motion data captured from an adult male. Note that the baby and the robot models have different body proportions and rest poses. Due to the limited space in this paper, other animations, including walking and pushing a box, are provided on our webpages. We use motion captured data instead of a hand-made animation to show that the extracted skeletons are robust enough to be used by arbitrarily selected motions and not only carefully designed motion. The motions, i.e., joints angles, are manually copied from the captured data to the skeletal joints.

The extracted skeletons can also help to plan motion, e.g., for navigating in the human colon or removing a mechanical part from an airplane engine. Sampling-based motion planners have been shown to solve difficult motion planning problems; see a survey in [Barraquand et al. 1997]. These methods approximate the free configuration space (C-space) of a movable object by sampling and connecting random configurations to form a graph (or a tree). However, they also have several technical issues limiting their success on

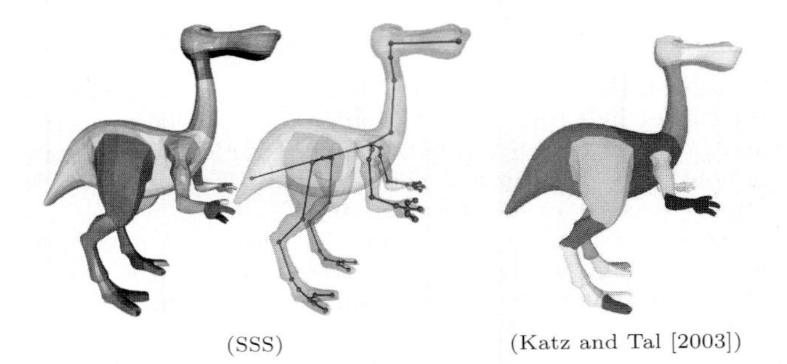

Figure 10: The decomposition with 0.7 convexity and the associated skeleton of the dino-pet model (with 6,564 triangles) are computed in 1.5 seconds whereas Katz and Tal's approach takes 57 seconds (on a P4 1.5 GHz CPU with 512 Mb RAM).

some important types of problems, such as the difficulty of finding paths that are required to pass through narrow passages [Wilmarth et al. 1999]. Using sampling biased toward the joints of the extracted skeleton, we can alleviate this so called "narrow passage" problem. Figure 13 shows that the graph constructed using the skeleton-based sampling can better represent the free C-space than using the uniform sampling [Kavraki et al. 1996] with the same number of samples. This is because more of the skeleton-based sampling samples are placed in the narrower regions. In addition, the connections between the samples in these narrow regions can be made easily because the components of the decomposition are nearly convex.

6 Discussion and Conclusion

In this paper, we propose a framework that simultaneously generates shape decompositions and skeletons. This framework is inspired by the observation that both operations share many common properties and applications but are generally considered as independent processes. This framework extracts the skeleton from the components in a decomposition and evaluates the skeleton by comparing it to the components. The process of simultaneous shape decomposition and skeletonization iterates until the quality of the skeleton becomes satisfactory.

We studied two simple skeleton extraction methods, using the centroids and the principal axis, and three quality evaluation measurements, that compute penetration, centeredness and convexity, respectively. In the experiments, we demonstrate that the proposed framework is efficient, ro-

Table 2: Robustness tests using perturbed and skeletal deformed meshes. D_O is the graph edit distance between a skeleton extracted from a perturbed or deformed mesh and a skeleton extracted from the original mesh. D_O^2 is D_O without counting operations on degree-2 nodes (which do not change the topology of the skeleton).

Shape Decomposition. 70% convexity

Original	Perturbed (random noise)	Deformed
female 16 components; horse 17 components; triceratop 9 components	female 16 components; horse 20 components; triceratop 9 components	female 18 components; horse 20 components; triceratop 9 components

Extracted Skeletons. 70% convexity

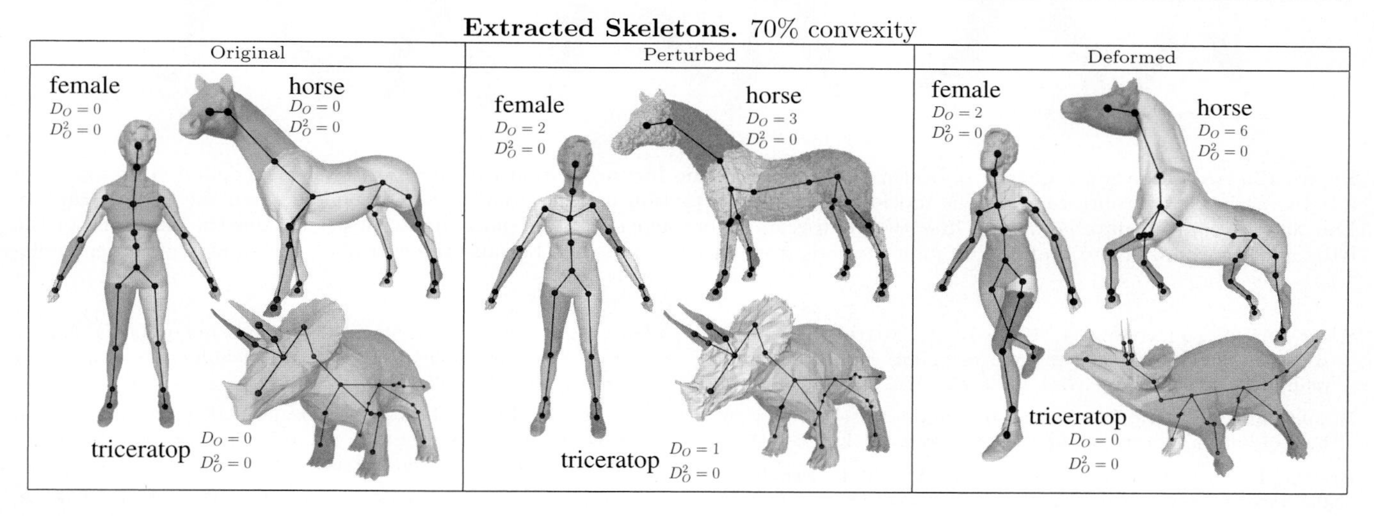

Original	Perturbed	Deformed
female $D_O = 0$, $D_O^2 = 0$; horse $D_O = 0$, $D_O^2 = 0$; triceratop $D_O = 0$, $D_O^2 = 0$	female $D_O = 2$, $D_O^2 = 0$; horse $D_O = 3$, $D_O^2 = 0$; triceratop $D_O = 1$, $D_O^2 = 0$	female $D_O = 2$, $D_O^2 = 0$; horse $D_O = 6$, $D_O^2 = 0$; triceratop $D_O = 0$, $D_O^2 = 0$

bust under perturbation and deformation, and can readily be used, e.g., to generate animations and plan motion.

There are several ways to extend the current work. First, there is a need to establish a systematic framework for comparing qualities of shape decomposition and skeletons using more quantitative measuring methods and benchmarks. Although the proposed quality measurements are based on a general idea of what a good skeleton should be, more studies are needed to investigate application-specific measurement criteria that should produce better and more "comparable" results. Second, not all models, such as a bowl, can have reasonable 1D skeletons. We are interested in using the same framework to extract the approximated medial axis from the components in a decomposition based on the idea that it is easier to extract the medial axis from a convex object than from a non-convex object. Finally, because the extracted skeletons and shape decompositions in our method co-evolve, we can provide more meaningful shape decompositions by using information from the extracted skeletons, e.g., merging components if the skeletons exacted from those components do not change the global skeleton made from the entire decomposition.

References

AMENTA, N., CHOI, S., AND KOLLURI, R. K. 2001. The power crust, unions of balls, and the medial axis transform. *Computational Geometry 19*, 2-3, 127–153.

AMENTA, N., CHOI, S., JUMP, M. E., KOLLURI, R. K., AND WAHL, T. 2002. Finding alpha-helices in skeletons. Tech. rep., Dept. of Computer Science, University of Texas at Austin.

ATTALI, D., AND LACHAUD, J.-O. 2001. Delaunay conforming iso-surface; skeleton extraction and noise removal. *Computational Geometry: Theory and Applications 19*, 2-3, 175–189.

ATTALI, D., BERTOLINO, P., AND MONTANVERT, A. 1994. Using polyballs to approximate shapes and skeletons. In *Proceedings of International Conference on Pattern Recognition (ICPR'94)*, 626–628.

ATTENE, M., BIASOTTI, S., AND SPAGNUOLO, M. 2001. Remeshing techniques for topological analysis. In *Proc. of the Shape Modeling International (SMI'01)*, 142–151.

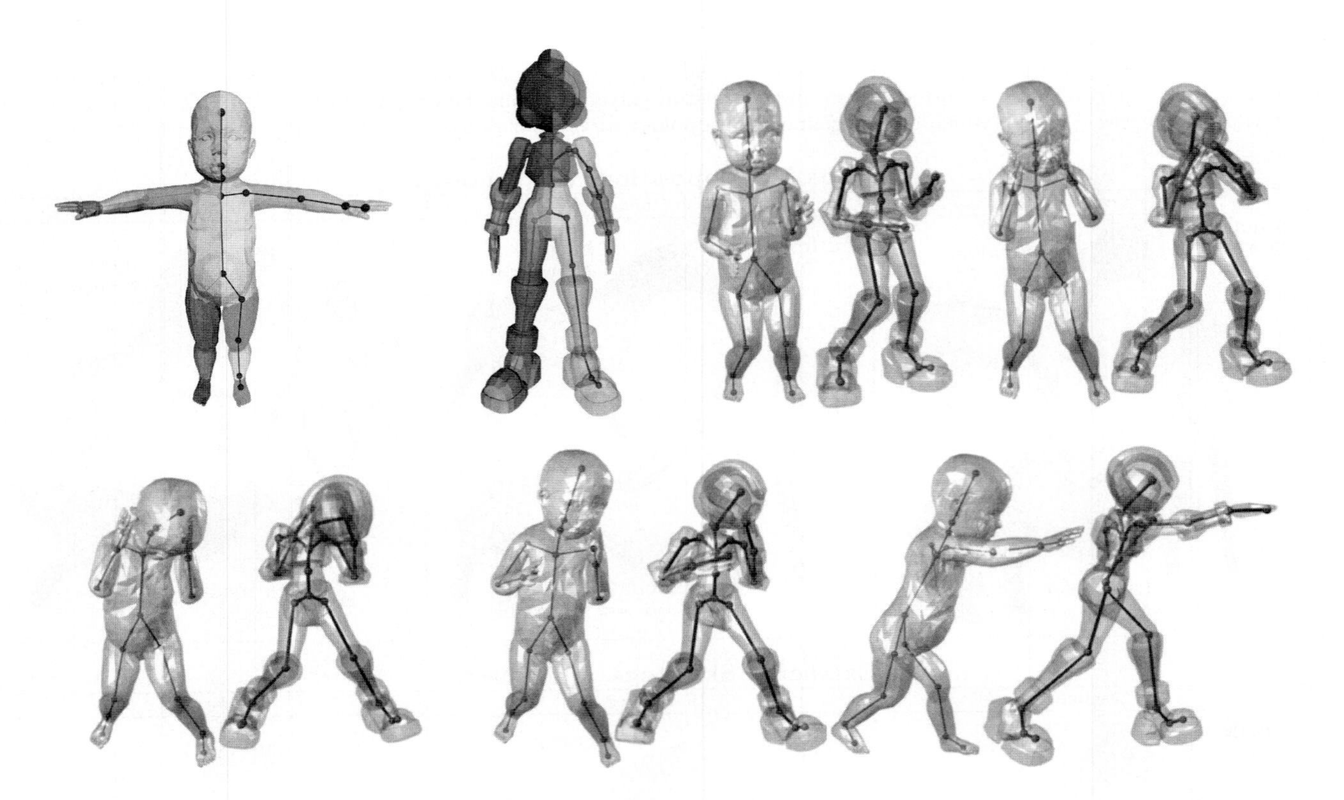

Figure 12: An animation sequence obtained from applying the boxing motion capture data to the extracted skeletons from a baby model and a robot model. The motion capture data (action number 13_17) are downloaded from the CMU Graphics Lab motion capture database. The first two figures in the sequence are the shape decompositions and the skeletons of the baby and the robot. Note that not all joint motions from the data are used because the extracted skeletons have fewer joints.

BARRAQUAND, J., LATOMBE, L. K. J., LI, T., MOTWANI, R., AND RAGHAVAN, P. 1997. A random sampling scheme for path planning. *Int. J. of Rob. Res 16*, 6, 759–774.

BIEDERMAN, I. 1987. Recognition-by-components: a theory of human image understanding. *Psychol. Rev. 94*, 2, 115–147.

BITTER, I., KAUFMAN, A. E., AND SATO, M. 2001. Penalized-distance volumetric skeleton algorithm. *IEEE Transactions on Visualization and Computer Graphics 7*, 3, 195–206.

BLUM, H. 1967. A transformation for extracting new descriptors of shape. In *Models for the Perception of Speech and Visual Form*, W. Wathen-Dunn, Ed. MIT Press, 362–380.

BRADSHAW, G., AND O'SULLIVAN, C. 2002. Sphere-tree construction using dynamic medial axis approximation. In *Proceedings of the ACM SIGGRAPH symposium on Computer animation*, ACM Press, 33–40.

BUNKE, H., AND KANDEL, A. 2000. Mean and maximum common subgraph of two graphs. *Pattern Recogn. Lett. 21*, 2, 163–168.

CAPELL, S., GREEN, S., CURLESS, B., DUCHAMP, T., AND POPOVIC, Z. 2002. Interactive skeleton-driven dynamic deformations. *ACM Transactions on Graphics 21*, 3, 586–593.

CHUANG, J.-H., TSAI, C.-H., AND KO, M.-C. 2000. Skeletonization of three-dimensional object using generalized potential field. *IEEE Transactions on Pattern Analysis and Machine Intelligence 22*, 11, 1241–1251.

CULVER, T., KEYSER, J., AND MANOCHA, D. 2004. Exact computation of the medial axis of a polyhedron. *Comput. Aided Geom. Des. 21*, 1, 65–98.

CYR, C. M., AND KIMIA, B. B. 2001. 3d object recognition using shape similarity-based aspect graph. In *ICCV'01.*

DEY, T. K., AND ZHAO, W. 2002. Approximate medial axis as a voronoi subcomplex. In *ACM Symposium on Solid Modeling and Applications*, 356–366.

DEY, T. K., GIESEN, J., AND GOSWAMI, S. 2003. Shape segmentation and matching with flow discretization. In *Proc. Workshop on Algorithms and Data Structures*, 25–36.

FUNKHOUSER, T., KAZHDAN, M., SHILANE, P., MIN, P., KIEFER, W., TAL, A., RUSINKIEWICZ, S., AND DOBKIN, D. 2004. Modeling by example. *ACM Trans. Graph. 23*, 3, 652–663.

HILAGA, M., SHINAGAWA, Y., KOHMURA, T., AND KUNII, T. L. 2001. Topology matching for fully automatic similarity estimation of 3d shapes. In *Proceedings of the 28th annual conference on Computer graphics and interactive techniques*, 203–212.

HOFFMAN, D., AND RICHARDS, W. 1984. Parts of recognition. *Cognition 18*, 65–96.

HOFFMAN, D., AND SINGH, M. 1997. Salience of visual parts. *Cognition 63*, 29–78.

HUBBARD, P. M. 1996. Approximating polyhedra with spheres for time-critical collision detection. *ACM Transactions on Graphics (TOG) 15*, 3, 179–210.

KATZ, S., AND TAL, A. 2003. Hierarchical mesh decomposition using fuzzy clustering and cuts. *ACM Trans. Graph. 22*, 3, 954–961.

KAVRAKI, L. E., SVESTKA, P., LATOMBE, J. C., AND OVERMARS, M. H. 1996. Probabilistic roadmaps for path planning in high-dimensional configuration spaces. *IEEE Trans. Robot. Automat. 12*, 4 (August), 566–580.

LEE, Y., LEE, S., SHAMIR, A., COHEN-OR, D., AND SEIDEL, H.-P. 2004. Intelligent mesh scissoring using 3d snakes. In *Proceed-*

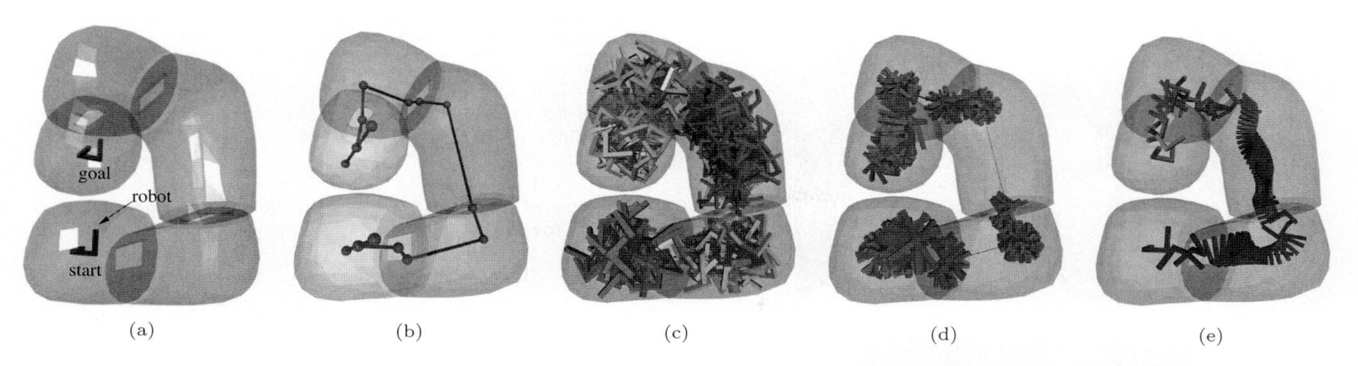

Figure 13: Narrow passage motion planning. (a) A difficult motion planning problem in which the robot is required to pass through four narrow windows to move from the start configuration (bottom) to the goal configuration (top). (b) A shape decomposition and a skeleton of the environment using the proposed method. In (c), uniform sampling results in a poor representation of the robot's free C-space. Uniform sampling samples 350 uniformly distributed collision-free configurations and connects these samples into a graph. In this example, there are five connected components in the graph. In (d), the skeleton-based sampling results in a better representation of the robot's free C-space. The skeleton-based sampling samples 350 collision-free configurations around the joints of the skeleton and connects these samples into a graph. In this example, the graph has one single connected component. (e) A collision-free path found by connecting the start and the goal configurations to the graph generated using the skeleton-based sampling.

ings of the 12th Pacific Conference on Computer Graphics and Applications (PG'04), 279–287.

LI, T.-Y., LIEN, J.-M., CHIU, S.-Y., AND YU, T.-H. 1999. Automatically generating virtual guided tours. In *Proc. IEEE Computer Animation (CA)*, 99–106.

LI, X., TOON, T. W., AND HUANG, Z. 2001. Decomposing polygon meshes for interactive applications. In *Proceedings of the 2001 symposium on Interactive 3D graphics*, 35–42.

LIEN, J.-M., AND AMATO, N. M. 2004. Approximate convex decomposition of polygons. In *Proc. 20th Annual ACM Symp. Computat. Geom. (SoCG)*, 17–26.

LIEN, J.-M., AND AMATO, N. M. 2006. Approximate convex decomposition of polyhedra. Tech. Rep. TR06-002, Parasol Lab, Dept. of Computer Science, Texas A&M University, Jan.

LIU, R., AND ZHANG, H. 2004. Segmentation of 3d meshes through spectral clustering. In *Proceedings of the 12th Pacific Conference on Computer Graphics and Applications (PG'04)*, 298–305.

LU, Y., GADH, R., AND TAUTGES, T. J. 1999. Volume decomposition and feature recognition for hexahedral mesh generation. In *Proc. 8th International Meshing Roundtable*, 269–280.

MANGAN, A. P., AND WHITAKER, R. T. 1999. Partitioning 3d surface meshes using watershed segmentation. *IEEE Transactions on Visualization and Computer Graphics 5*, 4, 308–321.

OGNIEWICZ, R., AND KUBLER, O. 1995. Hierarchic voronoi skeletons. *Pattern Recognition 28*, 3, 343–359.

PAGE, D. L., KOSCHAN, A. F., AND ABIDI, M. A. 2003. Perception-based 3d triangle mesh segmentation using fast marching watersheds. In *Proceedings of the 2003 Conference on Computer Vision and Pattern Recognition (CVPR '03)*, 27–32.

PALENICHKA, R. M., ZAREMBA, M. B., AND DU QUEBEC, U. 2002. Multi-scale model-based skeletonization of object shapes using self-organizing maps. In *16 th International Conference on Pattern Recognition (ICPR'02)*, 10143–10147.

ROM, H., AND MEDIONI, G. 1993. Hierarchical decomposition and axial shape description. *IEEE Transactions on Pattern Analysis and Machine Intelligence 15*, 10, 973–981.

ROM, H., AND MEDIONI, G. 1994. Part decomposition and description of 3d shapes. In *Proc. International Conference of Pattern Recognition*, 629–632.

SANDER, P. V., WOOD, Z. J., GORTLER, S. J., SNYDER, J., AND HOPPE, H. 2003. Multi-chart geometry images. In *Proceedings of the Eurographics/ACM SIGGRAPH symposium on Geometry processing*, 146–155.

SHEEHY, D. J., ARMSTRONG, C. G., AND ROBINSON, D. J. 1996. Shape description by medial surface construction. *IEEE Trans. Visualizat. Comput. Graph. 2*, 1 (Mar.), 62–72.

SHINAGAWA, Y., AND KUNII, T. L. 1991. Constructing a reeb graph automatically from cross sections. *IEEE Computer Graphics and Applications 11*, 6, 44–51.

SHINAGAWA, Y., KUNII, T. L., AND KERGOSIEN, Y. L. 1991. Surface coding based on morse theory. *IEEE Comput. Graph. Appl. 11* (Sept.), 66–78.

SIDDIQI, K., SHOKOUFANDEH, A., DICKINSON, S. J., AND ZUCKER, S. W. 1998. Shock graphs and shape matching. In *ICCV*, 222–229.

TEICHMANN, M., AND TELLER, S. 1998. Assisted articulation of closed polygonal models. In *Proceedings of the Eurographics Workshop*, 254–254.

VERROUST, A., AND LAZARUS, F. 1999. Extracting skeletal curves from 3d scattered data. In *International Conference on Shape Modeling and Applications*, IEEE Computer Society, 194–201.

WILMARTH, S. A., AMATO, N. M., AND STILLER, P. F. 1999. Motion planning for a rigid body using random networks on the medial axis of the free space. In *Proc. ACM Symp. on Computational Geometry (SoCG)*, 173–180.

WU, K., AND LEVINE, M. D. 1997. 3d part segmentation using simulated electrical charge distributions. *IEEE Transactions on Pattern Analysis and Machine Intelligence 19*, 11, 1223–1235.

WU, F.-C., MA, W.-C., LIOU, P.-C., LAING, R.-H., AND OUHYOUNG, M. 2003. Skeleton extraction of 3d objects with visible repulsive force. In *Computer Graphics Workshop 2003, Hua-Lien, Taiwan.*

WYVILL, G., AND HANDLEY, C. 2001. The "thermodynamics" of shape. In *Proc. of the Shape Modeling International (SMI'01)*, 2–8.

ZHOU, Y., AND TOGA, A. W. 1999. Efficient skeletonization of volumetric objects. *IEEE Transactions on Visualization and Computer Graphics 5*, 3, 196–209.

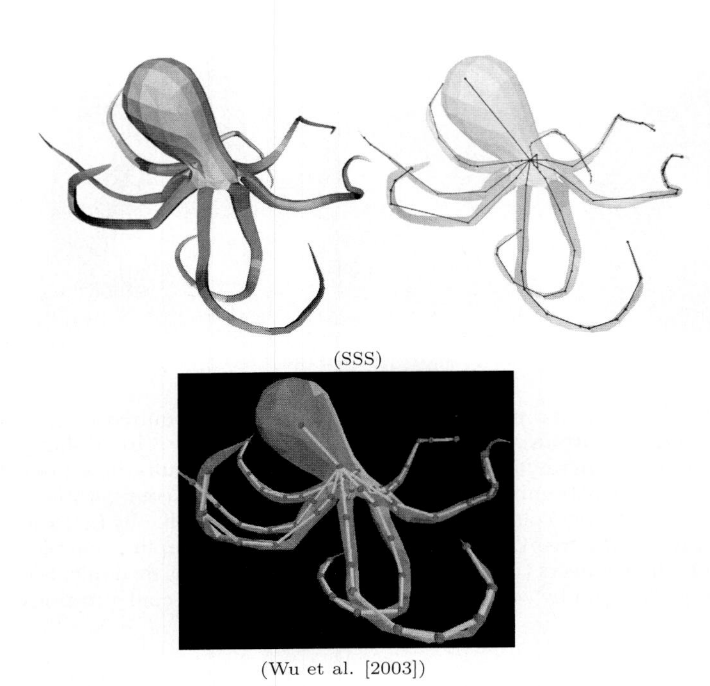

(SSS)

(Wu et al. [2003])

Figure 11: The decomposition with 0.7 convexity and the associated skeleton of the octopus model (with 8,276 triangles) are computed in 8.8 seconds whereas Wu et al.'s approach takes 53 seconds (on a P4 1.5 GHz CPU with 512 Mb RAM) using a simplified version of this model (with 2,000 triangles).

Appendix

A From a Principal Axis to a Skeleton

Here, we show how a local skeleton can be computed using the principle axis. Our goal is to find a match $M : O \rightarrow \mathcal{P}$ from the opening centroids O to the points $\mathcal{P}$ on the principle axis so that the total length of the match and the number of the matched points (joints) in $\mathcal{P}$ is minimized. We let the score function F of a match M be defined as

$$F(M) = s_1 \cdot |M| + s_2 \cdot J(M) \ , \tag{5}$$

where $|M|$ and $J(M)$ are the length and joint size of match M, and s_1 and s_2 are user specified scalars. In this paper, s_1 and s_2 are constants set to ten and one, resp. A brute force approach to find an optimal solution will take $O(|\mathcal{P}|^{|O|})$ time, where $|\mathcal{P}|$ and $|O|$ are the number of vertices in $\mathcal{P}$ and O, respectively. This exponential time complexity is in general impractical for most applications.

The main idea of finding the optimal match is to group opening centroids O and each group will connect to a point in $\mathcal{P}$. After knowing how O is grouped, it takes $O(|\mathcal{P}||O|)$ time to find solution.

Grouping O can be done using dynamic programming. An observation that enables us to group O is that two centroids are likely to be grouped when their closest points in $\mathcal{P}$ are close. Thus, we first sort O according to their closest points of $\mathcal{P}$ and then group sorted O. A dynamic programming is shown in Algorithm A.1 to group O. In Algorithm A.1, we use $G[i, j]$ to denote the optimal solution for the sub-problem $\{O_i, \cdots, O_j\}$ and use $< G_i G_j >$ and $G_i G_j$ to denote the joint of two groups G_i and G_j with and without merging G_i and G_j to one group.

Algorithm A.1 Optimal Matching(O, $\mathcal{P}$)

1: **for** $i \in \{1, \cdots, |O|\}$ **do**
2: $\quad G[i, i] = O_i$
3: **for** $l \in \{2, \cdots, |O|\}$ **do**
4: $\quad$ **for** $i \in \{1, \cdots, |O| - l + 1\}$ **do**
5: $\quad\quad j = i + l - 1$
6: $\quad\quad G[i, j] =< O_i \cdots O_j >$
7: $\quad\quad score = F(G[i, j], \mathcal{P})$ {F is defined in Eqn. 5}
8: $\quad\quad$ **for** $k \in \{i \cdots, j - 1\}$ **do**
9: $\quad\quad\quad s = F(G[i, k]G[k + 1, j], \mathcal{P})$
10: $\quad\quad\quad$ **if** $s_1 < score$ **then**
11: $\quad\quad\quad\quad G[i, j] = G[i, k]G[k + 1, j]$
12: $\quad\quad\quad\quad score = s_1$

B Compute level sets and centeredness

A level set of a component C in a decomposition is a set of points on the surface ∂C of the component with the same geodesic distance to the closest opening of C. A connected component in a level set is a list of connected edges, which usually forms a loop on ∂C. A level set can have one or multiple connected component(s). These level sets can be computed, similar to the construction process of a Reeb graph [Shinagawa and Kunii 1991], by flooding the entire ∂C from the boundaries of the openings of C. In each iteration of this flooding process, the wavefronts will propagate from the visited vertices to unvisited vertices via incident edges.

To compute centeredness, we need to know how a skeleton S intersects the level sets of C, i.e., we need the function $f(l_c, S)$ use in Eqn 3, which returns zero if S intersects the level set l_c. The function $f(l_c, S)$ can be implemented by simply checking the intersection between each line segment of S and the triangulation of l_c.

C ACD increases skeleton quality

In this section, we show that the error measurements of a skeleton described in Section 4.2, i.e., penetration, centeredness, and convexity, decrease as the input model is decomposed. This is a critical property, which allows the SSS framework to terminate.

Lemma C.1. *Let S be the skeleton of a polyhedron P and let S' be the skeleton of the components of the* ACD *of P. The error estimation of S' must be smaller than the error estimation of S measured using penetration, centeredness, and convexity defined in Section 4.2.*

Proof. We show that all error measurements become zero if the input model is convex. For penetration, because the segments connecting any two points inside the convex object must not intersect its boundary, a skeleton will never penetrate the object. For the same reason, the skeleton must not be 'outside' of any level set of a convex component. Finally, because the convexity of a convex object is one, its error must be zero.

Since ACD decomposes P into more convex components after each iteration, the error measurements of S' will be closer to zero than S for all three types of measurements. □

Experience in the Exchange of Procedural Shape Models using ISO 10303 (STEP)

Michael J. Pratt*
LMR Systems, UK

Junhwan Kim†
National Institute of Standards and Technology, USA

Abstract

The international standard ISO 10303 (STEP) is being extended to permit the exchange of procedurally defined shape models, with additional parameterization and constraint information, between CAD systems. The transfer of parameterized assembly models is an additional objective. Most of the essential new resources have already been published by ISO, and the remainder are well advanced in the standardization process. Because these are new capabilities, at present not quite complete, there are at present no commercial STEP translators making use of them. However, several proof-of-concept trials have been performed or are in progress, using development versions of the STEP documentation. This paper reports in some detail on one of those trials, and comments on the experience gained. The conclusion is that the standardized exchange of CAD models containing 'design intent' information has been successfully demonstrated, but that the development of translators for that purpose is not an easy task. One particular problem area is pinpointed, where further research is needed to find ways of improving the efficiency of such exchanges.

CR Categories: H.2.5 [Information systems]: Heterogeneous databases—Data translation; H.5.3 [Information systems]: Group and organization interfaces—computer-supported cooperative work; J.6 [Computer applications]: Computer-aided engineering—Computer-aided design (CAD).

Keywords: standard, product data exchange, construction history, design intent

1 Introduction

The earliest parts of the international standard ISO 10303 [Int 1994] for the exchange of product data in electronic form were published in 1994. Since then the standard has become widely used for the exchange of computer aided design (CAD) models between different systems within companies, and also between companies up and down the engineering supply chain. The most widely used part of the standard at present is the application protocol ISO 10303-203 (AP203: 'Configuration controlled design of mechanical parts and assemblies'), which provides for the exchange of wireframe, surface and boundary representation (B-rep) solid models, together with associated administrative data. As mentioned in earlier papers [Pratt 2004a; Pratt et al. 2005] such models cannot be effectively edited in a receiving system after a transfer because all of the information collectively known as 'design intent' is lost in the exchange. Design intent is considered to include the construction history of the model together with any parameterization schemes and constraint sets imposed upon it. AP203, whose development was started well before these kinds of information were widely used in CAD systems, makes no provision for the capture and transfer of design intent, which is why the models it exchanges have become known as 'dumb' models. On receipt after an exchange such a model may have additional detail defined upon it, but it is not in general possible to change the basic properties (e.g., dimensions) of the exchanged model itself.

The new parts of ISO 10303 mentioned above are as follows:

- ISO 10303-55: 'Procedural and hybrid representation' [Pratt et al. 2005];
- ISO 10303-108: 'Parameterization and constraints for explicit geometric product models' [Pratt 2004b; Pratt et al. 2005];
- ISO 10303-109: 'Kinematic and geometric constraints for assembly models [Pratt et al. 2005];
- ISO DIS 10303-111: 'Elements for procedural modelling of solid shapes' [Pratt et al. 2005];
- ISO DIS 10303-112: 'Modelling commands for the exchange of procedurally represented 2D CAD models'.

The first three of these have already been published by the International Organization for Standardization (ISO) as parts of ISO 10303; the last two are (at the time of writing) at the Draft International Standard (DIS) stage.

This paper does not address the exchange of assembly models, and all 2D profiles or sketches involved in the tests performed have been defined in terms of explicit geometric elements rather than procedurally. The only further mention of ISO 10303-109 and -112 will therefore be at the end of the paper where some information is given about other proof-of-concept translation tests that have been performed or are in progress.

In what follows, ISO 10303, whose official title is 'Industrial automation systems and integration — Product data representation and exchange', will usually be referred to briefly by its informal name of STEP (STandard for the Exchange of Product model data). The individual parts of the standard listed above will be cited as 'Part 55', 'Part 108', and so on. A single reference is given to ISO 10303 in the bibliography, and this covers all published and DIS parts of the standard. A brief overview of the standard is given in [Pratt 2001] and a more extended one in the book [Owen 1997].

An early suggestion for a method of exchanging CAD models in terms of their construction history was made by Hoffmann and Juan [Hoffmann and Juan 1992]. Their EREP (Editable REPresentation) was a specification for the representation of sequential feature-based design processes, which supported parameterization and constraints. It was suggested that during a design session a CAD system could be made to generate an EREP model in addition to its own internal model. It should then be possible to export the EREP model and process it using a different CAD system to

*e-mail: mike@lmr.clara.co.uk
†e-mail: junhwan.kim@nist.gov

SPM 2006, Cardiff, Wales, United Kingdom, 06–08 June 2006.
© 2006 ACM 1-59593-358-1/06/0006 $5.00

generate an equivalent model there. A trial implementation was made, but attention apparently later became focused on associated problems such as persistent naming [Capoyleas et al. 1996] and the solution of constraint systems [Bouma et al. 1995]. These are outside the scope of the STEP work, which has adopted a strategem (described later) for avoiding the persistent naming problem and addresses only the representation and transmission of constraint data, leaving the solution of constraint sets to individual CAD systems.

The closest recent parallel to the work reported here is that by Rappoport et al. [Rappoport 2003; Spitz and Rappoport 2004; Rappoport et al. 2005]. The primary differences are that the present research is aimed towards the development of an International Standard, and that it is based on the STEP philosophy of a standardized intermediate neutral representation. In the STEP approach the concept of 'feature rewrites' introduced in the cited papers occur during the preprocessing and postprocessing phases of the overall translation, rather than in a centralized processor. This may have the advantage of allowing the rewrites to be specifically tailored to the CAD systems concerned.

An alternative approach to the exchange of procedural models has been used at KAIST in Korea [Choi et al. 2002; Mun et al. 2003]. This is based on the capture and transfer of the journal file created by a CAD system, which contains a record of every action of the system user. However, the work of the team concerned is now directed to extending the STEP-related work described in this paper.

The basic assumptions made in this work are as follows:

1. Any design intent not present in the model in the sending system cannot be transmitted. This implies that the work carried out does not involve general mechanisms for such things as automated feature recognition or other implicitly defined model characteristics.
2. Even if it is present in a CAD model, any aspect of design intent that is not accessible via the applications programming interface (API) of a particular CAD system cannot be exchanged unless it can be inferred by indirect means.
3. STEP does not consider details of the behavior of the receiving system if the transmitted model is modified there, but only assumes that it is as far as possible intuitive for the system user.
4. The minimal criteria for a successful parametric model exchange are the correct transmission of parameters and constraints and their appropriate interpretation in the receiving system.

The exchange of parametric construction history models using STEP is based on the use of a dual model, consising of a primary procedural model and an associated secondary model of the B-rep or some closely related explicit type. The secondary model can be used in the receiving system as a check on the validity of reconstruction there, and to resolve ambiguities, e.g., to determine which of several valid solutions of a nonlinear constraint system was chosen in the original model. Spitz and Rappoport [Spitz and Rappoport 2004] similarly use pure geometric representations of individual features for verification purposes.

It is believed that the work described here illustrates several advances on previously published methods:

1. The transfer of parameters and mathematical relations between them has not been previously reported. Any attribute in the part model can be treated as a parameter in an exchange, so that it is possible (for example) to transfer a relation defining the number of hole instances in a circular hole pattern in terms of the radius of the pattern.
2. The method can handle several different types of constraints (algebraic, logical, dimensional) in a very general and easily extensible manner.
3. The method handles multiple simultaneous constraints in the model in addition to sets of independent constraints.
4. The method conforms to the international standard ISO 10303 through the use of new parts of that standard.

Having said this, there are some restrictions. The tests concentrate on the transfer of sketch-defined features, and all geometric constraints are two-dimensional. STEP permits the use of 3D constraints, which are mainly used in practice for inter-feature relationships and for constraining parts in an assembly. These capabilities have not so far been tested, but they are no different in principle from their 2D counterparts.

The methodology used for translator development has been based on a careful analysis of all the different types of information present in a procedural CAD model, and of the different usages made of each of those types of information. The structuring of information into optimal 'units of creation' has also been given detailed attention, as discussed below. Finally, semantic differences between CAD systems have been analyzed in the interests of maximizing interoperability and minimizing information loss in the exchange of CAD models.

2 Design intent and its representation in CAD systems

The most essential aspect of 'design intent' is the constructional history of the model. If this is recorded as the primary aspect of the model's representation, it is possible to replay that history, with modifications if desired, in the certainty that the designer's original methodology will be followed. Other important contributions to design intent are the presence of parameters in the model, representing values that it is permissible to change, and the presence of constraints, defining relationships that must be preserved in any change. Design intent, as defined in this paper, therefore corresponds to the way in which facilities provided by a CAD system are used in order to achieve intended design aims. A distinction is made between this and *design rationale*, which is concerned with the reasoning underlying the way those facilities are used (see Section 3.6).

Further details of some important aspects of design intent information are given in the following subsections.

2.1 Construction history

The construction history of a model is a procedural representation of that model, expressed in terms of the operations used to build it. Modern CAD systems provide users with a range of high-level constructional operations which shield them from having to work at the level of individual geometrical and topological elements. In CAD system terminology the configurations they create are referred to as 'features', though this is misleading. Strictly, a feature has some associated application semantics [Shah and Mäntylä 1995], but with the present level of CAD technology the intended design functionality of the features created in a design process is present only in the mind of the designer; it is not captured by the CAD system. These operations should therefore be regarded merely as 'shape macros' which, in B-rep terms, construct relatively complex subgraphs of the overall topological structure. However, because the use of the word 'feature' is prevalent in the CAD design context it will be used in the remainder of the paper, subject to the proviso made in this paragraph.

A construction history, then, is primarily a sequence of operations that create shape features. Part 111 of STEP, which provides represention of shape features, is based on a survey of major CAD systems. Its intention is to capture a range of the most widely used feature types. In general, the ordering of elements in a STEP exchange file is immaterial — in the exchange of a B-rep model, for example, the order in which elements are added to the model as it is rebuilt after a transfer does not matter provided all necessary elements are present in the file. As is well known, the ordering of operations in a construction sequence is crucial, however; a different ordering will in general lead to a different model. Part 55 of STEP therefore defines special structures for the capture of operation sequences.

Each operation is defined, as in the CAD systems themselves, in terms of the shape feature created (expressed in descriptive geometrical terms) and the size attributes of that configuration. For example, in the case of a simple rectangular pocket feature the pocket will be characterized by four planar walls and a planar floor. The values of its size attributes will define its length, width and depth. Implicit in the operation of the creation procedure will be the facts that opposite pairs of wall faces are parallel, that adjacent wall faces are perpendicular and that the walls are perpendicular to the floor, regardless of the values of the size attributes. The dimensional attributes themselves have no independent existence as elements of the model. In this case we refer to the constraints and dimensions of the feature as being *implicitly* defined.

Many feature creation operations require additional supporting information, though this varies from one CAD system to another. Typically, a round hole feature will need a centreline. In STEP terms this may be an unbounded line defined by a point and a direction, or some equivalent construct. The line and its defining point and direction need to be present explicitly in the exchange file for the hole feature to be reconstructed following a transfer. These elements do not occur in operation sequences, because provided they are present their place in the exchange file does not matter. They are therefore transmitted in 'traditional' STEP unordered mode.

In CAD systems, supporting elements for feature creation operations are usually stored within the data structures representing the features concerned. STEP, by contrast, transmits them at the model level rather than the feature level. The reasons for this are partly to maintain upwards compatibility with previous practice in STEP, and partly to keep the transferred model at the most general possible level so that it is compatible with a wide range of CAD modelling methodologies. There is also the distinction that a CAD system data structure is designed for efficiency and a neutral data structure such as STEP for informational completeness. It is therefore hardly surprising that they differ significantly, and this divergence leads to some problems for the STEP translator writer, as will be discussed below.

2.2 Parameters

Parameters represent values that may be changed in a part model to generate different members of a family of parts. They include the dimensional attributes of feature creation operations mentioned above. Since these have no independent existence as elements of the model they are here referred to as *implicit parameters*. Explicit parameters, by contrast, are model elements in their own right. Part 108 of STEP defines representations for their transmission in an exchange file. The primary uses of explicit parameters in CAD modelling are (i) for specifying dimensional relationships in the 2D sketches often used as the basis for created features, (ii) for specifying dimensional relations between features, and (iii) for positioning part models in assembly models.

CAD systems differ in the way they deal with parameters. Most have separate data structures for sketches in which explicit sketch parameters are represented. In some cases separate model-level parameter data structures are provided, and used as the basis for the generation of tabular displays of parameters that may be changed in the model. Some systems, under some circumstances, will generate explicit parameters coresponding to parameters that were initially implicitly defined.

Another important aspect of parameters is that they may have relationships defined between them. For example, it may be desired to model a family of rectangular blocks in which the length of a block is always twice the width. More complex algebraic relationships may also be defined. Further, it is possible to define parameters whose values do not correspond to any physical quantities in a model, as in the case where the length and width of a block are required to be given by t^2+1 and $t+3$, t being a parameter that is not directly associated with any specific dimension in the model. All CAD systems provide capabilities of these kinds, and Part 108 of STEP makes provision for their capture and transmission, though in that document they are treated as specialized constraints. The topic of constraints in general is covered below.

2.3 Constraints

As with parameters, constraints have implicit and explicit forms. Implicit constraints were mentioned above; they occur automatically as the result of creation operations. Explicit constraints, by contrast, are modelling elements in their own right, which make reference to other modelling elements and constrain them to satisfy specified relationships. For example, a sketch of a rectangle with rounded corners is made up of four line segments and four circular arcs. This collection of geometric elements may be supplemented by explicit constraints requiring (i) opposite pairs of sides of the rectangle to be parallel, (ii) adjacent pairs of sides to be perpendicular, and (iii) lines and arcs to be tangential where they adjoin. Some CAD systems will add constraints to ensure that the end points of lines and arcs are coincident where they adjoin, but others achieve this result in different ways.

It was mentioned above that constraints involved in feature creation are usually implicit. Explicit constraints have similar application areas to explicit parameters: (i) for specifying geometric relationships such as parallelism, perpendicularity or tangency between geometric elements of 2D sketches, (ii) for positioning and orienting features with respect to each other, or with respect to datum elements defined in the model, and (iii) for positioning and orienting part models in an assembly model.

Part 108 of STEP [Pratt 2004b] defines entities representing explicit constraints. Apart from constraints specifying explicit mathematical relationships between parameters, a wide range of 'descriptive' constraints is provided, expressing geometric relationships such as parallelism or tangency. The latter basically record no more than the nature of the constraint and the elements that are subject to it. All CAD systems implement geometric constraints of these types, their semantics are widely understood, and so it is best if their precise mathematical formulation is left to the systems concerned. Some descriptive constraints have dimensional subtypes. For example, the parallelism constraint, applying to lines and planes, has a dimensional subtype whose interpretation is 'parallel at a specified distance'; it only makes sense to assert the dimensional aspect of the constraint once the logical condition (parallelism) has been established.

CAD systems usually store explicit constraints in the datastructures of the individual sketches or features they relate to. However, Part 108 of STEP is part-oriented rather than feature-oriented, and explicit constraints may occur anywhere in an exchange file, in the same manner as the supporting information for feature operations mentioned earlier.

2.4 Sketches

CAD systems provide self-contained datastructures for sketches, as mentioned above. Parameters and constraints associated with sketches are usually stored in these datastructures. Part 108 of STEP provides representations for sketches, though it distinguishes between 2D sketches defined in neutral coordinate systems (such as might be stored in a library of sketches for multiple re-use) and the results of transforming them into 3D model space.

In the case of a sketch-based feature, the creation of the defining sketch is usually treated as a single operation, because although a sketch may contain many elements its parametric variation is determined by computation of a new solution to its constraint system rather than by a replay of its construction history.

2.5 Datums

Model elements are often positioned or dimensioned with respect to datums, which may or may not be geometric elements of the model. The centreline of an axisymmetric hole feature is an example of an element that is not a constituent of the B-rep of the model containing the hole. Despite this, it may be used as a reference element in positioning the hole in the model, perhaps with respect to a datum coprresponding to the axis of another hole. Most CAD systems store details of datums with the features or sketches that make use of them. Part 108 of STEP provides for the definition of datums under the general heading of 'auxiliary geometric elements', but as with parameters and constraints they are not associated in an exchange file with specific sketches or features.

3 Problems encountered in writing trial STEP translators for construction history models

This section outlines some of the primary problems encountered in writing translators for the exchange of CAD models of the construction history type with parameters and constraints. We start with a few general remarks.

The basic principle of the exchange of a feature-based construction history CAD model is that each successive creation operation is mapped onto a corresponding operation or combination of operations in the receiving system. This allows a natural decomposition of the overall process into a sequence of transfers of simpler components, which may be optimized for the transfer of each of those components. Ideally, the performance of the transferred sequence of operations in the receiving system will result in a correct reconstruction there. This process has the property that the partial models generated at various stages of the overall reconstruction process will also match the corresponding partial models that were created during the original design process. This is significant for the transfer of user-selected elements, as will be explained below.

In the tests performed, the translators read and wrote model information through the applications programming interfaces (APIs) of the CAD systems concerned. The success of such translations therefore depends crucially on the completeness of the functionality of those interfaces. In comparable experiments performed elsewhere it was found that not all CAD systems provided adequate access to the data required [Stiteler 2004]. The APIs of most CAD systems provide the translator developer with an entry point to the data structure of a represented model which, at the highest level, gives access to lists of features and sketches used in defining the model.

In the past, differences in the internal numerical tolerances used to judge coincidences etc. in CAD systems gave rise to major problems with geometry/topology incompatibilities in the STEP-based exchange of B-rep models [Gu et al. 2001]. These were largely overcome in the years following the initial publication of STEP, and currently there is a high success rate in the transfer of B-rep models. However, the introduction of design intent information into exchanged CAD models gives further scope for the occurrence of accuracy problems. In particular, constraints and dimensions may be satisfied by the accuracy criteria of the sending system but found to be unsatisfied by the more stringent criteria of a receiving system. So far, this has not been found to be a major problem in the tests performed. One can also take the view that the transfer of construction history information should alleviate accuracy incompatibilities, because the received model is always reconstructed according to the accuracy criteria of the receiving system, and mismatches are largely avoided. One potential area for mismatch remains, however, in the handling of elements selected by the user from the screen of the sending system. This will be discussed below, in Section 3.3.

3.1 Operation granularity

CAD systems differ in the complexity of the model substructures that are created by a single operation. We will use the term *granularity* in this context. An operation with coarse granularity in one system may need to be reformulated as a sequence of operations with finer granularity in another. The idea of a 'unit of construction' is a useful one; it is an operation or group of operations that results in the creation of a new geometric configuration in the model, but which may require different but equivalent sequences of one or more operations in different CAD systems. Two examples follow:

- Some CAD systems include positioning and orientation information in the basic definition of a feature (coarse granularity) while others allow the creation of the feature as an operation in its own right and then require the use of additional operations to position and orient it in the model (finer granularity).
- Some CAD systems allow the creation of underconstrained sketches or features, and permit later fine-tuning of the model in terms of lower-level operations. But other systems, with coarser granularity, only allow the creation of fully constrained constructs, possibly through the use of default options. The identification of such defaults, and their appropriate capture in an exchange file has proved to be one of the more difficult aspects of the work described.

The proof-of-concept tests described here attempt the automatic identification of units of construction with the same number of degrees of freedom in both the CAD system and the ISO 10303 neutral file. Degrees of freedom include dimensions and other parameters that may be defined either implicitly or explicitly. It is usually necessary to match a set of finer granularity system creation operations to a single ISO 10303 operation of coarser granularity, or vice versa. A one-to-one mapping would be ideal, but is rarely possible. The foregoing remarks apply both in the preprocessing phase (translation from the sending system to the ISO 10303 exchange file) and the postprocessing phase (translation from the exchange file to the receiving system).

The four possibilities for matching units of construction are clearly the following:

identity: there is a perfect match between operations;

aggregation: the translation must combine two or more finer-level operations to match an operation of coarser granularity;

decomposition: the translation must decompose a coarse granularity operation into two or more operations of finer granularity;

complex: it is necessary to use some combination of aggregation and decomposition.

To illustrate, we consider the case of a protrusion feature with a rectangular cross-section. Most CAD systems provide several means for the creation of such a feature, including

1. Creation of a block primitive and use of a Boolean union operation. These two operations map exactly onto operations that can be represented in STEP, and thus we have two cases of identity matches. The constraints on the form of the protrusion are implicit, being inherent in the definition of the block primitive.

2. Extrusion of a rectangular sketch defined on the part surface. The CAD system may provide rectangle creation as a single coarse-granularity operation in which the geometric constraints of the rectangle are defined implicitly, or may require the user to create a quadrilateral and impose the necessary constraints to make it a rectangle. In the latter case the constraints will be created explicitly in the CAD system by separate fine-granularity operations. At present STEP requires the constraints to be explicit, which implies the need for decomposition in the preprocessing phase if the sending CAD system is of the first type, and aggregation in the postprocessing phase if the receiving system is of the first type.

3. Creation of a general B-rep hexahedron, imposition of appropriate parallelism and perpendicularity constraints on its faces, association of explicit parameters with its dimensions, and use of a Boolean union as in Case 1 above. This method is possible in principle, and though it will rarely be used in practice it is useful for illustrative purposes. The protrusion will be editable in the receiving system because of its associated design intent information. However, in this case the block is defined entirely in terms of fine-granularity elements, and its mapping into the exchange file, for example as a block primitive, would require a high level of aggregation, together with the initial difficulty of recognizing automatically that the hexahedral B-rep does indeed have the form of a rectangular block. Alternatively, the exchange could be restricted to the transfer of the fine-granularity elements, without the added 'block' semantics. This would result in the exchange of the correct shape, but without the additional feature recognition process an important element of design intent would be lost.

As far as is known, feature recognition as suggested in Case 3 above has not yet been attempted for aggregation purposes in the the context of CAD model exchange, and it was ruled out of scope for the present work. If the model to be transferred contains explicit constraints and parameters these would provide a good basis for the use of the 'hint-based' apporach to feature recognition [Han et al. 2000].

In general, translation should be performed at the coarsest possible level of granularity, because this preserves the highest level of user intent. However, the translation software must be endowed with considerable intelligence to enable it to determine the most appropriate units of construction. In many cases, STEP feature definitions have a more general specification than the corresponding CAD system features, which allows the possibility of mapping the feature plus several additional constraints from the sending system to a single feature in the exchange file. This maximizes flexibility for interpretation of that feature in the receiving system. In postprocessing, it is best to select the coarsest granularity compatible option from the feature library of the receiving system. Any remaining differences in the representations can then be taken into account using additional finer-level elements.

3.2 Feature support information

Most modern CAD systems used in mechanical engineering allow design in terms of features. However, there are wide variations between systems regarding what is and what is not regarded as a feature, and this can lead to semantic mismatches. For example, a datum may be treated as a geometric element in one system but as a feature in its own right in another. Generally, CAD systems store design information as a collection of feature representations, each feature having associated with it all the supporting information needed to define that feature.

STEP, by contrast, is part-oriented rather than feature-oriented. This is partly for historical reasons and partly because of the need to cater for all types of systems, including any that may not be feature-based. Details of information supporting feature definitions is therefore present in the exchange file, but it is not identified in any way as being associated with specific features. Such information may include, for example, an explicitly defined line used as the centreline of a cylindrical hole, or an explicit direction specifying the direction of extrusion of a sketch. In these two cases the hole and the extrusion, respectively, will refer to the supporting elements, but there will be no references in the reverse sense. The translation process frequently requires these inverse references, and currently they can only be found by searching the entire exchange file until the feature referencing the supporting element in question is identified. At present this process is made more efficient by the generation of ephemeral data structures recording the inverse relationships as they are found. However, this has been found to be one of the most computing-intensive aspect of the translation, and it is felt desirable to amend the STEP resource that defines design features to make such searches more efficient.

One area where the part-oriented data structure of STEP causes problems is in the handling of explicit constraints. Such a constraint occurs as an instance in the exchange file, and it references the model elements that are the subject of the constraint. But it does not make any reference to the feature or sketch to which those elements belong. For translation into a feature-based CAD system, the translator must therefore identify the feature or sketch concerned indirectly, by a search process that compares the elements involved in the constraint with the elements of features and sketches represented in the file. For example, if an instance of a constraint in the exchange file references a line instance, and a sketch instance references the same line instance, then clearly the constraint belongs to the sketch, despite the fact that it is not directly referenced by it. The most robust exchanges will be those in which all constraints can be assigned to features or sketches in this way.

The proposed change to STEP is simply the addition of a new entity to Part 111, the STEP resource defining design features, whose instances will provide on the one hand a pointer to a construction operation and on the other the relevant sets of explicit parameters, explicit constraints and their supporting elements relating to that operation. Then by scanning all such instances in the exchange file the translator can create a structure that lists all sketches and features and the constraint elements that belong to them, and use this for the correct allocation of elements to the sketch and feature data structures of the receiving system. The use of this new entity will be optional; for example, it would not be appropriate to use it for translating a STEP file into a CAD system that was not feature-based. Subject to agreement by the relevant ISO technical subcommitee, this new entity will be added when the International Standard version of Part 111 is published. It will not represent a technical change in the way that models are represented and transmitted, but will provide a redundant additional construct to aid in the setting up of temporary data structures needed for the translation process.

3.3 Identifiers and user-selected elements

In some CAD systems each feature, topological element and geometric element has an identifier that is unique in the model. However, in most systems the data structures are based on individual feature or sketch elements, and identifiers are only unique within those subunits of the overall model. A STEP exchange file is model-based, individual instances of STEP entities being referred to uniformly by identifiers of the form #*n*, where *n* is an integer unique to the instance concerned.

During the CAD design process the system user frequently selects model elements by picking from the screen. Such selected elements may be used as datums, or may be the basis of further creation or modification operations. An example is provided by the selection of an edge, or a set of edges, to be blended or filleted in a subsequent operation. In the CAD system concerned the selected element is referred to by its internal system identifier, but such identifiers can be correctly interpreted only in the context of the system where they are created. In the preprocessing phase no attempt is made to associate the system identifiers of the sending system with STEP instance identifiers, and in the postprocesing phase the STEP instance identifiers will similarly be discarded as system identifiers are generated for the reconstructed elements created in the receiving system. That being so, some system-independent means is needed for indicating elements in the exchange file that correspond to user-selected elements in the sending system.

The method adopted in STEP exchanges is to write the selected element explicitly into the exchange file, but to mark it as a selected element so that it can be correctly interpreted in the receiving system. The reconstruction of the procedurally defined model in that system will give rise to an element that corresponds to the element selected in the sending system, and that element may be determined by matching all model elements of the appropriate type against the explicitly transferred selected element. In the absence of a perfect universal persistent naming method [Capoyleas et al. 1996] this has been found to be the most robust method of dealing with the selected element problem. Admittedly, geometric accuracy problems may in principle cause the matching process in the receiving system to fail, but so far this has not been found to happen in practice.

A similar approach to the handling of user-selected elements has been adopted by Rappoport et al. [Rappoport et al. 2005], who point out that the matching process is often complicated by the fact that different CAD systems use different topological structures in representing the same shapes. The cited paper deals with the matching process for vertices and edges. In the latter case matching may require identification of a pair of edges with different end-points but lying (to wihin some numerical tolerance) on the same curve.

Another area where identifiers need to be handled carefully is in the transmission of mathematical relations between the values of dimensions or other parameters. Again, each CAD system has its own internal method for allocating identifiers to dimensions and parameters. These are both treated in STEP as mathematical variables with associated semantics, and if they occur explicitly in the exchange file they are referenced in terms of their instance identifiers rather than by any other form of identifier (they may optionally have an additional name associated with them in the form of a text string, but this is not intended to play any part in the exchange process).

CAD systems generally store mathematical relationships as strings, in the manner of scientific programming languages. STEP, by contrast, for reasons of upwards compatibility with earlier parts of the standard, represents them in a parsed form in terms of sequences of entity instances defining individual operators and operands. Translation between the two forms presents little difficulty; the major requirement is the careful recording in a temporary data structure, for both the preprocessing and postprocessing phases, of correspondences between system names of variables or parameters and the identifiers of their representations in the STEP exchange file.

Part 108 of STEP provides two types of mathematical relationships, the *assignment*, where the value of one variable is required to be equal to the value of an expression involving other variables, and the *relationship*, which specifies a more general type of relation involving two or more variables. Both equality and inequality relationships are provided in the latter case.

3.4 The interplay between implicit and explicit data

This topic is related to that of the granularity of units of creation, previously discussed. To illustrate the connection, we will consider a CAD system that provides an operation of high granularity that creates, dimensions and positions a new feature on the model in a single operation. Then all the defining information, including the positioning information, will be input as arguments to the creation operation. This information will therefore be present implicitly in the model, to use the terminology introduced earlier. If transferred directly into a STEP exchange file it will occur as values of attributes of the feature instance in the file.

It was earlier mentioned that not all systems adopt this approach. In a system with lower granularity, the operation used to create the feature and the operations used to position and orient it in the model may be separate. Suppose that a dimensional constraint is created for positioning purposes. In this case a dimension which is implicitly represented in the first system is represented explicitly in the second system, as an entity instance in its own right. Then in an exchange of models between the first and the second system one or more implicitly defined items of information must be made explicit. Conversely, for exchange in the reverse direction explicit information must be made implicit. The place where the conversion is made depends upon the granularity of the representations of the two systems with respect to the granularity of the STEP representation. Even here matters are not totally clear-cut, because STEP often provides alternative ways of representing the same configuration, as illustrated in Section 3.1 above. The recommendation made there was that the highest possible level of granularity should always be used for maximal preservation of design intent.

Matters become more complex when implicit information is hidden in the sending system. For example, the creation of a constant radius blend feature will usually lead to the designer's chosen blend radius being stored by the system as an attribute of the feature representation. On the other hand, some systems provide the capability for defining a default value for blend radii, and in this case the value of the default radius may have to be accessed in a different manner through the system API. In either case, however, the value of the radius is accessible, and a corresponding explicit dimension can be created in the receiving system if that is appropriate.

3.5 Differences in modelling methodology

Attention is restricted in this section to the topic of explicit geometric constraints, to provide some illustrations of differences in CAD system modelling methodology that have to be taken into account in a successful inter-system model exchange.

An explicit geometric constraint has a specification describing its semantics, and refers to two or more constrained geometric elements. Some CAD systems only allow binary constraints, in which the number of geometric elements involved is limited to two. In this case if the designer selects $N > 2$ geometric elements, then $N - 1$ separate binary constraints are created.

Geometric constraints are often *directed*, in the sense that one or more model elements is constrained with respect to one or more reference elements. In such a case the configuration may only be modified by editing the reference elements(s), when the constrained elements will automatically change so that the constraint in question remains satisfied. Undirected constraints also exist, in which all pairs of members of a set of elements are required to satisfy a specified constraint; for example, a set of planes may be constrained to be parallel to each other but not with respect to any reference element.

The API of a CAD system, when queried for the geometric elements involved in a constraint, may return either the system names of those elements or direct pointers to the elements themselves.

Some CAD systems do not allow the definition of 3D constraints within a part model. In these cases, the effect of a 3D constraint may be achieved indirectly, for example, by the creation of a datum element based on an element of one feature that is used in the definition of a second feature. While this implies a 3D relationship between the two features concerned, it may not be represented explicitly as a constraint in the CAD system. In an ISO 10303 exchange file, however, the possibility exists for expressing this relationship via an equivalent explicit constraint.

As far as the STEP neutral file is concerned, all geometric constraints must reference the underlying geometry of constrained topological elements (points, curves, surfaces). Sometimes it is necessary to refer to the defining elements of geometrical entities rather than the entities themselves. For example, if it is desired to constrain a cylindrical surface to be perpendicular to a plane it is necessary to formulate the constraint in terms of the axial direction of the cylinder rather than the actual cylindrical surface. This approach allows the total number of constraint types to be reduced because specialized constraint types do not have to be defined for each specific type of geometrical entity. On the other hand, the referencing of defining elements of constrained curves and surfaces rather than the curves and surfaces themselves creates more difficulties in the implementation of translators because the references to constrained elements are indirect.

A problem frequently arises when a datum acting as reference element for a constraint is translated from the sending system to a STEP file. Datum elements are usually defined by the CAD system with default dimensions, though in some cases no dimension is specified for them. For example, a datum plane may be displayed by a CAD system with default dimensions of 100×100 units, and it may be represented internally with that precise size and the topology of a face, or as an unbounded plane that is displayed as finite for easier understanding by the designer. In either case the pre-processor should create appropriate geometry to enable the post-processor to reconstruct the relationship involving the datum correctly from the point of view of the receiving system. Initial experience suggests that the most appropriate type of geometry to transfer for a datum is the most general – for example, unbounded lines, planes, etc. rather than bounded ones.

CAD systems may also re-interpret the user's input in some cases. For example, in creating a sketch the user may select a plane as the reference element for a constraint. The CAD system will then often represent the reference element as a compatible line rather than the chosen plane, reducing everything to 2D terms. Such reinterpetations can cause dimensionality problems in a STEP exchange; STEP regards a positioned sketch in model space as a 3D construct composed of 3D elements, and therefore will require the reference element of the constraint to have dimensionality 3.

Finally, we give an example of how the same geometric situation may be represented in different ways. It is required to constrain a line in a 2D sketch to be vertical. Many CAD systems allow this to be achieved in several ways:

- by simply subjecting the line to a **vertical** constraint;
- by using a **same-coordinate** constraint, requiring the x-coordinates of the positions of the end-points of the line to be equal;
- by constraining the line to be **parallel** to some other line that is vertical;
- by constraining the line to be **perpendicular** to some other line that is horizontal.

STEP permits the constraint to be transferred in the second, third and fourth of these forms, and a constraint originally expressed in the first form will need to be reformulated appropriately for transmission. Similarly, a particular receiving system may not implement the **same-coordinate** constraint, for example, and in that case the postprocessor needs to be provided with the intelligence to output one of the corresponding forms in order to preserve the design intent.

3.6 Design rationale

This topic was not addressed in the tests described, but it is mentioned here as being important for the future. It has been pointed out by Ohtaka [Ohtaka 1999] that a constructional history, even if it can be transmitted effectively, may be difficult to work with in a receiving system. One reason for this is because the history alone lacks any information about the designer's motivation in choosing a particular design methodology (this motivation is referred to as *design rationale*). In its absence, it may be impossible to understand why the designer used his chosen approach to the construction of the model, why certain features are present, why certain constraints have been imposed, and so on. Another reason is that the construction history of a complex model may be large and have many embedded levels of detail, and consequently be difficult to understand and modify simply for that reason. Both difficulties could be overcome through the provision of design rationale information with the construction history transfer, and it is very desirable that such information is captured for long-term data archiving. However, no effective method has yet been found for capturing such information automatically during the design process, and the best that can be done at present is to make provision for the insertion of design rationale in text form at appropriate points in the history. That will require the designer to input the text as he proceeds (possibly by speech recognition rather than via the keyboard), and problems may still result from differences in the manner of description of the design process, which is likely to vary significantly between designers.

3.7 Nature of the tests performed

This paper being in the nature of a survey, and covering a wide spectrum of issues, it is not possible in a limited space to go into fine detail regarding the tests performed. More detailed information will be given in a forthcoming journal paper.

The primary systems used in the exchanges were SolidWorks and ProEngineer (both registered trade-marks of their respective vendor companies). However, other major CAD systems were also examined, to ensure that the approaches used were compatible with those systems and that the necessary information could be read and written via their APIs. The test parts used were fairly simple, and were chosen to exhibit a range of different aspects of parametric feature-based design. Two examples are given here.

3.7.1 Case Study 1

In Case Study 1 the base shape is the linear extrusion of a 2D sketch originally defined on a datum plane. The sketch contains geometric constraints and dimensions. Two additional features are defined upon this base shape, a circular protrusion and a circular depression. In the originating system (ProEngineer) the constraints defined are 5 point coincidences, 3 tangencies, 2 horizontal direction constraints, a 'same x-coordinate' constraint between the centre point of the arc R92 and the lower end-point of that arc, and 5 dimensions. The part is illustrated in Figure 1. The numbers in boxes are those of entity instances generated in the STEP exchange file.

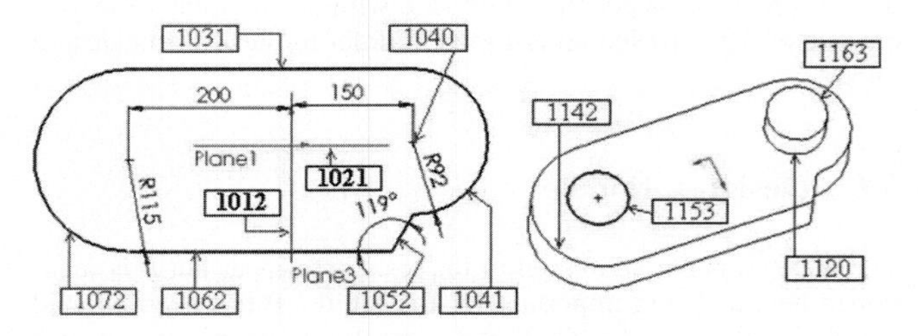

Figure 1: Extruded sketch with constraints and dimensions

In this case study the dimensions of the transferred model were editable in the receiving system, subject to the constraints imposed in the sending system.

3.7.2 Case Study 2

For Case Study 2 the base shape is a simple rotational shape with a circular pattern of four holes defined upon it, as shown in Figure 2. The position of the first hole is specified in terms of a datum plane and the distance between the axes of the part body and the hole. The pattern is then defined by a pattern creation operation, which automatically generates the angular dimension shown in the figure.

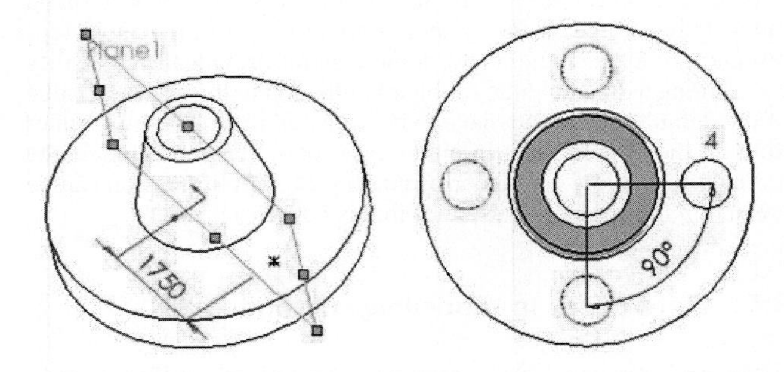

Figure 2: Rotational part with hole pattern

Two relations were defined between parameters for this part. The first simply expresses the diameter of its hole circle as twice the radius dimension (the value of the diameter is important in another context), and the second relates the number of holes in the circular pattern to the radius dimension. Variation of the salient dimensions of the part in the receiving system then allowed the generation of members of a family of parts as shown in Figure 3. The 'Master Model' is the case shown in Figure 2, where there are four holes. The model contains ten independent parameters; all ten degrees of freedom were tested, and no incorrect results were generated. Dependent parameters were also correctly evaluated, subject to transmitted constraint relationships.

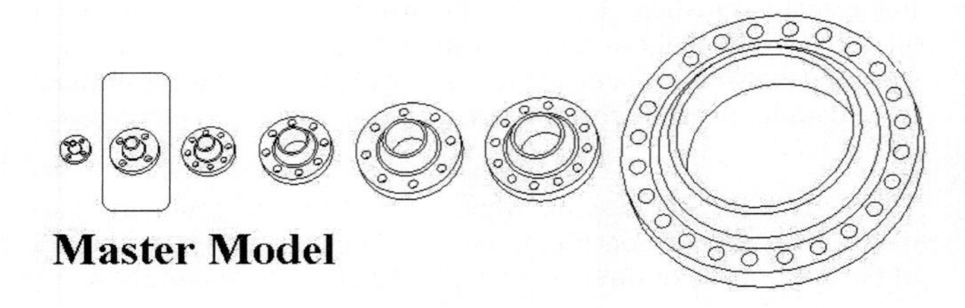

Figure 3: Members of a parametric part family

4 Other related tests

Other tests similar to those described here have been performed or are currently in progress, aimed at proving out different capabilities of the new STEP resources. A project coordinated by the organization PDES Inc. (http://pdesinc.aticorp.org) involved two major aerospace manufacturing companies and four of their supplier companies, and concentrated mainly on the transfer of pure construction history models. Some details are given in [Pratt et al. 2005], and the business case for exchanges of this type is made in [Stiteler 2004]. The rationale for pure construction history approach without parameterization and constraints is that many companies, in communicating designs with their suppliers, would prefer to suppress full details of the design intent in the original models because some of this information is regarded as proprietary. For data transfers within a company, however, the benefits of the more complete exchange are recognized.

The organization ProSTEP (http://www.prostep.de), based in Germany, is also engaging in tests of the new STEP facilities, though nothing has yet been published. The context here is automotive, and the Part 109 capability for representing parameterized assemblies is being evaluated. The data exchange experiments performed at KAIST in Korea have already been mentioned [Choi et al. 2002; Mun et al. 2003]; although they used a slightly different approach, there will soon be further demonstrations using Part 112 of STEP. Further related work, concentrating on the STEP-based exchange of sketches with parameterized geometry and constraints, has been reported from Troyes University of Technology in France [Charles et al. 2003].

5 Conclusion

The paper has described experiences in the testing of new STEP capabilities for the standard-based exchange of construction history CAD models with parameterization and constraints. Overall it has proved possible to transfer a range of different models, and to subject them to parametric variation in the receiving system. This is the primary criterion for success in such transfers; it demonstrates that design intent has been preserved in the exchange, a facility that was impossible until the recent development of Parts 55, 108 and 111 of STEP. However, certain difficulties and inefficiencies were identified in the course of the work, and one major potential improvement to Part 111 has been suggested that should significantly speed up the translation process in the future.

6 Acknowledgments

The first author is grateful for financial support from the US National Institute of Standards and Technology (NIST) under Contracts SB134102C0014 and SB134105W1299. As leader of the ISO TC184/SC4/WG12 Parametrics Group, he also acknowledges major recent contributions to the work of the group by Bill Anderson (Advanced Technology Institute, USA), Ray Goult (LMR Systems, UK), Soonhung Han (KAIST, Korea), Akihiko Ohtaka (Nihon Unisys, Japan), Vijay Srinivasan (IBM/Columbia University, USA), Tony Ranger (Theorem Solutions, UK), and Nobuhiro Sugimura (Osaka Prefectural University, Japan). He would also like to thank the many other people who made contributions during earlier phases of the work.

The second author gratefully acknowledges partial support through Korea Research Foundation Grant M01-2003-000-20351-0, funded by the Korean Government (MOEHRD).

7 Disclaimer

Certain commercial software systems are identified in this document. Such identification does not imply recommendation or endorsement by NIST; nor does it imply that the products identified are necessarily the best available for their purpose.

References

BOUMA, W., FUDOS, I., HOFFMAN, C. M., CAI, J., AND PAIGE, R. 1995. Geometric constraint solver. *Computer Aided Design* **27**, *6, 487 – 501.*

CAPOYLEAS, V., CHEN, X., AND HOFFMANN, C. M. 1996. Generic naming in generative constraint-based design. *Computer Aided Design* **28**, *1, 17 – 26.*

CHARLES, S., DUCELLIER, G., EYNARD, B., LI, L., AND RAKOTOMAMONJY, X. 2003. Standardisation des échanges de modèles géométriques 3D paramétrés non figés. *Revue Internationale de CFAO et d'informatique graphique* **18**, *4, 389 – 407.*

CHOI, G.-H., MUN, D.-W., AND HAN, S.-H. 2002. Exchange of CAD part models based on the macro-parametric approach. *International J. of CAD/CAM* **2**, *2, 23 – 31.* (Online at http://www.ijcc.org).

GU, H., CHASE, T. R., CHENEY, D. C., BAILEY, T. T., AND JOHNSON, D. 2001. Identifying, correcting, and avoiding errors in computer aided design models which affect interoperability. *Int. J. Computing and Information Science in Engineering* **1**, *1, 156 – 166.*

HAN, J.-H., PRATT, M. J., AND REGLI, W. C. 2000. Manufacturing feature recognition from solid models: A status report. *IEEE Trans. Robotics & Automation* **16**, *6, 782 – 797.*

HOFFMANN, C. M., AND JUAN, R. 1992. EREP — An editable, high-level representation for geometric design and analysis. In *Geometric Modeling for Product Realization*, North-Holland Publishing Co., P. R. Wilson, M. J. Wozny, and M. J. Pratt, Eds. (Proc. IFIP WG5.2 Workshop on Geometric Modeling, Rensselaerville, NY, Sept/Oct 1992).

INTERNATIONAL ORGANIZATION FOR STANDARDIZATION, GENEVA, SWITZERLAND. 1994. *ISO 10303:1994 – Industrial Automation Systems and Integration – Product Data Representation and Exchange.* (Information on ISO standards is given at http://www.iso.ch/cate/cat.html).

MUN, D.-H., HAN, S.-H., KIM, J.-H., AND OH, Y.-C. 2003. A set of standard modeling commands for the history-based parametric approach. *Computer Aided Design* **35**, *1171 – 1179.*

OHTAKA, A., 1999. Parametric representation and exchange: A sample data model for history-based parametrics and key issues. White Paper, ISO TC184/SC4/WG12/N295, International Organization for Standardization.

OWEN, J. 1997. *STEP: An Introduction*, 2nd ed. Information Geometers, Winchester, UK.

PRATT, M. J., ANDERSON, B. D., AND RANGER, T. 2005. Towards the standardized exchange of parameterized feature-based CAD models. *Computer Aided Design* **37**, *1251 – 1265.*

PRATT, M. J. 2001. Introduction to ISO 10303 — the STEP standard for product data exchange. *ASME J. Computing & Information Science in Engineering* **1**, *1, 102 – 103.*

PRATT, M. J. 2004a. Extension of ISO 10303, the STEP standard, for the exchange of procedural shape models. In *Procs. 2004 Shape Modelling and Applications Conf., Genova, Italy, June 2004*, IEEE Computer Society Press, F. Giannini and A. Pasko, Eds.

PRATT, M. J. 2004b. A new ISO 10303 (STEP) resource for modelling parameterization and constraints. *ASME J. Computing and Information Science in Engineering* **4**, *4, 339 – 351.*

RAPPOPORT, A., SPITZ, S., AND ETZION, M. 2005. One-dimensional selections for feature-based data exchange. In *Proc. 2005 ACM Solid & Physical Modeling Symposium, Cambridge, MA, USA*, ACM Press.

RAPPOPORT, A. 2003. An architecture for universal CAD data exchange. In *Proc. 2003 ACM Solid Modeling Symposium, Seattle, WA, USA*, ACM Press, 266 – 269.

SHAH, J. J., AND MÄNTYLÄ, M. 1995. *Parametric and Feature-based CAD/CAM*. Wiley.

SPITZ, S., AND RAPPOPORT, A. 2004. Integrated feature-based and geometric CAD data exchange. In *Proc. 2004 ACM Solid Modeling Symposium, Genova, Italy*, ACM Press.

STITELER, M. 2004. Construction History And ParametricS: Improving affordability through intelligent CAD data exchange. Tech. rep., CHAPS Program Final Report, Advanced Technology Institute, 5300 International Boulevard, North Charleston, SC 29418, USA.

Design For Manufacturing Feedback At Interactive Rates

Sara McMains

The contemporary engineering design curriculum embraces the concept of "concurrent engineering," in contrast to a more traditional, linear "over-the-wall" design process. In the old model, only after a design was all but complete would the manufacturing engineers become involved, typically requesting numerous design changes before compromising on a modified design that was both feasible and economical to manufacture. Not only does delayed manufacturability feedback increase time-to-market, it also increases cost. The flexibility to make design changes, particularly for interdependent components of a complex system, rapidly diminishes later in the design cycle. Since over 70% of manufacturing cost typically becomes fixed in the design stage, it is vital that designers be able to accurately assess the effect of design decisions on manufacturing costs early in the design process.

Thus the most effective design methodology must go beyond Design for Manufacturability (DFM) rules of thumb to a tighter integration between computer-aided design tools and computer-aided manufacturing tools. However, current software to aid in manufacturability checking rarely runs at interactive rates; furthermore, the considerable manual overhead needed to set it up it is an obstacle to its widespread adoption early in the design process. Thus, the ideal of being guided by detailed DFM analysis of candidate designs from the start, rather than re-designing for manufacturability later in the design cycle, continues to be rare in practice. In an ideal world, designers would be able to obtain continuous interactive feedback about relative manufacturing feasibility and cost with each design update, without even having to leave their design software.

In order to provide fast, automated, and relevant feedback, we focus on manufacturing cost drivers that are highly sensitive to changes in geometry. This talk will describe algorithms for analyzing complex part geometries to efficiently identify geometric configurations that influence such cost drivers for molding and casting processes, generating interactive feedback to guide the designer towards manufacturing-friendly design alternatives.

SPM 2006, Cardiff, Wales, United Kingdom, 06–08 June 2006.
© 2006 ACM 1-59593-358-1/06/0006 $5.00

Sara McMains is an Assistant Professor in the Department of Mechanical Engineering at the University of California, Berkeley. The focus of her research is efficient geometric algorithms for computer aided design and manufacturing, including applications in virtual reality design environments, design for manufacturing, and computer aided process planning for layered manufacturing and machining. She received her A.B. from Harvard University in Computer Science, and her M.S. and Ph.D. from UC Berkeley in Computer Science with a minor in Mechanical Engineering. She is the recipient of Best Paper Awards from Usenix (1995) and ASME DETC (2000). Recent research is funded in part by the Hellman Foundation, a Prytanean Alumnae Award, and an NSF CAREER Award.

Committees and Reviewers

Program Committee

George Allen
UGS, USA

Pierre Alliez
INRIA, France

Cecil Armstrong
Queen's University Belfast, UK

Hujun Bao
Zhejiang University, China

Mario Botsch
ETHZ, Switzerland

Wim Bronsvoort
TU Delft, Netherlands

Frederic Cazals
INRIA, France

Falai Chen
University of Science and Technology, China

Paolo Cignoni
CNR, Italy

Danny Cohen-Or
Tel Aviv University, Israel

Jonathan Corney
Heriot-Watt University, UK

Leila De Floriani
University of Genova, Italy

Tamal Dey
Ohio State University, USA

Gershon Elber
Technion, Israel

Craig Gotsman
Technion, Israel

Thomas Grandine
Boeing, USA

Xianfeng Gu
SUNY, USA

Baining Guo
Microsoft, China

Satyandra Gupta
University of Maryland, USA

Chris Hoffmann
Purdue University, USA

Ken Joy
UC Davis, USA

Bert Jüttler
University of Linz, Austria

Deok-Soo Kim
Hanyang University, Korea

Myung-Soo Kim
Seoul National University, Korea

Frank Langbein
Cardiff University, UK

Kunwoo Lee
Seoul National University, Korea

Bruno Levy
LORIA, France

Andre Lieutier
Dassault, France

Alan Liu
USUHS, USA

Ligang Liu
Zhejiang University, China

Takashi Maekawa
Yokohama National University, Japan

Dinesh Manocha
University of North Carolina, USA

Joe Mitchell
SUNY, USA

Valerio Pascucci
LLNL, USA

Helmut Pottmann
Technical University of Vienna, Austria

Enrico Puppo
University of Genova, Italy

Hong Qin
SUNY, USA

Ari Rappoport
Hebrew University of Jerusalem, Israel

Bill Regli
Drexel University, USA

Nickolas Sapidis
University of the Aegean, Greece

Vadim Shapiro
University of Wisconsin, USA

Alla Sheffer
University of British Columbia, Canada

Jack Snoeyink
University of North Carolina, USA

Michela Spagnuolo
CNR, Italy

Neil Stewart
University of Montreal, Canada

Robert Sumner
Georgia Tech, USA

Luiz Velho
IMPA, Brazil

Alvar Vinacua
Polytechnic University of Catalunya, Spain

Committees and Reviewers

Program Committee

Frank Zeilfelder
University of Mannheim, Germany

Caimin Zhang
Shandong University, China

Denis Zorin
New York University, USA

Additional Reviewers

Gill Barequet
Alexander Bobenko
Mathieu Desbrun
Shachar Fleishman
Fabio Ganovelli
Xiao-Shan Gao
Igor Guskov
Iddo Hanniel
Gero Hecklin
Ravi Janardan
Alexander Keller
Martin Marinov
Victor Milenkovic
Masha Nikolski
Mark Pauly
Nico Pietroni
Patrick Reuter
Martin Rumpf
Christian Rössl
Malcolm Sabin
Sagi Schein
Stefan Schirra
Ching-Kuang Shene
Jacob Subag
Vitaly Surazhsky
Jianmin Zheng

Cover Image Credits

Front Cover

Reference: "Computing Surface Hyperbolic Structure and Real Projective Structure," *Miao Jin, Feng-Luo, Xianfeng Gu,* pp. 105–116.

Back Cover

Reference: "Holoimages," *Xianfeng Gu, Song Zhang, Liangjun Zhang, Ralph Martin, Peisen Huang, Shing-Tung Yau,* pp. 129–138.

Author Index

(Page numbers in italic face are color plates.)

Color Plate Section

Figure 1: Illustration of the holomorphic one-form on the colon surface by texture-mapping a checker board image.

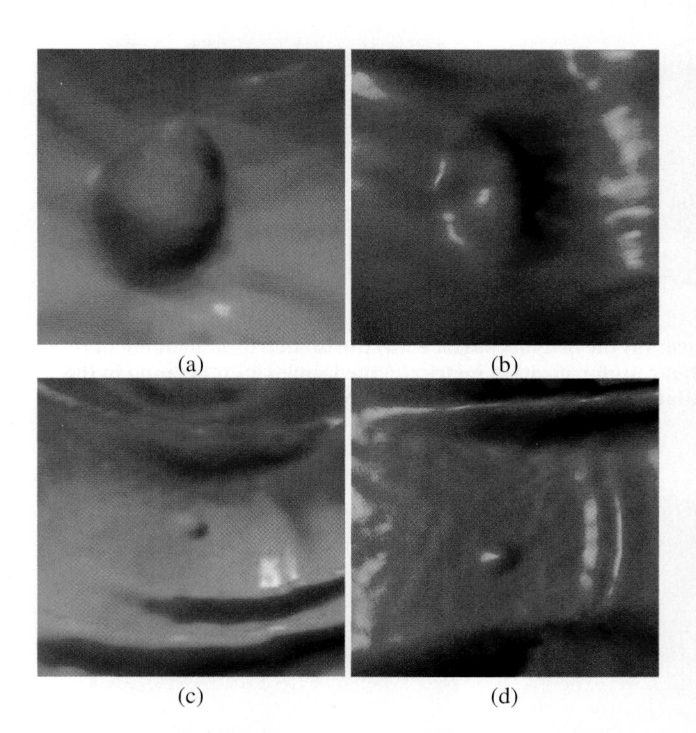

Figure 2: (a) A close view of a polyp rendered with volumetric ray casting; (b) A view generated from the flattened colon image showing the same polyp. (c) A view containing a small polyp generated from the navigation of a virtual colonoscopy system; (d) A view of the same smaller polyp generated from the flattened colon image.

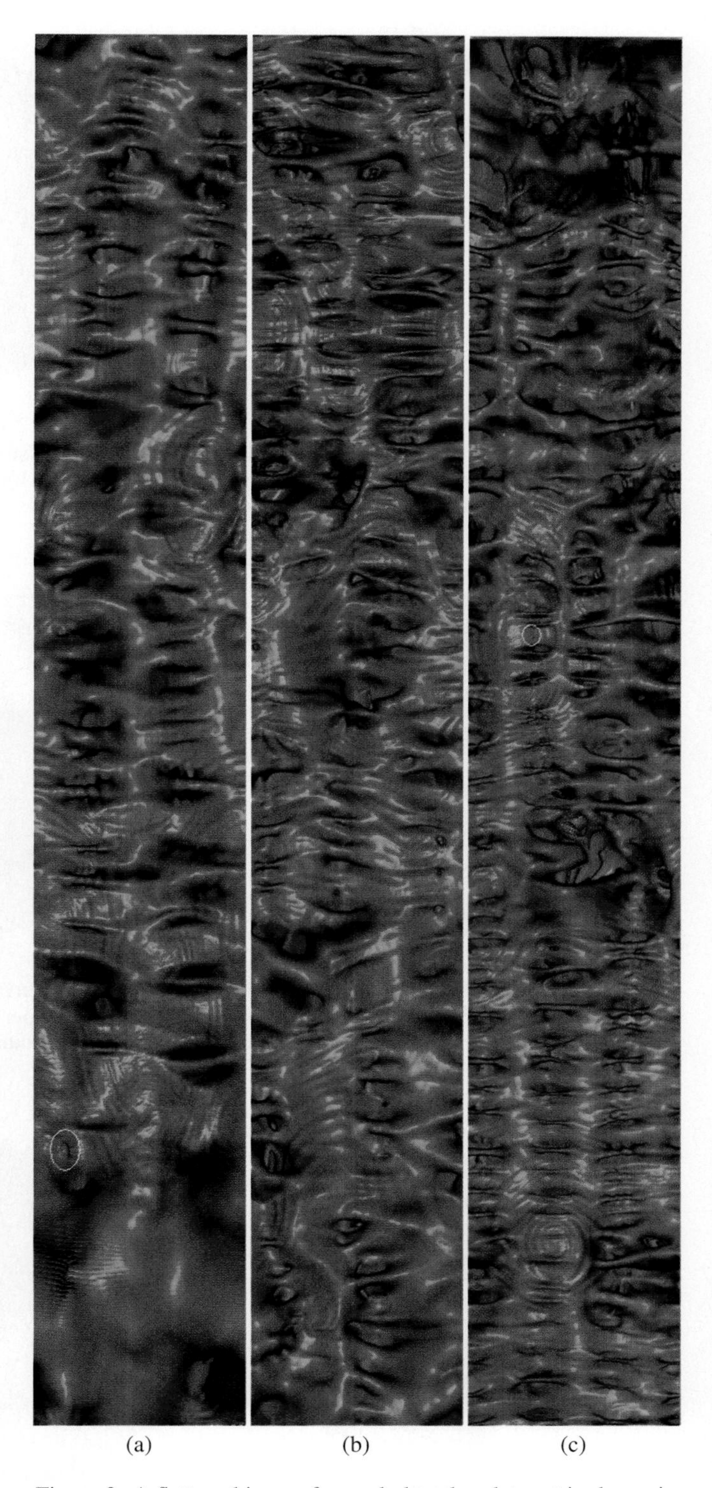

Figure 3: A flattened image for a whole colon data set is shown in three images. The bottom of image (a) is the rectum of the colon, and the top of image (c) is the cecum of the colon.

Figure 1: **The hyperbolic structure of a genus four closed surface: (a) the original surface ;(b) the isometric embedding of its universal covering space in the Poincaré disk using the hyperbolic uniformization metric;(c) the isometric embedding in the Klein disk which induces real projective structures**

Figure 2: **The hyperbolic structure of a genus two closed surface: (a) the original surface ;(b) the isometric embedding of its universal covering space in the Poincaré disk using the hyperbolic uniformization metric;(c) the isometric embedding in the Klein disk which induces real projective structures. The boundaries of fundamental domains are straightened to hyperbolic lines.**

Figure 3: **The hyperbolic structure of a genus four closed surface: (a) the original surface; (b) the isometric embedding of its universal covering space in the Poincaré disk using the hyperbolic uniformization metric;(c) the isometric embedding in the Klein disk which induces real projective structures.The boundaries of fundamental domains are straightened to hyperbolic lines.**

Figure 1: **Holoimage representing both geometry and shading.** The first image is a 24-bit holoimage with 512×512 resolution, with spatial frequency 64Hz, and 80° projection angle. All the other geometry and images are deduced from it, and are, in order, the phase map, the geometric surface, the shaded image, and the normal map.

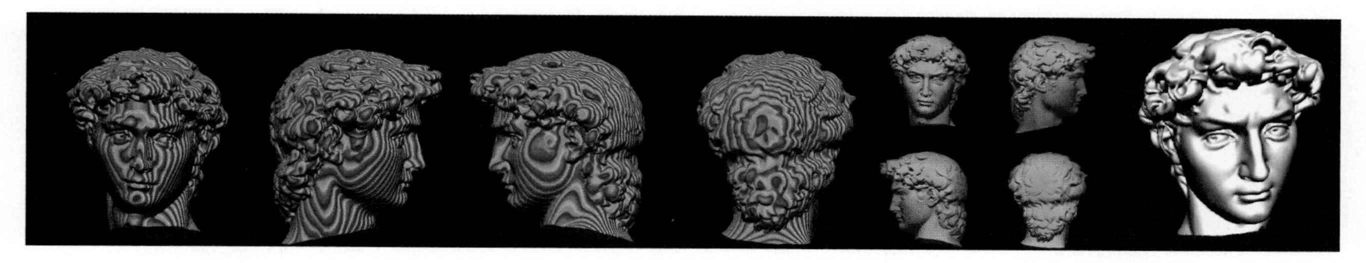

Figure 2: **Holoimage-based rendering.** The David head model represented as holoimages viewed from different directions at two wavelengths. The set of holoimages can be efficiently rendered by GPU to cover the whole surface. The last column is the rendered result.

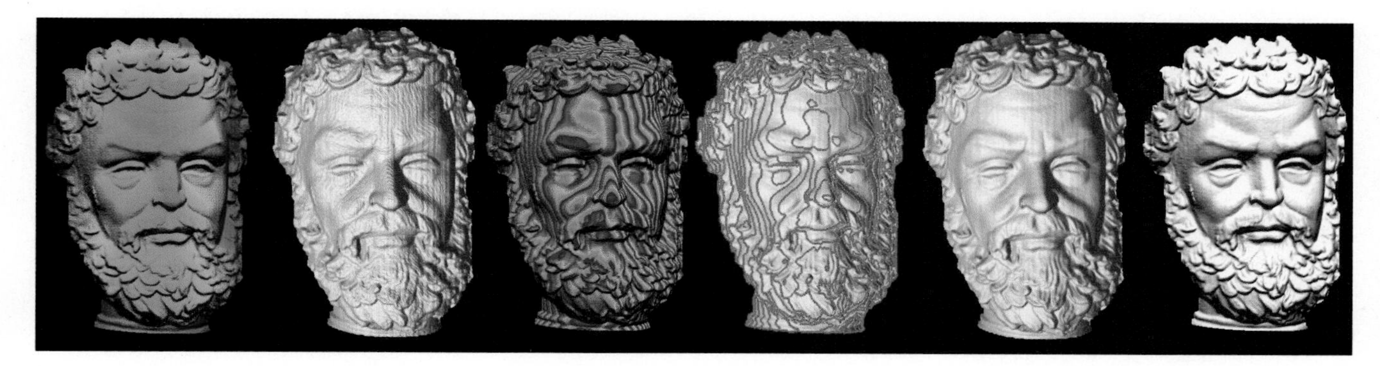

Figure 3: **Two-wavelength phase unwrapping.** The first image is a holoimage with spatial frequency 1Hz; the coarse reconstructed geometry is shown in the second image does not need phase unwrapping. This is used as a reference to phase unwrap holoimages with denser fringes. The 3rd image is a holoimage with spatial frequency 64Hz. The 4th image is the reconstructed geometry without phase unwrapping; the red contours show phase jumps. The 5th image is the reconstructed geometric surface after phase unwrapping to depth consistency with the 2nd image. The last image is a shaded image.

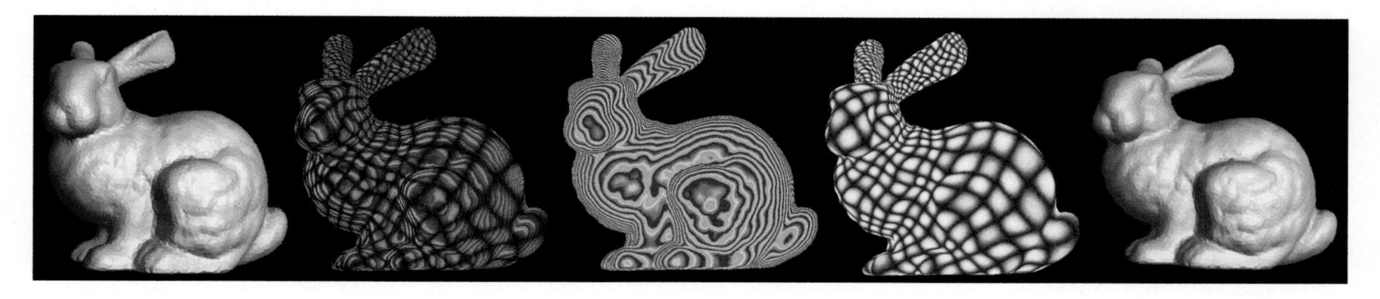

Figure 4: **Holoimage of texture mapped surfaces.** The original geometric surface is flat shaded in the first image, the flat triangles are visible. The second image is the 24-bit holoimage for the textured bunny surface. The third image shows the phase map. The fourth image shows the texture reconstructed from the holoimage. The reconstructed geometry is shown in the last image, the flat triangles are recovered.